Praise for
THE TELOMERE EFFECT

"Elizabeth Blackburn and Elissa Epel have discovered that telomeres, the capping structure at the end of your DNA that make up your chromosomes, do not simply carry out the commands issued by your genetic code. Telomeres are listening to you. They absorb the instructions you give them. They respond to you being stressed and to your being relaxed, to your being sad or your being happy. Thus, telomeres contribute to the state of your brain, to your mood, to the speed of your aging, and to your risk for neurodegenerative diseases. In other words, we can change the way we age at the most elemental cellular level. So if you want to keep your brain sharp, you have to know about your telomeres and stay in touch with them. This book shows you how you can do this, and it does so in a way that is both intellectually exhilarating yet easily accessible to all. This book is a classic. It is one of the most exciting health books to emerge in the last decade. It explains how we can slow the way we age at a fundamental level."

—Eric Kandel, Nobel laureate, author of *In Search of Memory: The Emergence of a New Science of Mind*

"Improving public health requires that people know the truth about their own lives. Blackburn and Epel reveal the discovery of how cells age and how certain forces in our lives cause us to get sick and age prematurely. *The Telomere Effect* explains the often-invisible things that affect all of our lives, giving us a fresh new level of awareness and helping us make better choices individually and socially for greater health and longevity. In short, it will change the way we think of aging and disease."

—David Kessler, MD, former FDA commissioner, author of the *New York Times* bestseller *The End of Overeating* and *Capture*

"Using both science and personal stories, Blackburn and Epel demonstrate that how we live each day has a profound effect not just on our health and well-being but how we age, as well. It's a manual for how to live younger and longer. Spoiler alert: sleep is a key element. *The Telomere Effect* is a book that will help you thrive at every level."

—Arianna Huffington

"Elizabeth Blackburn and Elissa Epel have discovered a revolutionary set of findings that can transform the way we live our lives, shaping the very health of our cells by how we use our minds. These pioneers of well-being unveil a story of the power of our interpersonal connections—in romance, friendship, and child-parent relationships—to slow the rate of cell aging.

"These powerful discoveries are made useful in your day-to-day life with a wealth of science-based suggestions, which will delight your mind, enrich your day, and improve your health."

—Daniel J. Siegel, MD,
author of *Mindsight* and *Brainstorm*

"Blackburn and Epel lay out a road map for thriving as we age by eloquently illuminating the intricate relationships between the psychology and the biology of aging. Drawing on telomere science, they empower and inspire readers to enhance their healthspans. The authors point to realistic possibilities long life affords, in language that is accessible, informative, and highly engaging."

—Laura L. Carstensen, PhD, professor of psychology,
founding director of the Center on Longevity at
Stanford University, author of
A Long Bright Future

"The Blackburn-Epel 'dream team' has condensed a massive body of complex scientific data into a highly readable, nontechnical 'how-to' manual on strategies that will help anyone who is human: a truly extraordinary gift to all of us who want to enhance our health, no matter at what stage of life we are in."

—Rita B. Effros, PhD, professor, David Geffen School of Medicine at UCLA, 2015 president, Gerontological Society of America

"The authors of this fascinating book show how the telomeres in our bloodstreams are responsive to many aspects of our daily existence. In these pages, telomeres become the nexus of an important discussion of vulnerability and resilience to influences of our social and physical environment and the important role of the mind-body connection. In our future, it may be that telomere monitoring will help us pursue better health—a new frontier waiting to be explored. Regardless, you will learn a lot that can benefit your 'healthspan.'"

—Bruce McEwen, PhD, professor of neuroscience, the Rockefeller University, author of *The End of Stress as We Know It*

"Dr. Elizabeth Blackburn is *the* expert on telomeres, which are the tips that protect our chromosomes and correlate remarkably with health and longevity. Her and Dr. Epel's scientific discoveries and their potential importance for our health, both individually and collectively, are profound, and their apparent relationship to stress opens up an exciting array of potential healthy lifestyle changes."

—Lee Goldman, MD, chief executive of Columbia University Medical Center, author of *Too Much of a Good Thing: How Four Key Survival Traits Are Now Killing Us*

"The breakthrough research that Drs. Elizabeth Blackburn and Elissa Epel have conducted has created a dramatic shift in our understanding of what is possible in terms of human health and longevity. Telomeres, the ends of your DNA, affect how quickly your cells age and die. As your telomeres get shorter, your life gets clouded with disease.

"Drs. Blackburn and Epel are the central researchers in the discovery of telomeres, their profound effect on health, and the myriad ways in which lifestyle choices can improve cellular aging. They have collaborated with researchers worldwide on studies from understanding cell-aging machinery to chemical exposures to mental-training classes aimed to improve cellular health. A study we collaborated on showed, for the first time, that comprehensive lifestyle changes may actually increase the length of our telomeres, thus beginning to reverse aging on a cellular level. This book is revolutionary, transforming the way our world thinks about health and living well, disease, and death. This work reveals a stunning picture of healthy aging—it's not simply about individuals; it's about how we are connected to each other, today and through future generations. It is hard to overstate its importance."

Dean Ornish, MD, founder and president, Preventive Medicine Research Institute, clinical professor of medicine, UCSF, *New York Times* bestselling author of *The Spectrum*

"Some of us emphasize social determinants of health; others of us emphasize behaviors such as diet and exercise; still others look at psychology and health. What if we had a coherent and readily comprehensible way of understanding the biology that links all of these to health and disease and to length and quality of life? We would not only gain better understanding of causes of health and disease, but what to do to improve things. As Blackburn and Epel lay out beautifully and clearly in this wonderful book, telomere length provides such a unifying biological mechanism. The authors take cutting-edge science and make it fascinating and intelligible for interested reader and expert alike. More than that, we warm to their humanity."

—Professor Sir Michael Marmot,
president of the World Medical Association, director,
University College London Institute of Health Equity, author
of *The Health Gap: The Challenge of an Unequal World*

"At last we are closing in on the intertwined biological, behavioral, and social influences on why some people thrive in good health while others are more likely to stumble and fall. Always educational and sometimes poetic, *The Telomere Effect* brings us a fascinating analysis from two of the world's best researchers on behavior, health, and longevity.

"Avoiding quick-fix exhortations like New Year's resolutions destined to be melting by spring, Blackburn and Epel explain the longer-term life patterns that play a role in longer telomeres, longer periods of good health, and longer life.

"This excellent book avoids the trap of seeing all stress and challenge as bad, and instead provides a nuanced understanding that trials and tribulations are not inevitably a health threat, because challenge can build resilience. The study of telomeres helps us understand what protects and toughens our cells. This is cutting-edge longevity science."

—Howard S. Friedman, PhD, distinguished professor
at the University of California, Riverside, author of
*The Longevity Project: Surprising Discoveries for Health and
Long Life from the Landmark Eight-Decade Study*

"*The Telomere Effect* gives us, in high relief and with exactly the practical level of detail we need, the long and the short of a new science revealing that how we live our lives, both inwardly and outwardly, individually and collectively, impinges significantly on our health, our well-being, and even our longevity. Mindfulness is a key ingredient, and importantly, issues of poverty and social justice are shown to clearly come into play as well. This book is an invaluable, rigorously authentic, and at its core, exceedingly compassionate and wise contribution to our understanding of health and well-being."

—Jon Kabat-Zinn, author of *Full Catastrophe Living* and *Coming to Our Senses*

THE
TELOMERE
EFFECT

THE TELOMERE EFFECT

A REVOLUTIONARY APPROACH TO
LIVING YOUNGER, HEALTHIER, LONGER

Elizabeth Blackburn, PhD
Elissa Epel, PhD

GRAND CENTRAL
PUBLISHING

NEW YORK   BOSTON

Copyright © 2017 by Elizabeth Blackburn and Elissa Epel

Cover design by Jeff Miller, Faceout Studio
Cover copyright © 2017 by Hachette Book Group, Inc.

Hachette Book Group supports the right to free expression and the value of copyright. The purpose of copyright is to encourage writers and artists to produce the creative works that enrich our culture.

The scanning, uploading, and distribution of this book without permission is a theft of the authors' intellectual property. If you would like permission to use material from the book (other than for review purposes), please contact permissions@hbgusa.com. Thank you for your support of the authors' rights.

Grand Central Publishing
Hachette Book Group
1290 Avenue of the Americas, New York, NY 10104
grandcentralpublishing.com
twitter.com/grandcentralpub

First Edition: January 2017

Grand Central Publishing is a division of Hachette Book Group, Inc. The Grand Central Publishing name and logo is a trademark of Hachette Book Group, Inc.

The publisher is not responsible for websites (or their content) that are not owned by the publisher.

The Hachette Speakers Bureau provides a wide range of authors for speaking events. To find out more, go to www.hachettespeakersbureau.com or call (866) 376-6591.

Illustrations by Colleen Patterson of Colleen Patterson Design

Library of Congress Cataloging-in-Publication Data

Names: Blackburn, Elizabeth H. (Elizabeth Helen), 1948– author. | Epel, Elissa S. (Elissa Sarah), 1968– author.

Title: The telomere effect : a revolutionary approach to living younger, healthier, longer / by Elizabeth Blackburn, Elissa Epel.

Description: New York : Grand Central Publishing, [2017] | Includes bibliographical references and index.

Identifiers: LCCN 2016028884| ISBN 9781455587971 (hardcover) | ISBN 9781455541713 (hardcover large-print) | ISBN 9781478940425 (audio cd) | ISBN 9781478940432 (audio download) | ISBN 9781455587964 (ebook)

Subjects: | MESH: Telomere—physiology | Aging—genetics

Classification: LCC QH600.3 | NLM QU 470 | DDC 572.8/7—dc23 LC record available at https://lccn.loc.gov/2016028884

ISBNs: 978-1-4555-8797-1 (hardcover), 978-1-4555-4171-3 (large print), 978-1-4555-8796-4 (ebook)

Printed in the United States of America

LSC-C

10 9 8 7 6 5 4

I dedicate this book to John and Ben,
the lights of my life, who simply make everything
for me worthwhile. —EHB

I dedicate this book to my parents, David and Lois,
who are an inspiration in how they live fully
and vibrantly, in their almost ninth decade of life,
and to Jack and Danny, who make my cells happy. —ESE

Contents

Authors' Note: Why We Wrote This Book

With a life span of 122 years, Jeanne Calment was one of the longest-living women on record. When she was eighty-five, she took up the sport of fencing. She was still riding a bike into her triple digits.[1] When she turned one hundred, she walked around her hometown of Arles, France, thanking the people who'd wished her a happy birthday.[2] Calment's relish for life captures what we all want: a life that is healthy right up to the very end. Aging and death are immutable facts of life, but how we live until our last day is not. This is up to us. We can live better and more fully now and in our later years.

The relatively new field of telomere science has profound implications that can help us reach this goal. Its application can help reduce chronic disease and improve wellbeing, all the way down to our cells and all the way through our lives. We've written this book to put this important information into your hands.

Here you will find a new way of thinking about human aging. One current, predominant, scientific view of human aging is that the DNA of our cells becomes progressively damaged, causing cells to become irreversibly aged and dysfunctional. But which DNA is damaged? Why did it become damaged? The full answers aren't known yet, but the clues are now pointing strongly toward telomeres as a major culprit. Diseases can seem distinct because they involve very different organs and parts of the body. But new scientific and clinical findings have crystallized into a new concept. Telomeres throughout the body shorten as we age, and this underlying mechanism contributes to most diseases of aging. Telomeres explain how we run out of the abilty to replenish tissue (called replicative senescence). There are other ways cells become dysfunctional or die early, and there are other factors that contribute to human aging. But telomere attrition is a clear and an early contributor to the aging process, and—more exciting—it is possible to slow or even reverse that attrition.

We've put the lessons from telomere research into the full story, as it is unfolding today, in language for the general reader. Previously this knowledge has been available only in scientific journal articles, scattered in bits and pieces. Simplifying this body of science for the public has been a great challenge and responsibility. We could not describe every theory or pathway of aging or lay out each topic in fine scientific detail. Nor could we state every qualification and disclaimer. Those issues are detailed in the scientific journals where the original studies were published, and we encourage interested readers to explore this fascinating body of work, much of it cited in this book. We have also written a review article covering the latest research on telomere biology, published in the peer-reviewed scientific journal *Science*, which will give you several good directions into the molecular-level mechanisms.[3]

Science is a team sport. We have been truly privileged to participate in research with a broad range of scientific collaborators from different disciplines. We have also learned from research teams from all over the world. Human aging is a puzzle made up of many pieces. Over several decades, new pieces of information have each added a critical part to the whole. The understanding of telomeres has helped us see how the pieces fit together—how aged cells can cause the vast array of diseases of aging. Finally a picture has emerged that is so compelling and helpful that we felt it was important to share it broadly. We now have a comprehensive understanding of human telomere maintenance, from cell to society, and what it can mean in human lives and communities. We are sharing with you the basic biology of telomeres, how they relate to disease, to health, to how we think, and even to our families and communities. Putting together the pieces, illuminated by knowledge of what affects telomeres, has led us to a more interconnected view of the world, as we share with you in the last section of the book.

Another reason we've written this book is to help you avoid potential risks. The interest in telomeres and aging is growing exponentially, and while there is some good information in the public domain, some of it is misleading. For example, there are claims that certain creams and supplements may elongate your telomeres and increase your longevity. These treatments, if they actually work in the body, could potentially increase your risk of cancer or have other dangerous effects. We simply need larger and longer studies to assess these potential serious risks. There are other known ways to improve your cell longevity, without risk, and we have tried to include the best of them here. You won't find any instant cures on these pages, but you *will* find the specific, research-supported ideas that could make the rest of your life healthy, long, and fulfilling. While some ideas may not be totally new to you, gaining a deep understanding of the behind-the-scenes reasons for them may change how you view and live your days.

Finally, we want you to know that neither of us has any financial interest in companies that sell telomere-related products or that offer telomere testing. Our wish is to synthesize the best of our understanding—as it stands today—and make it available to anyone who may find it useful. These studies represent a true breakthrough in our understanding of aging and living younger, and we want to thank all who have contributed to the research that we are able to present here.

With the exception of the "teaching story" that appears on the first page of the introduction, the stories in this book are drawn from real-life people and experiences. We are deeply grateful to the people who shared their stories with us. To protect their privacy, we have changed some names and identifying details.

We hope this book is helpful to you, your families, and all who can benefit from these fascinating discoveries.

A Tale of Two Telomeres

It is a chilly Saturday morning in San Francisco. Two women sit at an outdoor café, sipping hot coffee. For these two friends, this is their time away from home, family, work, and to-do lists that never seem to get any shorter.

Kara is talking about how tired she is. How tired she *always* is. It doesn't help that she catches every cold that goes around the office, or that those colds inevitably turn into miserable sinus infections. Or that her ex-husband keeps "forgetting" when it's his turn to pick up the children. Or that her bad-tempered boss at the investment firm scolds her—right in front of her staff. And sometimes, as she lies down in bed at night, Kara's heart gallops out of control. The sensation lasts for just a few seconds, but Kara stays awake long after it passes, worrying. *Maybe it's just the stress*, she tells herself. *I'm too young to have a heart problem. Aren't I?*

"It's not fair," she sighs to Lisa. "We're the same age, but I look older."

She's right. In the morning light, Kara looks haggard. When she reaches for her coffee cup, she moves gingerly, as if her neck and shoulders hurt.

But Lisa looks vibrant. Her eyes and skin are bright; this is a woman with more than enough energy for the day's activities. She feels good, too. Actually, Lisa doesn't think very much about her age, except to be thankful that she's wiser about life than she used to be.

Looking at Kara and Lisa side by side, you would think that Lisa really *is* younger than her friend. If you could peer under their skin, you'd see that in some ways, this gap is even wider than it seems. Chronologically, the two women are the same age. Biologically, Kara is decades older.

Does Lisa have a secret—expensive facial creams? Laser treatments at the dermatologist's office? Good genes? A life that has been free of the difficulties her friend seems to face year after year?

Not even close. Lisa has more than enough stresses of her own. She lost her husband two years ago in a car accident; now, like Kara, she is a single mother. Money is tight, and the tech start-up company she works for always seems to be one quarterly report away from running out of capital.

What's going on? Why are these two women aging in such different ways?

The answer is simple, and it has to do with the activity inside each woman's cells. Kara's cells are prematurely aging. She looks older than she is, and she is on a headlong path toward age-related diseases and disorders. Lisa's cells are renewing themselves. She is living younger.

WHY DO PEOPLE AGE DIFFERENTLY?

Why do people age at different rates? Why are some people whip smart and energetic into old age, while other people, much younger, are sick, exhausted, and foggy? You can think of the difference visually:

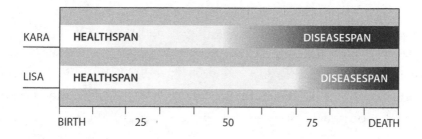

Figure 1: Healthspan versus Diseasespan. Our healthspan is the number of years of our healthy life. Our diseasespan is the years we live with noticeable disease that interferes with our quality of living. Lisa and Kara may both live to one hundred, but each has a dramatically different quality of life in the second half of her life.

Look at the first white bar in figure 1. It shows Kara's healthspan, the time of her life when she's healthy and free of disease. But in her early fifties, the white goes gray, and at seventy, black. She enters a different phase: the diseasespan.

These are years marked by the diseases of aging: cardiovascular disease, arthritis, a weakened immune system, diabetes, cancer, lung disease, and more. Skin and hair become older looking, too. Worse, it's not as if you get just one disease of aging and then stop there. In a phenomenon with the gloomy name *multi-morbidity*, these diseases tend to come in clusters. So Kara doesn't just have a run-down immune system; she also has joint pain and early signs of heart disease. For some people, the diseases of aging hasten the end of life. For others, life goes on, but it's a life with less spark, less zip. The years are increasingly marred by sickness, fatigue, and discomfort.

At fifty, Kara should be brimming with good health. But the graph shows that at this young age, she is creeping into the diseasespan. Kara might put it more bluntly: she is getting old.

Lisa is another story.

At age fifty, Lisa is still enjoying excellent health. She gets older as the years pass, but she luxuriates in the healthspan for a nice, long time. It isn't until she's well into her eighties—roughly the age that gerontologists call "old old"—that it gets significantly harder for her to keep up with life as she's always known it. Lisa has a diseasespan, but it's compressed into just a few years toward the end of a long, productive life. Lisa and Kara aren't real people—we've made them up to demonstrate a point—but their stories highlight questions that are genuine.

How can one person bask in the sunshine of good health, while the other suffers in the shadow of the diseasespan? Can you choose which experience happens to *you*?

The terms *healthspan* and *diseasespan* are new, but the basic question is not. *Why do people age differently?* People have been asking this question for millennia, probably since we were first able to count the years and compare ourselves to our neighbors.

At one extreme, some people feel that the aging process is determined by nature. It's out of our hands. The ancient Greeks expressed this idea through the myth of the Fates, three old women who hovered around babies in the days after birth. The first Fate spun a thread; the second Fate measured out a length of that thread; and the third Fate snipped it. Your life would be as long as the thread. As the Fates did their work, *your* fate was sealed.

It's an idea that lives on today, although with more scientific authority. In the latest version of the "nature" argument, your health is mostly controlled by your genes. There may not be Fates hovering around the cradle, but the genetic code determines your risk for heart disease, cancer, and general longevity before you're even born.

Perhaps without even realizing it, some people have come to believe that nature is *all* that determines aging. If they were pressed to explain why Kara is aging so much faster than her friend, here are some things they might say:

"Her parents probably have heart problems and bad joints, too."

"It's all in her DNA."

"She has unlucky genes."

4

The "genes are our destiny" belief is, of course, not the only position. Many have noticed that the quality of our health is shaped by the way we live. We think of this as a modern view, but it's been around for a long, long time. An ancient Chinese legend tells of a raven-haired warlord who had to make a dangerous trip over the border of his homeland. Terrified that he would be captured at the border and killed, the warlord was so anxious that he woke up one morning to discover that his beautiful dark hair had turned white. He'd aged early, and he'd aged overnight. As many as 2,500 years ago, this culture recognized that early aging can be triggered by influences like stress. (The story ends happily: No one recognized the warlord with his newly whitened hair, and he traveled across the border undetected. Getting older has its advantages.)

Today there are plenty of people who feel that nurture is more important than nature—that it's not what you're born with, it's your health habits that really count. Here's what these folks might say about Kara's early aging:

"She's eating too many carbs."

"As we age, each of us gets the face we deserve."

"She needs to exercise more."

"She probably has some deep, unresolved psychological issues."

Take a look again at the ways the two sides explain Kara's accelerated aging. The nature proponents sound fatalistic. For good or for bad, we're born with our futures already encoded into our chromosomes. The nurture side is more hopeful in its belief that premature aging can be avoided. But advocates of the nurture theory can also sound judgmental. If Kara is aging rapidly, they suggest, it's all her fault.

Which is right? Nature or nurture? Genes or environment? Actually, both are critical, and it's the interaction between the two that matters most. The real differences between Lisa's and Kara's rates of aging lie in the complex interactions between genes, social relationships and environments, lifestyles, those twists of fate, and especially how one responds to the twists of fate. You're born with a particular set of genes, but the way you live can influence how your genes express themselves. In some cases, lifestyle factors can turn genes on or shut them off. As the obesity researcher George Bray has said, "Genes load the gun, and environment pulls the trigger."[1] His words apply not just to weight gain but to most aspects of health.

We're going to show you a completely different way of thinking about your health. We are going to take your health down to the cellular level, to show you what premature cellular aging looks like and what kind of havoc it wreaks on your body—and we'll also show you not only how to avoid it but also how to reverse it. We'll dive deep into the genetic heart of the cell, into the chromosomes. This is where you'll find **telomeres (tee-lo-meres)**, repeating segments of noncoding DNA that live at the ends of your chromosomes. Telomeres, which shorten with each cell division, help determine how fast your cells age and when they die, depending on how quickly they wear down. The extraordinary discovery from our research labs and other research labs around the world is that the ends of our chromosomes can actually lengthen—and as a result, aging is a dynamic process that can be accelerated or slowed, and in some aspects even reversed. Aging need not be, as thought for so long, a one-way slippery slope toward infirmity and decay. We all will get older, but how we age is very much dependent on our cellular health.

We are a molecular biologist (Liz) and a health psychologist (Elissa). Liz has devoted her entire professional life to investigating telomeres, and her fundamental research has given birth to an entirely new field of scientific understanding. Elissa's lifelong work has been on psychological stress. She has studied its harmful effects on behavior, physiology, and health, and she has also studied how to reverse these effects. We joined forces in research fifteen years ago, and the studies that we performed together have set in motion a whole new way of examining the relationship between the human

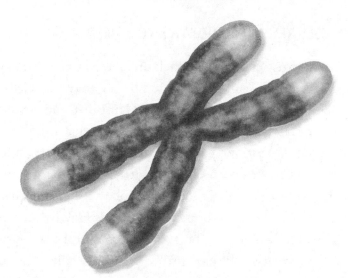

Figure 2: Telomeres at the Tips of Chromosomes. The DNA of every chromosome has end regions consisting of DNA strands coated by a dedicated protective sheath of proteins. These are shown here as the lighter regions at the end of the chromosome—the telomeres. In this picture the telomeres are not drawn to scale, because they make up less than one-ten-thousandth of the total DNA of our cells. They are a small but vitally important part of the chromosome.

mind and body. To an extent that has surprised us and the rest of the scientific community, telomeres do not simply carry out the commands issued by your genetic code. Your telomeres, it turns out, are listening to you. They absorb the instructions you give them. The way you live can, in effect, tell your telomeres to speed up the process of cellular aging. But it can also do the opposite. The foods you eat, your response to emotional challenges, the amount of exercise you get, whether you were exposed to childhood stress, and even the level of trust and safety in your neighborhood—all of these factors and more appear to influence your telomeres and can prevent premature aging at the cellular level. In short, one of the keys to a long healthspan is simply doing your part to foster healthy cell renewal.

HEALTHY CELL RENEWAL AND WHY YOU NEED IT

In 1961 the biologist Leonard Hayflick discovered that normal human cells can divide a finite number of times before they die. Cells reproduce by making copies of themselves (called mitosis), and as the human cells sat in a thin, transparent layer in the flasks that filled Hayflick's lab, they would, at first, copy themselves rapidly. As they multiplied, Hayflick needed more and more flasks to contain the growing cell cultures. The cells in this early stage multiplied so quickly that it was impossible to save all the cultures; otherwise, as Hayflick remembers, he and his assistant would have been "driven out of the laboratory and the research building by culture bottles." Hayflick called this youthful phase of cell division "luxuriant growth." After a while, though, the reproducing cells in Hayflick's lab stopped in their tracks, as if they were getting tired. The longest-lasting cells managed about fifty cell divisions, although most divided far fewer times. Eventually these tired cells reached a stage he called **senescence**: They were still alive but they had all stopped dividing, permanently. This is called the Hayflick limit, the natural limit that human cells have for dividing, and the stop switch happens to be telomeres that have become critically short.

Are all cells subject to this Hayflick limit? No. Throughout our bodies we find cells that renew—including immune cells; bone cells; gut, lung, and liver cells; skin and hair cells; pancreatic cells; and the cells that line our cardiovascular systems. They need to divide over and over and over to keep our bodies healthy. Renewing cells include some types of normal cells that can divide, like immune cells; progenitor cells, which can keep dividing even longer; and those critical cells in our bodies called stem cells, which can divide indefinitely as long as they are healthy. And, unlike those cells in Hayflick's lab dishes, cells don't always have a Hayflick limit, because—as you will read in chapter 1—they have telomerase. Stem cells, if kept healthy, have enough telomerase to enable them to keep dividing throughout our life spans. That cell replenishment, that *luxuriant growth*, is one reason Lisa's skin looks so fresh. It's why her joints move easily. It's one reason she can take in deep lungfuls of the cool air blowing in off the bay. The new cells are constantly renewing essential body tissues and organs. Cell renewal helps keep her feeling young.

From a linguistic perspective, the word *senescent* has a shared history with the word *senile*. In a way, that's what these cells are—they're senile. In one way it is definitely good that cells stop dividing. If they just keep on multiplying, cancer can ensue. But these senile cells are not harmless—they are bewildered and weary. They get their signals confused, and they don't send the right messages to other cells. They can't do their jobs as well as they used to. They sicken. The time of luxuriant growth is over, at least for them. And this has profound health consequences for you. When too many of your cells are senescent, your body's tissues start to age. For example, when you have too many senescent cells in the walls of your blood vessels, your arteries stiffen and you are more likely to have a heart attack. When the infection-fighting immune cells in your bloodstream can't tell when a virus is nearby because they are senescent, you are more susceptible to catching the flu or pneumonia. **Senescent cells can leak proinflammatory substances that make you vulnerable to more pain, more chronic illness. Eventually, many senescent cells will undergo a preprogrammed death.**

The diseasespan begins.

Many healthy human cells can divide repeatedly, so long as their telomeres (and other crucial building blocks of cells like proteins) remain functional. After that, the cells become senescent. Eventually, senescence can even happen to our amazing stem cells. This limit on cells dividing is one reason that there seems to be a natural winding down of the human healthspan as we age into our seventies and eighties, although of course many people live healthy lives much longer. A good healthspan and life span, reaching eighty to one hundred years for some of us and many of our children, is within our reach.[2] There are around three hundred thousand centenarians

9

worldwide, and their numbers are rapidly increasing. Even more so are the numbers of people living into their nineties. Based on trends, it is thought that over one-third of children born in the United Kingdom now will live to one hundred years.[3] How many of those years will be darkened by diseasespan? If we better understand the levers on good cell renewal, we can have joints that move fluidly, lungs that breathe easily, immune cells that fiercely fight infections, a heart that keeps pumping blood through its four chambers, and a brain that is sharp throughout the elderly years.

But sometimes cells don't make it through all their divisions in the way they should. Sometimes they stop dividing earlier, falling into an old, senescent stage before their time. When this happens, you don't get those eight or nine great decades. Instead, you get premature cellular aging. Premature cellular aging is what happens to people like Kara, whose healthspan graph turns dark at an early age.

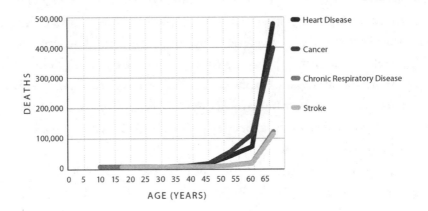

Figure 3: Aging and Disease. Age is by far the largest determinant of chronic diseases. This graph shows the frequency of death by age, up to age sixty-five and older, for the top four causes of death by disease (heart disease, cancer, respiratory disease, and stroke and other cerebrovascular diseases). The death rate due to chronic diseases starts to increase after age forty and goes up dramatically after age sixty. Adapted from U.S. Department of Health and Human Services, Centers for Disease Control and Prevention, "Ten Leading Causes of Death and Injury," http://www.cdc.gov/injury/wisqars/leadingCauses.html.

Chronological age is the major determinant of when we get diseases, and this reflects our biological aging inside.

At the beginning of the chapter, we asked, *Why do people age differently?* One reason is cellular aging. Now the question becomes, *What causes cells to get old before their time?*

For an answer to this question, think of shoelaces.

HOW TELOMERES CAN MAKE YOU FEEL OLD OR HELP YOU STAY YOUNG AND HEALTHY

Do you know the protective plastic tips at the ends of shoelaces? These are called aglets. The aglets are there to keep shoelaces from fraying. Now imagine that your shoelaces are your chromosomes, the structures inside your cells that carry your genetic information. Telomeres, which can be measured in units of DNA known as base pairs, are like the aglets; they form little caps at the ends of the chromosomes and keep the genetic material from unraveling. They are the aglets of aging. But telomeres tend to shorten over time.

Here's a typical trajectory for the life of a human's telomere:

Age	Telomere Length (in base pairs)
Newborn baby	10,000 base pairs
35 years old	7,500 base pairs
65 years old	4,800 base pairs

When your shoelace tips wear down too far, the shoelaces become unusable. You may as well throw them away. Something similar happens to cells. When telomeres become too short, the cell stops dividing altogether. Telomeres aren't the only reason a cell can become senescent. There are other stresses on normal cells that we don't yet understand very well. But short telomeres are one of the primary reasons human cells grow old, and they are one mechanism that controls the Hayflick limit.

Your genes affect your telomeres, both their length when you're born and how quickly they dwindle down. But the wonderful news is that our research, along with research from around the globe, has shown you can step in and take some control of how short or long—how *robust*—they are.

For instance:

- Some of us respond to difficult situations by feeling highly threatened—and this response is linked to shorter telomeres. We can reframe our view of situations in a more positive way.
- Several mind-body techniques, including meditation and Qigong, have been shown to reduce stress *and* to increase telomerase, the enzyme that replenishes telomeres.
- Exercise that promotes cardiovascular fitness is great for telomeres. We describe two simple workout programs that have been shown to improve telomere maintenance, and these programs can accommodate all fitness levels.
- Telomeres hate processed meats like hot dogs, but fresh, whole foods are good for them.
- Neighborhoods that are low in social cohesion—meaning that people don't know and trust one another—are bad for telomeres. This is true no matter what the income level.
- Children who are exposed to several adverse life events have shorter telomeres. Moving children away from neglectful circumstances (such as the notorious Romanian orphanages) can reverse some of the damage.
- Telomeres on the parents' chromosomes in the egg and sperm are directly transmitted to the developing baby. Remarkably, this means that if your parents had hard lives that shortened their telomeres, they could have passed those shortened telomeres on to you! If you think that might be the case, don't panic. Telomeres can build up as well as shorten. **You can still take action to keep your telomeres stable. And this**

news also means that our own life choices can result in a positive cellular legacy for the next generation.

MAKE THE TELOMERE CONNECTION

When you think about living in a healthier way, you may think, with a groan, about a long list of things you ought to be doing. For some people, though, when they have seen and understood the connection between their actions and their telomeres, they are able to make changes that last. When I (Liz) walk to the office, people sometimes stop me to say, "Look, I'm biking to work now—I'm keeping my telomeres long!" Or "I stopped drinking sugary soda. I hated to think of what it was doing to my telomeres."

WHAT'S AHEAD

Does our research show that by maintaining your telomeres you will live into your hundreds, or run marathons when you're ninety-four, or stay wrinkle free? No. Everyone's cells become old and eventually we die. But imagine that you're driving on a highway. There are fast lanes, there are slow lanes, and there are lanes in between. You can drive in the fast lane, barreling toward the diseasespan at an accelerated pace. Or you can drive in a slower lane, taking more time to enjoy the weather, the music, and the company in the passenger seat. And, of course, you'll enjoy your good health.

Even if you are currently on a fast track to premature cellular aging, you can switch lanes. In the pages ahead, you'll see how to make this happen. In the first part of the book, we'll explain more about the dangers of premature cellular aging—and how healthy telomeres are a secret weapon against this enemy. We'll also tell you about the discovery of telomerase, an enzyme in our cells that helps keep the protective sheaths around our chromosome ends in good shape.

The rest of the book shows you how to use telomere science to support your cells. Begin with changes that you can make to your mental habits and then to your body—to the kinds of exercise, food, and sleep routines that are best for telomeres. Then expand outward to determine whether your social and physical environments support your telomere health. Throughout the book, sections called "Renewal Labs" offer suggestions that can help you prevent premature cellular aging, along with an explanation of the science behind those suggestions.

By cultivating your telomeres, you can optimize your chances of living a life that is not just longer but better. That is, in fact, why we've written this book. In the course of our work on telomeres we've seen too many Karas—too many men and women whose telomeres are wearing down too fast, who enter the diseasespan when they should still feel vibrant. There is abundant high-quality research, published in prestigious scientific journals and backed by the best labs and universities, that can guide you toward avoiding this fate. We could wait for those studies to trickle down through the media and make their way into magazines and onto health websites, but that process can take many years and is piecemeal, and, sadly, information often gets distorted along the way. We want to share what we know now—and

THE HOLY GRAIL?

Telomeres are an integrative index of many lifetime influences, both the good, restorative ones like good fitness and sleep, and also malign ones like toxic stress or poor nutrition or adversities. Birds, fish, and mice also show the stress-telomere relationship. Thus it's been suggested that telomere length may be the "Holy Grail for cumulative welfare,"[4] to be used as a summative measure of the animals' lifetime experiences. In humans, as in animals, while there will be no one biological indicator of cumulative lifetime experience, telomeres are among one of the most helpful indicators that we know of right now.

we don't want more people or their families to suffer the consequences of unnecessary premature cellular aging.

When we lose people to poor health, we lose a precious resource. Poor health often saps your mental and physical ability to live as you wish. When people in their thirties, forties, fifties, sixties, and beyond are healthier, they will enjoy themselves more and will share their gifts. They can more easily use their time in meaningful ways—to nurture and educate the next generation, to support other people, solve social problems, develop as artists, make scientific or technological discoveries, travel and share their experiences, grow businesses, or serve as wise leaders. As you read this book, you are going to learn a lot more about how to keep your cells healthy. We hope you're going to enjoy hearing how easy it is to extend your healthspan. And we hope you're going to enjoy asking yourself the question: *How am I going to use all those wonderful years of good health?* Follow a bit of the advice in this book, and chances are that you'll have plenty of time, energy, and vitality to come up with an answer.

RENEWAL BEGINS RIGHT NOW

You can start to renew your telomeres, and your cells, right now. One study has found that people who tend to focus their minds more on what they are currently doing have longer telomeres than people whose minds tend to wander more.[5] Other studies find that taking a class that offers training in mindfulness or meditation is linked to improved telomere maintenance.[6]

Mental focus is a skill that you can cultivate. All it takes is practice. You'll see a shoelace icon, pictured here, throughout the book. Whenever you see it—or whenever you see your own shoes with or without laces—you might use it as a cue to pause and ask yourself what you're thinking. Where are your thoughts right now? If you're worrying or rehashing old problems, gently remind yourself to focus on whatever it is you're doing. And if you are not "doing" anything at all, then you can enjoy focusing on "being."

Simply focus on your breath, bringing all of your awareness to this simple action of breathing in and out. It is restorative to focus your mind inside (noticing sensations, your rhythmic breathing), or outside (noticing the sights and sounds around you). This ability to focus on your breath, or your present experience, turns out to be very good for the cells of your body.

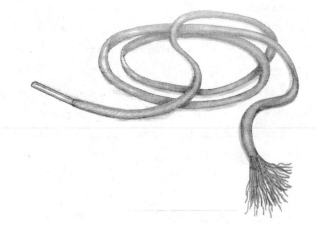

Figure 4: Think of Your Shoelaces. Shoelace tips are a metaphor for telomeres. The longer the protective aglets at the ends of the laces, the less likely the shoelace will fray. In terms of chromosomes, the longer the telomeres, the less likely there will be any alarms going off in cells or fusions of chromosomes. Fusions trigger chromosome instability and DNA breakage, which are catastrophic events for the cell.

Throughout the book, you will see a shoelace icon with long aglets. You can use that as an opportunity to refocus your mind on the present, take a deep breath, and think of your telomeres being restored with the vitality of your breath.

PART I

TELOMERES: A PATHWAY TO LIVING YOUNGER

CHAPTER ONE

How Prematurely Aging Cells Make You Look, Feel, and Act Old

Ask yourself these questions:

1. How old do I look?

- ▪ I look younger than my age.
- ▪ I look about my age.
- ▪ I look older than my age.

2. How would I rate my physical health?

- ▪ I'm in better health than most people my age.
- ▪ I'm about as healthy as most people my age.
- ▪ I'm less healthy than most people my age.

3. How old do I feel?

- ▪ I feel younger than my age.
- ▪ I feel about my age.
- ▪ I feel older than my age.

These three questions are simple, but your answers can reveal important trends in your health and aging. People who look older than their age can in fact be experiencing the early hair graying or skin damage associated with shorter telomeres. Poor physical health can happen for lots of reasons, but an early entry into the diseasespan is often a sign that your cells are aging. And studies show that people who feel older than their biological age also tend to be sicker earlier than people who feel younger.

When people say that they fear getting older, what they usually mean is that they fear a long, drawn-out diseasespan. They fear trouble getting up the stairs, trouble recovering from open-heart surgery, trouble wheeling around a tank of oxygen; they fear loss of bone, curved backs, and the dreaded loss of memory and of mind. And they fear a consequence of all these: loss of opportunities for healthy social connections and the need to replace those with dependency on others. But really, aging doesn't have to be so traumatic.

If your answers to our three questions suggest that you look and feel older than your age, perhaps it's because your telomeres are wearing down faster than they should. Those short telomeres could be sending your cells a signal that it's time to fast-forward the aging process. It's an alarming scenario, but take heart. There's plenty you can do to fight premature aging where it counts the most: at the cellular level.

However, you can't successfully fight your enemy until you really understand it.

In this section of the book, we'll give you the knowledge you need before you begin the battle. This first chapter scouts out what happens during premature cellular aging. You'll take a close-up look at aging cells and see why they are so damaging to your body and brain. You'll also discover why many of the most frightening and debilitating diseases are linked to short telomeres and thus cell aging. Then, in chapters 2 and 3, you'll see how telomeres and the fascinating enzyme telomerase (pronounced *tell-OMM-er-ase*) can either trigger an early diseasespan or work to keep your cells healthy.

HOW ARE PREMATURELY AGING CELLS DIFFERENT FROM HEALTHY CELLS?

Think of the human body as a barrel full of apples. A healthy human cell is like one of these fresh, shiny apples. But what if there is a rotten apple in that barrel? Not only can't you eat it, but worse, it will start to make the other apples around it rotten, too. That rotten apple is like an aged, senescent cell in your body.

Before we explain why, we want to return to the fact that your body is full of cells that need to constantly renew themselves to stay healthy. These renewing cells, which are called proliferative cells, live in places like your:

- immune system
- gut
- bones
- lungs
- liver
- skin
- hair follicles
- pancreas
- cardiovascular system lining
- heart's smooth muscle cells
- brain, in parts including the hippocampus (a learning and memory center of the brain)

For these crucial body tissues to stay healthy, their cells need to keep renewing. Your body has finely calibrated systems for assessing when a cell needs to be renewed; even though a body tissue can look the same for years, it is constantly being replaced by new cells in just the right numbers and at the correct rate. But remember that some cells have a limit to how many times they can divide. When cells can no longer renew themselves, the body tissues they supply will start to age and function poorly.

Cells in our tissues originate from stem cells, which have the amazing ability to become many different types of specialized cells. They live in stem cell niches, which are a kind of VIP lounge where stem cells are protected and lie dormant until they are needed. The niches are usually in or near the tissues that the stem cells will replace. Stem cells for skin live under the hair follicles, some stem cells for the heart live in the right ventricular wall, and muscle stem cells live deep in the fiber of the muscle. If all is well, the stem cells remain in their niche. But when there is a need to replenish tissues, the stem cell appears on deck. It divides and produces proliferative cells—sometimes called progenitor cells—and some of their progeny cells transform into whatever specialized cell is needed. If you get sick and need more immune cells (white blood cells), freshly divided stem cells for blood that were hiding out in the bone marrow will enter the bloodstream. Your gut lining is constantly being worn down by normal digestive processes, and your skin is being sloughed off, and stem cells keep these body tissues replenished. If you go jogging and tear your calf muscle, some of your muscle stem cells will divide, each stem cell creating two new cells. One of those cells replaces the original stem cell and remains comfortably in its niche; the other can become a muscle cell and help replenish the damaged tissue. Having a good supply of stem cells that are able to renew themselves is key to staying healthy and to recovering from sickness and injury.

But when a cell's telomeres become too short, they send out signals that put the cell's cycle of division and replication under arrest. An arrested cell stops in its tracks. The cell can no longer renew itself. It becomes old; it becomes senescent. If it is a stem cell, it goes into permanent retirement and will no longer leave its cozy niche when it is needed. Other cells that become senescent just sit around, unable to do the things they're supposed to do. Their internal powerhouses, the mitochondria, don't work properly, causing a kind of energy crisis.

An old cell's DNA can't communicate well with the other parts of the cell, and the cell can't keep house well. The old cell gets crowded inside, with—among other things—clumps of malfunctioning proteins and brown globs of "junk" known as lipofuscin, which can cause macular degeneration in the eyes and some neurological diseases. Worse still—and why they are like rotten apples in a barrel—senescent cells send out false alarms in the form of pro-inflammatory substances, reaching other parts of the body as well.

The same basic process of aging happens across the different types of cells in our bodies, whether they are liver cells, skin cells, hair follicle cells, or the cells that line our blood vessels. But there are some twists on the process that depend on the cell's type and its location in the body. Senescent cells in the bone marrow prevent blood and immune stem cells from dividing the way they're supposed to, or warp them into making blood cells in unbalanced amounts. Senescent cells in the pancreas may not correctly "hear" signals that regulate their production of insulin. Senescent cells in the brain may secrete substances that cause neurons to die. While the underlying process of aging is similar in most of the cells that have been studied, a cell's way of expressing that aging process can create different kinds of injury to the body.

Aging can be defined as the cell's "progressive functional impairment and reduced capacity to respond appropriately to environmental stimuli and injuries." Aged cells can no longer respond to stresses normally, whether the stress is physical or psychological.[1] This process is a continuum that often silently and slowly segues into the diseases of aging—diseases that can be traced, in part, to shorter telomeres and aging cells. To understand aging and telomeres a little better, let's go back to the three questions that we asked you at the beginning of this chapter:

How old do you look?
How would you rate your physical health?
How old do you feel?

OUT WITH THE OLD, IN WITH THE NEW: REMOVING SENESCENT CELLS IN MICE REVERSES PREMATURE AGING

One laboratory study followed mice that had been genetically altered so that many of their cells were senescent much earlier than usual. The mice began to age prematurely—they lost fat deposits, which made them look wrinkled; their muscles withered; their hearts weakened; and they developed cataracts. Some died early of heart failure. Then, in an experimental genetic trick that is not possible to replicate in humans, the researchers removed the mice's senescent cells. Taking out the senescent cells reversed many of the symptoms of premature aging. It cleaned up their cataracts and restored their wasted muscles, maintained their fat deposits (which reduced their wrinkles), and promoted a longer healthspan.[2]
Senescent cells control the aging process!

PREMATURELY AGING CELLS: HOW OLD DO YOU LOOK?

Age spots and blotches. Gray hair. The shrunken or stooped posture that comes with bone loss. These changes happen to all of us, but if you've been to a high school reunion recently, you've seen proof that they don't happen at the same time or in the same way.

Walk through the doors of your tenth high school reunion, when everyone is still in their twenties, and you'll spot classmates who sport expensive clothes—and classmates whose party outfits look a bit threadbare. Some classmates are parading their career successes, start-up companies, or productivity in offspring, and others are gulping down Scotch while commiserating about their latest heartbreaks. It may not seem fair. But in terms of the physical signs of aging, it's a level field. Almost everyone in the room—no matter whether they're rich, poor, successful, struggling, happy, or sad—will *look* like they're in their twenties. Their hair is healthy, their skin is clear, and a few of your classmates are an inch or two taller than when your class graduated ten years ago. They are in the radiant peak of young adulthood.

But show up for a reunion five or ten years later, and a different scene emerges. You'll notice that a few of your old classmates are starting to look like *old* classmates. They're a little gray around the ears or showing more forehead. Their skin looks speckled and cloudier; the crow's-feet around their eyes may be etched deeply. They may have protruding bellies and may even look a bit hunched over. These people are experiencing a rapid onset of outward physical aging.

Yet other classmates are graced with a slower aging trajectory. Over the years, as your twentieth, thirtieth, fortieth, fiftieth, and sixtieth reunions come and go, it's evident that the hair, faces, and bodies of these lucky classmates are changing—but these changes happen slowly and gradually, with elegance. Telomeres, as you'll see, play at least a small role in how quickly you develop an aged appearance, and whether you become one of those people who "age well."

Skin Aging

The skin's outer layer, the epidermis, is made up of proliferating cells that are constantly replenishing themselves. Some of these skin cells (the keratinocytes) make telomerase, so they don't wear out and become senescent cells, but most do slow down in their ability to replenish.[3] Underneath this visible layer of skin is the dermis, a layer of skin cells (skin fibroblasts) that creates the foundation for a healthy, plump epidermis—by producing collagen, elastin, and growth-promoting factors, for example.

With age, these fibroblasts secrete less collagen and elastin, which makes the outer layer of visible skin look old and loose. This effect translates upward through the skin layers to create a more aged outer appearance. Aged skin becomes thinner, as it loses fat pads and hyaluronic acid (which acts as a natural moisturizer for skin and joints). It becomes more permeable to the elements.[4] The aged melanocytes lead to age spots but also paleness. In short, the aging skin gets the all-too-familiar spotty, pale, saggy, and wrinkly look, mainly due to the aging fibroblasts that no longer support the outer cells.

In older people, skin cells often lose their ability to divide. Some older people do have skin cells that can still keep dividing. When researchers peer into their cells, they see the cells are better at fending off oxidative stress and have longer telomeres.[5] Although short telomeres don't necessarily cause aging skin, they play a role, especially when it comes to aging from the sun (also called photoaging). The UV rays from sun exposure can damage telomeres.[6] Petra Boukamp, a telomere skin researcher from the German Cancer Research Center in Heidelberg, and her colleagues have compared skin from a sun-exposed site—the neck—to a sun-protected site—the buttocks. The outer cells on the neck showed some telomere attrition from the sun, whereas the protected buttock cells showed almost no telomere attrition with aging! Skin cells, when protected from the sun, can withstand aging for a long time.

Bone Loss

Your bone tissue is remodeled throughout your life, and a healthy level of bone density results when you have a balance between the bone-building cells (osteoblasts) and the bone-busting cells (osteoclasts). Osteoblast cells need healthy telomeres in order to keep dividing and replenishing themselves—and when your telomeres are short, the osteoblasts get old and can't keep up with the osteoclasts. The balance tips, and the osteoclasts nibble away at your bones.[7] It doesn't help that after a person's telomeres wear down, the old bone cells become inflammatory. Lab mice bred to have extra-short telomeres suffer from early bone loss and osteoporosis;[8] so do people who have been born with a genetic disorder that causes their telomeres to be extraordinarily short.

Graying Hair

In a sense, we're all born with hair that's been colored. Each strand of hair begins inside its own follicle and is made from keratin, which produces a white hair. But there are special cells inside the follicle—melanocytes, the same kinds of cells responsible for skin color—that inject the hair with pigment. Without these natural hair-dye cells, hair color is lost. Stem cells in the follicle produce the melanocytes. When these stem cells' telomeres wear down, the cells can't replenish themselves fast enough to keep up with hair growth, and gray hair is a result. Eventually, when all the melanocytes have died, hair becomes pure white. Melanocytes are also sensitive to chemical stressors and to ultraviolet radiation; and in a study published in the journal *Cell*, mice who underwent X-rays developed damaged melanocytes and gray fur.[9] Mice with a genetic mutation that causes extremely short telomeres also develop early graying of their fur, and restoring telomerase turns their gray fur dark again.[10]

What's normal graying? Graying happens least in African Americans and Asians, and most in blonds.[11] Graying starts to happen in at least half of people by their late forties, and around 90 percent of people in their early sixties. The vast majority of cases of early graying are quite normal; only a very few people who find themselves with gray or white hair at an early age, in their thirties, may carry a genetic mutation that causes short telomeres.

What Does Your Appearance Say About Your Health?

Maybe you're thinking, "Well, I don't really mind having a few early gray hairs. And are a couple of age spots around my eyes really such a big deal? Aren't you asking me to focus on the wrong things—to value a youthful appearance instead of my health?" These are great questions. There is no contest here: health is what matters. But how much does aged appearance reflect inner health? Researchers have asked specially trained "raters" to estimate the age of a person just by looking at a photo.[12] As it turns out, the people who look older on average have shorter telomeres. This is not surprising, given the role that telomeres appear to play in skin aging and hair graying. Looking older is associated in small but worrisome ways with signs of poor physical health. People who look old tend to be weaker, to perform worse on a mental exam that tests memory, to have higher fasting glucose and cortisol levels, and to show early signs of cardiovascular disease.[13] The good news is that these are *very small effects*. What's on the inside of your body is what matters most, but looking older than your age—looking haggard—is a sign worth paying attention to. It may be an indicator that your telomeres need more protection.

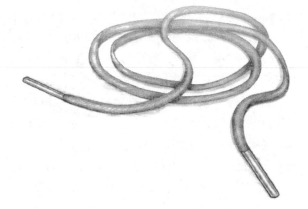

Remember what to do when you see this image? See page 16.

PREMATURE CELLULAR AGING: HOW IS YOUR PHYSICAL HEALTH?

You can see the real power of short telomeres to damage your cells and your health when you consider the next question: *How would you rate your physical health?*

Think again about your high school reunions. As you reach your twentieth or thirtieth reunion, you'll notice that many of your classmates are starting to suffer from the common diseases of aging. Yet they're only about forty or fifty. They're not chronologically old yet. So why are their bodies *acting* as if they are? Why are they entering the diseasespan at young ages?

Inflamm-Aging

Wouldn't it be interesting if you could peer into the cells deep inside each person at your reunion and measure the lengths of each person's telomeres? If you could, you would see that the people with the shortest telomeres are on average the ones who are sicker, weaker, or whose faces show the strain of coping with health problems like diabetes, cardiovascular disease, a weakened immune system, and lung diseases. You would probably also find that the ones with the shortest telomeres suffer from chronic inflammation. The observation that inflammation increases with age and is a cause of the diseases of aging is so important that scientists have a name for it: *inflamm-aging*. This is a persistent, low-grade inflammation that can accumulate with age. There are many reasons why this occurs, such as proteins becoming damaged. One other common cause of inflamm-aging involves telomere damage.

When a cell's genes are damaged or its telomeres are too short, that cell knows its precious DNA is in danger. The cell reprograms itself so that it emits molecules that can travel to other cells and call for help. These molecules, together called senescence-associated secretory phenotype, or SASP, can be useful. If a cell has become senescent because it's been wounded, it can send signals to neighboring immune cells and other cells with repair functions, to call in the squads that can get the healing process going.

And here's where things go terribly wrong. Telomeres have an abnormal response to DNA damage. The telomere is so preoccupied with protecting itself that even though the cell has called out for help, the telomere won't let the help in. It's like people who doggedly refuse assistance in the face of adversity because they're afraid to let their guard down. A shortened telomere can sit inside an aging cell for months, signaling and signaling for help but not allowing the cell to take action to resolve the damage. This unremitting but futile signaling can have devastating consequences. Because now that cell becomes like the rotten apple in the barrel. It starts affecting all the tissues around it. The SASP process involves chemicals like proinflammatory cytokines that, over time, travel through the body, leading to system-wide chronic inflammation. Judith Campisi of the Buck Institute of Aging discovered SASP, and she has shown that these cells create a friendly territory for cancer growth.

In the past decade or so, scientists have come to recognize that chronic inflammation (from SASP or other sources) is a crucial player in causing many diseases. Short-term, acute inflammation brings healing to injured cells, but long-term inflammation interferes with the normal functioning of body tissues. For example, chronic inflammation can cause cells of the pancreas to malfunction and not regulate insulin production properly, setting the stage for diabetes. It

Figure 5: A Rotten Apple in an Apple Barrel. Think of a barrel of apples. An apple barrel's health depends on each apple. One rotten apple sends out gases that rot the other apples. One senescent cell sends signals to surrounding cells, promoting inflammation and factors that promote what we might call "cell rot."

can cause plaque on an artery wall to burst. It can cause the body's immune response to turn on itself, so that it attacks its own tissues.

These are just a few of the most harrowing examples of inflammation's destructive power, but the list marches on to a deadly drumbeat. Chronic inflammation is also a factor in heart disease, brain diseases, gum disease, Crohn's disease, celiac disease, rheumatoid arthritis, asthma, hepatitis, cancers, and more. That's why scientists talk about inflamm-aging. It's real.

If you want to slow inflamm-aging, if you want to stay in the healthspan for as long as possible, you've got to prevent chronic inflammation. And a big part of controlling inflammation means protecting your telomeres. Since cells with very short telomeres

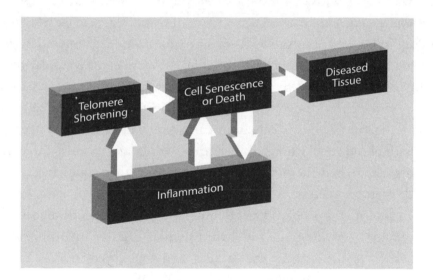

Figure 6: A Path from Short Telomeres to Disease. One early pathway to disease is telomere shortening. Shortened telomeres lead to senescent cells that either stick around or—if we are lucky—are removed from the scene early. While there are many factors that can cause senescence, telomere damage is a common one in humans. When the old senescent cells build up over decades to a critical mass, they become the foundation for diseased tissue. Inflammation is a cause of both telomere shortening and senescent cells, and senescent cells in turn create more inflammation.

send out constant inflammatory signals, you need to keep those telomeres a healthy length.

Heart Disease and Short Telomeres

Each of our arteries, from the large to the small, is lined with layers of cells called the endothelium. If you want your cardiovascular system to stay healthy, the cells of the endothelium need to replenish themselves, protect the lining, and keep immune cells from entering the arterial wall.

But in people with short telomeres in their white blood cells, the risk of cardiovascular disease goes up. (Usually, short telomeres in the blood reveal short telomeres in other tissues, like the endothelium.) People with common genetic variations that lead to shorter telomeres are also more prone to cardiovascular problems.[14] Just being in the bottom third of the population's blood telomere length means you are 40 percent more likely to develop cardiovascular disease in the future.[15] Why? We don't know all the pathways, but vascular senescence is one of them: When short telomeres tell cells to age prematurely, the endothelium can't renew itself to make strong, smooth blood vessel linings. It becomes weaker and more vulnerable to disease. When the actual vascular tissue with plaques is examined, short telomeres are indeed found.

In addition, short blood cell telomeres can also trigger inflammation, which sets the stage for cardiovascular disease. Inflammatory cells stick to the sides of the arteries and trap cholesterol to form plaques or make existing plaques unstable. If a plaque ruptures, a blood clot can form over the plaque, blocking the artery. And if this artery is a coronary artery, it chokes off the heart's blood supply and causes a heart attack.

Lung Diseases and Short Telomeres

People with asthma, chronic obstructive pulmonary disease (COPD), and pulmonary fibrosis (a very serious, irreversible disease in which scarred lung tissue leads to difficult breathing) have shorter telomeres in their immune cells and lung cells than people who are healthy. Pulmonary fibrosis in particular clearly results from not having good telomere maintenance; the proof for this comes from finding pulmonary fibrosis in those unlucky people with rare inherited telomere maintenance gene mutations. Along with this revealing fact, several other lines of smoking gun evidence exist. Together they point a strongly incriminating finger to inadequate telomere maintenance as a shared underlying problem that contributes to COPD, asthma, lung infections, and poor lung functioning—and this is true in everyone, not just people with a rare mutation in a telomere maintenance gene. Lacking robust telomere maintenance, lung stem cells and blood vessels of the lungs become senescent. They cannot keep the lung tissues replenished and supplied with their needs. Immune-cell senescence creates a proinflammatory environment that further taxes the lungs, so they function more and more poorly.

PREMATURE CELLULAR AGING: HOW OLD DO YOU FEEL?

Let's return to your high school reunion—this time, let's go to your fortieth reunion, when your classmates are closing in on age sixty. This is when the first people in your class start to show signs of cognitive slowing. It may be hard to put your finger on what exactly is different about these folks, but you'll notice that they seem a little fuzzier, a little out of it, a little less focused, and less tuned in to normal social cues. They may take a few extra seconds to remember your name. This mental loss, more than anything else, is what makes us feel really, truly old.

Cognitive Slowing and Alzheimer's Disease

You won't be surprised to hear that the people who have early cognitive problems also tend to have shorter telomeres. This effect may persist as people get older. In one study of otherwise healthy seventy-year-olds, shorter telomeres predicted general cognitive decline years later.[16] In young adults, there was no relationship between telomeres and cognitive function, but greater telomere shortening over approximately ten years predicted poorer cognitive function.[17] Researchers are fascinated by a possible relationship between the lengths of our telomeres and the sharpness of our thinking. Can short telomeres predict dementia or Alzheimer's disease?

A large, impressive Texas study set out to help answer this question.[18] Researchers imaged the brains of almost two thousand adults from Dallas County. The study controlled for age as well as other factors that affect the brain, such as smoking, gender, and the status of a gene, APOE-epsilon 4 (commonly called just APOE). A normal variant of APOE increases a person's risk of Alzheimer's. As expected, nearly everyone's brain showed some signs of shrinking with age. But then the researchers examined the parts of the brain that are specifically involved in emotions and memory. The hippocampus, for example, is a part of the brain that helps form, organize, and store memories; it also helps link those memories to your emotions and your senses. The hippocampus is a reason that the scent of a new box of erasers drops you back into the first day of grade school; it's a reason you can remember grade school at all. Amazingly, the Texas researchers found that when people had short telomeres in their white blood cells (which serve as a window into telomere length throughout the body), their hippocampuses were smaller than those of people with longer telomeres. The hippocampus is made up of cells that need to regenerate—and if you want to have good memory function, it's essential for your body to be able to replenish the hippocampus's cells.

It isn't just the hippocampus that is smaller in people with short telomeres. So are other areas of the brain's limbic system, including the amygdala and the temporal and parietal lobes. These areas, along with the hippocampus, help regulate memory, emotion, and stress—*and* they are the very same areas that atrophy in Alzheimer's disease. The Dallas study suggests that *short telomeres in the blood crudely indicate an aging brain.* It is possible that cellular aging, perhaps just in the hippocampus or perhaps throughout the body, may underlie an important pathway of dementia. Keeping telomeres healthy may be especially crucial for people who carry the variant of APOE gene that puts them at a higher risk for early Alzheimer's. One study found if you have this gene variant and also have short telomeres, your risk of dying earlier is nine times greater than if you have the same gene variant but your telomeres are long.[19]

Short telomeres may help cause Alzheimer's directly. There are common genetic variations (in genes called TERT and OBFC1) that can lead to short telomeres. Remarkably, people who have even just one gene with these common variations are statistically more likely to develop Alzheimer's.[20] This is not a big effect, but it demonstrates a causal relationship: telomeres are not just a marker for something else, or an epiphenomenon, but rather they are causing part of brain aging—putting us at greater risk for neurodegenerative disease processes. TERT and OBFC1 directly function to maintain telomeres in well-understood ways. The evidence keeps growing. If you want to keep your brain sharp, think about your telomeres. See notes at the back of this book for a research opportunity on brain aging.[21]

A Healthy "Felt Age"

If you went to your fortieth reunion, climbed up onto the stage, and asked this group of sixty-year-olds to raise their hands if they *feel* like sixty-year-olds, you'd get an interesting result. A majority of people—75 percent—would say that they feel younger than their age. Even as the years go by, and even as the date of birth on our driver's license tells us that we are getting older, many of us still *feel* young.[22] This response to aging is highly adaptive. Having a younger "felt age" is associated with more life satisfaction, personal growth, and social connections with others.[23]

Feeling younger is different from wishing to *be* younger. People who long to be a chronologically younger age (say, a man in his fifties who wishes he could be thirty again) tend to be unhappier and more dissatisfied with their lives. Wishing and longing for youth is really the opposite of our major developmental task as we age, which is to accept ourselves as we are, even while working toward maintaining our mental and physical fitness.

FOR A HEALTHIER OLD AGE, CHANGE HOW YOU THINK ABOUT IT

Be careful of how you think about old people. People who internalize and accept negative age stereotypes may *become* age stereotypes—they may develop more health challenges. This phenomenon, called stereotype embodiment, was identified by Becca Levy, a social psychologist at Yale University. Even when their current health status is taken into account, people who have negative beliefs about aging act differently from people who have a sunnier view of aging.[24] They believe they have less control over whether they develop disease, and they don't work as hard at health behaviors, such as taking prescribed medications. They're more than twice as likely to die of a heart attack, and as the decades go by they experience a steeper decline in memory. When they are injured or sick, they recover more slowly.[25] In another study, elderly people who were simply reminded of age stereotypes performed so poorly on a test that they scored as low as if they had dementia.[26]

If you have a negative view of aging, you can make a conscious effort to counter it. Here's a list of stereotypes we've adapted from Levy's Image of Aging Scale.[27] You might visualize yourself thriving in old age, embodying some of those positive traits. When you catch yourself thinking of old age negatively, remind yourself of the positive side of aging:

What's Your Image of Aging?	
grumpy	optimistic
dependent	capable
slow	full of vitality
frail	self-reliant
lonely	strong will to live
confused	wise
nostalgic	emotionally complex
distrustful	close relationships
bitter	loving

What is the profile of our emotional life as we age? Despite the image of older people as cranky or resentful of the young, Laura Carstensen, a researcher of aging at Stanford University, shows that our daily emotional experience is actually enhanced with age. Typically, older people experience more positive emotions than negative ones in daily life. The experience isn't purely "happy." Rather, our emotions grow richer and more complex over time. We experience more co-occurrence of positive and negative emotions, such as those poignant occasions when you get a tear in the eye at the same time you feel joy, or feeling pride at the same time you feel anger[28]—a capacity we call "emotional complexity." These mixed emotional states help us avoid the dramatic ups and downs that younger people have, and they also help us exercise more control over what we feel. Mixed emotions are easier to manage than purely positive or purely negative emotions. Thus, emotionally speaking, life just feels better. Better control over emotions and enhanced complexity means more enriched daily experiences.

People with more emotional complexity also have a longer healthspan.[29]

Gerontology researchers also know that we maintain interest in intimacy and sex as we age. Our social circles get smaller, but this is largely by choice. Over time, we shape our social circles to include the most meaningful relationships, and we weed away those more troublesome relationships. This leads to days with more positive feelings and less stress. We prioritize things better and focus our time on things that matter most to us. Perhaps that's one way to describe the wisdom of age.

Your efforts to imagine a better, healthier, more vibrant old age will pay off. Levy reminded older people of the benefits of aging—like enlightenment and accomplishment—and then gave them stressful tasks to perform. She found that they responded to the stress with less reactivity (lower heart rate and blood pressure) than a control group.[30] As the saying goes, "Age is an issue of mind over matter. If you don't mind, it doesn't matter."

TWO PATHS

Pause for a minute. Imagine what your future might look like if your telomeres shorten too quickly and your cells begin to age prematurely. This thought exercise is intended to make premature cellular aging vivid and real for you. Think about the kind of aging you *don't* want to experience in your forties, fifties, sixties, and seventies. Do you dread scenarios like these?

- "I've lost my sharpness. When I talk, my younger colleague's eyes glaze over because I'm rambling and unfocused."
- "I'm always in bed with a respiratory infection; I seem to catch every illness."

- "It is hard to breathe."
- "My legs feel numb."
- "My feet aren't steady underneath me. I'm scared of falling."
- "I'm too tired to do anything but sit on the couch watching TV all day."
- "I overhear my children saying, 'Whose turn is it to take care of Mom?'"
- "I can't travel the way I'd planned to, because I want to stay close to my doctors."

These statements reveal aspects of life with an early diseasespan—the kind of life you want to avoid. You may have parents or grand-parents who believed the old myth that everyone gets a few good decades and then it's time to get sick or give up. We all know people who turned sixty or seventy and quietly declared that their lives were finished. These are the folks who pull on their sweatpants, sit back in the recliner, and watch television until disease takes over.

Now envision a different future, one with long, healthy telomeres and with cells that renew. What do these decades of good health look like? Do you have a role model you can picture?

Aging is often portrayed in such negative ways that most of us try not to even think about it. If you had parents or grandparents who got sick early, or who simply gave up once they hit a certain age, it may be hard for you to imagine that it's possible to be old, healthy, and energetically engaged with life. But if you can form a clear, positive picture of how you'd like to age, you suddenly have a goal to work toward while aging—and a compelling reason to keep your telomeres and cells healthy. If you think of aging in a positive way, odds are that you'll live seven and a half years longer than someone who doesn't, at least according to one study![31]

One of our favorite examples of a person who is constantly renewed in spirit is my (Liz's) friend Marie-Jeanne, a delightful molecular biologist who lives in Paris. Marie-Jeanne is about eighty years old; she has white hair and wrinkles, and her back is slightly stooped, but her face is lively and intelligent. Marie-Jeanne and I met up for the afternoon recently. We had lunch. We visited the Petit Palais art museum, walking upstairs and downstairs and seeking out most of the exhibits. We explored the Latin Quarter on foot and visited bookshops. Six hours later, Marie-Jeanne was looking fresh with no sign of slowing down. I was ready to drop from exhaustion. I proposed heading back ("so Marie-Jeanne could rest"). As Marie-Jeanne suggested yet another venue to visit, I, ashamed to admit how desperate I was to put up my aching feet, conjured up a previous engagement so my tired legs could get home to collapse.

Marie-Jeanne checks many of the boxes that define healthy aging for us:

- She's stayed interested in her work over many years. Although she's officially past retirement age, she still goes into the office at her research institution.
- She socializes with all sorts of people. She hosts monthly dinner discussions (held in many languages) for her younger colleagues.
- She lives in a fifth-floor walkup apartment. At times, her younger friends have to forgo a dinner party there because they are too stiff or fatigued to get up all those flights of stairs—but Marie-Jeanne navigates them as deftly as she has for many years.
- She is always interested in new experiences, like visiting museum exhibits that come to town.

You might have your own role model, or your own goals for getting older. Here are some others we've heard:

- "When I'm older, I want to be like the actress Judi Dench, especially the way she played M in the Bond movies:

40

white-haired but completely in command, and the smartest person in the room."

- "I'm inspired by the idea of life's 'third act.' The first act of my life was all about educating myself; the second act was about growing my teaching career; and for my third act I am planning to work with not-for-profits to help teen parents stay in school and complete their degrees."
- "My grandfather took us kids cross-country skiing well into his seventies and showed us how to build fires in the snow. I want to do the same for my own grandchildren."
- "As I picture getting older, I imagine that the kids are grown up and out of the house. I miss them, but I've got more time. I can finally accept the offer to chair my department."
- "If I'm still intellectually curious, and actively working on writing projects or a philanthropic project, I'll be happy. I want to be giving back in more than one way, appreciating our beautiful planet and the best in others, including myself."

Your cells are going to age. But they don't need to age before their time. What most of us really want is to have a long, satisfying life, with advanced cellular aging pushed toward the very end.

The chapter you've just read showed you how prematurely aging cells can hurt you. Next, we're going to show you exactly what telomeres are and how they can give you the best shot at a long, good life.

CHAPTER TWO

The Power of Long Telomeres

It's 1987. Robin Huiras is twelve years old and standing on her school's playing field, waiting to begin a timed mile-long run. The weather is good for running—it's a chilly Minnesota morning— and Robin is fit and slender. Although she doesn't enjoy being put through her paces by the gym teacher, she expects to do well.

She doesn't. The gym teacher fires the starting pistol and almost immediately the other girls in the class are ahead of Robin. She tries to catch them, but the pack recedes along the red-dirt running path. Robin is no slacker—she gives everything she's got, but as the race goes on she falls farther and farther behind. Her final time is one of the slowest in the class, almost as if she'd stopped partway through the course and taken a leisurely stroll across the finish line, but long after the race is finished Robin is still doubled over from the exertion, gulping for air.

The next year, when Robin is thirteen, she spots a gray strand threading its way through her brown hair. Then another gray hair appears, and another, until her hair takes on the light salt-and-pepper appearance that's common among women in their forties or fifties. Her skin changes, too—there are days when normal activities leave deeply colored bruises on her arms and legs. Robin is only a teenager, but her energy is low, her hair is turning gray, and her skin is fragile. It's as if she's growing old before her time.

In a very real way, that was exactly what was happening. Robin has a rare telomere biology disorder, an inherited disorder that causes extremely short telomeres and, in turn, early cell aging. Well before people with telomere biology disorders are chronologically old, they can experience rapidly accelerated aging. Outwardly, it shows up in the skin. Melanocytes, for example, which are the skin's color cells, lose their ability to keep the skin evenly toned. The result is age blotches and spots, along with gray or white hair, even at a young age. Fingernails and toenails look old, too. Because nails have cells that turn over quickly, they become ridged and split. Bones grow older, too: osteoblasts—cells that your bones need to stay solid and strong—can stop renewing themselves. Robin's father, who had the same telomere biology disorder, had so much bone loss and muscle pain that he needed both of his hips replaced twice before the disorder took his life at age forty-three.

But an aged appearance and even bone loss are some of the milder effects of telomere biology disorder. The more devastating ones can include scarred lungs, unusually low blood counts, a weakened immune system, bone-marrow disorders, digestive problems, and certain cancers. People with telomere disorders do not tend to live full life spans, though the precise symptoms and the average length of life varies; one of the oldest known telomere disorder patients alive right now is in her sixties.

Severe inherited forms of telomere biology disorders such as Robin's are an extreme form of much more common conditions, which we now collectively call "telomere syndromes." We understand which genes accidentally go wrong to cause these inherited, severe forms, and what these genes do in cells. (Eleven such genes are known to date.) Thankfully, these extreme, inherited telomere syndromes are rare; they affect about one in a million people. And thankfully, Robin was eventually able to take advantage of medical advances and undergo a successful stem cell transplant (one which contained a donor's blood-forming stem cells). One testimony to the transplant's success is Robin's platelet count. Because Robin's blood stem cells could not effectively repair their telomeres, or make new cells, her platelets had plummeted to alarmingly low numbers, with counts as low as 3,000 or 4,000. (Low blood counts are a reason she couldn't keep up during the mile run.) Six months after the transplant, Robin's counts shot up to more normal levels of almost 200,000. Robin, who is now in her thirties and runs an advocacy organization for people with telomere biology disorders, has more wrinkles around her mouth and eyes than other people her age. Her hair is almost entirely gray, and she sometimes experiences severe joint and muscle pain. But habitual exercise helps keep the pain at bay, and the transplant has restored much of her energy.

Severe inherited telomere syndromes carry a powerful message for all of us, because what is happening inside Robin's cells is also happening inside your own. It's just happening to her faster than it's happening to you. In all of us, telomeres shrink with age. And premature cellular aging can happen—in a slower way—to basically healthy people. We can think of all of us as being susceptible, to some extent, to telomere syndromes of aging, although to much lesser degrees than Robin and her father. Patients with the inherited telomere syndromes are powerless to stop the premature aging process, because it takes place with overwhelming speed in their bodies, but the rest of us are luckier. We have much more control over premature cellular aging, because—to a surprising extent—we have some real control over our telomeres.

That control begins with knowledge—knowledge about telomeres and how their length corresponds to your daily habits and health. To understand the role telomeres play in your body, we need to turn to an unlikely source. We need to spend some time with pond scum.

POND SCUM SENDS A MESSAGE

Tetrahymena is a single-celled organism that swims valiantly through bodies of freshwater, searching for food or a mate. (There are seven sexes of *Tetrahymena*, a curious fact to ponder next time you are splashing around in a lake.) *Tetrahymena* is, literally, pond scum. Yet it's almost adorable. Seen under a microscope, it boasts a plump little body and hairlike projections that make it look like a fuzzy cartoon creature. Look at it long enough, and you might notice a resemblance to Bip Bippadotta, the wild-haired Muppet who scats the famously infectious song "Mahna Mahna."

Inside *Tetrahymena*'s cell is its nucleus, its central command center. Deep within that nucleus is a gift to molecular biologists: twenty thousand tiny chromosomes, all identical, linear, and very short. That gift makes it relatively easy to study *Tetrahymena*'s telomeres, those caps at the ends of chromosomes. That gift is the reason that in 1975, I (Liz) was standing in a laboratory at Yale, cultivating millions of tiny *Tetrahymena* in big glass lab jars. I wanted to collect enough of their telomeres to understand just what they were made of, at the genetic level.

For decades, scientists had theorized that telomeres protect chromosomes—not just in pond scum but in humans, too—but no one knew exactly what telomeres were or how they worked. I thought that if I could pinpoint the structure of the DNA in telomeres, I might be able to learn more about their function. I was driven by my desire to understand biology; at this point, no one knew that telomeres would prove to be one of the primary biological foundations of aging and health.

By using a mixture of what was essentially dish detergent and salts, I was able to release *Tetrahymena*'s DNA from its surrounding matter, out of the cell. Then I analyzed it, using a combination of the chemical and biochemical methods that I'd learned during my PhD graduate years in Cambridge, England. Under the dim, red, and warm safelight of the lab's darkroom, I reached my goal. The darkroom was quiet; only a trickle of water sounded as it ran next to the old-fashioned developing tanks. I held a dripping X-ray film up to the safety light, and excitement surged through me as I understood what I was seeing. At the ends of chromosomes was a simple, repeated DNA sequence. The same sequence, over and over and over. I had discovered the structure of telomere DNA. And, in the ensuing months, as I toiled over pinpointing its details, an unexpected fact rose up: Remarkably, these tiny chromosomes were not as identical as they had seemed. Some had ends with more, and some had ends with fewer numbers of the repeats.

No other DNA behaves in this strangely variable, sequential, repeating way. The telomeres of pond scum were sending a message: There

Figure 7: _Tetrahymena_. This tiny one-celled creature, which Liz studied to decode the DNA structure of telomeres and to discover telomerase, provided the first precious information about telomeres, telomerase, and a cell's life span. This foreshadowed what was later learned in humans.

is something special here at the ends of chromosomes. Something that would turn out to be vital for the health of human cells. That variability in the lengths of the ends turns out to be one of the factors that explains why some of us live longer and healthier than others.

TELOMERES: THE PROTECTORS OF OUR CHROMOSOMES

It became clear from that dripping X-ray film that telomeres are composed of repeated patterns of DNA. Your DNA consists of two parallel, twisting strands that are made up of just four building blocks ("nucleotides") that are represented by the letters A, T, C, and G. Remember grade school field trips, when you had to hold hands with a buddy as you walked through a museum? The letters of DNA operate on the buddy system, too. A always pairs with T, and

Figure 8: Telomere Strands Up Close. At the tips of the chromosome are the telomeres. The telomere strand is made up of repeating sequences of TTAGGG that sit across from their base pair partners, AATCCC. The more of these sequences we have, the longer our telomeres. In this diagram we depict just the DNA of telomeres, but it is not bare like this—it is covered by a protective sheath of proteins.

C always pairs with G. The letters on the first strand of DNA pair up with their partners on the second strand. The two make up a "base pair," which is the unit we measure telomeres in.

In human telomeres (as would later be discovered), the first strand consists of repeating sequences of TTAGGG, and they are coupled with their pairs, AATCCC, on the second strand, twisted into the helix shape that is DNA.

These are the base pairs of telomeres that, repeated thousands of times, offer a way of measuring their length. (Note some of our graphs measure telomere length in a unit called a t/s ratio, instead of base pairs, which is just another way to measure telomeres.) The repeating sequence highlights the differences between telomeres and other DNA. Genes, which are made of DNA, live within a chromosome. (Inside a cell we have twenty-three pairs of chromosomes, for a total of forty-six.) This genetic DNA is what forms your body's blueprint, its instruction manual. Its paired letters create complicated "sentences" that send instructions for building the proteins that make up your body. Genetic DNA can help determine how quickly your heart beats, whether your eyes are brown or blue, and whether you're going to have the long legs and arms of a distance runner. The DNA of telomeres is different. First of all, it doesn't live inside any gene. It sits outside of all the genes, at the very edges of the chromosome that contains genes. And unlike genetic DNA, it doesn't act like a blueprint or code. It's more like a physical buffer; it protects the chromosome during the process of cell division. Like beefy football players who surround a quarterback, absorbing the hardest blows from the onrush of opposing players, telomeres take one for the team.

This protection is crucial. As cells divide and renew, they need their precious chromosome cargoes of genetic instruction manuals (the genes) to be delivered intact. Otherwise, how would a child's body know to grow tall and strong? How would your cells know to produce the body traits that make you feel like *you*? Yet cell division is a potentially dangerous time for chromosomes and the genetic material inside. Without protection, chromosomes and the genetic material they carry could easily become unraveled. The chromosomes can break, can fuse with others, or can mutate. If your cell's genetic instruction manuals were scrambled like this, the result would be disastrous. A mutation can lead to cell dysfunction, cell death, or even proliferation of a now-cancerous cell, and as a consequence you probably wouldn't live very long.

Telomeres, which seal off the ends of the chromosomes, keep this unthinkable event from happening. That is the message sent to us by the special repeating sequences of telomere DNA. Jack Szostak and I (Liz) discovered this function in the early 1980s, when I isolated a telomere sequence from *Tetrahymena* and Jack put it into a yeast cell. The *Tetrahymena* telomeres protected the chromosomes of the yeast during cell division by donating some of their own base pairs.

Every time a cell divides, its precious "coding DNA" (which makes up the genes) is copied so it can stay safe and whole. Unfortunately, with each division, telomeres lose base pairs from the sequences at the two ends of each chromosome. Telomeres tend to shorten as we get older, as our cells experience more and more divisions. But the trend is not just a straight line. Take a look at the graph on the next page.

In the Kaiser Permanente Research Program on Genes, Environment, and Health study of one hundred thousand people's salivary telomere length, telomeres on average grew shorter and shorter as people progressed from their twenties, hitting rock bottom at around age seventy-five.[1] In an interesting coda, telomere length appears to stay the same or even go *up* as people live past seventy-five. This trend is probably not true lengthening happening; it just looks that way because the folks with shorter telomeres have passed away by this age (which is called survival bias—in any aging study, the oldest people are the healthy survivors). It's the people with longer telomeres who are living into their eighties and nineties.

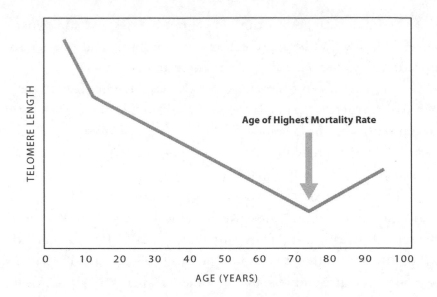

Figure 9: Telomeres Shorten with Age. Telomere length declines with age, on average. It declines fastest during early childhood and then has a slower average rate of decline with age. Interestingly, many studies find telomere length is not shorter in those who live to be a lot older than seventy years. This is thought to be due to "survival bias," meaning that those still alive at this age tended to have been those people with longer telomeres. Their telomeres probably had been longer all along, starting from birth.

TELOMERES, THE DISEASESPAN, AND DEATH

Telomeres shorten with age. But can our telomeres really help determine how long we'll live or how soon we enter the diseasespan?

Science says *yes*.

Short telomeres don't predict death in every study, since there are many other factors that predict when we die. They do predict time of death in around half of studies, including the largest study yet. A 2015 Copenhagen study of more than sixty-four thousand people shows that short telomeres predict earlier mortality.[2] The shorter your telomeres, the higher your risk of dying from cancer, cardiovascular disease, and of dying at younger ages generally, known as all-cause mortality. Look at figure 10, and you'll see that telomere length is broken out by percentiles in ten groups. People in the 90th percentile of telomere length (with the longest telomeres) are at the left; people in the 80th percentile are just next to them; and so on all the way to the right side, where the people in the lowest percentile are represented. There is a graded response: people with the longest telomeres are the healthiest, and as telomeres get shorter, people get sicker and are more likely to die.

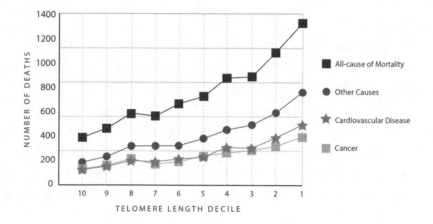

Figure 10: Telomeres and Death. Telomere length predicts mortality overall, and from different diseases. Those with the longest telomeres (90th percentile) have the lowest rate of death from cancers, heart disease, and all causes added up. (Figure is from the data in Rode et al., 2015.[3])

The Kaiser Permanente study previously mentioned measured telomere length in one hundred thousand research volunteers who happened to be members of Kaiser's health coverage plan. In the three years after their telomeres were measured, the people with shorter telomeres were more likely to die when all causes of death were combined.[4] The study was controlled for differences among the subjects that might likely lead to differences in health and longevity, including age, gender, race and ethnicity, education, smoking, physical activity, drinking, and body mass index (BMI). Why did the scientists control for so many variables? Because any one, some, or all of these factors might in theory have been the real reasons contributing to the increased mortality, not the shortened telomeres. For example, a clear relationship exists between tobacco smoking histories and all-cause mortality rates. And many studies have found a relationship between more smoking and more telomere shortening. Yet even after correcting for all those potential explanations, the relationship between telomere shortness and all-cause mortality still held true. It does indeed look as though telomere shortness itself is a real contributor to our overall risks for mortality.

Over and over and over, telomere shortness has also been linked to the major diseases of aging. Many large studies have shown that people with shorter telomeres are more likely to have a chronic disease, such as diabetes, cardiovascular disease, lung diseases, impaired immune function, and certain types of cancers, or to develop one of these diseases over time.[5] Many of these associations have now been reinforced by large reviews (called meta-analyses) that give us confidence that the relationships are accurate and reliable. Flip these findings, and the optimistic opposite is true: one study of a healthy elderly U.S. sample (the Health ABC study) showed that in the general population, people with longer telomeres in their white blood cells had more years of healthy life without any major diseases—a longer healthspan.[6]

TURN THE TIDE IN HEALTH

People like Robin Huiras, whose rare inherited disorder leads to telomeres that are drastically short, show us the power of telomeres. Sometimes, as in Robin's case, it's a kind of dark, corroding power that speeds up the cellular aging process. The good news is that we have learned a great deal about the nature of telomeres. By donating blood and tissue samples, for example, Robin and her family have helped researchers pinpoint one of the gene mutations that caused her disorder. That knowledge is a first step to better diagnoses, treatments, and, one day, a cure.

And you can use our knowledge about telomeres to turn the tide in health—in your health, the health of people in your community, and the health of generations to come. Because as you're about to see, telomeres can change. *You* have the power to influence whether your telomeres are going to shorten early, or whether they are going to stay supported and healthy. To show you what we mean, we need to take you back to Liz's lab. There, *Tetrahymena* telomeres began to behave in a strange, unexpected way.

CHAPTER THREE

Telomerase, the Enzyme That Replenishes Telomeres

Not long after I (Liz) had read the X-ray revealing the DNA of telomeres, I was hired by the University of California, Berkeley, where in 1978 I set up my own laboratory to continue my research into telomeres. There I began to notice something that shocked me. I was still growing *Tetrahymena,* that hairy, Muppet–like pond scum, and now I was able to tell the sizes of their telomeres from the length of their DNA. And mysteriously, under some conditions, the *Tetrahymena*'s telomeres would sometimes *grow.*

This was a shock because I expected that if telomeres were going to change at all, they were going to get shorter, not longer: with each cell division, the number of DNA sequences in the telomeres would more likely shrink. Yet it looked to me as if *Tetrahymena* was creating new DNA. But this was not supposed to happen. DNA is not supposed to change. You've probably heard that the DNA we are born with is the DNA we die with, and that DNA is produced solely through a kind of biochemical photocopying. I checked, and double-checked, and confirmed that what was supposed to be impossible was, in fact, happening. Next, we saw the same thing happening in yeast cells, too. ("We" here included my student Janice Shampay in my lab, working on the experiments that Harvard researcher Jack Szostak and I had dreamed up together.) Then reports from other scientists trickled in that suggested these changes might happen in other tiny, *Tetrahymena*-like creatures, too. The organisms were, in fact, producing new DNA at the ends of their telomeres. Their telomeres were growing.

No other element of DNA behaves in this way. For decades, genetic scientists believed that any stretch of chromosomal DNA existed only because it had been copied from preexisting DNA. The accepted wisdom was that DNA could not be created from whole cloth where there had not been DNA before. The discovery of this odd behavior told me *there was something going on here that no one had seen before.* For scientists, that's one of the most exciting kinds of discoveries to make. It's thrilling when a strange finding suggests that there are new, unknown street corners of the universe, ripe for exploration. As it turned out, this behavior of telomeres led to more than just a new street corner of the universe; this was a whole new neighborhood, one that no one had known existed.

TELOMERASE: THE SOLUTION TO TELOMERE SHRINKAGE

I kept pondering this strange behavior of the telomere, its apparent ability to grow. I wanted to look for an enzyme in a cell that might add DNA onto telomeres—an enzyme that might replenish telomeres after they'd lost some of their pairs of letters. It was time for me to roll up my sleeves and make more *Tetrahymena* cell extracts. Why *Tetrahymena*? Because it's such a good source of plentiful telomeres. I reasoned that it might be a good source of enzymes that could form telomeres, if such an enzyme existed.

In 1983 I was joined in this quest by Carol Greider, a new graduate student in my lab. We began devising experiments, and then refining those experiments, and on Christmas Day in 1984, Carol developed an X-ray film called an autoradiograph. The patterns on that film showed the first clear signs of a new enzyme at work. Carol went back home and danced with excitement in her living room.

The next day, her face alight with suppressed glee in her anticipation of my reaction, she showed me the X-ray film. We looked at each other. Each of us knew this was it. Telomeres could add DNA by attracting this previously undiscovered enzyme, which our lab named telomerase. Telomerase creates new telomeres patterned on its own biochemical sequence.

But science does not work only by the exhilaration of one single eureka moment. We had to be sure. As the weeks stretched into months, we experienced surges of doubt followed by thrills of joy as we painstakingly conducted follow-up experiments. Step by step, we ruled out every possible reason that our exciting first moments in 1984 could have been just a false lead. Eventually, a deeper understanding of telomerase emerged: Telomerase is the enzyme responsible for restoring the DNA lost during cell divisions. Telomerase makes and replenishes telomeres.

Here's how telomerase works. It includes both protein and RNA, which you can think of as a copy of DNA. That copy includes a template of the telomere's DNA sequence. Telomerase uses that sequence in the RNA as its own inbuilt biochemical guide to create the right sequence of brand-new DNA. The right sequence is needed to make a DNA scaffold perfectly shaped to attract a sheath of telomere-protective proteins that cover the telomeric DNA. This new segment of DNA is added by telomerase to the end of the chromosome, guided by the RNA template sequence and the DNA's buddy partner system of pairing up its letters. This ensures that the right sequence of building blocks of telomeric DNA is added. In this way, telomerase re-creates new endings at the chromosome's tips and replaces ones that have been worn down.

The mystery of the growing telomeres was solved. Telomerase replenishes telomeres by adding telomeric DNA to them. Each time a cell divides, telomeres gradually shorten until they reach a crisis point that signals cell division to stop. But telomerase counteracts this telomere shortening by adding DNA and building back the chromosome end each time a cell divides. This means that the chromosome itself is protected, and an accurate copy of it is made for the new cell. The cell can continue to renew itself. **Telomerase can slow, prevent, or even reverse the shortening of telomeres that comes with cell division.** Telomeres can, in a sense, be renewed by telomerase. We had found a way to get around the Hayflick limit of cell division... in pond scum.

TELOMERASE: NO ELIXIR OF IMMORTALITY

After these discoveries, both the scientific world and the global media buzzed with hopeful speculation. What if we could increase our supply of telomerase? Could we be like *Tetrahymena*, with cells that renew forever? (This may have been the first recorded instance of humans fervently wishing to be more like pond scum.)

People wondered if telomerase could be distilled and served up as an elixir of immortality. In this wishful scenario, we'd visit our local telomerase bar every now and then for a hit of the enzyme, which would let us live healthy lives all the way to the very end of the known maximum human life span—or beyond it.

These dreams are perhaps not as ridiculous as they might seem. Telomeres and telomerase form a crucial biological foundation for cell aging. The demonstration of the relationship between telomerase and cell aging first came from *Tetrahymena*. Guo-Liang Yu, then a graduate student in my Berkeley lab, performed a simple but surgically precise experiment. He replaced the normal telomerase in *Tetrahymena* cells with a precisely inactivated version. If you feed them properly, *Tetrahymena* cells are normally immortal in the laboratory. Like the Energizer Bunny, *Tetrahymena*'s cell divisions normally just keep going and going and going. But this inactivated telomerase caused the telomeres to become shorter and shorter as the *Tetrahymena* cells divided. Then when the telomeres had become too short to protect the genes inside the chromosome, the cells stopped dividing. Think again of a shoelace. It's as though the shoelace tip wore down and the shoelace—with all that vital genetic material—became frayed. Inactivating telomerase made the *Tetrahymena* cells mortal.

Abundant Telomerase as Cell Divides

Insufficient Telomerase as Cell Divides

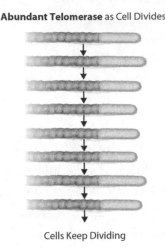

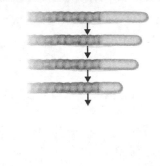

Cells Keep Dividing

Cell Division Stops Prematurely

Figure 11: Consequences of Enough, or Not Enough, Telomerase Action. Telomere DNA shortens because the enzymes to duplicate the DNA don't work at the telomere ends (incomplete DNA replication). Telomerase elongates telomeres and thus counterbalances the inexorable attrition of telomeric DNA. With abundant telomerase, telomeres are maintained and cells can keep dividing. With insufficient telomerase (due to genetics, lifestyle, or other causes) telomeres shorten rapidly, cells stop dividing, and senescence soon follows. Reprinted with permission from AAAS (Blackburn, E., E. Epel, and J. Lin., "Human Telomere Biology: A Contributory and Interactive Factor in Aging, Disease Risks, and Protection," *Science* (New York) 350, no. 6265 (December 4, 2015): 1193–98).

Without telomerase, the cells stop renewing.

And then, at other labs around the world, the same was found for nearly all cells, except bacteria (whose chromosomes are circles of DNA instead of lines and thus have no ends to protect). Longer telomeres and more telomerase delayed premature cellular aging, and shortened telomeres and less telomerase sped it up. The telomerase–health connection was nailed when clinician Inderjeet Dokal and his colleagues in the United Kingdom and the United

States first discovered that when people have a genetic mutation that slashes telomerase levels in half, they develop severe inherited telomere syndromes.[1] This is the same category of disease that was diagnosed in Robin Huiras. Without sufficient telomerase, the telomeres quickly shorten, and the body succumbs to early disease.

Tetrahymena cells have telomerase in sufficient quantities so that they can constantly rebuild their telomeres. This allows *Tetrahymena* to perpetually renew itself and to forever avoid cell aging. But we humans normally don't have enough telomerase to accomplish this feat. We are very miserly when it comes to telomerase. Our cells are reluctant to hand out telomerase willy-nilly to their telomeres all the time. We produce telomerase in sufficient quantities to rebuild telomeres...but only up to a point. As we age, the telomerase in most of our cells generally becomes less active, and telomeres get shorter.

TELOMERASE AND THE CANCER PARADOX

It's natural to wonder if we could extend human life through artificial methods of increasing telomerase. Ads for telomerase-boosting supplements abound on the Internet claiming that we can. Telomerase and telomeres have wonderful properties that can allow us to avoid horrible diseases and feel more youthful. But they are not magical life extenders—they don't let us live past the normal human life span as we know it. In fact, if you try to extend your life by using artificial methods of increasing telomerase, you are putting yourself in danger.

That's because telomerase has a dark side. Think of Dr. Jekyll and Mr. Hyde—they are the same person, but one with a drastically different character depending on whether it is day or night. We need our good Dr. Jekyll telomerase to stay healthy, but if you get too much of it in the wrong cells at the wrong time, telomerase takes on its Mr. Hyde persona to fuel the kind of uncontrolled cell growth that is a hallmark of cancer. Cancer is, basically, cells that won't stop dividing; it's often defined as "cell renewal run amok."

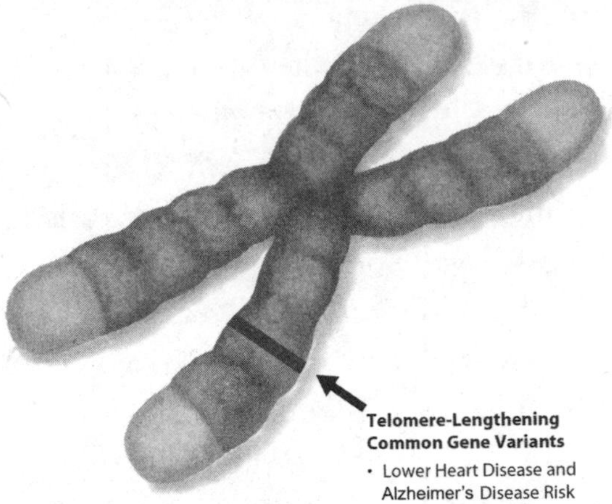

Telomere-Lengthening
Common Gene Variants
• Lower Heart Disease and
  Alzheimer's Disease Risk
• Greater Cancer Risk

Figure 12: Telomere-Related Genes and Disease. Telomere maintenance genes can protect us from common diseases, but can put us at risk for some cancers. Having gene variants for more telomerase and telomere proteins means longer telomeres. This natural genetic way of making telomeres longer lowers risks for most diseases of aging, including heart disease and Alzheimer's disease, but the high telomerase also means that cells that are prone to become cancerous can keep dividing unchecked, causing a greater risk for certain types of cancer (brain cancers, melanoma, and lung cancers). Bigger isn't always better!

You don't want to bomb your cells with artificial telomerase that may goad them into taking the road toward becoming cancerous. Unless the telomerase supplement field comes up with more thorough demonstrations of safety in large—and *long-term*—clinical trials, in our view it's sensible to skip any pill, cream, or injection that claims it will increase your telomerase. Depending on your individual propensity for different types of cancer, you may be potentially increasing the chance of developing any of a number of different cancers (such as melanomas or brain and lung cancers). Knowing this, it comes as no surprise that our cells keep their telomerase on a tight rein.

Given these scary-sounding findings, you may be wondering, why are we suggesting activities that boost telomerase? The answer is that there is a big difference between the body's normal physiological responses to the lifestyle suggestions we make for your health in this book and taking an artificial substance (no matter how "natural" its plant source—remember that plants are some of nature's biggest chemical warfarers, having evolved an armamentarium of strong chemicals to fend off hungry animals and marauding pathogens). The suggestions we include in this book for increasing your telomerase action are gentle and natural—and they increase telomerase in safe amounts. You do not need to worry about an increased cancer risk with these strategies. They simply don't increase telomerase to the levels or in the ways that would be harmful.

Paradoxically, we do need to keep our telomeres healthy to ward off cancer, too. Some types of cancers are more likely to develop when *too little available telomerase* makes telomeres too short—blood cancers like leukemia, skin cancers besides melanoma, and some gastrointestinal cancers such as pancreatic cancer. This was proved with the discovery that people born with a mutation that precisely inactivated a telomerase gene had much higher risk for these cancers. Such cancers arise because losing telomere protection allows our genes to become more easily damaged—and altered genes can eventually lead to cancers. Furthermore, too little telomerase weakens the telomeres in our immune cells. Our immune system usually keeps a sharp eye out for anything perceived as "foreign," and that includes harmful cancer cells as well as pathogenic invaders from the outside such as bacteria and viruses. Without telomeres long enough to act as buffers, the cells of the immune system will eventually become senescent.

Some of these immune cells are like surveillance cameras posted at every corner of the body. If they become senescent, their lenses then act as though they are steamed up, and they miss the "foreign" cancer cells. So the teams of immune cells that would normally be called up fail to leap into action. The result of weakened telomeres is that the immune defenses of the body are more likely to lose the fight against a cancer (or a pathogen).

TELOMERASE AND HOPE FOR NEW CANCER TREATMENTS

Too much telomerase, spurred by the actions of even normal variants of telomerase genes, can increase risks of developing several forms of cancer. And overactive telomerase fuels most cancers once they turn malignant. But even this "dark side" of telomerase may not always be dark. Researchers have learned that telomerase is hyperactive in roughly 80 to 90 percent of malignant human cancers, with levels that are turned up ten to hundreds of times as high as in normal cells. This discovery may one day turn out to be a potent weapon in our fight against the disease. If telomerase is necessary for cancers to grow so relentlessly, perhaps we can treat cancer by turning off the telomerase in just the cancer cells. Researchers are at work on this idea.

The key is to well regulate the action of telomerase on telomeres— in the right cells and at the right times, because only that will keep telomeres and us healthy. **The body knows how to do this, and we can help it with a lifestyle full of renewal strategies.**

YOU CAN INFLUENCE YOUR TELOMERES AND TELOMERASE

By the turn of the millennium, scientists had become accustomed to thinking about both telomeres and telomerase as foundations of cell renewal. But the telomere syndromes, starting with the shocking finding that cutting telomerase by merely half could have such drastic effects, had galvanized everyone into thinking only in terms of *genes* that determined whether our telomeres were long or short, and whether we had enough telomerase to replenish worn-down telomeres.

That was when I (Elissa) began a postdoctoral fellowship in

health psychology at the University of California, San Francisco. Susan Folkman, the now-retired director of the Osher Center for Integrative Medicine and a pioneer in the study of stress and coping, invited me to join a team that was interviewing mothers of children with chronic conditions, a group under tremendous psychological strain.

I felt a profound empathy for these caregiving mothers, who seemed extraordinarily worn out and older than their chronological ages. By then, Liz had moved to the San Francisco campus of the University of California, and I was aware of her work on biological aging. I approached Liz and told her about the caregiving mothers we were studying. If I could come up with the funds, would it be possible to test the mothers' telomeres and telomerase? Was it worth investigating whether stress could shorten telomeres and lead to early cell aging?

Like most other molecular biologists at the time, I (Liz) was peering down at telomeres from one particular mountaintop. I was thinking about our telomere maintenance in terms of the cellular molecules specified by the genes that control telomeres. When Elissa asked me about studying caregivers, however, it was as though I suddenly saw telomeres from a whole new viewpoint, from a completely different mountain. I responded as both a scientist and a mother. "We need another ten years just to more fully understand the genetics of telomeres," I mused, somewhat doubtfully, but I could also well imagine the tremendous stresses these women were under. I thought about the way we describe exhausted, stressed people: *careworn*. Mothers of chronically sick children are women who are worn down. Was it possible that their telomeres were worn down, too? "Yes," I agreed. "Let's do this study, if we can find a scientist in my lab who will help with the measurements." My postdoctoral fellow, Jue Lin, raised her hand. She proceeded to refine a way to sensitively and carefully measure telomerase in healthy human cells, and the work began.

We selected a group of mothers who were each caring for a chronically ill biological child. A research subject who might have an outside "issue" could warp the results, so any mother with a major health problem was screened out of the study. We used a similar process to select a control group of mothers whose children were healthy. This process took several years of careful selection and assessments.

We took a sample of each woman's blood and measured the telomeres in the white blood cells. We recruited the help of Richard Cawthon at the University of Utah, who had recently devised a new easier way to measure the length of telomeres in white blood cells (applying a method called polymerase chain reaction).

One day in 2004, the assay results came in. I (Elissa) was sitting in my office as the numerical analysis came out of the printer. I looked down at the scatter plot and gasped. There was a pattern to the data, the exact gradient we thought might exist indeed was right there, on the page. It showed that the more stress you are under, the shorter your telomeres and the lower your telomerase levels.

I immediately picked up the phone and called Liz. "The results are in," I said, "and the findings are even more striking than we'd thought they might be."

We'd asked the question *Can the way we live change our telomeres and telomerase?* Now we had an answer.

Yes.

Yes, the mothers who perceived themselves to be under the most stress were the ones with the lowest telomerase.

Yes, the mothers who perceived themselves to be under the most stress were the ones with the shortest telomeres.

Yes, mothers who had been caregiving for the longest time had shorter telomeres.

This triple yes meant that our results weren't just a coincidence or a statistical blip. **It also meant that our life experiences, and the way we respond to those events, can change the lengths**

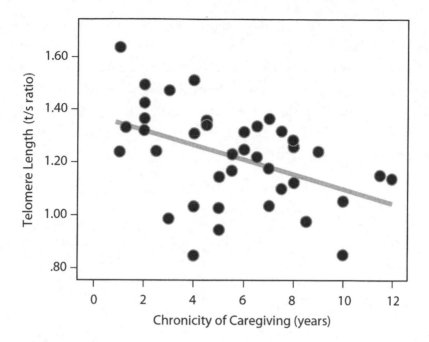

Figure 13: Telomere Length and Chronic Stress. The more years since the child had been diagnosed (thus the more years of chronic stress), the shorter the telomeres.[2]

of our telomeres. In other words, we can change the way that we age, at the most elemental, cellular level.

Whether aging can be sped up, slowed, or reversed has been a topic of medical debate for centuries. What we have learned since this first study of caregivers is wholly new. We, as a field, have learned that by our actions, we can keep our telomeres—and hence our *cells*—from aging prematurely. We may even be able to partly reverse the cellular aging process caused by telomere wear and tear. Over the years, the results of our initial study have held up, and many additional studies that you will read about here have taken this first finding much further, showing that many different life factors can affect our telomeres.

In the rest of this book you will hear us talking about how you can increase telomerase and protect your telomeres. Our recommendations are based on studies, some that measure telomeres, some that measure telomerase activity, and some that do both. You can join us on the journeys of exploration that have followed from those first mountaintop views. Use this research as a North Star to help you change the way you use your mind, take care of your body, and even how you interact with your community, to protect your telomeres and enjoy your healthspan.

RENEWAL LABS: A Guide

Life is full of little experiments we can learn from. Throughout the rest of this book, you will find a Renewal Lab at the end of each chapter. There, you become the researcher if you wish. Your mind, your body, and your life become your personal laboratory where you can try out practical applications of telomere science or behavioral science and learn ways to change your daily life to enhance your cellular health. In most cases, the Renewal Labs have been directly linked to better telomere length, and in all cases they are associated with better physical or mental health. (You can find the relevant studies in the book's Notes section, which begins on page 337.)

When we say "laboratory," we really mean it. They're experiments, not written-in-stone commandments. What works best for you depends on your individual mind and body, your preferences, and your stage of life. So give them a try—perhaps only one or two at a time. If you find one that works for you, focus on it for a while, until it becomes a habit. If you practice any of these Labs regularly, they should enhance your cellular health as well as your daily wellbeing. Studies have found that lifestyle changes can have an effect on telomere maintenance (that means increased telomerase or telomere length) as soon as three weeks to four months. Remember, as Ralph Waldo Emerson said, "Don't be too timid or squeamish about your actions. All life is an experiment. The more experiments you make the better."

PART II

YOUR CELLS ARE LISTENING TO YOUR THOUGHTS

ASSESSMENT: Your Stress Response Style Revealed

Part Two, "Your Cells Are Listening to Your Thoughts," offers insights into how you experience stress and how you can shift that experience to be healthier for your telomeres and more beneficial in your daily life. To get you started, here's a quick self-test. It assesses your underlying sources of stress reactivity and stress resilience, some of which have been linked to telomere length.

Think of a situation that bothers you a great deal and that is ongoing in your life. (If you cannot think of a current situation, think of your most recent difficult problem.) Circle your numerical response to each question.					
1. When you think about dealing with this situation, how much do you feel hope and confidence vs. feelings of fear and anxiety?	**0** hopeful, confident	**1**	**2** same amount of each	**3**	**4** fearful, anxious
2. Do you feel you have whatever it takes to cope effectively with this situation?	**4** not at all	**3**	**2** somewhat	**1**	**0** extremely
3. How much are you caught up in repetitive thoughts about this situation?	**0** not at all	**1**	**2** somewhat	**3**	**4** extremely
4. How much do you avoid thinking about the situation or try not to express negative emotions?	**0** not at all	**1**	**2** somewhat	**3**	**4** extremely
5. How much does this situation make you feel bad about yourself?	**0** not at all	**1**	**2** somewhat	**3**	**4** extremely
6. How much do you think about this situation in a positive way, seeing some good that could come from it, or telling yourself statements that feel comforting or helpful, such as that you are doing the best you can?	**4** not at all	**3**	**2** somewhat	**1**	**0** extremely
TOTAL SCORE (Add up the numbers; notice questions 2 and 6 are positive responses so the scale is reversed.)					

The point of this informal test (not a validated research measure) is to raise awareness of your own tendencies to respond in a certain way to chronic stress. It is not a diagnostic scale. Also know that if you're dealing with a severe situation, your response style score will naturally shift to be higher. This is not a pure measure of response style, because our situations and our responses inevitably get a bit mixed together.

Total score of 11 or under: Your stress style tends to be healthy. Instead of feeling threatened by stress, you tend to feel challenged by it, and you limit the degree to which the situation spills over into the rest of your life. You recover quickly after an event. This stress resilience is positive news for your telomeres.

Total score of 12 or over: You're like most of us. When you're in a stressful situation, the power of that threat is magnified by your own habits of thinking. Those habits are linked, either directly or indirectly, to shorter telomeres. We'll show you how to change those habits or soften their effects.

★ ★ ★

Here's a closer look at the habits of mind associated with each question:

Questions 1 and 2: These questions gauge how threatened you feel by stress. High fear combined with low coping resources turn on a strong hormonal and inflammatory stress response. **Threat stress** involves a set of mental and physiological responses that can, over time, endanger your telomeres. Fortunately, there are ways to convert threat stress into a feeling of challenge, which is healthier and more productive.

Question 3: This item assesses your level of **rumination**. Rumination is a loop of repetitive, unproductive thoughts about something that's bothering you. If you're not sure how often you ruminate, now you can start to notice. Most stress triggers are short-lived, but we humans have the remarkable ability to give them a vivid and extended life in the mind, letting them fill our headspace long after the event has passed. Rumination, also known as brooding, can slip into a more serious state known as depressive rumination, which includes negative thoughts about oneself and one's future. Those thoughts can be toxic.

Question 4: This one's about **avoidance and emotion suppression**. Do you avoid thinking about the stressful situation or avoid sharing feelings about it? Is it so emotionally loaded that the thought of it makes your stomach clench? It's natural to try to push difficult feelings away, but although this strategy may work in the short term, it doesn't tend to help when the situation is chronic.

Question 5: This question addresses "**ego threat**." Does it feel as if your pride and personal identity could be damaged if the stressful situation doesn't go well? Does the stress trigger negative thoughts about yourself, even to the extent that you feel worthless? It's normal to have these self-critical thoughts sometimes, but when they are frequent, they throw the body into an overly sensitive, reactive state characterized by high levels of the stress hormone cortisol.

Question 6: This question asks whether you're able to engage in **positive reappraisal**, which is the ability to rethink stressful situations in a positive light. Positive reappraisal lets you take a less than ideal situation and turn it to your benefit or at least take the sting out of it. This question also measures whether you tend to offer yourself some healthy **self-compassion**.

If the assessment revealed that you struggle with your stress responses, take heart. It's not always possible to change your automatic response, but most of us can learn to change our responses *to our responses*—and that's the secret sauce of **stress resilience**. Now let's get to work understanding how stress affects your telomeres and cells, and how you can make changes that will help protect them.

CHAPTER FOUR

Unraveling: How Stress Gets into Your Cells

We explore the stress–telomere connection, explain toxic stress versus typical stress, and show how stress and short telomeres affect the immune system. People who respond to stress by feeling overly threatened have shorter telomeres than people who face stress with a rousing sense of challenge. Here, you'll learn how to move from harmful stress responses to helpful ones.

Nearly fifteen years ago, my husband and I (Elissa) were driving across the country. We had just finished graduate school at Yale and were taking on postdoctoral fellowships in the Bay Area. San Francisco is an expensive city, and so we had arranged to live with my sister and her family. We expected that when we arrived in San Francisco, we would meet our new nephew, who was supposed to be born at any moment. In fact, he was quite overdue. I called every day for news, but I'd had trouble reaching anyone in the family for days.

About halfway through the trip, just after we'd passed Wall Drug Store in South Dakota, my cell phone finally rang. Tearful voices wavered on the other end. The baby had been born, but something had gone terribly wrong during an induced delivery. The baby was on life support and being fed through a gastric tube to his stomach.

He was a beautiful healthy boy, but an MRI showed his brain had been profoundly damaged. He was paralyzed, blind, and wracked with seizures.

Eventually, after several months, the baby left the intensive care ward and came home. We joined the family team to help take care of this little guy, who had extraordinary needs. We became intimate with both the demands and sorrows that come with a life of caregiving. We were accustomed to pressure and hard work, but this had nothing in common with the types of stresses we had known. Now there were new feelings of constant vigilance, intermittent urgency, worry about the future, and most of all, a heavy weight on the heart. One of the hardest parts was seeing and feeling the pain my sister and brother-in-law were experiencing every day. On top of the emotional suffering there was, all of a sudden, a new, unexpected, and demanding life centered around medical caregiving.

Caregiving is one of the most profound stresses a person can experience. Its tasks are emotionally and physically demanding, and one reason caregivers get so worn out is that they don't get to go home from their caregiving "jobs" and recover. At night, when we all need to biologically check out and refresh body and mind, caregivers are on call. They may be repeatedly woken from sleep to respond to someone in need. Caregivers rarely have time to take care of themselves. They skip their own doctor appointments as well as opportunities for exercise and going out with friends. Caregiving is an honorable role that is taken on out of love, loyalty, and responsibility, but it is not supported by society or recognized for its value. In the United States alone, family caregivers perform an estimated $375 billion in unpaid services each year.[1]

Caregivers often feel unappreciated and become isolated. Health researchers have identified them as one of the most chronically stressed groups of people. This is why we often ask caregivers to volunteer for our studies on stress. Their experiences can tell us a lot about how telomeres react to serious stress. In this chapter, you'll learn what our groups of caregivers have taught us—that chronic, long-lasting stress can erode telomeres. Fortunately for all of us who cannot escape chronic stress (and for all of us who scored higher than 12 on the stress assessment on page 71), we've also learned that we can protect our telomeres from some of stress's worst damage.

"LIKE THERE IS AN ASSAILANT, WAITING FOR ME": HOW STRESS HURTS YOUR CELLS

In our very first study together we looked at some of the most highly stressed caregivers of all: mothers who were taking care of their chronically ill children. This is the study we've told you about. It's the one that first revealed a relationship between stress and shorter telomeres. Now we want to show you a close-up look at the extent of that damage. More than ten years later we still find it sobering.

We learned that the years of caregiving had a profound effect, grinding down the women's telomeres. The longer a mother had been looking after her sick child, the shorter her telomeres. This held true even after we took into account other factors that might affect telomeres, like the mother's age and body mass index (BMI), which are related to shorter telomeres themselves.

There was more. The more stressed out the mothers felt, the shorter their telomeres. This was true not just for the caregivers of sick children, but also for *everyone* in the study, including the control group of mothers who had healthy children at home. The high-stress mothers also had almost half the levels of telomerase than the low-stress mothers, so their capacity to protect their telomeres was lower.

People experience stress in many different ways: "like a fifty-pound weight on my chest," "like a knot in my stomach," "like a vacuum in my lungs that doesn't let me take a full breath," "my heart pounds like there is an assailant, waiting for me." These metaphors are grounded in the body, because stress is as present in the body as in the head. When the stress–response system is on high alert, the body produces more of the stress hormones cortisol and epinephrine. The heart beats faster and blood pressure increases. The vagus nerve, which helps modulate the physiological reaction to stress, withdraws its activity. That's why it's harder to breathe, harder to stay in control, harder to imagine that the world is a safe place. When you suffer from chronic stress, these responses are on a low but constant alert, keeping you in a state of physiological vigilance.

In our caregivers, several aspects of the physiological stress response, including lower vagus activity during stress, and higher stress hormones while sleeping, were linked to shorter telomeres or to less telomerase.[2] These responses to stress appeared to be accelerating the biological aging process. We had discovered a new reason that stressed–out people look haggard and get sick: their heavy stresses and cares are wearing down their telomeres.

SHORT TELOMERES AND STRESS: CAUSE OR EFFECT?

When a scientific finding suggests a cause-and-effect relationship, you have to ask whether the relationship really runs in the direction you think it does. For example, people used to think that fevers caused sickness. Now we know that the relationship is the reverse: sickness causes fevers.

As the results of our first study of caregivers came in, we were careful to ask ourselves *why* shorter telomeres appeared in people with higher stress. Does stress really lead to short telomeres? Or can short telomeres somehow predispose a person to feeling more stress? Our caregiving mothers provided the first convincing data about this question. The relationship between the years of caregiving stress and telomere length is a strong indicator that the stress exposure happens over time, causing telomeres to shorten.

Short telomere length (after correcting for age) could not have determined how many years a mother had been a caregiver, so it had to be the other way around—that the years of caregiving were the cause of the shorter telomeres. We also tested whether an older age of the child was related to shorter telomeres. If the years of difficult caregiving were wearing down telomeres more than the years of parenting by the control mothers, we would see the relationship between the child's age and the mother's telomeres in the caregivers but not in the control moms. Indeed, this was what we found. Now there are animal studies showing that inducing stress can actually cause telomere shortening.

The depression story is more complicated. The findings above were not enough to rule out the possibility that cell aging could cause depression. In humans, depression runs in families. Not only are girls whose mothers have depression more prone to depression themselves, but even before any depression has developed, these girls have shorter blood telomeres than girls who are not depressed.[3] Also, the more stress reactive the girls are, the shorter the telomeres. So the arrow likely points in both directions with depression—short telomeres may precede depression, and depression may speed up telomere shortening.

HOW MUCH STRESS IS TOO MUCH?

Stress is unavoidable. How much of it can we handle before our telomeres are damaged? A consistent lesson from the past decade of studies—and a lesson that echoes what the caregivers taught us—is that stress and telomeres have a dose-response relationship. If you drink alcohol, you're familiar with dose and response. An occasional glass of wine with dinner is rarely harmful to your health and may even be beneficial, as long as you're not drinking and driving. Drink several glasses of wine or whiskey, night after night, and the story changes. As you "dose" yourself with more and more alcohol, the poisonous effects of alcohol take over, damaging your liver, heart, and digestive system and putting you at risk for cancer and other serious health problems. The more you drink, the more damage you do.

Stress and telomeres have a similar relationship. A small dose of stress does not endanger your telomeres. In fact, short-term, manageable stressors can be good for you, because they build your coping muscles. You develop skills and confidence that you can handle challenges. Physiologically, short-term stress can even boost your cells' health (a phenomenon called hormesis, or toughening). The ups and downs of daily life are usually not wearing to your telomeres. But a high dose of chronic stress that wears on for years and years will take its toll.

We now have evidence that links particular kinds of stress to shorter telomeres. These include long-term caregiving for a family member and burnout from job stress. As you may imagine, more serious traumas, both recent and in childhood, have also been linked to damaged telomeres. These traumas include rape, abuse, domestic violence, and prolonged bullying.[4]

Of course, it's not the situations themselves that produce the short telomeres; it's the stress responses that many people feel when they're in these situations. And even under these stressful circumstances, dose matters. A monthlong crisis at work can be stressful, but there's no reason to think your telomeres will take a hit. They are more robust than that; otherwise, we'd all be falling apart. (A recent review showed that there is a relationship between short-term stress and shorter telomeres, but that effect is so tiny that we don't think it will have a meaningful effect on an individual person.[5] And even if short-term stress shortens your telomeres, the effect is likely temporary, with telomeres quickly recovering their lost base pairs.) But when stress is an enduring, defining feature of your life, it can act as a slow drip of poison. The longer the stress lasts, the shorter your telomeres. It is vitally important to get out of long-term, psychologically toxic situations if it's at all possible.

But fortunately for the many of us who live with stressful situations we cannot change, that's not the whole story. **Our studies have shown that being under chronic stress does not *inevitably* lead to telomere damage.** Some of the caregivers we've studied were weathering enormous burdens without losing telomere length. These stress-resistant outliers have helped us understand that you do not necessarily have to escape difficult situations to protect your telomeres. Incredible as it sounds, you can learn to use stress as a source of positive fuel—and as a shield that can help protect your telomeres.

DON'T THREATEN YOUR TELOMERES—CHALLENGE THEM

When we looked at the data for our first caregiver study, we realized we had a mystery on our hands. Some of the caregiving mothers in the group reported less stress, and these mothers had longer telomeres. We wondered: *Why* would they feel less stress? After all, they had been caregiving for just as long as the other mothers in the group. They had a similar number of daily duties and spent just as many hours in the day performing those duties (appointments, administering injections and other treatments, managing tantrums, having to hand- or tube-feed, diaper, and bathe older disabled children).

To understand what was protecting these mothers' telomeres, we wanted to see people respond to stress in real time, before our eyes. We decided to bring more women into the lab and, essentially, stress them out. Research volunteers who arrive at our stress lab are told something like, "You're going to perform some tasks in front of two evaluators. We want you to try hard to do your best. You are going to prepare a five-minute speech and then deliver it, and perform some mental arithmetic. You can make some notes for your speech, but you will have to do all the math in your head." Sound easy? Not really, and especially not in front of an audience.

One by one, the volunteers are escorted into a testing room. Each study volunteer stands at the front of the room and faces two researchers sitting at a desk. The researchers look at the volunteer in a manner best described as stony-faced. No smiling, no nodding, no encouragement. Technically, a stony-faced expression is neutral, neither positive nor negative, but most of us are used to seeing other people smile at us, nod as we talk, or at least make an effort to seem pleasant. When compared to our usual interactions, a stony expression can come across as disapproving or strict.

The researchers explain the task, saying something like, "Please take the number 4,923 and subtract the number 17 from it, out loud. Then take your answer and subtract 17 from it, and so on, as many times as you can in the next five minutes. It is important that you perform this task quickly and accurately. We will judge you on various aspects of your performance. The clock starts now."

As each volunteer begins the math task, the researchers stare at her, pencils poised to record her answer. If she fumbles (and almost everyone fumbles), the researchers turned toward each other and whisper.

Then the volunteer goes on to her five-minute speech with the same researchers evaluating her and behaving in a similar way. If she finishes before the five minutes are up, the researchers point to the timer and say, "Please continue!" As she talks, the researchers glance at each other and slightly furrow their brows and shake their heads.

This lab stressor test, developed by Clemens Kirschbaum and Dirk Hellhammer, is a staple of psychology research, and its point is definitely *not* to test math and speech skills. Instead, it's designed to induce stress. What makes it so stressful? Mental math and on-the-fly public speaking are tricky to perform well. The most stressful element, though, is what's called social evaluative stress. Anyone who tries to perform a task in front of an audience will probably feel increased stress about their performance. When that audience appears judgmental, the stress is intensified. Even though our volunteers' physical survival was not at risk and they were safe in a clean, well-lighted university lab, this test was capable of eliciting a full-blown stress response.

We've put caregivers and noncaregivers through this protocol. We assessed their thoughts at two different times during the lab stressor: just after they'd learned what they were going to do, and just after they'd finished the two tasks. What we found was that although all the women felt *some* stress, not everyone had the *same* type of stress response. And only one kind of stress response went hand in hand with unhealthy telomeres.[6]

The Threat Response: Anxious and Ashamed—and Aging

Some of the women had what's known as a threat response to the lab stressors. The threat response is an old, evolutionary response, a kind of switch to be flipped in case of dire emergency. Basically, the threat response was designed to surge when we are face-to-face with a predator who is probably going to eat us. The response prepares our body and mind for the trauma of being attacked. As you might guess, if it keeps on happening without letup, this is *not* the response associated with telomere health.

If you already suspect that you have an exaggerated threat response to stress, don't worry. In a moment, we'll show you some lab-tested ways to convert a habitual threat response into one that is healthier for your telomeres. First, though, it's important to know what a threat response looks and feels like. Physically, the threat response causes your blood vessels to constrict so that you'll bleed less if you're wounded, but also less blood flows to your brain. Your adrenal gland releases cortisol, which gives you energizing glucose. Your vagus nerve, which runs a direct line from your brain to your viscera and normally helps you feel calm and safe, withdraws its activity. As a result, your heart rate accelerates, and your blood pressure increases. You may faint or even release your bladder. A branch of the vagus innervates the muscles of facial expression, and when that nerve isn't active, it becomes even harder for someone to interpret your facial expressions accurately. If others are wearing a similarly ambiguous expression, one that leaves lots of room for your interpretation, you in turn may view them as more hostile. You tend to freeze, you are unable to run or fight—and your hands and feet get colder, making movement more difficult.

A full-throttle threat response unleashes some uncomfortable physical reactions, but there are psychological ones, too. As you might expect, the threat response is associated with fear and anxiety. Shame, too, if you're worried about failing in front of other people. People with a strong habitual threat response tend to suffer from anticipatory worry; they imagine a bad outcome to an event that hasn't happened yet. That was exactly what happened to many of the caregivers in our lab. They felt high levels of threat—not just after they had finished the tasks but *before* the tasks had even begun. This group of caregivers became fearful and anxious when they heard the somewhat vague information about having to give a speech and do mental math. They anticipated a bad outcome, and they felt failure and shame.

As a group, our caregivers had a stronger threat response. The chronic stress of being a caregiver had made them more sensitive to a lab stressor. The ones with the strongest threat responses also had the shortest telomeres. The noncaregivers were less likely to have an exaggerated threat response, but those who did had shorter telomeres, too. Having a large anticipatory threat response—meaning that they felt threatened at the mere thought of the lab stressor before it even happened—was what mattered most.[7] Here was some vital information about how stress gets into our cells. **It's not just from experiencing a stressful event, it's also from feeling threatened by it, even if the stressful event hasn't happened yet.**

Excited and Energized: The Challenge Response

Feeling threatened is not the only way to respond to stress. It's also possible to feel a sense of challenge. People with a challenge response may feel anxious and nervous during a lab stressor test, but they also feel excited and energized. They have a "bring it on!" mentality.

Our colleague, Wendy Mendes, a health psychologist at the University of California, San Francisco (UCSF), has spent over a decade examining the body's responses to different types of stressors in the lab, and has mapped out the differences that occur in the brain, in the body, and in behavior during "good stress" compared to "bad stress." Whereas the threat response prepares you to shut down and tolerate the pain, the challenge response helps you muster your resources. Your heart rate increases, and more of your blood is oxygenated; these are positive effects that allow more blood to flow where it's needed, especially to the heart and brain. (This is the opposite of what happens when you're threatened. Then, the blood vessels constrict.) During the challenge response, your adrenal gland gives you a nice shot of cortisol to increase your energy—but then your brain quickly and firmly shuts off cortisol secretion when the stressful event is over. This is a robust, healthy kind of stress, similar to the kind you may have when you exercise. The challenge response is associated with making more accurate decisions and doing better on tasks, and is even associated with better brain aging and a reduced risk of developing dementia.[8] Athletes who have a challenge response win more often, and a study of Olympic athletes has shown that these highly successful folks have a history of seeing their life problems as challenges to be surmounted.[9]

The challenge response creates the psychological and physiological conditions for you to engage fully, perform at your best, and win. The threat response is characterized by withdrawal and defeat, as you slump in your seat or freeze, your body preparing for wounding

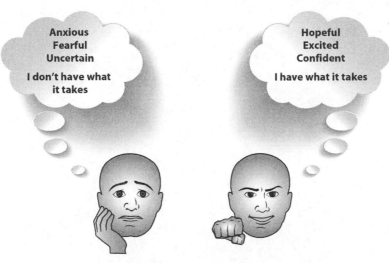

Figure 14: Threat versus Challenge Responses. People tend to have many thoughts and feelings when facing a stressful situation. Here are two different types of responses: One is characterized by feeling threatened, by a fear of losing, or possibly being shamed. The other is characterized by feeling challenged and confident about achieving a positive outcome.

and shame as you anticipate a bad outcome. A predominant habitual threat response can, over time, work itself into your cells and grind down your telomeres. A predominant challenge response, though, may help shield your telomeres from some of the worst effects of chronic stress.

People don't generally show responses that are *all* threat or *all* challenge. Most experience some of both. In one study, we found that it was the proportion of these responses that mattered most for telomere health. The volunteers who felt more threat than challenge had shorter telomeres. Those who saw the stressful task as more of a challenge than a threat had longer telomeres.[10]

What does this mean for you? It means you have reason to be hopeful. We do not mean to trivialize or underestimate the potential that very tough, difficult, or intractable situations have for harm

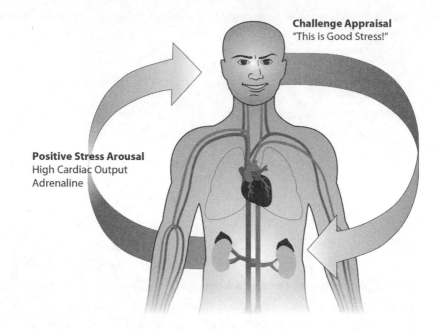

Figure 15: Positive Stress (Challenge Stress) Energizes. Our body automatically reacts to a stressful event within seconds and also reacts to our thoughts about the event. When we start to notice the stress response in our muscle tension, heart rate, and breathing, we can relabel it by saying, "This is good stress, energizing me so I can perform well!" This can help shape the body's response to be more energizing, bringing more dilation to the vessels and more blood to the brain.

to your telomeres. But when you can't control the difficult or stressful events in your life, you can still help protect your telomeres by shifting the way you view those events.

WHY DO SOME PEOPLE FEEL MORE THREAT THAN OTHERS?

Reflect on incidents in your life that have been difficult. Ask yourself: Do you tend to respond by feeling more threatened or challenged? Do you borrow trouble, feeling anticipatory threat about events that haven't happened yet—and that may not ever happen?

When you're stressed, do you feel ready for action, or do you feel like diving under the covers and hiding?

If you tend to feel more of a threat response, don't waste your time feeling bad about it. Some of us are simply wired to be more stress reactive. It has been critical to human survival for some of us to respond in a robust way to changes in our environment, and for others to be more sensitive. After all, someone's got to alert the tribe to dangers and warn the more gung-ho members against taking foolhardy risks.

Even if you weren't strongly wired at birth to feel threat, conditions in your life may have altered your natural response. Teenagers who were exposed to maltreatment when they were children respond to stressful tasks with blood-flow patterns characteristic of a threat response, experiencing vasoconstriction rather than strong blood flow out of the heart.[11] (On the other hand, people who experienced moderate adversity in childhood tend to show more of a challenge response than people who had it easy as children—more evidence that small doses of stress can be healthy, provided that resources are available to help you cope.) As we described earlier, prolonged stress can wear down emotional resources, making people more prone to feeling threatened.[12]

Either by birth or by the circumstances of your life, you may have a strong threat response. The question is: Can you learn to feel challenged instead? Research says the answer is *yes*.

DEVELOPING A CHALLENGE RESPONSE

What happens as an emotion arises? Scientists used to believe it was a more linear process—that we experience events in the world, our limbic system reacts with an emotion, like anger or fear, which causes the body to respond with an increased heart rate or sweaty palms. But it's more complicated than that. The brain is wired to *predict things ahead of time*, not just *react after things have happened*.[13] The brain uses memories of past experiences to continually anticipate what will happen next, and then corrects those predictions with both the current incoming information from the outside world, and from all the signals within our body. *Then* our brain comes up with an emotion to match all of this. Within seconds, we patch all this information together, without our awareness, and we feel some emotion.

If our "database" of past experience has a lot of shame in it, we are more likely to expect shame again. For example, if you feel high arousal and jittery, maybe from that morning's strong coffee, and if you see two people who could be talking about you, your mind may quickly cook up the emotions of shame and threat. Our emotions are not pure reactions to the world; they are our own fabricated constructions of the world.[14]

Knowing how emotions are created is powerful. Once you know this, you can have more choice over what you experience. Instead of feeling your body's stress responses and viewing them as harmful, a common experience in your brain's database, you can think about your body's arousal as a source of fuel that will help your brain work quickly and efficiently. And if you practice this enough, then eventually your brain will come to predict feelings of arousal as helpful. Even if you're one of those people whose brain is hardwired to feel more threat, you can feel that immediate instinctive survival response—and then revise the story. You can choose to feel challenged.

Sports psychologist Jim Afremow, PhD, who consults with professional and Olympic athletes, was once approached by a sprinter who was struggling with her hundred-meter time. She had already diagnosed the reason she wasn't running as well as she wanted to. "It's the stress," she said. "Before every race, my pulse races. My heart is about to jump out of my chest. You've got to help me stop it!"

Afremow laughed. "Do you really want to stop your heart?" The worst thing athletes can do, he says, is try to get rid of their stress.

"They need to think of stress as helping them get ready to perform. They need to say, 'Yes! I need this!' Instead of trying to make the butterflies in their stomach go away, athletes need to make those butterflies line up and fly in formation." In other words, they need to make the stress work for them.

The sprinter took Afremow's advice. By viewing her physical responses as tools that would help her rise to the challenge of a race, she was able to shave milliseconds off her time (a big deal for a hundred-meter runner) and set a personal record.

It sounds unbelievably simple, but research backs up this efficient method of converting threat to challenge. When research volunteers are told to interpret their body's arousal as something that will help them succeed, they have a greater challenge response. One study found that students who are encouraged to view stress in this way score higher on their GREs.[15] And when researchers put people through lab stressors, the ones who are told to think of stress as useful are able to maintain their social equilibrium. Instead of looking away, playing with their hair, or fidgeting—all signs of feeling somewhat threatened—the challenge participants make direct eye contact. Their shoulders are relaxed, and their bodies move fluidly. They feel less anxiety and shame.[16] All these benefits happened simply because people were told to think of their stress as good for them.

A challenge response doesn't make you less stressed. Your sympathetic nervous system is still highly aroused, but it is a positive arousal, putting you in a more powerful, more focused state. To channel your stress so that it gives you more good energy for an event or performance, say to yourself, "I'm excited!" or "My heart is racing and my stomach is doing cartwheels. *Fantastic*—those are the signs of a good, strong stress response." Of course, if you are under the kind of emotionally depleting stress that our caregiving mothers experienced, this language could feel too glib. Instead, talk to yourself in a gentler way. You could say, "My body's responses are trying to help me. They're designed to help me focus on the tasks at hand. They're a sign that I care." The challenge response is not a falsely chipper, gee-I'm-so-happy-that-stressful-things-are-happening-to-me attitude. It is the knowledge that even though times may be very difficult, you can shape stress to your purpose.

For those who feel addicted to "good stress"—the achievement stress involved in the constant excitement of, for example, working in a start-up company and never having downtime, know that even good stress can be overdone. It's healthy to have times when your cardiovascular system is mobilized and your psyche is primed for action. But our bodies and minds aren't built to sustain this kind of high stimulation on a consistent basis. Being able to relax, although it's been overrated as a sole source of stress management, is still necessary. We recommend that you regularly engage in an activity that brings you deep restoration. There is high-quality evidence that meditation, chanting, and other mindfulness practices can reduce stress, stimulate telomerase, and perhaps even help your telomeres to grow. See page 153 to learn more about these cell-protecting strategies.

Even in chronically stressful situations like caregiving, the stress is not a monolith or a blanket of darkness that cannot be lifted. Stress and stressful events do not live in each little moment, although they can visit. There is some freedom in each moment, because we can have a choice about how we spend this moment. We can't rewrite the past and we can't dictate what happens in the future, but we can choose where to place our attention in the moment. And although we can't always choose our immediate reactions, we can shape our subsequent responses.

Some clever studies have shown that merely anticipating a stressful event has almost the same effect on the brain and body as experiencing the stressful event.[17] When you worry about events that haven't happened yet, you're letting stress flow over its time boundaries the way a river can overflow its banks, flooding the minutes, hours, and days that could otherwise be more enjoyable. It is almost

OUR FEATHERY FRIENDS: STRESSED BIRDS, STRESSED TELOMERES

Is the stress-telomere relationship really causal? To test this, researchers have done experiments on birds. When wild European shag chicks were given water laced with the stress hormone cortisol, or were stressed out by being held, they developed shorter telomeres compared to controls.[18] Not good, since in this species, short telomeres early in life predict early death! When parrots are housed alone and can't have their usual social chats with each other, they develop shorter telomere length.[19] We know humans are sensitive to their social environments, and it seems birds are, too.

always possible to find something to worry about and therefore possible to keep the stress response engaged on an almost constant basis. When you anticipate a bad outcome before an event has even begun, you increase your dose of threat stress, and that's the last thing you need. But rather than avoiding thinking about stressful things, it's how we think about them that matters.

A SHORT PATH TO A LONG DISEASESPAN: STRESS, AGING IMMUNE CELLS, AND INFLAMMATION

It never fails. Just after you've met an important work deadline, or as you're boarding a plane for a long-overdue beach vacation, you come down with the mother of all colds: sneezing, runny nose, sore throat, fatigue. Coincidence? Probably not. While your body is actively fighting stress, your immune system can be bolstered for a time. But that effect can't last forever. Chronic stress suppresses aspects of the immune system, leaving us more vulnerable to infections, causing us to produce fewer antibodies in response to vaccinations, and making our wounds heal more slowly.[20]

There is an unsavory relationship between stress, immune suppression, and telomeres. For years, scientists were unsure just how stress, which lives in the mind, could damage the immune system. Now we have an important part of the answer: telomeres. People with chronic stress have shorter telomeres, and short telomeres can lead to prematurely aging immune cells, which means worse immune function.

Shorter Telomeres, Weaker Immune System

Certain immune cells are like SWAT teams that fight viral infections. These cells are known as T-cells, because they are stored in the thymus gland, which sits under the sternum bone in the chest. Once T-cells mature, they leave the thymus and circulate continuously throughout the body. Each T-cell has a unique receptor on its surface. The receptor acts like a searchlight on a police helicopter, sweeping the body and looking for "criminals"—cells that are either infected or cancerous. Of particular interest to aging is the type of T-cell called a CD8 cell.

But it isn't enough for the T-cell to simply spot a villainous cell. In order to complete the job, the T-cell needs to receive a second signal from a surface protein, called CD28. When the T-cell kills its target, the cell develops "memory" so that if the same virus infects the body again in the future, the T-cell can multiply into thousands and thousands of progeny cells just like itself. Together they can mount a rapid, efficient immune response against that specific virus. This is the basis of vaccination. The vaccine is typically a piece of a viral protein or a killed virus; the immunity lasts for years, since the T-cells that have responded to the initial vaccination remain in the body for a long time (sometimes for life) and are available to fight off an infection if the virus should work its way into the body again.

We have a tremendously large repertoire of T-cells, each with the capacity to recognize just one particular antigen or virus. Because we have such a huge variety of different T-cells, when we become infected with a particular virus, the few T-cells that have the correct receptor for the virus must create many progeny in order to combat the infection. During this massive process of cell division, telomerase is ratcheted up to high levels. However, it simply can't keep up with the speedy rate of telomere shortening, and eventually the telomerase response weakens to a whisper, and the telomeres in those responding T-cells keep getting shorter. So they pay for those heroic responses. When a T-cell's telomeres grow short, the cell becomes old, and it loses the CD28 surface marker that is necessary for mounting a good immune response. The body becomes like a city that's lost its budget for police helicopters and searchlights. The city looks normal from the outside, but lies vulnerable to criminal infestation. The antigens on bacteria, viruses, or cancerous cells are not cleared from the body. That's a reason people with aging cells— including the elderly and the chronically stressed—are so vulnerable to sickness, and why it's hard for them to weather diseases like the flu or pneumonia. It's partly why HIV progresses to AIDS.[21]

When telomeres in these aging T-cells are too short, even young people are more vulnerable. Sheldon Cohen, a psychologist at Carnegie Mellon University, asked young, healthy volunteers to live isolated in hotels so he could study the effects of giving them a noseful of the virus that causes the common cold. First, he measured their telomeres. The people with shorter telomeres in their immune cells, and especially in their near-senescent CD8 cells, developed colds faster, with more severe symptoms (which were measured by weighing their used tissues).[22]

What's Stress Got to Do with It?

Our CD8 T-cells (the fighters in the immune system) appear to be especially vulnerable to stress. In another of our family caregiver studies, we took blood samples from mothers who had a child with autism living at home. We found that these caregiving mothers had lower telomerase in their CD8 cells that had lost the critical CD28 surface marker, suggesting they would be in danger of developing critically short telomeres over the years. Rita Effros, an immunologist from University of California, Los Angeles, and a pioneer in understanding aging immune cells, has created "stress in a dish"—she has shown that exposing immune cells to the stress hormone cortisol dampens their levels of telomerase.[23] A compelling reason to learn how to respond to stress in a healthier way.

Shorter Telomeres, More Inflammation

Unfortunately, the news gets worse. When the telomeres of aging CD8 cells wear down, the aging cells send out proinflammatory cytokines, those protein molecules that create systemic inflammation. As the telomeres continue to shorten and the CD8 cells become fully senescent, they refuse to die and they accumulate in the blood over time. (Normally CD8 T-cells gradually die by a natural type of cell death called apoptosis. Apoptosis rids the body of old or damaged immune cells so they do not overwhelm the body or develop into the types of blood cancers called leukemias.) These senescent T-cells are the rotten apples in the barrel, with their bad effects spreading outward. They pump out slightly more inflammatory substances each year like a slow drip. If you have too many of these aging cells in your bloodstream, you're at risk for rampant infections *and* all the diseases of inflammation. Your heart, your joints, your bones, your nerves, and even your gums can become diseased. When stress makes your CD8 cells grow old, you grow old, too—no matter what your chronological age is.

Experiencing stress and pain is unavoidable. It is part and parcel of being involved with life, of loving and caring for people, caring about issues, and taking risks. Use the challenge response to protect your cells while you engage fully with life. The Renewal Lab at the end of this chapter offers some specific techniques to help you cultivate this response. The challenge response is not the only tool in your box, though. For powerful stress-relievers that are great for your telomeres, check out "Stress-Reducing Techniques Shown to Boost Telomere Maintenance" at the end of Part Two. And if stress tends to lead you into destructive thinking patterns—maybe you suppress painful thoughts or ruminate excessively about them, or perhaps you begin to anticipate negative responses from other people—turn to the next chapter. We'll help you protect your telomeres from this harmful thinking.

TELOMERE TIPS

- Your telomeres don't sweat the small stuff. Toxic stress, on the other hand, is something to watch for. Toxic stress is severe stress that lasts for years. Toxic stress can dampen down telomerase and shorten telomeres.
- Short telomeres create sluggish immune function and make you vulnerable even to catching the common cold.
- Short telomeres promote inflammation (particularly in the CD8 T-cells), and the slow rise of inflammation leads to degeneration of our tissues and diseases of aging.
- We cannot rid ourselves of stress, but approaching stressful events with a challenge mentality can help promote protective stress resilience in body and mind.

RENEWAL LAB

REDUCE "EGO THREAT" STRESS

If you feel that an important aspect of your identity is on the line, you are probably going to feel a strong threat response. This is why a final exam can be so stressful if your main identity is as a "good student," or why a sports competition can feel terrifying if you strongly identify as an athlete. If you do poorly, you don't just suffer a bad grade or a loss. The experience takes a bite out of your sense of self-worth. A challenge to your identity leads to threat stress, which can lead to poor performance, which can wound your identity. It's a vicious cycle, one that may have a negative effect on your telomeres. Break the cycle by reminding yourself that your identity runs wide and deep:

Instructions for defusing ego threat: Think of a stressful situation. Now in your mind or on a piece of paper, make a list of the things that you value (it's best to choose things unrelated to the stressful situation). For example, you may think about some social roles that are important to you (being a parent, good worker, community member, etc.) or values you believe are particularly important (such as your religious beliefs, community service). Next, think about a specific time in your life when one of these roles or values was particularly salient for you.

There are many studies documenting this effect; typically in these studies, volunteers are asked to write for ten minutes about their personal values. This small manipulation (called value affirmation) reduces stress responses in the lab, and in real life, and helps people engage in stressful tasks with a challenge mind-set.[24] Identifying values translates into better performance and higher grades on science tests.[25] It activates the reward area of the brain that may help buffer stress reactions.[26]

The next time a threat looms, pause and list what's most important to you. One caregiving mother we know stops and reminds herself that one of her highest priorities is helping her son who has autism, which seems to absorb her tension and protect her from caring about what other people think. When he has a meltdown in a public space, she ignores the judgmental stares from other people and simply does what her son needs. "It's like I'm in a protective bubble," she says. "It's a lot less stressful in there." When you see just how broad your values run, you validate your sense of self-worth, so there's less of your identity riding on the outcome of a single event.

DISTANCING

Create some space between your feeling self and your thinking self. Researchers Ozlem Ayduk and Ethan Kross and their colleagues have conducted several lab studies to manipulate the emotional stress response, in order to see what amps it up and what allows emotions to dissipate quickly. They've discovered that by distancing your thoughts from your emotions, you can convert a threat response into a positive feeling of challenge. Below are the methods Ayduk and Kross have identified to create this distance:

Linguistic self-distancing. Think about an upcoming stressful task using the third person, as in "What is making Liz nervous?" Thinking in the third person "puts you in the audience," so to speak, or makes you a fly on the wall. You don't feel so caught up in the drama. Moreover, research shows that frequent self referencing ("I," "me," and "mine") is a sign of being self-focused and is related to feeling more negative emotions. Ayduk and Kross have found that thinking in the third person and not using "I" leads people to feel less threatened, anxious, and ashamed, and to engage in less rumination. They perform better at stressful tasks, and raters view them as more confident.[27]

Time distancing. Think about the immediate future, and you will have a bigger emotional response than if you take a longer-term view. Next time you are in the grip of a stressful event, ask yourself, *In ten years, will this event still have an impact on me?* In studies, people asked to pose this question to themselves had more challenge thoughts. Recognizing the impermanence of an event helps you get over it faster.

Visual self-distancing. Distancing is a trick you can play on the threat response after the fact. If you have experienced a stressful event that you still feel emotional about, visual distancing allows you to emotionally process it in a way that will help put it to rest. Rather than just relive the event straight up, which can induce the same emotions you felt in those moments, *step back and view the event from afar, as if it's happening in a movie that you're watching.* That way, you won't reexperience the event in your emotional brain. Instead, you'll view it with greater separation and clarity. Distancing takes some of the power away from a negative memory. This technique is also known as cognitive defusion, and it's been shown to immediately reduce the brain's neural stress response,[28] probably because it activates the brain's more reflective, analytical areas instead of its emotional ones. Here is a modified version of the script Ayduk and Kross use to help their research volunteers create distancing (we combined visual, linguistic, and time distancing):[29]

> *Instructions for distancing:* Close your eyes. Go back to the time and place of the emotional experience and see the scene in

your mind's eye. Now take a few steps back. Move away from the situation to a point where you can now watch the event unfold from a distance and see yourself in the event, the distant you. Now watch the experience unfold as if it were happening to the distant you all over again. Observe your distant self. As you continue to watch the situation unfold for your distant self, try to understand his [or her] feelings. Why did he [or she] have those feelings? What were the causes and reasons? Ask yourself, "Will this situation affect me in ten years?"

If you suffer from retrospective stress—if you feel a lot of negative emotions and shame after an event is over—the visual distancing strategy can be especially useful. You can also try this strategy while you're actually in the stressful moment. By mentally stepping outside your body, you can bypass its sense of imminent threat and attack.

CHAPTER FIVE

Mind Your Telomeres: Negative Thinking, Resilient Thinking

We are largely unaware of the mental chatter in our minds and how it affects us. Certain thought patterns appear to be unhealthy for telomeres. These include thought suppression and rumination as well as the negative thinking that characterizes hostility and pessimism. We can't totally change our automatic responses—some of us are born ruminators or pessimists—but we can learn how to keep these automatic patterns from hurting us and even find humor in them. Here we invite you to become more aware of your habits of mind. Learning about your style of thinking can be surprising and empowering. To see what your own tendencies are, take the personality assessment at the end of this chapter (page 128).

One day several years ago, Redford Williams came home from a difficult day at the office and headed for the kitchen. Then he stopped. There was a pile of catalogs sitting on the counter—a pile that his wife, Virginia, had agreed to get rid of the day before. Yet there Virginia stood, calmly stirring a pot on the stove. The catalogs remained exactly where she'd left them.

Redford exploded. "Get the damn catalogs off the counter!" he ordered. It was the first thing he'd said to her since walking in the door.

What was he thinking? It's a natural question when we hear about bewildering, out-of-proportion hostility like this. Because Redford Williams is now a renowned professor of psychology and neuroscience at Duke University and an expert in anger management, he can provide some answers. "I was thinking that I was exhausted, surprised, and angry. I was thinking that Virginia was being lazy, and that she was deliberately avoiding a task she'd promised to do," he said. "I was impugning her motives." He discovered later that Virginia had not moved the catalogs because she was busy cooking him a meal that would be good for his heart.

Scientists are learning that certain thought patterns are unhealthy for telomeres. Cynical hostility, which is characterized in part by the kind of suspicious, angry thoughts that gripped Williams when he saw a less than perfectly tidy kitchen, is linked to shorter telomeres. So is pessimism. Other thought patterns, including mind wandering, rumination, and thought suppression, may also lead to telomere damage.

These thought patterns, unfortunately, can be automatic and hard to change. Some of us are born cynics or pessimists; some of us have been ruminating about our problems practically since we were old enough to talk. In this chapter, we'll describe each of these automatic patterns, but you'll also discover that you can learn to laugh at your negative thoughts and keep them from hurting you so much.

CYNICAL HOSTILITY

In the 1970s, the best-selling book *Type A Behavior and Your Heart* made *type A personality* a household term. The book claimed that type A behavior—characterized by hard-charging impatience, an emphasis on personal achievement, and hostility toward others—was a risk factor for heart disease.[1] You still see the idea of type A lingering in online assessments and casual conversation. ("Oh, I hate to stand in long lines—I'm so type A.") Actually, subsequent research showed that being a quick-on-the-draw high achiever is not necessarily harmful to your health. It's the hostility component of type A that is so damaging.

Cynical hostility is defined by an emotional style of high anger and frequent thoughts that other people cannot be trusted. A person with hostility doesn't just think, "I hate to stand in long lines at the grocery store." A person with hostility thinks, "That other shopper deliberately sped up and beat me to my rightful position in the line!"—and either seethes or makes a nasty expression or comment to the unsuspecting person who is standing in front of him. People who score high on measures of cynical hostility often cope passively by eating, drinking, and smoking more. They tend to get more cardiovascular disease, metabolic disease,[2] and often die at younger ages.[3]

They also have shorter telomeres. In a study of British civil servants, men who scored high on measures of cynical hostility had shorter telomeres than men whose hostility scores were low. The most hostile men were 30 percent more likely to have a combination of short telomeres and high telomerase—a profile that is worrisome, because it seems to reflect unsuccessful attempts of telomerase to protect telomeres when they are too short.[4]

The men who had this vulnerable cell aging profile had the opposite of a healthy response to stress. Ideally, your body responds to stress with a spike in cortisol and blood pressure, followed by a quick return to normal levels. You're prepared to meet whatever challenge is facing you, and then you recover. When these men were exposed to stress, their diastolic blood pressure and cortisol levels were blunted, a sign that their stress response was, basically, broken from overuse. Their systolic blood pressure increased, but instead of returning to normal levels after the stressful event was over, it stayed elevated for a long time afterward. The men also had few of the resources that usually buffer people from stress. In addition to the greater hostility, they had fewer social connections and less optimism, for example.[5] In terms of their physical and psychosocial health, these men were highly vulnerable to an early diseasespan.

Women tend to have lower hostility, and it is less related to heart disease for them, but there are other psychological culprits affecting women's health, such as depression.[6]

PESSIMISM

One of the brain's chief jobs is to predict the future. The brain is constantly scanning the environment and comparing it to past experience, looking for upcoming threats to your safety. Some people have brains that are faster to spot danger. Even in ambiguous or neutral situations, these people tend to think, "Something bad is going to happen here." These folks are the first to prepare for a worst-case scenario, the first to expect a bad outcome. In other words, they're pessimists.

I (Elissa) am reminded of pessimism when I am out hiking with my friend Jamie. I see off-trail paths as an adventure; she sees them as potential for poison oak. When we see a house in the woods or in the middle of nowhere, I feel some delight and anticipation. Someone may invite us in for tea! Or maybe we'll at least get a smile and a hello if someone steps out onto the porch. But Jamie has a different set of thoughts. She's sure that if someone steps out onto the porch, it will be with a furrowed brow, gruff words, and maybe even a rifle. Jamie has a more pessimistic style of thought.

When our research team conducted a study on pessimism and telomere length, we found that people who scored high on a pessimism inventory had shorter telomeres.[7] This was a small study, of about thirty-five women, but similar results have been found in other studies, including a study of over one thousand men.[8] It also fits with a large body of evidence that pessimism is a risk factor for poor health. When pessimists develop one of the diseases of aging, like cancer or heart disease, the disease tends to progress faster. And, like cynically hostile people—and people with short telomeres generally—they tend to die earlier.

We already know that people who feel threatened by stress tend to have shorter telomeres than people who feel challenged by it. Pessimists, by definition, feel more threatened by stressful situations. They are more likely to think they won't do well, that they can't handle the problem, and that the problem is going to linger. They tend *not* to get pumped up about a challenge.

Although some people are born pessimists, some kinds of pessimism are forged by early environments in which a child learns to expect deprivation, violence, or distress. In these situations, pessimism can be seen as a healthy adaption, a protection against the pain of repeated disappointment.

MIND WANDERING

As you sit, holding this book or your e-reader in your hands, are you thinking about what you're reading? If you are thinking about something else, are your thoughts pleasant, unpleasant, or neutral? And how happy are you, right now?

Harvard psychologists Matthew Killingsworth and Daniel Gilbert used a "track your happiness" iPhone app to ask thousands of people questions like these. At random times across the day, the app prompts people to respond to similar questions about what activity they are engaged in, what their minds are doing, and how happy they are.

As the data came in, Killingsworth and Gilbert discovered that we spend half of the day thinking about something other than what we're doing. This is true almost no matter what activity is at hand. Having sex, engaging in conversation, or exercising are the activities that produce the least mind wandering, but even these have a 30 percent mind-wandering rate. "The human mind is a wandering mind," they concluded. Emphasis on "human": they noted that we are alone among the animals in our ability to think about something that's not happening right now.[9] This power of language lets us plan, reflect, and dream—but it's a power that comes with a price tag.

The iPhone mind-wandering study showed that when people are

not thinking about what they're doing, they're just not as happy as when they're engaged. As Gilbert and Killingsworth also observed, "A wandering mind is an unhappy mind." In particular, *negative* mind wandering (thinking negative thoughts, or wishing you were somewhere else) was more likely to lead to unhappiness in their next moments—no surprise there. (To gauge how often your own mind wanders, you can download the app at https://www.trackyourhappiness.org.)

Together with our colleague Eli Puterman, we studied close to 250 healthy, low-stress women who ranged from fifty-five to sixty-five years old, and we assessed their tendency to mind-wander. We asked them two questions to assess their presence in the moment and negative mind wandering:

How often in the past week have you had moments when you felt totally focused or engaged in doing what you were doing at the moment?

How often in the past week have you had any moments when you felt you didn't want to be where you were, or doing what you were doing at the moment?

Then we measured the women's telomeres. The women with the highest levels of self-reported mind wandering (which we defined as low present-oriented focus along with wanting to be somewhere else) had telomeres that were shorter by around two hundred base pairs.[10] This was regardless of how much stress they had in their lives. That's why it's a good habit to notice if you are having thoughts of wanting to be somewhere else. That thought reveals an internal conflict that creates unhappiness. This type of negative mind wandering is the antithesis of a mindful state. As Jon Kabat-Zinn, founder of the worldwide program Mindfulness Based Stress Reduction (MBSR), has said, "When we let go of wanting something else to happen in this moment, we are taking a profound step toward being able to encounter what is here now."[11]

Splitting your attention by multitasking is a low-grade source of noxious stress, even if you are not aware of it. We naturally mind-wander much of the time, and some kinds of mind wandering can be creative. But when you are thinking negative thoughts about the past, you are more likely to be unhappy, and you may possibly even experience higher levels of resting stress hormones.[12] It's becoming increasingly clear that *negative* mind wandering may be an invisible source of strife.

UNITASKING

We all have pressure on our limited attention these days and are inclined to multitask, to check e-mail, to use our time efficiently. It turns out the most efficient use of time is to do one thing and to pay full attention to it. This "unitasking," sometimes termed "flow," is also the most satisfying way to spend moments. We allow ourselves to be content and absorbed. When I (Elissa) have a day of meetings, I can easily and frenetically split myself between giving divided bits of attention to the meeting, my phone, e-mail, and intrusive thoughts about what else I need to be doing, or I can decide to be focused fully on the person in front of me. The latter is a simple pleasure, and the person in front of me has a different experience as well.

And I (Liz) felt the same contrasting sets of pulls and tugs on my attention when I was an active research scientist and mother while also serving an administrative position as the chair of my department at UCSF. On a day when I would allow myself to become absorbed in my lab doing experimental manipulations with molecules and cells in tiny test tubes, hours of productive work would fly by before I was aware of their passing. Any weekend at home just spending time with the family seemed to finish almost as soon as it had begun. Those times felt very different from juggling many kinds of time-constrained work duties. Of course, sometimes such tight schedules with multitasking cannot be avoided. But whatever you are doing, whether it is in the form of "flow" or various rapidly transitioning activities, you can try to eliminate other distractions, and be fully present, at least for part of the day.

RUMINATION

Rumination is the act of rehashing your problems over and over. It's seductive. Rumination's siren call sounds something like this: *If you keep mulling things over, if you think some more about an unresolved issue or why a bad thing happened to you, you'll have some kind of cognitive breakthrough. You'll solve problems, you'll find relief!* But rumination only *looks* like the act of problem solving. Being caught in rumination is more like getting sucked into a whirlpool that hurtles you through increasingly negative, self-critical thoughts. When you ruminate, you are actually less effective at solving problems, and you feel much, much worse.

How do you tell rumination from harmless reflection? Reflection is the natural curious, introspective, or philosophical analysis about why things happen a certain way. Reflection may cause you some healthy discomfort, especially if you are thinking about something you wish you hadn't done. But rumination feels *awful*. You can't stop yourself, even if you try. And it doesn't lead to a solution, only to more ruminating.

If for some reason you wanted to prolong the bad effects of stress long after a difficult event was past, rumination would be an effective way to do it. When you ruminate, stress sticks around in the body long after the reason for the stress is over, in the form of prolonged high blood pressure, elevated heart rate, and higher levels of cortisol. Your vagus nerve, which helps you feel calm and keeps your heart and digestive system steady, withdraws its activity—and it remains withdrawn long after the stressor is over. In one of our most recent studies, we examined daily stress responses in healthy women who were family caregivers. The more the women ruminated after a stressful event, the lower the telomerase in their aging CD8 cells, the crucial immune cells that send out proinflammatory signals when they are damaged. People who ruminate experience more depression and anxiety,[13] which are in turn associated with shorter telomeres.

THOUGHT SUPPRESSION

The final dangerous thought pattern we'll describe is actually a kind of antithought. It's a process called thought suppression, the attempt to push away unwanted thoughts and feelings.

The late Daniel Wegener, a Harvard social psychologist, was reading one day when he came across this line from the great nineteenth-century Russian writer Fyodor Dostoevsky: "Try to pose for yourself this task: not to think of a polar bear, and you will see that the cursed thing will come to mind every minute."[14]

Wegener, feeling that this idea rang true, decided to put it to the test. Through a series of experiments, Wegener identified a phenomenon he called ironic error, meaning that the more forcefully you push thoughts away, the louder they will call out for your attention. That's because suppressing a thought is hard work for your mind. It has to constantly monitor your mental activity for the forbidden item: *Is there a polar bear around here anywhere?* The brain can't sustain that work of monitoring. It fatigues. You try to push the polar bear behind an ice floe, and it pops back up, poking its head above the water and bringing a few friends along for good measure. You get *more* thoughts of polar bears than if you didn't try to suppress them in the first place. Ironic error is one reason smokers who are trying to quit will constantly think of cigarettes and why dieters, trying desperately not to think of food, are tortured by images of sweet Frappuccinos.

Ironic error may also be harmful to telomeres. We know that chronic stress can shorten telomeres—but if we try to manage our stressful thoughts by sinking the bad thoughts into the deepest waters of our subconscious, it may backfire. The chronically stressed brain's resources are already taxed (we call this cognitive load), making it even harder to successfully suppress thoughts. Instead of less stress, we get *more*. A classic example of suppression's dark power comes from people with post-traumatic stress disorder (PTSD), who—understandably—don't want to remember events that have caused them terrible distress. But their ghastly memories barge into their daily lives in an unexpected, jarring way or enter their dreams at night. Often they will judge themselves harshly for letting the intrusive thought into their minds—for not being strong enough to hold it back—and for having an emotional response to the thought.

Take a moment to absorb the links here. We push away our bad feelings, which inevitably roar back, and then we feel bad, and *then* we feel bad about feeling bad. That additional layer of negative judgment—the layer of feeling bad about feeling bad—can be like a heavy blanket that smothers that last bit of energy you had available for coping. It's one reason people fall into a serious depressive state. In one small study, greater avoidance of negative feelings and thoughts was associated with shorter telomeres.[15] Avoidance alone is probably not enough to shorten telomeres. But as you'll see in the next chapter, there's a body of evidence showing that untreated clinical depression is extremely bad for telomeres. In short: Thought suppression is a royal road to chronic stress arousal and depression, both of which shorten your telomeres.

THE ANATOMY OF A STRESSFUL DAY

In a recent study, we followed mothers caring for a child with an autism spectrum disorder. We wanted to understand the emotional anatomy of their days. Not surprisingly, the caregivers woke with more dread about the day than a control group of mothers with typical children. As the day unfolded, they viewed its stressful events as more threatening. The caregiving mothers ruminated more about the stressful things that had happened. They also reported more negative mind wandering. It appears that the chronic stress of caregiving creates a **hyperreactive stress syndrome**, in which stressful events are more often anticipated, worried about, overreacted to, or ruminated upon.

When we looked at these caregivers' cells, we found that telomerase was significantly lower in their aged CD8 cells. And for all the women in the study, negative thinking was associated with the lower telomerase. On the positive side, there were many caregivers who awoke with joy, who had a challenge response to stress, and who managed to avoid rumination—and these habits were all associated with higher telomerase.

RESILIENT THINKING

If you suffer from any of the painful mental habits we've just described (pessimism, rumination, negative mind wandering, and the thinking that characterizes cynical hostility) you probably want to make some changes. But it's unlikely that you'll end negative thinking just by ordering yourself to stop. People who scold themselves about needing to change their thoughts remind us of the *Seinfeld* episode when Frank Costanza, worked into a lather about the seating arrangements in George's car, raises his hands in the air and shouts, "Serenity now! Serenity now!" Frank explains that this is what he's supposed to say to calm himself when his blood pressure gets too high. George gazes pointedly into the rearview mirror at his father, who is red-faced and practically foaming at the mouth, the very opposite of serene.

"Are you supposed to *yell* it?" he asks.

Yelling at yourself doesn't work. For one thing, personality traits such as cynical hostility and pessimism have a genetic component— they're baked in. And if you had a lot of trauma in your childhood, you may have negative thoughts frequently. Those thoughts are life-long habits, and it's possible they won't ever completely go away. So scolding yourself is unlikely to be effective. Fortunately, you can shield yourself from some of the effects of negative thought patterns by using resilient thinking.

Resilient thinking is encompassed in a new generation of therapies based on acceptance and mindfulness. These therapies don't try to alter your thoughts. Instead, they help you change your relationship to them. You don't need to believe your negative thoughts, or act on them, or have a lot of bad feelings because the thoughts crossed your mind. Below are some suggestions for responding to negative thought patterns in a more resilient way. These suggestions will help you feel better—and we believe, based on preliminary clinical trials that have been conducted so far, that improving your stress resilience is good for your cellular health in general.

Thought Awareness: Loosen the Grip of Negative Thought Patterns

The negative thought patterns we've described here are automatic, exaggerated—and controlling. They take over your mind; it's as if they tie a blindfold around your brain so that you can't see what is really going on around you. When your negative thought patterns are in control, you really believe that your wife is lazy; you can't see that she is working hard to make sure you have a healthy dinner. You believe that the stranger will come out of the house with a rifle; you can't see how exaggerated this scenario is. But when you become more aware of your thoughts, you take off the blindfold. You don't necessarily stop the thoughts, but you have more clarity.

Activities that directly promote better thought awareness include most types of meditation, especially mindfulness meditation, along with most forms of mind–body exercises. Even long-distance running, with its repetitive footfalls, can help with thought awareness and present orientation. You can notice the rhythm you create when your foot hits the ground, notice details of the trees and leaves you pass, notice your thoughts passing. Engaging in any kind of mind–body practice regularly allows you to be less focused on negative thoughts about yourself; you get better at noticing your surroundings and other people. And in times of reactivity, you are able to notice you are experiencing negative thoughts, and they dissolve sooner. Thought awareness promotes stress resilience.

To become aware of your thoughts, close your eyes, take some relaxed breaths, and focus on the movie screen of your mind. Take a mental step back and watch your thoughts go by, as if you're watching traffic on a busy street. For some of us, that street is like the New Jersey Turnpike during a thunderstorm—slick, crowded, and heart-thumpingly fast. That's fine. As you become more aware of your thoughts, including the ones that are distressing, you can label them, accept them, and even laugh at them. ("Oh, I'm criticizing myself again. I do that so often it's funny.") Instead of pushing your thoughts under the surface or letting them control your behavior, you let the negative thoughts pass by.

Thought awareness can reduce rumination.[16] It can help with automatic negative thinking by putting some distance between your instinctive thought and your reaction to that thought. You realize that you don't have to follow the story line inside your head—because, as you'll notice, the story line doesn't usually lead to productive thinking. We apparently have around sixty-five thousand thoughts a day. We're not really in control of *generating* our thoughts; they come no matter what we do. And this includes thoughts we would never invite in. But when you practice thought awareness, you notice that about 90 percent of your thoughts are repeats of thoughts that came before. You feel less compelled to grab on to them and take them wherever they lead you. They're just not worth following. With time, you learn to encounter your own ruminations or problematic thoughts and say, "That's just a thought. It'll fade." And that is a secret about the human mind: We don't need to believe everything our thoughts tell us. (As the wise bumper sticker says, "Don't believe everything you think.") The only thing we can be sure of is that our thoughts are constantly changing. Thought awareness helps us perceive the truth of this statement.

I (Liz) went on a mindfulness meditation retreat several years ago in order to learn about and experience this technique, since some of my collaborative studies on telomeres involved meditation interventions. With other interested scientists, and psychologists, I spent a week at a quiet location in southern California to learn from Alan Wallace, an experienced teacher of Tibetan meditation techniques. As a newcomer to mindfulness, I was surprised to learn how much emphasis was placed on training the mind to focus its attention. I found that the mindfulness meditation techniques produced a calm mind, along with pleasant, unbidden feelings like gratitude.

Now, years later, the ability to focus better on whatever is at hand has stayed with me. To keep this ability topped up, I use micro-meditations at times that could otherwise leave me bored, antsy, or impatient: when I'm waiting for the airplane to take off; during a shuttle ride in San Francisco en route to a meeting; waiting for my computer to boot up; or even just waiting for the microwave to heat a cup of tea.

Next time you notice that unwelcome thoughts are racing through your head, you might want to try this: *Close your eyes. Let yourself breathe normally, but pay attention to your breath. When thoughts come into your mind, imagine you are simply a witness to them and watch them gently waft away. Try not to judge the thoughts or yourself for having the thoughts. Bring your attention back to your breath, focusing on the natural feel of it as you breathe in and breathe out.*

With practice, the thoughts that are buzzing in your mind will settle down, and you'll be in a more focused state. Picture your mind like a snow globe. Minds are often in an unsettled state, and the globe is cloudy with thoughts. But taking a mini-meditation break allows the thoughts to eventually settle, allowing you more mental clarity. You won't be at the mercy of following your thoughts.

Of course, it is wonderful if you can practice for a longer time, or to attend a mindfulness retreat to learn this new skill more easily.

But don't let the perfect be the enemy of the good. Shorter periods of mindfulness will also help you develop thought awareness and reduce the power of your negative thought patterns.

Mindfulness Training, Purpose in Life, and Healthier Telomeres

In one of the most dramatic and comprehensive studies of meditation ever performed, experienced meditators headed up to the Colorado Rockies for a mountain retreat with Buddhist teacher Alan Wallace. For three months, they followed an intense practice of concentration meditation aimed to cultivate a relaxed, vivid, and stable attentional focus. The meditators also engaged in practices designed to foster beneficial aspirations for themselves and others, such as compassion.[17] The meditators also put up with a lot of experimentation, including blood draws. The intrepid University of California, Davis, researcher Clifford Saron and his colleagues decided to measure the meditators' telomerase, too, so they built a wet lab right on the mountain, complete with a refrigerated centrifuge and a dry ice freezer for storing the meditators' cells at the necessary temperature of eighty degrees below zero degrees Celsius—meaning that they had to haul five thousand pounds of dry ice up the mountain over the course of the project.

The results were what you might expect from three months sitting in a beautiful locale, listening to an inspiring teacher, and meditating among like-minded indviduals every day. After the retreat, the meditators felt better—less anxious, more resilient, empathetic. They had longer sustained focus and could better inhibit their habitual responses.[18] The researchers checked in with the meditators five months after the retreat, and these effects were still strong. They found that the enhanced ability to inhibit responses, gained from the retreat, predicted the longer-term improvements in emotional wellbeing.[19] A control group of experienced meditators who waited back at home for their turn on the mountain (but who were flown to the retreat center for testing) did not experience these effects until they went on the retreat themselves.

The meditators also experienced an increased sense of purpose in life. When you have a sense of purpose, you wake up in the morning with a sense of mission, and it's easier to make decisions and plans. In a study led by neuroscientist Richard Davidson of the University of Wisconsin, volunteers were exposed to distressing pictures, which typically enhances one's startle response to loud noise. The eyeblink startle response reflects an automatic defense response in the brain. The people with the strongest sense of purpose in life had a more resilient stress response, less reactivity, and faster recovery of their eyeblink startle response.[20]

Stronger feelings of life purpose are also related to reduced risk of stroke and improved functioning of immune cells.[21] Life purpose is even linked to less belly fat and lower insulin sensitivity.[22] In addition, having a higher purpose in life may inspire us to take better care of ourselves. People with greater purpose tend to get more lifesaving tests to detect early disease (such as prostate exams and mammograms), and when they do get sick, they stay in the hospital fewer days.[23] The writer Leo Rosten said, "The purpose of life is not to be happy—but to matter, to be productive, to be useful, to have it make some difference that you lived at all." But it doesn't have to be a competition between being happy and being productive with purpose—they come together.

Life purpose is what brings us eudaemonic happiness, the healthy feeling that we are involved in something bigger than ourselves. Eudaemonic happiness is not the transitory happiness we experience when eating or buying something we really want; it is enduring wellbeing. A strong sense of our values and purpose can serve as a bedrock foundation that helps us feel stability through life events, those earthquakes both minor and major. In hard times, we can bring them to mind over and over again. They may even protect us from threat stress at an unconscious automatic level. With a strong sense of purpose, the vicissitudes of life, including both joy and sorrow, can then fit more easily into a meaningful context or container.

What about cell aging? Saron had used the blood draws and wet lab to spin, separate, and save the white blood cells for later lab analysis by Liz and our colleague Jue Lin, who examined the meditators' telomerase activity. (Back then, we didn't think telomeres could change quickly, so we did not measure them in studies that followed people for only a few months.) Tonya Jacobs carefully analyzed telomerase in relation to self-reported psychological changes in wellbeing, such as purpose in life. Overall, the retreat group had 30 percent more telomerase than the group that had been wait-listed. And the more that the meditators improved on scores of purpose in life, the higher their telomerase.[24] Meditation, if it is of interest to you, is obviously one important way to enhance your own purpose in life. There are innumerable ways to achieve a greater sense of purpose, and which one you choose depends on what is most meaningful to you.

A New Purpose During Retirement? The Experience Corps

Imagine you've been retired for years. You have your routine, and you know what to expect each day. Then you are approached and asked to tutor an at-risk child in your neighborhood. What would you say? What is it like for someone not used to a daily job anymore, and certainly not used to working in a low-income school with little kids? So just what does happen when retired people join a tutoring program, volunteering fifteen hours a week?

The Experience Corps is a remarkable program that matches retired men and women as tutors in public schools for low-income, urban, young children. It is also a high-intensity volunteer experience and comes with its own stresses. A group of gerontology researchers wanted to see if this intergenerational program could improve health for all involved, so they have been examining the benefits of the program for both the children and the adults. So far the results are profound.

First, let's take a closer look at the volunteers' experiences of stress. Many of them were interviewed about the stresses and rewards of volunteering. They dealt with the kids' behavioral issues and sometimes couldn't get to their lesson. They saw the children's personal problems, sometimes including parental neglect, up close. They didn't always get along with the teachers. However, the rewards were numerous, and on balance the benefits outweighed the stressful aspects. They enjoyed helping the children and seeing them improve and developed special relationships.[25] This sounds like a form of positive stress!

To examine effects on health, researchers built a controlled trial into the Experience Corps, randomizing older adults to either volunteering in the Corps or a control condition. Two years later, the volunteers felt more "generative" (more accomplished from helping others).[26] The volunteers had some physiological transformations as well: While the control group had declines in brain volume (cortex and hippocampus), the volunteers had increases, especially the men. The men showed a reversal of three years of aging over two years of volunteering. This increase means better brain function—the larger the brain-volume increase, the better their increase in memory.[27] These increases in wellbeing and brain volume remind us that "life shrinks or expands in proportion to one's courage," as the writer Anaïs Nin has said.

A TELOMERE-HEALTHY PERSONALITY TRAIT

Personality traits like cynical hostility and pessimism may damage your telomeres, but there's one personality trait that appears to be good for them: conscientiousness. Conscientious people are organized, persistent, and task oriented; they work hard toward long-term goals—and their telomeres tend to be longer.[28] In one study, teachers were asked to rank their young students according to their conscientiousness. Forty years later, the students who'd scored highest on conscientiousness had longer telomeres than the ones who were the least conscientious.[29] This finding is important, because conscientiousness is the personality trait that is the most consistent predictor of longevity.[30]

Part of conscientiousness is having good impulse control, being able to delay the lure of immediately rewarding (and often dangerous) things like overspending money, driving too fast, excess eating, or alcohol use. Having high levels of impulsivity is associated with shorter telomeres as well.[31]

Conscientiousness in childhood predicts longevity decades later, and in a study of Medicare patients, those with high self-discipline lived 34 percent longer than their less conscientious counterparts.[32] Perhaps that's because conscientious people are better able to control impulses, engage in healthy daily behaviors, and follow medical advice. They also tend to have healthier relationships and find better work environments, all of which mutually reinforce wellbeing and thriving.[33]

Trade Pain for Self-Compassion

Another technique for resilient thinking is self-compassion. Self-compassion is nothing more than kindness toward yourself, the knowledge that you are not alone in your suffering, and the ability to turn toward and face difficult emotions without getting lost in them. Instead of beating yourself up, you treat yourself with the same warmth and understanding you'd extend to a friend.

To gauge your self-compassion, answer these questions, based on Kristin Neff's Self-Compassion Scale:[34] Do you try to be patient and tolerant toward aspects of your personality you don't like? When something painful happens, do you try to take a balanced view of the issue? Do you remind yourself that everybody has flaws and that you are not alone? Do you give yourself the care you need? Yes answers indicate that you're high in self-compassion, and you probably recover quickly from most stresses.

Now try these questions: When you fail at something important to you, do you berate yourself? Do you become consumed by feelings of inadequacy? Are you judgmental about your flaws? Do you feel isolated and alone, separate from other people?

If you've answered yes to these, it's a sign that you struggle to feel compassionate toward yourself. Self-compassion is a skill you can develop. And it's a skill that will help you develop a resilient response to your negative thoughts. (See the Renewal Lab on page 122 for a few ideas.)

When people who are high in self-compassion have a flood of negative thoughts and feelings, they do things differently from the rest of us. They don't criticize themselves for having faults. They can observe their negative thoughts without getting swept up in them. This means that they don't have to push away negative feelings; they just let those feelings happen and then fade. This kind attitude has positive effects on their health. People high in self-compassion react to stress with lower levels of stress hormones,[35] and they have less anxiety and depression.[36]

You may be objecting to the idea of self-compassion. Some people think it is more honest and more honorable to be self-critical. Of course, it's wise to have an accurate sense of your strengths and shortcomings, but that's different from judging yourself harshly. It's different from carping at yourself when you think you're not measuring up to the competition. Self-criticism cuts like a knife. It hurts, and those invisible knife wounds don't make you stronger or better. In fact, self-criticism is a particularly painful form of self-pity, not self-improvement.

Self-compassion *is* self-improving, because it cultivates the inner strength to cope with the troubles of life. By teaching us to rely on ourselves for encouragement and support, self-compassion makes us more resilient. Depending on other people to make us feel good about ourselves is fraught with peril. When we need other people to think well of us, the thought of their disapproval is so painful that we try to beat them to it—and that's when we jump to criticize ourselves. We cannot overrely on others for comfort. Developing self-compassion is not weak or wimpy at all. It is self-reliance, and a part of stress resilience.

Wake Up Joyfully

We have found that women who wake up with feelings of joy have more telomerase in their CD8 immune cells, and their waking cortisol peak is less exaggerated than women who wake up without joy or with dread. We don't know if this is causal, of course, but let's hedge our bets and talk about those first moments of waking. They can shape the rest of your day. Regardless of which day of your life it happens to be, you can start the day with gratitude. Upon waking—and before mentally jumping into your to-do list—see what it feels like to think "I am alive!" and welcome in the day. Even though you can't know or control what the future holds, you can turn your attention to the beauty of having a fresh new day and acknowledge some small thing you are grateful for.

I (Elissa) was struck by hearing how the fourteenth Dalai Lama wakes up: "Every day, think as you wake up, today I am fortunate to be alive, I have a precious human life, I am not going to waste it." It's too easy to never have this thought and to miss this life-affirming perspective.

As you've now read, there are many ways to promote stress resiliency. A handful of more formal techniques have been studied in relation to telomere maintenance (telomerase or telomere length). Some of these are studies that have compared people cross-sectionally. For example, people practicing Zen meditation,[37] or loving-kindness meditation,[38] have longer telomeres than nonmeditators. But we don't know if a third factor (a "confound") might be causing that effect: Meditators have different values and behaviors. They may eat more kale chips, and fewer potato chips, than nonmeditators. The highest type of scientific evidence is controlled trials, where people are randomized to an active treatment or control group. You already heard about the meditators on a three-month mountain retreat. Good news: There have been more controlled trials that show that you don't have to leave home. A range of mind-body activities—Mindfulness Based Stress Reduction, yogic meditation, Qigong, and intensive lifestyle change—all promote better telomere maintenance. We describe these research studies in the "Master Tips" section at the end of Part Two (page 153).

TELOMERE TIPS

- Getting to know our habits of thinking is an important step toward wellbeing. Negative styles of thinking (hostility, pessimism, thought suppression, rumination) are common but cause us unnecessary suffering. Fortunately, they can be tempered.
- Increasing our stress resiliency—through purpose in life, optimism, unitasking, mindfulness, and self-compassion—combats negative thinking and excessive stress reactivity.
- Telomeres tend to be shorter with negative thinking. But they may be stabilized or even lengthened by practicing habits that promote stress resiliency.

RENEWAL LAB

TAKE A SELF-COMPASSION BREAK

Whenever you run into a situation that is difficult or stressful, try taking a self-compassion break. Kristin Neff, a psychologist at the University of Texas at Austin, has conducted extensive research on self compassion. Her early trials suggest that practicing self compassion can reduce rumination and avoidance, and increase optimism and mindfulness.[39] Here is a modified description of how to do it:[40]

Instructions: Recall a situation in your life that is bothering you, such as a health issue, a relationship conflict, or perhaps a work problem.

1. Say whatever word or expression that feels true to your situation:

"This is painful." "This is stressful." "This is really hard right now."

2. Acknowledge the reality of suffering: "Suffering is a part of life." Say something that reminds you of our common humanity and that this pain is not unique to you: "I'm not alone." "Everyone feels this way sometimes." "We all struggle in our lives." "This is part of being human."

3. Put your hands over your heart, or any other place that feels soothing and comforting, perhaps your belly or gently over your eyes. Take a deep breath, and say to yourself, "May I be kind to myself."

You can use a different statement that reflects your needs in the moment, including any of the following:

I accept myself as I am, a work in progress.
May I learn to accept myself as I am.
May I forgive myself.
May I be strong.
I will be as kind to myself as possible.

The first few times you take a self-compassion break, you may feel awkward, you may feel only a slight relief of pain. Keep at it anyway. When you feel pain, acknowledge it; remind yourself that you are not alone in your suffering; and put your hand on your heart, with kindness. Eventually, you will become proficient at giving yourself compassion, and you will find that these mini-breaks restore your resilient thinking.

MANAGE YOUR EAGER ASSISTANT

Most of us have been told at one time or another to beware the internal critic, that inner voice that whispers dark words into your psyche, telling you that you're not good enough, that everyone is against you, that you're thinking the wrong way. But that's counterproductive. The inner critic is part of you; get angry at it, and you're getting angry at yourself. Ultimately, you're just getting trapped in more negative thought patterns and causing yourself more discomfort.

Instead of fighting the critic, or trying to banish the critic, try accepting the critic. You can do this by thinking of the internal voice in friendlier terms. Darrah Westrup is a clinical psychologist and the author of several books about ACT, a therapy that's based on accepting life—and your mind—as it is. She suggests that you think of that voice in your head as an eager assistant. Your eager assistant isn't evil or cruel. You don't need to fire her or scold her or send her to the basement filing room. Your eager assistant is like a bright-eyed young intern, one who desperately wants to prove her worth by providing you with a steady stream of *well-intentioned but often misguided advice.*

It's unlikely you will ever get the eager assistant to stop offering a barrage of suggestions and comments about what you are doing, could have done better, or should do in the future. But you can manage your eager assistant. Be aware of her. Understand that what she's saying isn't necessarily "truth." Treat her in the same way you might manage an overly "helpful" young staff member at the office: Smile, nod, and tell yourself, "Oh, there goes my eager assistant again. She means well, but doesn't know what she's talking about here." That way, you're not in a battle with your own thoughts. By letting them be, they'll have much less influence over you.

WHAT'S ON YOUR TOMBSTONE?

The study of meditators in the Colorado Rockies found that a strong purpose in life appears to increase telomerase. Mindfulness meditation can increase your sense of purpose, but so can other activities. The following exercise may sound a bit ghoulish, but it can be clarifying:

Instructions: Write down the epitaph you'd like to see on your tombstone, the few words that you'd like the world to remember you by. To get ideas flowing, first ask yourself, what are you deeply passionate about? Here are examples we've heard:

- "Devoted father and husband."
- "Patron of the arts."
- "A friend to everyone."
- "Always learning, always growing."
- "An inspiration to all."
- "No one spread more love in one lifetime."
- "We make a living by what we get but we make a life by what we give."
- "If you don't climb the mountain, you can't view the plain."

There's not a lot of room on a tombstone! That's the whole point of the exercise; it forces you to articulate the one or two principles that are the most important to you. After doing this exercise, some people realize that they've been distracted by things that aren't all that important to them, and that it's time to attend to the priorities at the top of their list. Other people begin the exercise believing that they have a somewhat humdrum existence—but when they write their epitaph, they realize, joyfully, that they have been living in accordance with their highest goals.

SEEKING STRESS? YES, POSITIVE STRESS!

Is there something in your life that makes you nervous or excited? Is daily life too filled with predictable routine and not enough novelty to stretch your problem-solving, creative, or socializing skills? Maybe you could add more "challenge stress" to enliven your day. Doing cognitive exercises like crossword puzzles may be a good thing for maintaining your mental sharpness,[41] but they don't do much for living with vitality and purpose. You might consider stepping outside the box of daily routine and adding a new activity that is meaningful, fulfilling, and...antiaging. And as we saw with the Experience Corps, positive stress may even improve brain aging.

To pursue a new dream, we may need to stretch and get out of our comfort zone. New situations may make us anxious, but if we avoid them, we are missing opportunities to grow and thrive. Positive stress for you may be doing something you've wanted to try but were apprehensive about.

Instructions: If you say yes to positive stress, close your eyes and think of what is at the top of your list. Take some time to think of something both exciting and feasible, a mini-adventure. Choose a small step toward that goal, something you can look into *today*. Bolster yourself with affirmations of your values and reappraisals to remind yourself that challenge stress is good stress.

ASSESSMENT: How Does Your Personality Influence Your Stress Responses?

Some personality traits can lead to bigger stress responses. To determine whether your personality could affect how your mind responds when stress comes your way, take the assessment on the next page. Whatever you learn about your personality, celebrate it. Personality is the spice of life, and knowledge about it is power. There is no right or wrong way to be. The point is to know yourself and be aware of your tendencies, not to change your personality. In fact, personality cannot change easily. It tends to be stable. Both genetics and life experiences have shaped our temperament. The more we are aware of our general tendencies, the more we can notice and live better with our natural habits of reacting to stress. And that can help us improve our telomere health.

A note to the skeptical: Some magazines or books contain personality assessments that are made up. They're fun, but they're not necessarily accurate. The personality assessments here include the actual measures used in research, reprinted with permission. (The hostility questions are an exception because those questions aren't available for public use. We've done our best to write our own questions that we feel will give you a good sense of your hostility level.) They're validated, meaning that they have been tested to see if they really measure the personality trait in question. (Note: These are shorter versions, but longer versions, those that include more questions, are more reliable.)

Instructions: For each question, circle the number that best describes how much you agree or disagree with the statement. As you take the assessments, pay attention to the words rather than the numbers. There are no right or wrong answers. Be as honest as you can.

WHAT'S YOUR THINKING STYLE?

How Pessimistic Are You?

1. I hardly ever expect things to go my way.	**4** Strongly Agree	**3** Agree	**2** Neutral	**1** Disagree	**0** Strongly Disagree
2. I rarely count on good things happening to me.	**4** Strongly Agree	**3** Agree	**2** Neutral	**1** Disagree	**0** Strongly Disagree
3. If something can go wrong for me, it will.	**4** Strongly Agree	**3** Agree	**2** Neutral	**1** Disagree	**0** Strongly Disagree
TOTAL SCORE					

Now calculate your total score by adding up the numbers you circled.

- If you scored between 0 and 3, you are **low** in pessimism.
- If you scored between 4 and 5, you are **average** in pessimism.
- If you scored 6 or above, you are **high** in pessimism.

How Optimistic Are You?

1. In uncertain times, I usually expect the best.	**4** Strongly Agree	**3** Agree	**2** Neutral	**1** Disagree	**0** Strongly Disagree
2. I'm always optimistic about my future.	**4** Strongly Agree	**3** Agree	**2** Neutral	**1** Disagree	**0** Strongly Disagree

3. Overall, I expect more good things to happen to me than bad.	**4** Strongly Agree	**3** Agree	**2** Neutral	**1** Disagree	**0** Strongly Disagree
TOTAL SCORE					

Now calculate your total score by adding up the numbers you circled.

- If you scored between 0 and 7, you are **low** in optimism.
- If you scored 8, you are **average** in optimism.
- If you scored 9 and above, you are **high** in optimism.

How Hostile Are You?

1. I usually know more than people I have to listen to or follow.	**4** Strongly Agree	**3** Agree	**2** Neutral	**1** Disagree	**0** Strongly Disagree
2. Most people cannot be trusted.	**4** Strongly Agree	**3** Agree	**2** Neutral	**1** Disagree	**0** Strongly Disagree
3. I am easily annoyed or irritated by other people's habits.	**4** Strongly Agree	**3** Agree	**2** Neutral	**1** Disagree	**0** Strongly Disagree
4. I get angry at other people easily.	**4** Strongly Agree	**3** Agree	**2** Neutral	**1** Disagree	**0** Strongly Disagree
5. I can be harsh or rough to people who are disrespectful or annoying.	**4** Strongly Agree	**3** Agree	**2** Neutral	**1** Disagree	**0** Strongly Disagree
TOTAL SCORE					

Now calculate your total score by adding up the numbers you circled.

- If you scored between 0 and 7, you are **low** in hostility.
- If you scored between 8 and 17, you are **average** in hostility.
- If you scored 18 and above, you are **high** in hostility.

How Much Do You Ruminate?

1. My attention is often focused on aspects of myself I wish I'd stop thinking about.	**4** Strongly Agree	**3** Agree	**2** Neutral	**1** Disagree	**0** Strongly Disagree
2. Sometimes it is hard for me to shut off thoughts about myself.	**4** Strongly Agree	**3** Agree	**2** Neutral	**1** Disagree	**0** Strongly Disagree
3. I tend to ruminate or dwell on things that happen to me for a really long time afterward.	**4** Strongly Agree	**3** Agree	**2** Neutral	**1** Disagree	**0** Strongly Disagree
4. I don't waste time rethinking things that are over and done with.	**0** Strongly Agree	**1** Agree	**2** Neutral	**3** Disagree	**4** Strongly Disagree
5. I never ruminate or dwell on thoughts about myself for very long.	**0** Strongly Agree	**1** Agree	**2** Neutral	**3** Disagree	**4** Strongly Disagree
6. It is hard for me to put unwanted thoughts out of my mind.	**4** Strongly Agree	**3** Agree	**2** Neutral	**1** Disagree	**0** Strongly Disagree
7. I often reflect on episodes in my life that I should no longer concern myself with.	**4** Strongly Agree	**3** Agree	**2** Neutral	**1** Disagree	**0** Strongly Disagree
8. I spend a great deal of time thinking back over my embarrassing or disappointing moments.	**4** Strongly Agree	**3** Agree	**2** Neutral	**1** Disagree	**0** Strongly Disagree
TOTAL SCORE					

Now calculate your total score by adding up the numbers you circled (be extra careful when adding your scores for questions 4 and 5—the numbers are reversed).

- If you scored between 0 and 24, you are **low** in rumination.
- If you scored between 25 and 29, you are **average** in rumination.
- If you scored 30 and above, you are **high** in rumination.

How Conscientious Are You?

I see myself as someone who . . .

1. Does a thorough job.	**4** Strongly Agree	**3** Agree	**2** Neutral	**1** Disagree	**0** Strongly Disagree
2. Can be somewhat careless.	**0** Strongly Agree	**1** Agree	**2** Neutral	**3** Disagree	**4** Strongly Disagree
3. Is a reliable worker.	**4** Strongly Agree	**3** Agree	**2** Neutral	**1** Disagree	**0** Strongly Disagree
4. Tends to be disorganized.	**0** Strongly Agree	**1** Agree	**2** Neutral	**3** Disagree	**4** Strongly Disagree
5. Tends to be lazy.	**0** Strongly Agree	**1** Agree	**2** Neutral	**3** Disagree	**4** Strongly Disagree
6. Perseveres until the task is finished.	**4** Strongly Agree	**3** Agree	**2** Neutral	**1** Disagree	**0** Strongly Disagree
7. Does things efficiently.	**4** Strongly Agree	**3** Agree	**2** Neutral	**1** Disagree	**0** Strongly Disagree
8. Makes plans and follows through with them.	**4** Strongly Agree	**3** Agree	**2** Neutral	**1** Disagree	**0** Strongly Disagree
9. Is easily distracted.	**0** Strongly Agree	**1** Agree	**2** Neutral	**3** Disagree	**4** Strongly Disagree
TOTAL SCORE					

Now calculate your total score by adding up the numbers you circled (be extra careful when adding your scores for questions 2, 4, 5, and 9—the numbers are reversed).

- If you scored between 0 and 28, you are **low** in conscientiousness.
- If you scored between 29 and 34, you are **average** in conscientiousness.
- If you scored 35 and above, you are **high** in conscientiousness.

How Much Purpose in Life Do You Feel?

1. There is not enough purpose in my life.	**0** Strongly Agree	**1** Agree	**2** Neutral	**3** Disagree	**4** Strongly Disagree
2. To me, the things I do are all worthwhile.	**4** Strongly Agree	**3** Agree	**2** Neutral	**1** Disagree	**0** Strongly Disagree
3. Most of what I do seems trivial and unimportant to me.	**0** Strongly Agree	**1** Agree	**2** Neutral	**3** Disagree	**4** Strongly Disagree
4. I value my activities a lot.	**4** Strongly Agree	**3** Agree	**2** Neutral	**1** Disagree	**0** Strongly Disagree
5. I don't care very much about the things I do.	**0** Strongly Agree	**1** Agree	**2** Neutral	**3** Disagree	**4** Strongly Disagree
6. I have lots of reasons for living.	**4** Strongly Agree	**3** Agree	**2** Neutral	**1** Disagree	**0** Strongly Disagree
TOTAL SCORE					

Now calculate your total score by adding up the numbers you circled (be extra careful when adding your scores for questions 1, 3, and 5—the numbers are reversed).

- If you scored between 0 and 16, you are **low** in life purpose.

132

- If you scored between 17 and 20, you are **average** in life purpose.
- If you scored 21 and above, you are **high** in life purpose.

SELF-ASSESSMENT SCORING AND INTERPRETATION

This assessment is simply meant to raise your awareness of your personal style. It is not meant to diagnose you or make you feel bad about being a certain way. Self-awareness of tendencies that make us more vulnerable to stress reactivity (and possibly telomere shortening, in several studies) is valuable! Awareness can help us notice unhealthy thought patterns and choose different responses. It can also help us know and accept our tendencies. As Aristotle reportedly said, "Knowing yourself is the beginning of all wisdom."

Dimensions that make us more vulnerable to stress	Score (Circle)		
Pessimism	High	Medium	Low
Hostility	High	Medium	Low
Rumination	High	Medium	Low

Dimensions that may help us be more stress resilient	Score (Circle)		
Optimism	High	Medium	Low
Conscientiousness	High	Medium	Low
Purpose in Life	High	Medium	Low

HOW DID WE DECIDE WHAT DETERMINES HIGH OR LOW SCORES?

In general, we determined the high, medium, and low score categories by looking at the data from large representative samples of people who had taken the test. We divided the population into thirds based on their scores. If you are in the top third (33 percent)

of scores, you scored "high." If you are in the bottom third
(33 percent) of scores, you scored "low." If you are in the middle, you
scored "average." The actual studies used are described below.

The cutoff points should not be taken too literally. First of all, the
comparisons are made to some large samples, but any one sample
is never representative of everyone. There are always differences
in how people score based on their race/ethnicity, sex, culture,
and even age, that we could not take into account. Second, we
assumed that there is a statistically "normal distribution" for scores
for each measure, which means the same number of people score
high as low, in the same symmetrical distribution pattern. In reality,
few measures have scores that are perfectly normally distributed.
Therefore, our cutoff points are not statistically perfect, nor are
they perfectly accurate when applied to individuals.

THE PERSONALITY TYPES AND SCALES
USED IN THIS ASSESSMENT

Optimism/Pessimism

Optimism is the tendency to expect or anticipate positive events
and outcomes rather than negative ones. Optimism is characterized
by a sense of hope and positivity about the future. **Pessimism** is
the tendency to expect or anticipate negative events and outcomes
rather than positive ones. Pessimism is characterized by a lack of
hope and positivity about the future.

We used the "Life Orientation Test—Revised" (LOT-R) developed
by Professors Charles Carver and Michael Scheier.[1] Optimism and pessimism are strongly related, but not totally overlapping, which means
they are different aspects of personality. So it is helpful to examine them
separately.[2] Two studies assessed the relationship with telomere length,

and both found correlations with pessimism but not optimism.[3] That is not to say that optimism doesn't matter for health. It absolutely does, especially for mental health. It's just that with stress-related health outcomes, negative traits are often stronger predictors than positive traits, and they are more directly tied to stress physiology. Positive traits can buffer you from stress, and are weakly related to positive restorative physiology.

For scoring, we used the average levels on each LOT-R subscale from a study that tested over two thousand men and women who varied in age, gender, race, ethnicity, level of education, and socioeconomic class.[4]

Hostility

Hostility is thought to have cognitive, emotional, and behavioral manifestations.[5] The cognitive component, possibly the most important part of hostility, is characterized by negative attitudes toward others, colored by cynicism and mistrust. The emotional component ranges from irritation to anger to rage. The behavioral component is the tendency to act out verbally or physically in ways that could hurt others.

Hostility scales are not free to the public, so this scale includes items we created that should roughly measure hostility in the same way as the standardized research scales, particularly the most common one, the Cook-Medley Hostility Questionnaire, which is part of the MMPI Personality measure. We estimated the cutoffs based on the mean scores from a study of men from the Whitehall study, which used a short version of the Cook-Medley Hostility Questionnaire. This study found high hostility is related to shorter telomeres in men.[6]

Rumination

Rumination is "self-attentiveness motivated by perceived threats, losses, or injustices to the self."[7] In other words, rumination is the act of spending a significant amount of time thinking about and perseverating on past negative events in one's life and one's role.

We used the eight-item rumination subscale from the "Rumination-Reflection Questionnaire," developed by Professor Paul Trapnell.[8] To determine cutoffs, we used the item mean for the eight-item version.[9] While no studies have directly linked rumination to telomere length, we think it is an important part of the stress process. That's because it keeps stress alive in the mind and body long after the event has passed. In our daily diary study of caregivers, we have found that daily rumination is associated with lower telomerase.

Conscientiousness

Conscientiousness is the measure of the degree to which a person is organized, how careful a person is in certain situations, and how disciplined he or she tends to be.

We used the conscientiousness subscale from the "Big Five Inventory" developed by Professor Oliver John.[10] This scale was used in a study that found a positive correlation between higher conscientiousness and longer telomeres.[11] For scoring, we used means from a large study that examined conscientiousness scores across ages.[12]

Purpose in Life

Purpose in life is not a typical personality dimension, but rather how much we are aware of having some explicit purpose or goal for our life. It is something that can change based on life experiences and personal growth. An individual who scores high on the Purpose in Life scale is characterized as having a strong sense of meaning in life, having aims and engaging in activites he or she strongly values, or having an outlook that gives life meaning.[13]

We used the "Life Engagement Test," a six-item scale developed by Professor Michael Scheier and colleagues.[14] For scoring, we used normative data from a study of 545 older adults (adjusted to a 0-to-3 scale).[15] No studies have directly linked purpose in life to telomere length. However, in a meditation retreat study, increased purpose in life was associated with higher telomerase. As reviewed in the previous chapter, purpose in life is linked to better health behaviors, physiological health, and stress resiliency.

CHAPTER SIX

When Blue Turns to Gray: Depression and Anxiety

Clinical depression and anxiety are linked to shorter telomeres—and the more severe these disorders are, the shorter your telomeres. These extreme emotional states have an effect on your cells' aging machinery: telomeres, mitochondria, and inflammatory processes.

Dave had been suffering from a viral infection—sneezing, cough, stuffy nose—for several days when suddenly he had trouble breathing. At first it was uncomfortable to draw a deep breath, and then it was agonizing. "I'm hyperventilating," Dave thought, and tried breathing into a paper bag. When that didn't help, he called his wife at work, and she agreed to pick him up at the street corner and drive him to an urgent care clinic. As he walked outside, the landscape seemed to darken, even though the day was bright. It was as if a deep shadow was overtaking his vision. His skin prickled. All this time, he continued to hyperventilate. When Dave arrived at the clinic, the nurses had to give him a mild sedative so that he could breathe well enough to describe his symptoms.

He was diagnosed with a panic attack, an intense episode of fear and anxiety. For Dave, the panic attack was actually a change from the depressive symptoms that have dogged him for most of his life. When he's depressed, he feels as if he has no possibilities, no future.

Every activity, even cracking an egg for a breakfast omelet, even just looking through his bedroom window, feels overwhelmingly strenuous and even physically painful. "I squint like I'm facing a strong wind," he says.

There are still people out there who don't take depression and anxiety seriously, who don't understand the phenomenal breadth and depth of suffering they cause. A global view helps put these problems into perspective: Mental disorders and substance abuse are the top causes of disability (defined as "productive days of life lost") worldwide, and the biggest player in this mix of disorders is depression, the "common cold" of psychiatry.[1] Heart disease, high blood pressure, and diabetes all develop earlier and faster in people with depression and anxiety. It's now much harder to write off depression and anxiety as "all in your head," because research has demonstrated that these states reach past your mind and soul, past your heart, past your bloodstream, and all the way into your cells.

ANXIETY, DEPRESSION, AND TELOMERES

Anxiety is characterized by excessive dread or worry about the future. It's not necessarily as dramatic as Dave's panic attack; often it's more like a steady, low-level thrum of unease. "I was standing at the edge of my driveway," said one woman we know, "waiting for my son to come home from a late hockey practice. I felt a little shaky, and my heart was beating fast. At first I thought that I was just worried about my son coming home safely. Then I realized that I felt this way most of the time. I finally asked myself, 'Is this normal?'" It isn't. The next week, she was diagnosed with generalized anxiety disorder.

Anxiety is a relatively recent subject of telomere research. People who are in the throes of clinical anxiety tend to have significantly shorter telomeres. The longer the anxiety persists, the shorter the telomeres. But when the anxiety is resolved and the person feels better, telomeres eventually return to a normal length.[2] This is a grand argument in favor of identifying and treating anxiety. Sometimes, though, anxiety is hard to spot. As our friend came to realize, anxiety can seem normal when you're accustomed to feeling it, when it's the air that you breathe.

The depression–telomere connection has a more solid scientific literature behind it, possibly because depression is so pervasive: More than 350 million worldwide suffer from it. An impressive large-scale study of almost twelve thousand Chinese women by Na Cai and colleagues (at Oxford University and Chang Gung University in Taiwan) found that women who are depressed have shorter telomeres.[3] Depressed people, like anxious ones, show that dose-response we've talked about before. The more severe and prolonged the depression, the shorter the telomeres.[4] (See the bar chart in Figure 16.)

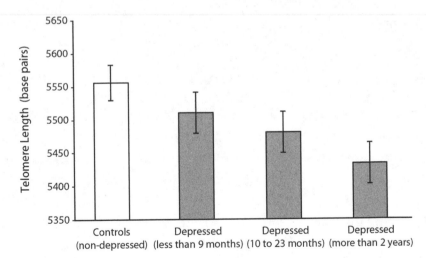

Figure 16: Duration of Depression Matters. The Netherlands Study of Depression and Anxiety has followed almost three thousand people, including those with depression and nondepressed controls. Josine Verhoeven and Brenda Penninx found those with current depression lasting less than ten months did not have significantly shorter telomeres than controls, but those with depression for more than ten months did.

A few studies suggest that short telomeres may directly lead to depression. People with depression have shorter telomeres in the hippocampus, an area of the brain that plays an important role in the disorder.[5] (They don't have shorter telomeres in other parts of the brain, just this part that is so crucial for mood.) Rats put under stress have less telomerase in the brain's hippocampus and less brain cell growth (neurogenesis), and they are more likely to develop depression.[6] However, when their telomerase is increased, the rats have more neurogenesis and do not become depressed. Cell aging in the brain may be one pathway to depression.

Here's an apparently odd phenomenon: depressed people have shorter telomeres but more telomerase in their immune cells. *What?* How could depression lead to shorter telomeres but more telomerase? This paradoxical combination appears in other situations, too: in people who are overloaded with stressful circumstances, people who haven't graduated from high school, men who are cynically hostile, and people who are at high risk for coronary disease. We believe that in these situations, the cells are producing more telomerase in response to telomere shortening, in a (sadly) ineffectual effort to build back the telomere segments that are being lost.

More support for this idea: Our colleague Owen Wolkowitz, a psychiatrist at UCSF, has been examining how telomerase might help with depression. Give depressed people an antidepressant (an SSRI), and their high telomerase levels climb up even further! The more that telomerase goes up, the more likely it is that their depression will lift.[7] It is possible that the efforts of the immune cells to replenish their lost telomeres reflect what is happening in the brain, with neurons doing the same thing. There may be a rejuvenation of sorts, in which more effective telomerase action (as opposed to ineffectual attempts by telomerase to elongate telomeres) could be promoting neurogenesis, the birth of new brain cells.

TRAUMA, DEPRESSION, AND THE REVERSAL OF STRESS EFFECTS

So far, most psychiatric disorders studied have been linked to telomere shortness, as shown by a meta-analysis.[8] Part of this could be due to the underlying stress leading up to onset of the disorders or from having the disorders. One of the most hopeful messages from the neuroscience of stress is that there is tremendous potential for brain plasticity, especially for the reversal of stress effects. We can overcome the effects of severe stress with antidepressants, exercise and other healthy buffers, and the passage of time. Telomere maintenance also shows plasticity. For example, in humans and rats it looks like telomeres likely shorten a small amount at the time of a stressful event, but in most cases they can eventually repair themselves.[9] Researcher Josine Verhoeven has examined patterns of recovery over time in the large Netherlands Study of Depression and Anxiety (NESDA) cohort: Major events in the last five years are associated with telomere shortness, but events in the distant past, more than five years ago, are not.[10] Similarly, having a current anxiety disorder is associated with shorter telomere length, but having one in the past is not, a finding that suggests telomeres can recover when an episode of anxiety is over. And the more years since the episode, the longer the telomere length.[11] Depression, however, appears to have a stronger imprint than stressful events or anxiety, as often people with past depression still have shorter telomeres.[12]

Cai's large Chinese study found a pattern suggesting that telomeres tend to rebound in people with past trauma—unless the person develops a serious depression. Then the telomeres remain short. It's as if trauma plus depression is a weight too heavy to bear. The good news is that even though telomeres can carry scars of past severe trauma plus depression, they can also be stabilized, and possibly lengthened, through activities that help boost telomerase. Telomeres can recover, thanks to telomerase.

Inside the cell, mitochondria are another important target of stress damage. Can mitochondria recover from stress as well? Mitochondria are critical to aging but have only recently been studied in terms of mental health. Mitochondria are the energy plants of the cell. Give them fuel, in the form of food molecules, and they'll process it into nutrient-rich molecules that power the cell. Some cells, like nerve cells, have one or two mitochondria; other cells need many more to keep up with their energy needs. Muscle cells, for example, typically have thousands of mitochondria. When you're in certain states of physical stress—if you have diabetes or heart disease—mitochondria can malfunction and cells won't receive enough energy. This can affect brain function, because the neurons don't have enough energy to fire. Your muscles may be weaker. The liver, heart, and kidneys—all organs that consume mass quantities of energy—suffer. One way to tell if cells are under major stress is to examine their mitochondrial DNA copy numbers, which tell us how hard the body is working to produce additional mitochondria to supplement the ones that are weary and damaged. In the Chinese study, it appeared that the greater the childhood adversity or depression, the shorter the telomeres and the higher the mitochondrial DNA copy numbers.

If you take mice and do some not very nice things to them (such as hanging them by their tails or forcing them to swim), they will, quite naturally, be stressed. Like humans, mice who are under stress develop an excessive number of mitochondria. It appears that their mitochondria are faulty and not working efficiently. Their cells are thus working desperately to increase their energy supply, with limited success. As you may imagine, stressed-out mice with high mitochondrial DNA copy numbers are not superenergetic. Further, their telomeres are 30 percent shorter. But given a month to recover from the stress, their telomeres and mitochondrial DNA become normal again. There is no lingering sign of accelerated aging.[13]

Biology can be molded by experience, then remolded. Cells can renew themselves. In a mouse's life, time-limited adversity can be mostly erased. Fortunately this appears true for many types of adversity in humans, too.

PROTECTING YOURSELF FROM DEPRESSION AND ANXIETY

Mental health is not a luxury. If you want to protect your telomeres, you need to protect yourself from the effects of depression and anxiety. Some proclivity toward these disorders is partially influenced by genes. But that does not mean everything is out of your control.

Depression is a complicated illness that lives in the emotions, in the thoughts, and in the body, and it's beyond the scope of this book to describe depression (or anxiety) in full. But here is one beautifully clear idea that drives some successful treatments: Depression is partly a dysfunctional response to stress. Instead of simply feeling the stress, depressed people tend to cope by using some of the negative thought patterns we've talked about. They try to suppress the bad feelings so that they can't be deeply felt, or they keep their problems alive by ruminating about them over and over and over. They criticize themselves. They feel irritable and angry, not just at whatever circumstances have caused their sorrow and stress, but at the fact of feeling sorrow and stress.

As we've said, this is a set of dysfunctional responses. Completely understandable, but dysfunctional nevertheless. Over time, this cycle can pull a person down past stress and into depression. Negative thoughts are like microtoxins—relatively harmless when your exposure is low, but in high quantities they are poisonous to your mind. Negative thoughts are not signs that you are truly unworthy or a failure. They are the substance of depression itself.

These counterproductive mental reactions are part of anxiety, too. Imagine this: You're at a cocktail party and you accidentally call the hostess by the wrong name. She gasps a little, recovers with a tight smile, and corrects your error. You're embarrassed. Who wouldn't be? But for most of us, it's a pretty mild stress. Our cheeks may flush a little, we apologize, and we move on. But some people have what's known as anxiety sensitivity, and their bodies will produce an outsized physical response to the same event. Put these folks at a party, and if they make a mistake like forgetting a name, their hearts will race, they'll feel light-headed, and they may even think they're going to have a heart attack. It's a really uncomfortable state. A person with anxiety sensitivity naturally thinks, "Well, that was awful. From now on, I'll just avoid parties."

The problem with avoiding whatever is making you anxious is that the avoidance actually perpetuates the feelings of anxiety. You avoid the things you want and need to do, and you never learn that it's possible to tolerate the discomfort. In psychology terms, you never *habituate* to the stressful situation. Your life becomes smaller and smaller and tenser and tenser. Those anxious feelings bloom into a full-grown clinical disorder that interferes with your life. Just as depression is an intolerance of being sad, anxiety is an intolerance of feeling anxious. That's the reason that treatment for anxiety disorders often involves exposure to the triggers and sensations that make you the most anxious. You learn that you can ride through the waves of anxiety and survive it.

Stress plus this kind of avoidant coping style can lead to both anxiety and depression. Understanding how the mind works, why and how it gets stuck in these cycles of thoughts, is a key part of overcoming these disorders. If you have frequent painful feelings and thoughts that prevent you from living fully, it's important to protect your telomeres and seek help. Don't be one of the millions suffering untreated. Coping skills take a while to develop and embody as habits, so give yourself time to learn them with the help of a therapist, and don't give up.

WHERE YOU PLACE YOUR ATTENTION MATTERS

What if nothing is really wrong with you, except your thoughts that are insisting otherwise? When we're feeling sad, we naturally try to think our way out of it. We notice the gap between how we feel and how we want to feel. We start to live in that gap, wishing things could be different, trying hard to find an escape.

Mindfulness-based cognitive therapy, or MBCT, helps people out of that gap. It combines traditional strategies of cognitive therapy with mindfulness practices. Cognitive therapy helps you change distorted thoughts; mindfulness, as we've mentioned, helps you change how you relate to your thoughts in the first place. MBCT is potent against that great threat to your telomeres, major depression. It's been shown to be *as effective as an antidepressant*.[14] One of the bleakest aspects of depression is that it can become chronic; 80 percent of its sufferers experience a recurrence. John Teasdale, formerly of the University of Cambridge; Zindel Segal of the University of Toronto Scarborough; and Mark Williams of Oxford University have found that in people with three or more recurring depressions, MBCT slashes the risk of depression's return in half.[15] It's also becoming clear that MBCT helps with anxiety, and it's useful for any of us who struggle with difficult thoughts and emotions.

There are two basic modes of thinking, MBCT teaches. There is the "doing mode," which is what we do when we're trying to get out of the gap between how life is and how we want it to be. But there is another mode, and that is the "being mode." In the being mode, you can more easily control where you put your attention. Instead of frantically striving to change things, you can choose to do little things that bring you pleasure, and things that help you feel masterful and in control. Because "being" also allows you to pay more attention to people, you can more fully connect with them—a state that typically brings humans the most joy and contentment. Have you ever experienced the contentment of focusing all your attention on a small task, such as cleaning out a messy drawer? That's what being mode feels like.

Doing Mode versus Being Mode[16]

	Doing Mode (Automatic)	Being Mode
Where is your attention?	**Not noticing what you are doing**	**Paying attention to the moment**
What time period are you living in?	**Past or future**	**Now**
What are you thinking about?	**Absorbed in stressful ideas** Thinking about where I wish I was, not where I am right now. Nothing feels satisfactory.	**Absorbed in current experience** Able to fully taste, smell, touch, and feel. Able to fully connect with others. Radical acceptance of self, unconditional kindness.
Level of metacognition (thoughts about thoughts)	**Believe thoughts are true** Cannot observe the mind's workings. Mood is controlled by thoughts.	**Freedom from believing the thoughts** Understand the transient nature of thoughts, can observe thoughts as they come and go. Can tolerate unpleasantness.

This chapter may have been a bit disturbing. Many of us have suffered from one of these common maladies of the mind or know someone in our circles who has. But the larger story is that telomeres can recover from episodes of adversity and depression, and when they don't, they can still be protected going forward. You can fortify your resources to prepare for the next challenge that appears. You can adopt a resilient mind-set to allow for more peace in mind and body, as we've reviewed earlier, such as building an awareness of both your stress response style and your habits of thinking. You might also adopt the breathing break or heart-focused meditations at the end of this chapter.

The telomere scars we may carry from adversity are also evidence of a state that we might call "worn wise." Dealing with adversity can leave us wiser and stronger. One of my (Elissa's) favorite scales measures how much one has grown from trauma in various ways (feeling closer relationships, feeling more self-reliant, increasing faith or spirituality). We used this scale in our first study of caregivers. We were confused at first when we saw that the caregivers who had shorter telomeres also experienced more psychological growth. A closer look at this pattern revealed what was going on—it was all about duration of struggle. Those who had been providing care the longest had more telomere wear and tear but also had more life-enriching changes.[17] As Elisabeth Kübler-Ross, a Swiss psychiatrist who studied grief and mourning, once said, "The most beautiful people we have known are those who have known defeat, known suffering, known struggle, known loss, and have found their way out of the depths. These persons have an appreciation, a sensitivity, and an understanding of life that fills them with compassion, gentleness, and a deep loving concern. Beautiful people do not just happen."

TELOMERE TIPS

- Major stress, depression, and anxiety are linked to shorter telomeres in a dose-response fashion. But in most cases, these personal histories can be erased, thank goodness. For example, major events have no residue five years later.
- Mitochondria function is also impacted by severe stress and depression, but at least in mice there is recovery with time.
- The cognitive machinery that drives depression and anxiety includes exaggerated forms of negative thinking—intolerance of negative feelings and excessive avoidance that doesn't really work. Depression is characterized by being stuck in the "doing mode" mind-set, including ruminative thoughts, which create a vicious cycle.
- Mindfulness-based interventions can help move us from the common overdoing mode to a being mode and reduce rumination. See the "Three-Minute Breathing Break" in this chapter's Renewal Lab.

RENEWAL LAB

THREE-MINUTE BREATHING BREAK

The pioneers of MBCT (mindfulness-based cognitive therapy)—John Teasdale, Mark Williams, and Zindel Segal—have developed training programs to help people attain the being mode. It is best to work with a practitioner to help you fully learn MBCT, but you can easily take advantage of a core activity of MBCT, which is a quick, three-minute "time in."[18] This breathing break is like practicing thought awareness. You might recognize that you are feeling something painful. You label your thoughts, allowing them to exist in your mind, and know that they will pass. The lifetime of an emotion, even a very unpleasant one, is no longer than ninety seconds—unless you try to chase it away or engage with it. Then it lasts longer. The breathing break is a way to keep negative emotions from living past their natural life spans. You can make it a habit, so it helps anchor you at any time, not just during hard moments. You can picture this exercise like an hourglass—invite whatever is present in your mind broadly, then focus narrowly on the breath, and then expand awareness out to your full surroundings. Here's our modified version:

1. Becoming aware: Sit upright and close your eyes. Connect with your breathing for a long inhalation and exhalation. With this awareness, ask yourself, "What is my experience right now? What are my thoughts? Feelings? Bodily sensations?" Wait for the responses. Acknowledge your experience and label your feelings, even if they are unwanted. Notice any pushing away of your experience, and soften around it, allowing space for all that comes up in your awareness.

2. Gathering your attention: Gently direct your full attention to your breathing. Notice each inhalation and each long exhalation. Follow each breath, one after another. Use your breathing as an anchor into this present moment. Tune in to a state of stillness that is always there right below the surface of your thoughts. This stillness allows you to come from a place of being (versus doing).

3. Expanding your awareness: Sense your field of awareness expanding around you, around your breathing, around your whole body. Notice your posture, your hands, your toes, your facial muscles. Soften any tension. Befriend all of your sensations, greet them with kindness. With this expanded awareness connect with your whole being, encompassing all that is you in the present moment.[18]

This breathing break calms your body and offers you more control over your stress reactions. It shifts your thinking away from self-focus and the doing mode and moves it toward the peaceful being mode.

A HEART-FOCUSED MEDITATION: RELEASE MENTAL PRESSURE, RELEASE YOUR BLOOD PRESSURE

Our breath is a window into knowing and regulating our mind–body. It is an important switch influencing the communication between brain and body. It's sometimes easier to change our breath to relax than to change our thoughts. When we breathe in, our heart rate goes up. When we exhale, our heart rate goes down. By having a longer exhalation than inhalation, we can slow our heart rate more, and we can also stimulate the vagus nerve. Breathing into our lower belly (abdominal breathing) stimulates the sensory pathways of the vagus nerve that go directly to our brain, which has an even more calming effect. Dr. Stephen Porges, an expert in understanding the vagus nerve, has shown why there is a strong link between the vagus nerve, breath, and feelings of social safety. Many mind–body techniques naturally stimulate the vagus nerve, sending our brain those critical safety signals.

Exercises that slow breathing, such as mantra meditation or paced breathing, are a reliable way to lower blood pressure.[19] You are slowing down your body's need to be aroused. You are turning up the volume on your vagus nerve activity, suppressing the sympathetic nervous system and slowing your heart rate even more. The vagus also turns on growth and restorative processes.

For some, focusing on the heart can be more peaceful than focusing on the breathing, and can still slow the breathing rate. The heart has such a complex and responsive nervous system that it is thought of as the "heart brain." Below we provide a script for a short heart-focused meditation. It also has some words from loving-kindness meditations in it. This has not been tested to examine any telomerase effect, but as you can see above, breathing is the basis of relaxation.

If you'd like, try this script now:

HEART-FOCUSED MEDITATION

Sit comfortably. Take some long and slow inhales and even longer exhales.

Continue to breathe in, and breathe out, repeating a calming word or hold a beautiful image each time you breathe out slowly. Notice the pause between the breaths.

Become aware of your thoughts: "Where are my thoughts right now?" Smile at each of them as they pass through your mind; then return to your exhalation word or image.

Place your hands (palms or fingers) on your heart. You might say to yourself "Ahhh" as you exhale. Let the burdens you are holding release and flow out of your body.

"May I be in peace."

"May my heart be filled with kindness."

"May I be a source of kindness for others."

Picture your heart radiating love. Picture a pet or person that you feel complete love for. Let that love radiate out toward others in your life.

Continue to breathe in, and breathe out slowly. Notice where you are holding in tension. As you exhale, let yourself feel enveloped in safety, warmth, and kindness. ♥

MASTER TIPS FOR RENEWAL: Stress-Reducing Techniques Shown to Boost Telomere Maintenance

The mind–body techniques and practices here have been shown in at least one study to increase telomerase in immune cells or lengthen telomeres. These effects are healthy for everyone, but they are especially important if you have high stress levels. Mind-body techniques, including meditation, Qigong, tai chi, and yoga, have been shown in clinical trials to improve wellbeing and reduce inflammation.[1] Many types of meditation also promote the mental skills for metacognition, changing how we see and respond to stressful events. While a very small number of people have negative experiences from meditation, in general there are minimal negative side effects of these practices and an abundance of positive benefits. The evidence so far does not suggest superiority of one type of meditation mind-body practice over another for telomere health.

We provide brief instruction or further resources on several of the methods below on our website, telomereeffect.com

MEDITATION RETREATS

The benefits of meditation on mental and physical health have been widely covered. When practiced regularly, it can help soothe negative thought patterns, help you connect more deeply with other people, and in some cases increase your sense of life purpose. Emerging research suggests it may even help your telomeres grow.

Researcher Cliff Saron of the University of California, Davis, has been studying the effects of residential retreats on experienced meditators. He found higher telomerase at the end of a three-month Shamatha retreat compared to a control group, and especially if the meditators had developed more purpose in life. A new study he conducted with researcher Quinn Conklin found that after three weeks of an intensive residential insight meditation retreat, experienced meditators had longer telomeres in their white blood cells than when they began, whereas the control group showed little change.[2]

As part of a collaborative group, we had an opportunity to conduct a highly controlled exploratory meditation study where both the retreat and control group lived at a resort. We examined the biological effects of a weeklong mantra-like meditation retreat led by Deepak Chopra and his colleagues at the Chopra Center, in Carlsbad, California. Women who had never or rarely meditated were randomized to either being on vacation at the resort or being in the meditation retreat. We compared them to women who were regular meditators and had already signed up for the same retreat. We found that after the week, everyone felt fantastic, showing dramatic improvements on all of our scales of wellbeing, regardless of what group they were in. Gene-expression patterns showed large changes— reductions in inflammation and stress pathways. Since these psychological and gene-expression improvements occurred in all groups, we think of this as the powerful vacation effect of unplugging from daily demands and staying at a resort. There appeared to be a meditation effect as well: Telomerase increased, but only in the experienced meditators, a finding that was marginally significant. And some other telomere-protective genes seemed more active.[3] Such intriguing findings point to greater benefits to cell aging for those already trained, but clearly have to be replicated.

MINDFULNESS-BASED STRESS REDUCTION (MBSR)

Mindfulness-based stress reduction, or MBSR, is a program created by Jon Kabat-Zinn at the University of Massachusetts Medical School for people with no or little meditation experience. Since 1979, around twenty-two thousand people have taken this program, and its benefits, such as reducing stress and physical symptoms like pain, have been well established.[4] MBSR includes training in the nature of the mind, mindful breathing, a mindful body scan (in which you slowly move your attention from your toes to the top of your head), and yoga. Taking a class in a group is a unique live experience, but for those who don't have access to MBSR locally, the University of Massachussetts Medical School's Center for Mindfulness offers an online course (http://www.umassmed.edu/cfm/stress-reduction/mbsr-online/). Their website also has a registry of trained MBSR teachers globally so you can see if you are near one.

In one study, people practicing MBSR increased their telomerase by 17 percent in a three-month period compared to a control group.[5] In another study, distressed breast-cancer survivors in the control group lost telomere base pairs—whereas the ones assigned to a form of MBSR focused on cancer recovery were able to maintain telomere length. A third group, one that received therapy based on emotional expression and support (supportive expressive group therapy), also maintained telomere length, a finding that provides encouraging news that the benefits of stress reduction on cell aging work across many different types of practices, not just meditation.[6] MBSR is great for anyone who wants to reduce stress; it's an especially good match for people suffering from chronic physical pain.

YOGIC MEDITATION AND YOGA

There are many types of meditation from many different traditions. Kirtan Kriya is a more traditional form of meditation from yoga principles that involves chanting and tapping of the fingers (called yoga mudras). Helen Lavretsky and Michael Irwin of UCLA conducted a study of people who were caring for family members with dementia, most of whom had at least mild symptoms of depression. Our lab measured their telomerase. When the caregivers practiced Kirtan Kriya for twelve minutes a day for two months, they increased their telomerase by 43 percent and decreased their gene expression related to inflammation.[7] (A control group listened to relaxing music; their telomerase increased, too, but only by 3.7 percent.) They were also less depressed, and their cognitive abilities improved.[8]

Unlike mindfulness meditation, which can help you develop metacognition and tolerate negative emotions, Kirtan Kriya works by putting you into a state of deep concentration and producing a calm, integrated state of body and mind. Afterward, your mind feels sharper and refreshed, as if you've just awakened from a great night's sleep.

A brief description can be found here: http://alzheimerspreven tion.org/research/12-minute-memory-exercise.

You may be wondering about Hatha Yoga, the type we commonly think of as exercise. It is a moving meditation that integrates physical postures, breathing, and a present mental state. Yoga has not been studied yet in relation to telomeres, but there is a tremendous research literature now on many health benefits of yoga. (For full disclosure, it is Elissa's personal favorite, and so hard not to mention.) Yoga improves quality of life and mood for people across different types of illnesses,[9] reduces blood pressure, and possibly inflammation and lipids.[10] Yoga has recently been shown to increase spine bone density if practiced long term.[11]

QIGONG

Qigong is a series of flowing movements; with its emphasis on posture, breathing, and intention, it is a kind of moving meditation.

Qigong is part of the wellness program of ancient Chinese medicine, a practice that has been developed and refined for more than five thousand years. Much like Kriya Yoga, Qigong induces a state of concentration and relaxation by integrating the body and mind. It is supported by thousands of years of practice but also the best kind of scientific evidence—randomized, controlled trials. For example, Qigong reduces depression[12] and may improve diabetes.[13] In a trial of Qigong on cell aging, researchers examined people with chronic fatigue syndrome. They found that people who practiced Qigong for four months had significantly greater increases in telomerase, and reductions in fatigue, than people who were assigned to a wait list.[14] A teacher showed the volunteers how to do Qigong for the first month, and then they practiced on their own at home, thirty minutes a day.

I've (Elissa) learned Qigong from Roger Jahnke, a doctor of Oriental medicine and an expert in medical Qigong. He recommends the practice both to prevent illness and to address particular health problems. The exercises are easy for anyone to do and can provide a strong sense of placid calm and wellbeing within minutes. (See examples on our website.) Many people are sensitive to how the body changes during this meditative activity, and they can feel a tingling sensation in the tips of their fingers (called chi/Qi sensation). This is partly due to the now-well-understood mechanisms of the relaxation response, which involves the activation of the parasympathetic nervous system and the dilation of blood vessels, creating new blood flow. This feeling is attributed to something from Chinese medicine that we have no concept for in our Western knowledge: chi/Qi energy flow.

INTENSIVE LIFESTYLE CHANGE: STRESS REDUCTION, NUTRITION, EXERCISE, AND SOCIAL SUPPORT

Dean Ornish, MD, president of the nonprofit Preventive Medicine Research Institute and clinical professor of medicine at UCSF, was the first to show that intensive lifestyle changes can reverse the progression of coronary heart disease. His program integrates stress-management techniques with other lifestyle changes. He wanted to see how this program might affect cell aging, so he studied this in men with low-risk prostate cancer. The men ate a diet high in plants and low in fat; they walked for half an hour, six days a week; they attended weekly support group sessions. They also practiced stress management on their own, with gentle yoga stretches, breathing, and meditation. In a prior randomized, controlled trial, this program was shown to slow or stop the progression of early-stage prostate cancer. At the end of the three months, the men's telomerase had also increased. Further, those who had greater reductions in their distressing thoughts about prostate cancer had the larger increases in telomerase, suggesting that stress reduction contributed to the improvements seen.[15] He followed a subgroup of these men for five years, and the ones who adhered to the program had significantly lengthened telomeres by 10 percent. His program for reversing heart disease is one of the very few behavioral programs that is now paid for by Medicare and many health insurance companies. You can find a certified provider for the heart disease program: https://www.ornish.com/ornish-certified-site-directory/.

PART III

HELP YOUR BODY PROTECT ITS CELLS

ASSESSMENT: What's Your Telomere Trajectory? Protective and Risky Factors

Next we are going to focus on the body—activity, sleep, eating. But before reading further, you are probably wondering how your telomeres are doing, and how you can find out. We pause here for a mini-assessment. We have telomeres in every cell of our body, in our different tissues, organs, and blood. They are correlated loosely—if we have short telomeres in our blood, we tend to have short telomeres in other tissues. A few commercial labs offer tests that measure the length of telomeres in your blood, but for individuals the usefulness of this is limited (see "Information about Commercial Telomere Tests" on page 333 and our website for a discussion of blood measures). It is more useful to assess the factors that are known to protect or damage telomeres and then, with the results of the assessment in mind, try to shift aspects of our daily lives so that our telomeres are more protected. That leads us to the Telomere Trajectory Assessment.

TELOMERE TRAJECTORY ASSESSMENT

You can assess the personal wellbeing and lifestyle factors that we know are related to telomere length. This assessment takes around ten minutes to complete and will help you identify the main areas you may want to improve.

When possible, we reprinted the actual scales used in the research

161

described in this book. Research details for each scale are described after each section.

You will be asked about these areas:

Your Wellbeing

- current major stressful exposures
- clinical levels of emotional distress (depression or anxiety)
- social support

Your Lifestyle

- exercise and sleep
- nutrition
- chemical exposures

Do You Have Any Severe Stress Exposures?

Enter *1* next to any questions that apply to you and *0* next to any questions that don't apply. The situations must have been ongoing for at least several months for a score of 1.

Are you experiencing severe ongoing job stress, where you feel emotionally exhausted, burned out, cynical about your work, and fatigued, even when you wake up?	
Are you serving as a full-time caregiver for an ill or disabled family member and feeling overwhelmed by it?	
Do you live in a dangerous neighborhood and regularly feel unsafe?	
Are you experiencing severe extreme stress almost every day due to some chronic situation or a recent traumatic event?	
TOTAL SCORE	

Calculate your TOTAL SCORE by adding up items 1–4: _____

Circle the telomere points below that relate to your score.

Severe Stress Exposure Score	Telomere points (Circle)
If you scored 0, you have **low risk**.	2
If you scored 1, you have **some risk**.	1
If you scored 2 or higher, you have **high risk**.	0

Explanation: This Severe Stress Exposure Checklist is not a standardized scale. Instead, it measures whether you are experiencing an extreme situation that has been linked to shorter telomeres. For example, work-related emotional exhaustion,[1] being a caregiver for a family member with dementia,[2] and regularly feeling unsafe where you live[3] are related to shorter telomeres in at least one study, after controlling for factors such as BMI, smoking, and age. Any severe event has the potential to contribute to telomere shortening, if it occurs over years. Exposure alone is not a determinant—your response is important, too, as we discuss in chapter 4. Last, having one situation may be manageable, but having more than one severe chronic situation is more likely to exhaust one's coping resources. Multiple severe chronic situations are categorized here as a higher risk.

Any Mood Disorders?

Have you been currently diagnosed with depression or an anxiety disorder (such as posttraumatic stress disorder or generalized anxiety)?

Circle the telomere points below that relate to your score:

Clinical Distress Score	Telomere Points (Circle)
If you do not have a diagnosable condition, you are at **low risk**.	2
If you have been diagnosed with a severe condition, you are at **high risk**.	0

Explanation: From various studies, it appears the symptoms of moderate distress alone are not related to shorter telomeres, but actual diagnoses—which means symptoms that are severe enough to interfere with your daily life—are related.[4]

How Much Social Support Do You Have?

Answer the following questions about social support you typically receive from significant others, family, friends, and community members.

	1	2	3	4	5
1. Is there someone available to give you **good advice** about a problem?	None of the time	A little of the time	Some of the time	Most of the time	All of the time
2. Is there someone available whom you can **count on to listen to you** when you need to talk?	None of the time	A little of the time	Some of the time	Most of the time	All of the time
3. Is there someone available to you who shows you **love and affection**?	None of the time	A little of the time	Some of the time	Most of the time	All of the time
4. Can you count on anyone to provide you with emotional support (**talking over problems** or **helping you make a difficult decision**)?	None of the time	A little of the time	Some of the time	Most of the time	All of the time
5. Do you have as much contact as you would like with someone you feel close to, someone in whom you can trust and confide in?	None of the time	A little of the time	Some of the time	Most of the time	All of the time
TOTAL SCORE					

Now calculate your total score by adding up the numbers you circled.

Circle the telomere points below that relate to your score.

Social Support Score	Telomere Points (Circle)
If you scored 24 or 25, you are **high** in social support.	2
If you scored between 19 and 23, you are **average** in social support.	1
If you scored between 5 and 18, you are **low** in social support.	0

Explanation: This questionnaire is the five-question version of the ENRICHD Social Support Inventory (ESSI), originally created to assess social support of post–heart attack patients and used in epidemiological studies.[5] Versions of this questionnaire have been used in studies relating telomere length to social support.[6]

Cutoffs for the social support categories are approximations from data from a large study, and the effects in this study were found in the oldest age group only.[7] The ENRICHD trial used the score of 18 as a lower cutoff point to define people who were low on social support.

How Much Physical Activity Do You Do?

During the past month, which statement best describes the kinds of physical activity you usually did?

1. I did not do much physical activity. I mostly did things like watching television, reading, playing cards, or playing computer games, and I took one or two walks.

2. Once or twice a week, I did **light activities** such as getting outdoors on the weekends for an easy walk or stroll.

3. About three times a week, I did **moderate activities** such as brisk walking, swimming, or riding a bike **for about 15–20 minutes each time**.

4. Almost daily (five or more times a week), I did **moderate activities** such as brisk walking, swimming, or riding a bike **for 30 minutes or more each time**.

5. About three times a week, I did **vigorous activities** such as running or riding hard on a bike **for 30 minutes or more each time**.

6. Almost daily (five or more times a week), I did **vigorous activities** such as running or riding hard on a bike **for 30 minutes or more each time**.

Circle the telomere points that relate to your score.

Exercise Score	Telomere Points (Circle)
If you chose options 4, 5, or 6, you are at **low risk**.	2
If you chose option 3, you are at **average risk**.	1
If you chose option 1 or 2, you are at **high risk**.	0

Explanation: This questionnaire is the Stanford Leisure-Time Activity Categorical Item (L–CAT) (permissions granted by Nature Publishing Group).[8] The L–CAT assesses six different levels of physical activity. Scores of 4, 5, or 6 meet the CDC recommendations for aerobic activity (150 minutes of moderate exercise, like brisk walking, or 75 minutes of vigorous exercise like jogging; note the CDC also recommends muscle-strengthening activities at least two days a week). As we explain in chapter 7 ("Training Your Telomeres"), if you are fit and do regular exercise, there doesn't appear to be an upper limit to its benefits as long as you don't overdo it during workouts and you give yourself recovery time after big workouts. Think "regular exerciser," not "weekend warrior."

People who are more physically active appear to be better buffered from the telomere shortening that occurs due to extreme stress than people who are less active.[9] Additionally, an intervention showed that exercising forty-five minutes three times a week led to increases in telomerase.[10]

What Are Your Sleep Patterns?

During the past month, how would you rate your sleep quality overall?	**0** Very good	**1** Fairly good	**2** Fairly bad	**3** Very bad
How many hours of sleep do you get on average each night (not including lying in bed awake)?	**0** 7 hours or more	**1** 6 hours	**2** 5 hours	**3** Less than 5 hours

Circle the telomere points that relate to your score.

Sleep Score	Telomere Points (Circle)
If you scored a 0 or 1 on both questions, you're at **low risk**.	2
If you scored a 2 or 3 on one question, you're at **some risk**.	1
If you scored 2 or 3 on both questions, or you have poorly treated sleep apnea, you have **high risk**.	0

Explanation: The item on sleep quality is from the Pittsburgh Sleep Quality Index (PSQI), which assesses quality and disturbances of sleep.[11] Several studies relating telomere length to sleep have used the PSQI to measure sleep quality.[12] Duration of sleep is also important. If you report sleeping at least six hours per night and describe your sleep as good or very good, you're at low risk. If you report either poor sleep quality or shorter sleep durations, this adds risk. And if you report both poor sleep quality and shorter sleep durations, this is categorized as high risk. Since studies have not tested for an additive effect of both short and poor sleep, we are making an assumption that having both is worse.

If you have sleep apnea and do not treat it nightly, you are also at risk.

What Are Your Nutrition Habits?

How often do you have the following? Circle *1* or *0* for each question.

1. Omega-3 supplements, seaweed, or fish that contains high omega-3 oils:	
3 servings or more a week of these products?	1
Less than 3 times a week	0
2. Fruits and vegetables:	
At least daily?	1
Not every day	0
3. Sugared sodas or sweetened beverages (not including when you add your own sugar to coffees or teas, which typically adds up to substantially less sugar than in commercially sweetened drinks):	
At least one 12-ounce drink on most days	0
Not regularly	1
4. Processed meat (sausage, lunch meats, hot dogs, ham, bacon, organ meats):	
Once a week or more	0
Less than once a week	1
5. How much of your diet is whole foods (whole grains, vegetables, eggs, unprocessed meats) versus processed food (packaged or processed with salts and preservatives)?	
Mostly eat whole foods	1
Mostly eat processed foods	0

Add your total points from all five nutrition questions, creating a score between 0 and 5.

TOTAL SCORE (sum of items 1–5): _____

Circle the telomere points that relate to your score.

Telomere Nutrition Score	Telomere Points (Circle)
If you scored a 4 or 5, you have excellent protection from diet.	2
If you scored a 2 or 3, you have **average risk** from diet.	1
If you scored 0 or 1, you have **high risk** from diet.	0

Explanation: The frequencies were extrapolated from telomere studies.

For omega-3s, food sources are best. If you rely on supplements, try algae-based products instead of fish for sustainability reasons. People with higher blood levels of omega-3 fatty acids (DHA, or docosahexaenoic acid, and EPA, or eicosapentaenoic acid) have slower attrition over time.[13] Those who ate a half serving of seaweed each day had longer telomeres later in life.[14] An omega-3 supplement study found that dose didn't matter as much as what's absorbed in your blood: Taking either a 1.25 gram or 2.5 gram omega-3 supplement decreased the ratio of omega-6s to omega-3s in the blood at least to some extent for most people, which in turn was associated with an increase in telomere length.[15] It's hard to know how much your body will absorb, but it should be sufficient to have fish several times a week, or take a gram of omega-3 oils a day.

While supplements are also associated with longer telomere length, real foods with antioxidants and vitamins are superior if available (i.e., lots of vegetables and some fruit).

Sugar carbonated beverages are linked to shorter telomeres in three studies,[16] and it is prudent to assume that daily consumption would be a sufficient dose to have an effect, as suggested in one of these studies. Most sweetened beverages have over 10 grams of sugar, typically 20 to 40 grams.

For processed meat, one study showed that those in the highest quartile of the sample—those who ate processed meats once a week (or a tiny portion each day)—had shorter telomeres.[17]

How Much Are You Exposed to Chemicals?

Circle either *Yes* or *No* for the following questions.

Do you regularly smoke cigarettes or cigars?	Yes	No
Do you do regular agricultural work with pesticides or herbicides?	Yes	No

Do you live in a city with very heavy traffic-related pollution?	Yes	No
Do you work in a job with heavy exposure to chemicals listed on the Telomere Toxins table (see page 266), such as hair dyes, household cleaners, lead or other heavy metal exposure (for example, in a car mechanic's shop)?	Yes	No

Telomere Chemical Exposure Score	Telomere Points (Circle)
If you answered all nos, you have **low risk** from chemical exposures.	2
If you answered yes to one or more, you have **high risk**.	0

Explanation: Here we listed exposures that have been linked to telomere shortness in at least one study. The exposures include smoking,[18] pesticide exposure,[19] chemical exposures in dyes and cleaners,[20] pollution,[21] lead exposure,[22] and exposures in a car mechanic shop.[23]

How Did You Score Overall?

Area	Telomere Points (Circle)		
WELLBEING:	high risk	average	low risk
Stress exposures	0	1	2
Clinical emotional distress	0	1	2
Social support	0	1	2
LIFESTYLE:			
Exercise	0	1	2
Sleep	0	1	2
Nutrition	0	1	2
Chemical exposures	0	1	2
Total Score (range 0 to 14) _____			

How to Understand Your Total Telomere Trajectory

The summary score is a way to show overall risk and protection of your telomere rate of decline. If you have a high score, you likely have great telomere maintenance. Keep up the good work! The most useful way to use this assessment is to focus on individual areas rather than your total score. **If you scored a 2 on any area in the summary grid, you are doing a great job at telomere protection. You are doing more than simply dodging risk. Typically, this score means you are performing protective behaviors every day, engaging in the daily work creating the foundation of a good healthspan.**

If you scored a 0 (high-risk category), you are likely to experience the typical age-related telomere decline, made worse by risk factors, but ones that you hopefully can gain more control over.

Choose an Area to Work On

The best way to use this chart is to notice the areas in which you scored 0, and then decide which will be the easiest for you to change. If you don't have any 0s, choose a category in which you scored a 1. Wherever you begin, **we suggest you choose only one area to work on at a time**. Make a commitment to improve one small thing in this area. Put a reminder of the change you're trying to make on your bedside table, or set a reminder alarm on your phone to go off at a helpful time of day. At the end of Part Three, you will see some tips to get you started on your new goal.

CHAPTER SEVEN

Training Your Telomeres: How Much Exercise Is Enough?

Exercise reduces oxidative stress and inflammation, so it's not surprising certain exercise programs have also been shown to increase telomerase. However, weekend warriors should beware: overdoing a workout can actually promote oxidative stress, and overdoing it on a chronic basis (overtraining) may cause serious damage to you and your telomeres.

In May 2013, Maggie ran her first ultramarathon. She had been a strong contender in shorter races, and she liked the idea of pushing herself to run very long distances, like this hundred-mile race through the desert. She didn't even let herself hope that she might take a top spot; she just wanted to finish. Halfway through the ultramarathon, one of Maggie's friends met up with her and said, "Did you know you're in thirteenth place? You could finish in the top ten!"

Maggie decided to dig deeper. Over the next several hours, she overtook the twelfth-place runner, and then the eleventh, and then the tenth. She crossed the finish line in tenth place, guaranteeing that she would be invited back the next year to run in a place of honor.

That summer, Maggie ran three more ultramarathons: another

hundred-miler in June, and two more in July and August. She felt great. In September, she decided that instead of taking a long recovery period from her grueling workout schedule, she'd train for another ultra in December. Then, suddenly, a few weeks into training, Maggie virtually stopped sleeping. She spent entire nights wide awake; she would sit up in bed and watch as her phone lit up in the morning, its alarm ringing. "I've never done drugs, but I imagine this is what being on speed is like," Maggie says. "I couldn't sleep, and I wasn't tired. I had a ton of energy. It was *very* weird."

Maggie continued to train. Then the illnesses set in: colds, the flu, other viruses. She tried cutting back on her workouts, but she didn't notice an improvement in her symptoms, so she resumed her schedule. Then, in early winter, her body broke down. She couldn't complete her workouts. She could barely get to work. She could barely get out of bed.

Maggie had nearly all the signs of overtraining syndrome, an unofficial diagnosis that is characterized by sleep changes, fatigue, moodiness, vulnerability to illness, and physical pain.

When Maggie reminisces about her "grand slam summer" of ultramarathons, the people around her have mixed reactions. Some are judgmental—they declare, almost gleefully, that such intense exercise is bound to be bad for the human body. Others feel guilty. Despite Maggie's troubles, they feel that something is wrong with them if they're not working out at this elite level. Others use Maggie's experience as an excuse not to work out at all.

Exercise can be a confusing topic; it can also be an emotional one. But telomeres offer some clarity. Telomeres do not need extreme fitness regimens to thrive, and that is good news for all of us who feel discouraged when we meet people like Maggie, who spent her grand slam summer taking her body to the breaking point and then past it. Another piece of good news is that telomeres appear to respond powerfully to many different levels and types of exercise. In this chapter, we'll show what the range of healthy exercise looks like and how you can gauge whether you are doing too little—or, as in Maggie's case, too much.

TWO PILLS

Let's pretend you're at a drugstore of the future. You consult with the pharmacist, who gives you a choice between two pills. You point to the first one and ask what it does.

The pharmacist ticks the benefits off on her fingers. "Lowers your blood pressure, stabilizes your insulin levels, improves your mood, increases your calorie burn, fights osteoporosis, and cuts your risk of stroke and heart disease. Unfortunately, its side effects include insomnia, skin rash, heart problems, nausea, gas, diarrhea, weight gain, and lots of others."

"Hmmm," you say. "How about the second pill? What does it do?"

"Oh, it's got the same benefits," the pharmacist says brightly.

"And the side effects?" you ask.

She beams. "There aren't any."

The first pill is imaginary, a fantasy synthesis of beta-blockers to control high blood pressure, statins to reduce cholesterol, diabetes drugs to regulate insulin, antidepressants, and osteoporosis medications.

The second pill is real, sort of. It's called exercise. People who exercise live longer and have a lowered risk of high blood pressure, stroke, cardiovascular disease, depression, diabetes, and metabolic syndrome. *And* they avoid dementia for longer.

If exercise is like a drug that pumps wondrous effects through your entire physiology, how does it work? You already know the macro view of exercise. It increases blood flow to your heart and your brain, builds muscle, and strengthens your bones. But if you could take a powerful microscope to exercise's effects, and peer into the heart of human cells when they are regularly exercised, what would you see?

Calmer, Slimmer, and Better at Fighting Free Radicals: The Cellular Benefits of Exercise

People who exercise spend less time in the toxic state known as oxidative stress. This noxious condition begins with a free radical, a molecule that is missing an electron. A free radical is rickety, unstable, incomplete. It craves the missing electron, so it swipes one from another molecule—which is now unstable itself and needs to steal a replacement electron of its own. Like a dark mood that is passed from one person to another, each person feeling a little better the bad feelings are dumped onto someone else, oxidative stress is a state that can shear through a cell's molecular population. It's associated with aging and onset of the diseasespan: cardiovascular disease, cancer, lung problems, arthritis, diabetes, macular degeneration, and neurodegenerative disorders.

Fortunately, our cells also contain antioxidants, which offer natural protection against oxidative stress. Antioxidants are molecules that can donate an electron to a free radical but still remain stable. When an antioxidant gives an electron to a free radical, the chain reaction ends. An antioxidant is like a wise friend who says, "Okay, tell me all your bad feelings; I'll listen and you'll feel better, but I'm not going to let you make me feel bad, too. And I'm definitely not going to pass your black mood on to someone else."

In an ideal situation, your cells have enough antioxidants to keep up with the need to neutralize free radicals in your body. Free radicals will never be completely eradicated from our bodies. They are continuously being made by the very processes of life—they occur normally through metabolism. In fact, very small numbers of free radicals are important for the normal communication processes in our cells. But radicals can also be created in excess when you're exposed to environmental stresses like radiation and smoking, or to severe depression. The danger seems to occur when free radicals build up. And when you have more free radicals than antioxidants, you enter an imbalanced state of oxidative stress.

That's one reason exercise is so valuable. In the short term, exercise actually causes an increase in free radicals. One reason is that you're taking in more oxygen. Most of those oxygen molecules are used to create energy from special chemical reactions in the mitochondria in your cells, but an unavoidable by-product of these vital processes is that some of them also form free radicals. But that short-term response creates a healthy counterresponse: The body steps up by producing more antioxidants. Just as short-term psychological stress can toughen you up and increase your ability to handle difficulty, the physical stress of moderate-intensity regular exercise ultimately improves the antioxidant–free radical balance so that your cells can stay healthier.

Your cells soak up the benefits of exercise in other ways, too. When you exercise regularly, the cells in your adrenal cortex (located inside your adrenal glands) release less cortisol, the notorious stress hormone. With less cortisol, you feel calmer. With regular exercise, cells throughout your body become more sensitive to insulin, which means your blood sugar is more stable. If you want to avoid the common midlife trifecta of stress, belly weight gain, and high blood sugar, you need to exercise.

Immunosenescence: Exercise Can Keep You in the Healthspan Longer

Immunosenescence is an important process underlying increased sickness and malignancy as we age. As a result of immunosenescence, you experience higher circulatory levels of proinflammatory cytokines, molecules that can spread inflammation through the body like a fire whipped along by gusts of wind. This hastens more of your T-cells toward senescence so they can't do their work of fighting off sickness. Some senescent immune cells, as you know from earlier in the book, can even go rogue. These aging immune cells can leave you more vulnerable to the kind of nasty bugs that can put you into a hospital bed. If you have a lot of immunosenescent cells and you get a vaccine for pneumonia or this year's strains of flu, there's a good chance the vaccine won't "take" and that you'll end up feverish and coughing anyway.[1] Your aging cells make it harder for you to enjoy the benefits of preventive medicine.

However. Compared to couch potatoes, people who exercise regularly have lower inflammatory cytokine levels, respond more successfully to vaccinations, and enjoy a more robust immune system. Immunosenescence is a natural process that happens with age…but people who exercise may be able to delay it until the end of life. As the exercise and immunology researcher Richard Simpson has said, these and other signs "indicate that habitual exercise is capable of regulating the immune system and delaying the onset of immunosenescence."[2] Consider exercise an excellent bet for keeping your immune system biologically young.

WHAT KIND OF EXERCISE IS BEST FOR TELOMERES?

Exercise helps protect your cells by warding off inflammation and immunosenescence. Now there is an additional explanation of exercise's cellular benefit: exercise helps you maintain your telomeres. This was true even in a study of 1,200 twin pairs, which allowed exercise's effects to be teased apart from effects of genetics: the active twin had longer telomeres than the less active twin.[3] After adjusting for age and other factors that affect telomeres, by statistically removing their effect, the relationship between telomeres and activity was laid bare. And it's not just that exercise is helpful; we also know that sedentariness itself is terrible for metabolic health. Now several studies have found that sedentary people have shorter telomeres than people who are even a little more active.[4]

But are all types of exercise equal when it comes to cell aging? Researchers Christian Werner and Ulrich Laufs of Saarland University Medical Center in Homburg, Germany, tested three types of exercise in a small but exciting study. Their results hint that exercise really may increase telomerase's replenishing action—and they help us understand which kinds of exercise are best for keeping our cells healthy. Two kinds of exercise stood out. Moderate aerobic endurance exercise, performed three times a week for forty-five minutes at a time, for six months, increased telomerase activity twofold. So did high-intensity interval training (HIIT), in which short bursts of heart-pounding activity are alternated with periods of recovery. Resistance exercise had no significant effect on telomerase activity (although it had other benefits; the researchers concluded that "resistance exercise should be complementary to endurance training rather than a substitute"). And all three forms of exercise led to improvements in telomere-associated proteins (such as telomere-protecting protein TRF2) and reduced an important marker of cellular aging known as p16.[5] They also found that regardless of exercise type, those who increased their aerobic fitness the most had greater increases in telomerase activity. This tells us it's the underlying cardiovascular fitness that matters most.

So try to do moderate cardiovascular exercise or HIT. Either is great. Our Renewal Lab at the end of this chapter will show you these evidence-based workouts for strengthening your telomeres. You may not want to restrict yourself to just one type of workout, though. We benefit from variety. In a study of thousands of Americans, the more categories of exercise—from walking to biking to strength training—that people engaged in, the longer their telomeres.[6] And this is a reason to perform strength training. Even though strength training doesn't appear to be significantly related to longer telomeres, it helps maintain or improve bone density, muscle mass, balance, and coordination—all of which are vital for aging well.

How, exactly, does exercise strengthen telomeres?

Maybe the wonderful cellular effects of exercise, including less inflammation and oxidative stress, are good for telomeres. Or maybe exercise is good for telomeres because it prevents stress from causing some of its usual damage. The stress response can leave cellular damage and debris in its wake—but exercise switches on autophagy, the housekeeping activity in the cell that eats up those damaged molecules and recycles them.

It's also possible that exercise improves telomeres directly. For example, getting on the treadmill induces an acute stress response, which increases the expression of TERT, a telomerase gene.[7] Athletes have higher expression of TERT than sedentary people.[8] Exercise releases a newly identified hormone, irisin, that boosts metabolism and in one study was associated with longer telomeres.[9]

But no matter how the exercise–telomere connection works, what's most significant is that exercise is essential for your telomeres. To keep your telomeres healthy, you need to work them out. For the workouts that have been shown to improve telomere maintenance, see the Renewal Lab.

EXERCISE AND INTRACELLULAR BENEFITS

Exercise leads to myriad beautiful intracellular changes. Exercise causes a brief stress response, which triggers an even bigger restorative response. Exercise damages molecules, and damaged molecules can cause inflammation. However, early on in a bout of exercise, exercise induces autophagy, the Pac-Man-like process that eats up damaged molecules. This prevents inflammation. Later in the same exercise session, when there are too many damaged molecules, and autophagy can no longer keep them under control, the cell dies a quick death (called apoptosis), in a cleaner way that doesn't lead to debris and inflammation.[10] Exercise also increases the number and quality of those energy-producing mitochondria. In this way, exercise can reduce the amount of oxidative stress.[11] After exercise, when your body is recovering, it is still cleaning up cell debris, making cells healthier and more robust than before exercise.

GAUGE YOUR TELOMERE FITNESS

It's not simply working out that is crucial for telomere health. As we hinted earlier, fitness, the ability to perform physical tasks, is important, too. It is very possible for someone to perform light regular exercise, but not be fit. And some lucky people can be fit without exercising, especially when they're young. (Think of twentysome-things who can successfully complete a long, arduous hike even if they haven't worked out since high school.) For telomere health, you need to get regular exercise *and* you need to be fit.

But how fit do you need to be? Do you need to be capable of running ultramarathons, like Maggie? Swim five miles in the open water? Do you have to be like one of our Midwestern friends, who spends Saturday mornings in October at races featuring "zombies" who chase her through fields of corn? Our cultural standards for fitness are getting higher and higher, and it can be hard to know whether you are fit enough to stay healthy.

Fitness is, in fact, crucial for telomere health.[12] But you may be relieved to know that significant telomere benefits are gained by having a very moderate, achievable level of fitness. Our colleague Mary Whooley, of UCSF, put a group of adults, all of whom have heart disease, on a treadmill. They began by walking, with the incline and speed gradually increased until they could do no more. The results were clear: The less exercise capacity they had, the shorter their telomeres.[13] The people with the lowest cardiovascular fitness couldn't even sustain a brisk walk, whereas those most fit sustained the equivalent pace to taking a hike. Those with low fitness had fewer base pairs by an amount that translates to about four years of extra cell aging, compared to the fit group.

Are you able to mow your lawn? Shovel snow? Carry your own clubs while playing golf? If not, you are in this category of low fitness. There are easy ways to slowly and safely build your capacity. Check with your doctor and then consider our walking plan, in the Renewal Lab. On the other hand, if you can walk vigorously or maintain a light jog for forty-five minutes, three times a week, you are fit enough to support your telomere health. Remember that fitness and exercise are related, but separate. Even if you are naturally fit, you still need an exercise program to keep your telomeres healthy.

TOO MUCH EXERCISE?

Moderate exercise and fitness are clearly wonderful for telomeres, but what about Maggie, the ultramarathon runner? Are her telomeres longer because she took exercise to the extreme? Are they shorter? Few of us run ultramarathons, but as more people participate in endurance sports, questions like these become more urgent.

Most extreme exercisers can breathe a sigh of relief. One remarkable study of ultrarunners found that their cells were the equivalent of sixteen years younger than those of their sedentary counterparts.[14] Does this mean that we should all sign up for the next hundred-mile race? Not at all. Those ultrarunners were compared to *sedentary* people. When endurance athletes are compared to more ordinary runners, who might run around ten miles a week, you find that both groups have nice, healthy telomeres compared to the more sedentary group—and that there appears to be no additional benefit for the ultra-long-distance endurance group in terms of telomeres.[15]

Endurance athletes sometimes worry about whether it's safe to continue their extreme training year after year, as opposed to training for a single endurance event and then returning to a more typical exercise routine. One study looked at older men who had been elite athletes when they were younger. Their telomeres were similar in length to other men their age, so their many years of extremely vigorous training didn't appear to have had a cumulative wearing effect.[16] Another German study examined a group of senior "master athletes" who had been competing in endurance races since their

181

youth. Most of them still compete, just at a slower rate (such as taking eight hours instead of two to complete a marathon). The long-term athletes both looked younger and had less telomere shortening compared to matched controls.[17] Another study examined years of exercise and found longer telomeres in people who had actively exercised for the past ten years or more.[18] It appears important to start exercising when you are young—but don't be discouraged. It's never too late to start, and benefits always await you.

Maggie, however, may be in some trouble. One study of extreme exercisers found that they had shorter muscle telomere length— but only if the exercisers suffered from a fatigue-overtraining syndrome.[19] When athletes develop a fatigue syndrome, as Maggie did, it is a sure sign they have overtrained and damaged their muscles to the extent that they can't be easily repaired. Progenitor cells (also known as satellite cells) repair muscle tissue that has been damaged—but it is thought that overtraining damages those crucial cells, leaving them less able to do their repair work. It appears to be overtraining, not extreme exercise, that is damaging to telomeres, at least in the muscle cells.

Overtraining is defined by too much training time relative to rest and recovery. It can happen to anyone, from beginning runners to professional athletes, and it happens when you don't support your body with enough rest, nutrition, and sleep. Psychological stress can contribute, too. Some warning signs of overtraining include fatigue, moodiness, irritability, trouble sleeping, and susceptibility to injuries and illness. The cure is rest—which sounds easy but is hard for athletes who are accustomed to pushing themselves.

Any discussion of overtraining is complicated, because there is no set point that constitutes "too much exercise." That point is different for everyone, and it depends on individual physiology and level of training. If telomeres tell us anything, they remind us of how context-dependent health can be. What is good for one person may be harmful to another. If you are an extreme athlete, make sure you're working closely with a trainer or physician so that any signs of overtraining can be spotted early.

In general, it's a good idea to begin *any* exercise program slowly, gradually working up to better fitness. Weekend warriors who sit in the office for five days and then overdo it on the weekend, pushing themselves to break down a lot of muscle at once, will feel fatigued and sometimes even nauseated. They are not doing their bodies any favors. Remember that exercise initially creates additional oxidative stress in the body, and then there is a healthy counterresponse that reduces that stress. But if you overdo it, that counterresponse can be overwhelmed. You'll end up with more oxidative stress, not less.

STRESSED OR DEPRESSED? EXERCISE IS RESILIENCE TRAINING FOR YOUR CELLS

"I don't have time to exercise. I'm already overcommitted and overscheduled."

"I'll exercise when I feel better. I am so stressed right now that I can't push myself to do one more hard thing."

Sound familiar? Yet it turns out that the most important time to exercise is exactly when you might not want to—when you are feeling overwhelmed. Exercise can improve your mood for up to three hours after working out[20] and can reduce stress reactivity.[21] Stress can shorten telomeres, but exercise shields telomeres from some of stress's damage. Our colleague Eli Puterman, a psychologist and exercise researcher at the University of British Columbia, has studied high-stress women, including many stressed-out caregivers. The more the women exercised, the less their stress ate away at their telomeres (see figure 17). The exercise actually buffered their telomeres from the insidious and telomere-shortening effects of stress. Even if your schedule is packed, even if you feel too exhausted to do a hard workout, find a way to slip in some exercise. For example, the two of us maintain busy schedules, but while working on this book we took walks together, thinking through the chapters aloud as we went up and down the hills of San Francisco.

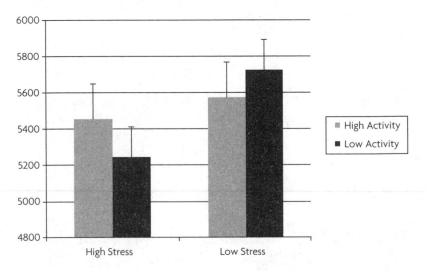

Figure 17: Physical Activity May Buffer Stress-Related Telomere Shortening. Women high on perceived stress had shorter telomeres, but only if they were relatively sedentary. If they exercised, they did not show the stress-telomere relationship.[22] The raw values (unadjusted) for telomere length in base pairs are shown here on the vertical axis.

You can probably exercise more often than you realize. But on the days when you just can't make it happen, take heart. In psychology, resilience is a kind of Holy Grail. Resilience is what keeps you bouncing back after being knocked down and lets stress slide off your shoulders without damaging your mind and body. Eli Puterman's stress research shows that telomeres can be resilient, too. The more you can practice good health habits—effective emotion regulation, strong social connections, good sleep, and good exercise—the less that stress hurts your telomeres. This is especially true if you have depression.[23] Exercise is a potent way to make your telomeres resilient, but when you can't exercise, step up other resiliency behaviors. Everything you do will help, and that's an encouraging piece of news.

TELOMERE TIPS

- People who exercise have longer telomeres than those who don't. This is true even for twins. It's the increased aerobic fitness that is most tied to good cell health.
- Exercise charges up the cell clean-up crew, so that cells have less junk buildup, more efficient mitochondria, and fewer free radicals.
- Endurance athletes, who have the best fitness and metabolic health, have long telomeres. But those telomeres are not much longer than those with moderate exercise. We don't need to aspire to extremes.
- Athletes who overexercise and burn out develop many physical issues, including risk of shorter telomeres in their muscle cells.
- If you have a high-stress life, exercise is not just good for you. It is essential. It protects you from stress-shortened telomeres.

RENEWAL LAB

IF YOU LIKE A STEADY, CARDIOVASCULAR WORKOUT...

Here is the cardiovascular workout tested in the German study, the one that showed a significant increase of telomerase.[24] It's pretty straightforward: Simply walk or run at about 60 percent of your maximum ability. You should be breathing somewhat hard, but you should still be able to maintain a conversation. Do this for forty minutes, at least three times a week.

IF YOU PREFER HIGH-INTENSITY INTERVAL TRAINING (HIT)...

This interval workout has been associated with the same gains in telomerase as the cardiovascular workout above. Plan to do it three times a week:

Cardiovascular Workout (Running)	
Warm up (easy)	10 minutes
Interval (repeat 4 times):	
Run (fast)	3 minutes
Run (easy)	3 minutes
Cool down (easy)	10 minutes

IF YOU WANT AN INTERVAL WORKOUT THAT'S LESS INTENSE...

Runners shouldn't have a monopoly on interval training. This plan is less intense but still incorporates some doable intervals. If you are out of shape, add a ten-minute warmup and cool down:

Walking Workout	
Interval (repeat 4 times):	
Walk fast (on a exertion scale of 1 to 10, be at a 6 or 7)	3 minutes
Stroll gently	3 minutes

This walking plan hasn't been tested specifically for its effects on telomeres or telomerase in any studies yet, but it certainly falls into the category of healthy exercise. One study tested this plan and found that it had a much more beneficial effect on multiple measures of fitness than just moderate, steady walking. More important, over two-thirds of the adults in this study, who were midlife or older, stayed with this walking regime for years afterward.[25]

SMALL STEPS COUNT

In addition to planned exercise, it's important to keep moving throughout the day. Activity that is woven into your daily life lifts you out of the dreaded "sedentary" category, which is linked to shorter telomeres and causes metabolic changes that can lead to more insulin resistance and inflammation.[26] So add little walks all day: Park farther away from your destination, take the stairs, or have a walking meeting. Some apps (and the iWatch) have programs that remind you to stand up every hour. Or a simple pedometer can be our daily reminder that our steps can add up.

CHAPTER EIGHT

Tired Telomeres: From Exhaustion to Restoration

Poor quality sleep, sleep debt, and sleep disorders are all linked to shorter telomeres. Of course, most of us already know that we need more sleep—the problem is figuring out how to get it. Here, we'll draw on the newest research that goes beyond the standard sleep hygiene advice and shows how both cognitive changes and mindfulness can help you get more restorative sleep. Even when you can't get more sleep, these techniques help you suffer less from the effects of sleeplessness.

Maria's sleep problems began more than fifteen years ago. She and her husband had been fighting a lot, and she'd wake up in the dark of the night, unable to stop replaying their arguments in her mind. When she consulted with a marriage and family therapist, that first bout of insomnia departed. Unfortunately, it left a door ajar, and several times a year, Maria's sleep troubles would walk back in. When they did, she'd feel too alert and aroused at night to fall asleep. She'd drift off and then wake again, often worrying about financial problems and how her sleeplessness would affect her work the next day. During the day, she felt depleted and exhausted, but her mind was racing too fast for sleep. When she attended a sleep program for insomnia, Maria was asked to track how much sleep she was actually getting. Her average minutes of sleep per night: 124.

Are *you* getting enough sleep? A quick gauge, one used by sleep researchers, is to ask yourself whether you're sleepy during the day. If you are, you need more sleep, even if your sleep loss isn't nearly as dramatic as Maria's. A better test is to ask yourself whether you fall asleep unintentionally while watching television or a movie, or while you're a passenger in a car. Many people just don't sleep enough, whether it's because of diagnosable sleep disorders, common lifestyle-related sleep problems, or being too busy; according to the National Sleep Foundation's 2014 Sleep Health Index, 45 percent of Americans say that poor or insufficient sleep affected their daily activities at least once in the last week.[1]

Telomeres need their sleep. We now know that getting enough sleep is important for healthy telomeres in all adults. Chronic insomnia is associated with shorter telomeres, particularly for people over

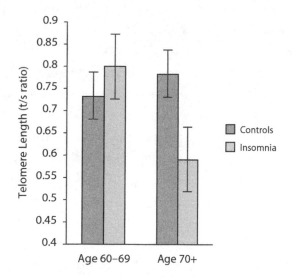

Figure 18: Telomeres and Insomnia. In men and women ages 60 to 88 years old, insomnia was related to shorter telomere length, but only in those 70 years and older. This graph shows the average telomere length from peripheral blood mononuclear cells.

seventy years old (see Figure 18).[2] In this chapter, we'll show you how good sleep protects your telomeres, buffers some of the effects of aging, regulates your appetite, and soothes the pain of some of your most stressful memories. For the newest techniques to help you get more sleep—and to help you feel better when sleep just isn't possible—read on.

THE RESTORATIVE POWER OF SLEEP

We don't usually think of sleep as an activity, but that's exactly what it is. In fact, it's the most restorative activity you can perform. You need that rejuvenating time to set your internal biological clock, regulate your appetite, consolidate and heal your memories, and refresh your mood.

Set Your Biological Clock

Do you struggle to wake up and feel alert in the morning?

Are you wide awake at bedtime?

Do you feel hungry at strange hours?

If you've answered yes to any of these questions, or your body's timing simply feels "off," you may suffer from at least slight dysregulation in a brain structure known as the suprachiasmatic nucleus, or SCN.[3] A structure of a mere fifty thousand cells, the SCN snuggles like a tiny egg within the larger nest of the brain's hypothalamus. Don't be fooled by its size, because the SCN is incredibly important. It's your body's central internal clock. It tells you when to feel tired, when to feel alert, and when to get hungry. It also drives the nightly task of cellular housekeeping, when damaged parts are swept away and DNA is repaired.[4] When your SCN is working well, you'll have more energy when you need it, deeper rest at night, and cells that are functioning more efficiently.

Like a delicate, handmade timepiece, the SCN is highly sensitive. It needs information from you to keep itself well tuned. Light signals, which are transmitted directly to the SCN through the optic nerve, allow the SCN to set itself to a proper day/night cycle. By getting light exposure during the day, and by dimming the lights at night, you keep your SCN on schedule. If you keep to regular eating and sleeping times, you also give your SCN the information it needs to inhibit the sleep drive during the day and unleash that drive throughout the night.

Control Your Appetite

Your body also depends on deep, restorative REM sleep to regulate your appetite. (REM sleep is characterized by rapid eye movements, higher heart rate, faster breathing, and more dreaming.) During REM sleep, cortisol is suppressed, and your metabolic rate increases. When you don't sleep well, you get less REM in the second half of the night, and that results in higher levels of cortisol and insulin, which stimulate appetite and lead to greater insulin resistance. In plain terms, this means that *a bad night of sleep can throw you into a temporary prediabetic state.* Studies have shown that even one night of partial sleep, or one night without enough REM sleep, can lead to elevated cortisol the next afternoon or evening, along with changes in the hormones and peptides that regulate appetite and lead to greater feelings of hunger.[5]

Good Memories, Bad Memories, and Emotions

"We sleep to remember, and we sleep to forget," says Matt Walker, a sleep researcher at the University of California, Berkeley. When you are well slept, you are better at learning and remembering. Tired people just aren't as successful at focusing their attention, so they don't take in new information as well. And sleep itself creates new connections between brain cells, which means that you're both learning and stabilizing your memory of what you've learned.

Sometimes, though, memories are painful. Sleep works its healing powers on these memories, reducing their emotional charge.

Walker has found that most of this work is accomplished during REM sleep, which shuts off some of the stimulating chemicals in your brain and allows you to split off your emotions from the content of the memory. With time, this action allows you to remember a painful experience but without an intense jolt to your mind and body.[6]

And of course, we need sleep to refresh emotionally. If you don't already know that sleep loss makes you more irritable, ask your family or colleagues. They'll quickly confirm this for you. When you are not well slept, you have a physiological and emotional stress response that's measurably bigger.[7] You can even get giddy or giggly more easily.[8] Sleep deprivation makes *all* emotions more intense. This is, perhaps, a reason Maria felt so hyperaroused and jumpy.

HOW MANY HOURS OF SLEEP DO TELOMERES NEED?

As scientists have realized that sleep is crucial to your mind, your metabolism, and your mood, they have increasingly included measurements of telomeres in their sleep studies. Researchers have looked at how sleep length affects telomeres in different populations, and the same answer keeps coming up: Long sleep means long telomeres.

Getting at least seven hours of sleep or more is associated with longer telomeres, especially if you are older.[9] The famous Whitehall study of British civil servants found that men who slept five hours or fewer most nights had telomeres that were shorter than men who slept more than seven hours.[10] This finding was after adjusting for other factors such as socioeconomic status, obesity, and depression. Seven hours of sleep appears to be the cutoff point for telomere health. Get less than seven, and telomeres start to suffer. If you're one of those rare people who need very little sleep (about 5 percent of the population needs only five or six hours of sleep per night), this cutoff point doesn't apply to you. Then again, if you feel terrible without eight or nine hours of sleep, don't try to scrape by with seven. Get those extra hours. And remember that rule of thumb, which offers highly customized sleep advice: *If you feel sleepy during the day, you need more sleep at night.*

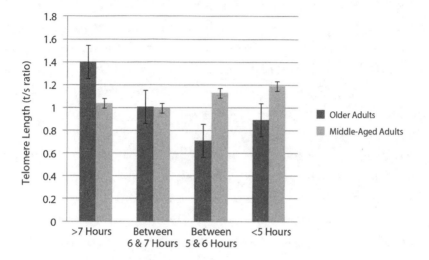

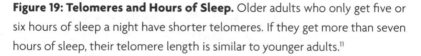

Figure 19: Telomeres and Hours of Sleep. Older adults who only get five or six hours of sleep a night have shorter telomeres. If they get more than seven hours of sleep, their telomere length is similar to younger adults.[11]

It's Not Just Hours in Bed: Sleep Quality, Regularity, and Rhythm

As you keep the seven-hour goal in mind, try not to become obsessed with it, because it's not just hours that matter. Think back to the past week and how well you've slept. How would you rate your sleep quality for the last seven days? Was it very good, fairly good, fairly bad, or very bad? The answers to this straightforward question have been scientifically linked to your telomere health. The closer you are toward the "very good" end of the scale, the healthier your telomeres are likely to be. In several studies examining sleep quality, people who gave themselves high ratings for sleep quality had longer telomeres.

Good sleep appears to be especially protective as we get older, buffering the natural age-related decline in telomere length. In one study age was not related to shorter telomere length in those with great sleep quality.[12] When sleep quality remains good, telomeres stay pretty stable across the decades.

Good sleep quality also protects the telomeres in your immune system's CD8 cells. When these cells are young, they attack viruses, bacteria, and other foreign invaders. Your body is constantly fighting off threats, but when you're protected by a vigorous army of immune cells, including CD8 cells, you hardly notice those threats. That's because the invaders are being surrounded and destroyed. These CD8 cells are part of an incredibly effective defense system. Until, that is, their telomeres shorten and they start to grow old. Then they are less able to fight off foreign bodies in your bloodstream; that's why people with shorter telomeres in CD8 cells are more likely to catch a cold virus. Short telomeres in CD8 cells can, with time, lead to systemic inflammation, as we've mentioned earlier. Aric Prather, PhD, a sleep researcher at the University of California, San Francisco, has found that women who rated their quality of sleep as poor were more likely to have short telomeres in their CD8 cells; excessive daytime sleepiness was also a predictor of shorter telomere length. Women with high stress were the most vulnerable to the effects of poor-quality sleep.[13]

Sleep length and sleep quality are important. Now add sleep rhythm to the list. Keeping a good sleep-wake rhythm—going to bed and waking up at regular times—may be critical to your cells' ability to regulate telomerase. In one study, scientists removed the "clock genes" from mice. Although normal mice show higher telomerase in the morning and lower telomerase at night, the mice without the clock gene did not show this telomerase diurnal rhythm, and their telomeres shortened. Then the same investigators turned to humans whose work schedules had, effectively, broken their internal clocks. Emergency room physicians who pulled night-shift duty also lacked this normal rhythm of telomerase.[14] This study was small, but it suggests that good sleep-wake rhythms may be critical in helping to maintain the rhythm of telomerase activity best for keeping your telomeres replenished.

HELP FOR SLEEP PROBLEMS: COGNITION AND METACOGNITION

Some of us need to be convinced that sleep is vital for health, but Maria did not. Driven by desperation, she attended a clinic that was experimenting with a novel approach to sleep problems.

Insomnia is characterized by some universal experiences: feeling too alert to fall sleep; trying too hard to sleep; and, especially, the common habit of rehearsing the past or worrying about the future. In order to sleep, we need to feel physically and psychologically safe. But at night, small worries can morph into large, looming threats, making it hard to feel secure enough to sleep. Usually these threats are, as Elissa's father used to say, "mere demons of the night" that disappear in daylight. He was right. Nighttime can turn manageable worries, problems that can be solved by day, into a chain of catastrophes that are replayed in a stupor of tired rumination.

But there is a second layer of worries that can arise. This tricky layer is made up of worries about insomnia and its effects. They include:

- "I can't function tomorrow without a good's night sleep."
- "I should be able to sleep as soundly as my bed partner does."
- "I'm going to look terrible tomorrow."
- "I'm going to have a nervous breakdown."

These thoughts can push an episode of tossing and turning into full-blown insomnia, and they can color the negative emotions you might feel the next day with an even darker tint.

One method that's been shown to help with this second layer of thoughts is to examine them directly. Like the demons of the night, your thoughts about sleep are usually much less foreboding and dramatic when you examine them in the light of day. They're what we call "cognitive distortions," and they mostly aren't true. Challenge these thoughts, and you'll find more accurate statements emerging:

- "Although I don't function as well without sleep, I can still manage to get my tasks done."
- "My partner's sleep needs are not the same as mine."
- "I look pretty good" or "Thank goodness for makeup!"
- "I'm going to be okay."

Dr. Jason Ong directed the sleep program that Maria attended. Cognitive behavioral therapy is the best known treatment for insomnia so far, as it challenges your thoughts about insomnia. At the same time, Jason also noticed that when sleep therapists challenged their patients' thoughts, some patients felt a bit bullied, as if the doctor was telling them what to think. Or they felt as if they were on opposite sides of a debate, with opposing arguments flying back and forth.

So in Dr. Ong's workshops, patients practice the usual good sleep behaviors that most doctors prescribe—getting out of bed when they are not able to sleep, waking up at the same time every morning, not trying to make up for lost sleep with naps—but instead of telling patients to think differently, the therapists encourage them to watch their thoughts from a distance. This is, again, a form of mindfulness. At the clinic, patients like Maria learn different forms of meditation, including moving meditations (for example, walking slowly while paying attention to each step) and more traditional meditations (sitting quietly and noticing the breath). They are encouraged to accept their thoughts about insomnia, and then to let those thoughts go. Meditation isn't used as a way to make people feel sleepy—it's a method for promoting awareness of that second layer of thoughts that make insomnia so much worse. It defuses those thoughts.

It can take a while to change your relationship to your thoughts. Maria stuck with the meditation program for six weeks without seeing much improvement. Finally she expressed her frustration. She said, "During my meditations, I have been trying to clear my mind and sometimes I can keep it blank for a while, but [the thoughts] always come back."

Dr. Ong suggested that Maria stop trying to exercise so much power over her mind. He asked her to consider what would happen if she simply let her thoughts run their course. "It is not the thoughts you are trying to control, but instead it is letting go of the effort of forcing these thoughts to go in a certain direction," he explained.

Maria mulled this over, and tried the meditation again with this new, less forceful approach. The following week, her anxiety levels downshifted. She felt less worked up before going to bed at night, and at the next workshop she was noticeably more relaxed. "For a long time I thought I had to get rid of my thoughts to sleep better. It's funny that once I stopped trying to make that happen, my sleep seemed to get better," she reported. Over the next few weeks, she nearly doubled her average sleep time—not a total cure, but a significant improvement. Her doctors predicted that as she continued to practice mindfulness, she would make further gains.[15]

Ong tested out his eight-week mindfulness-based treatment for insomnia. The program, officially known as MBTI, was compared to a group who simply wrote down their sleep times and levels of arousal. Those in the MBTI group had greater decreases in insomnia, and within six months, 80 percent showed improvement in their sleep.[16]

FRESH STRATEGIES FOR MORE SLEEP

What about everyone else, including those of us who don't have chronic insomnia but who could use help getting a little more sleep? Following are some suggestions.

Give Yourself the Gift of Protected Transition Time

Your mind is not a car engine. You can't run it high speed right up until bedtime, doing work, exercise, chores, or tending to children, and then expect to switch it off and drop into slumber. It doesn't work that way. Biologically, *your brain is more like an airplane.* You need a slow descent into sleep, landing as gently as possible. So give yourself the gift of transition time between work and sleep, a sleep routine or ritual that lets you wind down. The smoother the transition, the less jolted you'll feel when you land.

Even five minutes of transition time can make a difference. Begin by unplugging. Turn off your phone or set it to airplane mode; let your body have a break from instant responding. If you have the willpower, leave your phone in a different room entirely. By setting aside phones and other screens, you minimize the number of stressors that can feature on your mind's IMAX screen of nighttime worries. You already have enough stress to contend with, given the natural human tendency to ruminate and rehearse worries at night. (In the next section, you'll see that screens are also sources of blue light, which keeps you awake.) After you've turned off the screens, perform a quiet, pleasant activity—not to make yourself sleepy but to create a transition period of calm and comfort. Some people like to read or knit or even open up a stress-relieving coloring book designed for adults. (You can find an adult coloring page in this chapter's Renewal Lab.) You can listen to an audio meditation or music that relaxes you.

Blue Light Suppresses Melatonin

There was a worldwide sleep debt even before our current addiction to screens. But now there are extra challenges to sleep. Do you bring smartphones, tablets, or other screens into your bedroom? The blue light from screens can suppress melatonin, the sleepiness hormone. In a study by sleep researcher Charles Czeisler and colleagues, people who used an e-reader immediately before bed released around 50 percent less melatonin than people who read from a print book.[17] The people using an e-reader took longer to fall asleep, had less REM sleep, and felt less alert in the morning.

Try to avoid screens for an hour before bedtime. If you can't, try using smaller screens, and holding them farther from your eyes, to help minimize blue light exposure. Liz uses a free software program called f.lux that matches a screen's light to the time of day, so that the blue light fades into yellow as you head into evening. You can download it at https://justgetflux.com. Apple computer's new operating system 9.3 has Night Shift, a program that automatically shifts from blue to yellow at night.

All light suppresses melatonin, though, so keep things as dark as you can. Look around your room at night. Where do you see light? Minimize light from windows and digital clocks. Wear an eye mask, and let the melatonin flow.

NOISE, HEART RATE, AND SLEEP

We all come with different settings for sleep. Some just aren't bothered by noise and some are. People with a particular pattern of brain activity, whose EEGs show the bursts of brain waves known as spindles, appear to be more resilient to nighttime noises.[18] For the rest of us, hearing sounds like car horns or sirens speeds up the heart rate and disrupts the sleep cycle.[19] If you are highly sensitive to your surroundings, you need to control your exposures. The more completely you can tune out of your environment, the more you can feel safe from the intrusion of noise, and the deeper you'll sleep. Earplugs are a good place to begin.

Synchronize Your Brain with Your Internal Clock

Your suprachiasmatic nucleus, the brain's clock, is trying to keep circadian rhythms on track. Help it along by eating and sleeping at regular times. This regularity will help your brain know when to release melatonin, and it will help your cells know when it's time to repair DNA and perform other restorative functions. Regular mealtimes and sufficient sleep also leads to greater insulin sensitivity, which helps you burn fat more efficiently.

Sleep Debt Is Not a Blame Game

People lose sleep at predictable times: after a baby is born, when their partner goes through a snoring phase, when they feel depressed or stressed, when hot flashes hit, or when first adapting to age-related changes in sleep. These events are usually temporary. They happen, and they pass. But today's epidemic levels of sleep loss are not caused by these events. Most sleep loss is caused by "voluntary sleep curtailment," otherwise known as sleep procrastination, otherwise known as not getting yourself into bed early enough.

You might be responding as I (Elissa) did when I heard that term: "I'm not volunteering to lose sleep—it's just that I have too much to do." But instead of mentally preparing your defense, remind yourself that sleep loss can't be resolved by playing a blame game. Just gently remind yourself that unless you're a new parent or a caregiver, your bedtime is one of the few areas of sleep you *can* control. Take advantage of this power and go to sleep earlier. (An exception: Severe insomnia and age-related sleep changes don't respond to an earlier bedtime. In these cases, going to bed earlier can boomerang on you and make it harder to get quality sleep all the way through the night.)

Treat Sleep Apnea and Snoring

Severe sleep apnea, the repeated cessation of breathing during sleep, has been linked to shorter telomeres in adults.[20] The cellular effects of sleep apnea may even be transmitted in the womb. In a sample of pregnant women, 30 percent responded to a sleep assessment with answers suggesting symptoms of apnea. When these women's babies were born, the telomeres in the babies' cord blood were shorter.[21] The same was true for women who snored. And here's some bad news for the many snorers out there: More time spent snoring is also linked to telomere shortness, at least in a large sample of Korean adults.[22] If you suspect you have sleep apnea, get tested and then take advantage of the effective new treatments, which are more comfortable than traditional CPAP machines, the ones that apply air pressure through a mask.

SLEEP IS A GROUP PROJECT

You probably know a few people who get enough sleep. You can spot them easily: They're the ones with bright eyes and skin, the ones who don't constantly complain about how tired they are, who don't always have a Starbucks Grande in one hand, the ones who don't wonder why they feel hungry at strange times of the day. What do these people have that the rest of us don't? Well, a few things. They may have a partner who encourages good sleep—and who suggests leaving the phone in the kitchen to recharge overnight. They may have colleagues who don't drop emergency e-mails at 10:00 p.m. They may have children who go to bed and stay there!

What we're saying here is that sometimes sleep is a group project. We have to support one another in reducing sleep procrastination, in going to bed earlier, in not doing business late at night. As the saying goes, be the change you wish to see in the world. Make a pact with your spouse to leave a few minutes at night to transition out of the stress mind-set. Make another pact with your colleagues not to send late-night messages (if you have to write them at night, save them in your drafts folder until morning). You can't tell your children not to have the kinds of nightmares that send them running to you at 2:00 a.m., but you can set an example of what good sleep habits look like in adulthood.

TELOMERE TIPS

- With sufficient sleep you will feel less hungry, less emotional, and lose fewer telomere base pairs.
- Telomeres like at least seven hours of sleep. Many strategies can help us boost our sleep quality, some as simple (but hard) as removing electronic screens from our bedroom.
- Try to minimize the effects of sleep apnea, snoring, and insomnia. These problems are common later in life. And when insomnia visits, use comforting thoughts to soften the alarming ones. If you have severe insomnia, cognitive-behavioral therapy can help.

RENEWAL LAB

FIVE BEDTIME RITUALS

Bringing restfulness into your bedtime space promotes a better night's sleep. Begin by making the next day's to-do list. Then put the list aside. That way, you'll feel more peaceful about tomorrow, and you'll leave behind some of the mental effort that keeps you in vigilance-anticipation mode. After that, you're ready for a bedtime ritual. Here are five rituals to support maximum tranquility and relaxation:

1. Spend five minutes in transition: breathing, meditating, or reading. The centuries-old practice of reading a book before bed can also help transition from an overly active mind to a state of absorbed attention. The transfer of focus away from self and to the content of the book can quiet the mind—provided that the book is not too exciting, of course.

2. Listen to soothing music. Soothing music calms your nervous system and your mind, and it sends a signal to start transitioning into a state of rest. The Spotify app offers several bedtime playlists, including "Bedtime Bach" (for lovers of classical music), "Best Relaxing Spa Music" (if you prefer New Age), and many soporific options under "sleep" such as "Sleep: Into the Ocean" (if you like nature sounds).

3. Set a mood for relaxation. Use essential oils, light a candle, and dim the lights. When our environments are restful and peaceful, so are we. Calming scents like lavender, cedar, or sandalwood are soothing for your entire system and brain. Reducing artificial light, and then turning off the lights completely, is a must for becoming restful enough to drift to sleep.

4. Brew warm herbal tea an hour or more before bed. A warm, scented mug of tea will help you wind down from your day. Try making your own blend of herbal tea from chamomile, lavender, rose petals, and a slice of fresh lemon or ginger. Don't drink the tea right before bed, or your sleep might be disrupted by a bathroom break.

5. Perform bedtime stretches or do some gentle yoga. Simple head and neck rolls will help melt the tension and anxiety from the day. For a more structured bedtime yoga routine, try this one. You can do it on a yoga mat or right on your bed:

Gentle Rolling of the Head and Neck: Start by slowly and gently rolling your head and neck in a clockwise motion while inhaling and exhaling long, deep breaths. Bring your attention particularly to the exhale, as this helps you let go of any stress incurred during your day. After one minute, gently switch directions. Roll your head and neck counterclockwise for one minute.

Forward Bend: Sit with a straight spine and your legs fully stretched out before you and parallel with your mat or bed. Pause here and take a long, deep inhale. On the exhale, start to bend or hinge at your waist, stretching your hands toward your feet. You can rest your hands on your calves, on the bed or mat beside your thighs, or on the tops of your feet. Allow yourself to be in this modified forward bend for at least three breaths or longer. When you're ready, slowly and mindfully activate your core to roll your spine into an upright position, long and straight, right where you began.

Child's Pose: The perfect send-off to bed is simply lying and breathing in child's pose (see Figure 20). Child's pose is a traditional resting posture in yoga and allows your entire system and body to relax. Start in a seated kneeling position. Take one long inhale, and on the exhale hinge or bend forward, bringing your head down to the mat or bed. Rest feeling fully supported in child's pose for several minutes, mindfully following your breath. When you're ready, return to the original kneeling position.

You are now ready for a good night's sleep.

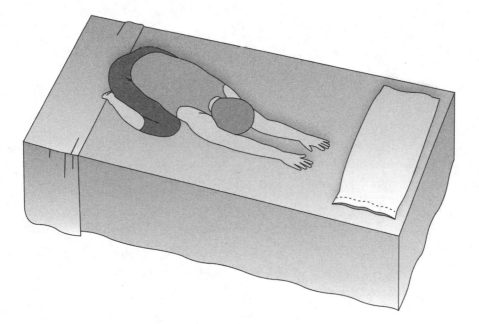

CHAPTER NINE

Telomeres Weigh In: A Healthy Metabolism

Your telomeres care how much you weigh—but not as much as you might imagine. What really appears to matter to telomeres is your metabolic health. Insulin resistance and belly fat are your real enemies, not the pounds on the scale. Dieting affects telomeres, both for good and for ill.

My (Elissa's) friend Peter is a genetic researcher and athlete who competes in Olympic-distance triathlons. He is muscular and burly, and his handsome face glows from his daily exercise. Peter has a huge appetite, but he works hard to keep himself from eating too much. I have spent a lot of time studying the psychology of eating, so I asked him what it's like to think so much about *not* eating:

I would have been an awesome hunter-gatherer. I can sniff out food in a second, especially sweets. At work, it's a joke: when food appears, so does Peter. I know where people will put food out—one person has a candy jar she fills periodically, another puts a plate of food out on a counter near her office, and lots of people put snacks or leftovers from parties or their kids' Halloween stash on the table in the kitchen.

I try to avoid seeing the food. When I meet with the woman who has the candy bowl, I try hard not to look at it (she's my boss, and I should be listening to her, but sometimes I'm thinking about not looking at the candy). When I get up to go to the bathroom, I choose a route that won't take me near the kitchen. But that means I can't even pee without thinking about food: Will I go by the kitchen to see if something is there? Or will I be strong and take a different route? I have to answer that question pretty much every time I leave my desk, because it's so easy to choose a route that will take me by a place where there might be food.

My plans to eat well don't always work. For instance, I often bring a healthy salad to work, but I don't always eat it, because I have to store it in the kitchen. I'll be on my way to get a salad, and get intercepted by the pound cake that someone put on the kitchen table. I end up eating a pound of cake—isn't that why they call it a pound cake?—while the salad sits, wilting and forgotten.

As Peter has discovered, it's hard work to think about food all the time, and even harder to lose weight. However, there is hopeful news for Peter and everyone else who struggles with weight, diet, and stress: It is not necessary, or even healthy, to think so much about food and caloric intake. That's because your telomeres care about your weight, but not as much as you might think.

IT'S THE BELLY, NOT THE BMI

Does eating too much shorten your telomeres? The quick and easy answer is *yes*. The effect of excess weight on telomeres is real—but it's not nearly as striking as the relationship between, say, depression and telomeres (which is around three times larger).[1] The weight effect is small and probably not directly causal. This finding may come as a surprise to people like Peter, who devote a huge chunk of their mental resources to the effort of eating less. It may be a bit shocking to *everyone* who's heard the message that weight loss is the most urgent goal in public health. Yet being overweight (and not obese) is, surprisingly, not linked strongly to shorter telomeres (nor is it strongly linked to mortality). Here's the reason: weight is a crude stand-in measure for what really matters, which is your metabolic health.[2] Most obesity research relies on the measure of body mass index (BMI, a measure of weight by height), but this does not tell us much about what really matters—how much muscle versus body fat we have, and where the fat is stored. Fat stored in the limbs (subcutaneously, so under the skin but not in the muscle) is different and maybe even protective, while fat stored deep inside, in the belly, liver, or muscles, is the real underlying threat. We are going to show you what it means to have poor metabolic health and show you why dieting may not be the way to get healthier.

Growing up, Sarah impressed her friends and family with her appetite. "I'd eat an Italian sub sandwich as an after-school snack, washed down by two glasses of sweet iced tea, and I'd never gain weight," she recalls wistfully. Sarah ate her way through high school and college; throughout a charmed early adulthood, she was slim. Until, suddenly, she wasn't. She was eating the same things and exercising the same amount (which was very little). Her upper body and legs were still trim, but her pants stopped fitting. Sarah had developed a belly. "I look like a strand of spaghetti with a meatball in the middle," she says now. She's worried, because both her parents take medication for high levels of bad cholesterol. After three decades of feeling effortlessly healthy, Sarah is wondering if she is going to join her parents in line at the pharmacy.

She's right to be worried, and it's not just her cholesterol levels that are at stake. Sarah's body type, where the weight is overrepresented at the belly, is closely associated with poor metabolic health. This is true *no matter how much you weigh.* It's true for people who carry a huge beer belly, and it's true for Sarah, whose BMI is normal but whose waist circumference is bigger than her hips.

When we say a person has poor metabolic health, we generally mean that he or she has a package of risk factors: belly fat, abnormal cholesterol levels, high blood pressure, and insulin resistance. Have three or more of these risk factors and you get labeled with "metabolic syndrome," a precursor to heart problems, cancer, and one of the greatest health threats of the twenty-first century: diabetes.

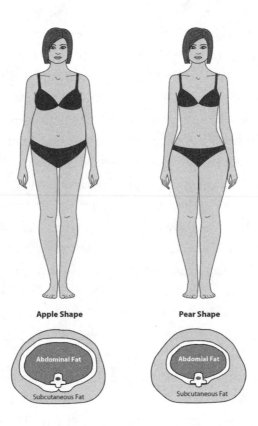

Figure 22: Telomeres and Belly Fat. Here you see what it means to have excessive fat around the waist, an apple shape (reflecting high intra-abdominal fat, measured by a greater waist-to-hip ratio, or WHR), versus more fat in the hips and thighs, a pear shape (smaller WHR). Subcutaneous fat, found under the skin and in limbs, carries fewer health risks. High intra-abdominal fat is metabolically troublesome and indicates some level of poor glucose control or insulin resistance. In one study, greater WHR predicts 40 percent greater risk for telomere shortening over the next five years.[3]

BELLY FAT, INSULIN RESISTANCE, AND DIABETES

Diabetes is a global public health emergency. The list of its long-term effects is long and chilling: heart disease, stroke, vision loss, and vascular problems that can require amputations. Worldwide, more than 387 million people—that's nearly 9 percent of the global population—have diabetes. That includes 7.3 million in Germany, 2.4 million in the United Kingdom, 9 million in Mexico, and a colossal 25.8 million people in the United States.[4]

Here's how type 2 diabetes develops: in a healthy person, the digestive system breaks food down into glucose. The beta cells in the pancreas make a hormone, insulin, which is released into the bloodstream and allows glucose to enter the body's cells to be used as fuel. In a wonderfully tidy system, insulin binds to receptors on the cells, like a key fitting into a lock. The lock turns, the door opens, and glucose can enter the body's cells. But too much belly or liver fat can cause your body to become insulin resistant, meaning that cells don't respond to insulin the way they should. Their "locks"—the insulin receptors—gum up and stick; the key no longer fits as well. It's harder for glucose to enter the cells. The glucose that can't get in through the door remains in the bloodstream. Glucose builds up in the blood even as your pancreas churns out more and more insulin. Type 1 diabetes is related to failure of the beta cells in the pancreas; they can't produce enough insulin. You're at risk for metabolic syndrome. And if your body can't keep glucose in the normal range, diabetes results.

HOW SHORT TELOMERES AND INFLAMMATION CONTRIBUTE TO DIABETES

Why do people with belly fat have more insulin resistance and diabetes? Poor nutrition, inactivity, and stress are all associated with belly fat and higher levels of blood sugar. But people with belly fat develop shorter telomeres over the years,[5] and it's very possible that these short telomeres worsen the insulin resistance problem. In a Danish study of 338 twins, short telomere length predicted increases in insulin resistance over twelve years. Within twin pairs, the twin with shorter telomeres developed higher insulin resistance.[6]

There's also a well-established connection between short telomeres and diabetes. People afflicted with inherited short telomere syndromes are much more likely to develop diabetes than the rest of the population. Their diabetes comes on early and strong. Other evidence comes from Native Americans, who are at a high risk of diabetes for a variety of reasons. When an Native American has short telomeres, he or she is twice as likely to develop diabetes over the course of five years than other members of this ethnic group with longer telomeres.[7] A meta-analysis across studies of around seven thousand people shows that short blood cell telomeres predict future onset of diabetes.[8]

We even have a glimpse into the mechanism that causes diabetes and can see what's happening in the pancreas. Mary Armanios and her colleagues have shown that when a mouse's telomeres are shortened throughout its body (through a genetic mutation), its pancreas's beta cells are not able to secrete insulin.[9] And the stem cells in the pancreas become exhausted; they run out of telomere length and can't replenish the damaged pancreatic beta cells that should have been doing the work of insulin production and regulation. These cells die off. Type 1 diabetes steps in and begins its malevolent work. In the more common type 2 diabetes, there is some beta cell dysfunction, and so short telomeres in the pancreas may play some role there as well.

In an otherwise healthy person, the pathway from belly fat to diabetes may also be traveled via our old enemy, chronic inflammation. Abdominal fat is more inflammatory than, say, thigh fat. The fat cells secrete proinflammatory substances that damage the cells of the immune system, making them senescent and shortening their telomeres. (Of course, one hallmark of senescent cells is that they can't stop sending out proinflammatory signals of their own. It's a vicious cycle.)

If you have excess belly fat (and more than half the adults in the United States do), you may be wondering how you can protect yourself—from inflammation, from short telomeres, and from metabolic syndrome. Before you go on a diet to reduce belly fat, read the rest of this chapter; you may decide that a diet will only make things worse. And that's fine, because soon we will suggest some alternate ways to improve your metabolic health.

DIETING IS DISAPPOINTING (WHAT A RELIEF)

There is a relationship between dieting, telomeres, and your metabolic health. But as in all things related to weight, it's complicated. Here are some results of research into weight loss and telomeres:

- Weight loss leads to a slowdown in the normal attrition rate of telomeres.
- Weight loss has no effect on telomeres.
- Weight loss encourages telomeres to lengthen.
- Weight loss leads to shorter telomeres.

It's a mind-bending set of findings. (In that final study, people who underwent bariatric surgery had shorter telomeres one year after their procedure, though this effect was possibly from the physical stress of the surgery.)[10] We think that these mixed results are telling us that, once again, it's not really the weight that matters. Weight loss is only a crude stand-in for positive changes to underlying metabolic health. One of those changes is the loss of belly fat. Lose weight overall, and you'll inevitably take a bite out of that "apple," and this may be more true if you are increasing your exercise rather than just reducing calories.

Another positive change is improved insulin resistance. One study followed volunteers for ten to twelve years; as the people in the study gained weight (as people tend to do), their telomeres got shorter. But then the researchers examined which mattered more, weight gain or the insulin resistance that often comes with it. It was insulin resistance that carried the weight, so to speak.[11]

This idea—that improving your metabolic health is more important than losing weight—is vital, and that's because repeated dieting takes a toll on your body. There are some internal "push back" mechanisms that makes it hard for us to keep weight off. Our body has a set point that it defends, and when we lose weight, we also slow our metabolism in an effort to regain the weight ("metabolic adaptation"). While this is well known, we didn't know how dramatic this adaptation could be. There is a tragic lesson here from the brave volunteers who have joined the reality TV show *The Biggest Loser*. For this show, very heavy people compete to lose the most pounds over 7.5 months, using exercise and diet. Dr. Kevin Hall and his colleagues from the National Institutes of Health decided to examine how this rapid massive weight loss affected their metabolism. At the end of the show, they had lost 40 percent of their weight (around 58 kg). Hall checked their weight and metabolism again six years later. Most had regained weight, but they kept an average of 12 percent weight loss. Here is the hard part: at the end of the program, their metabolism had slowed so that they were burning 610 fewer calories per day. By six years, despite weight regain, their metabolic adaptations had become even more severe, where they were burning around 700 calories fewer than their baseline.[12] Ouch. While this is an example of extreme weight loss, this metabolic rate slowing happens to a lesser extent whenever we lose weight, and, apparently, even when we regain it.

In the phenomenon known as weight cycling (or "yo-yo dieting"), dieters gain pounds and lose them, and gain and lose, and so on. Fewer than 5 percent of people who are trying to lose weight can stick to a diet and maintain the weight loss for five years.

The remaining 95 percent either give up or become weight cyclers. Weight cycling has become a way of life for many of us, especially women; it's what we talk about; it's how we laugh together. (An example: "Inside me there's a skinny woman crying to get out, but I can usually shut her up with cookies.") Yet weight cycling appears to shorten our telomeres.[13]

Weight cycling is so unhealthy, and also so common, that we feel strongly that everyone should understand it. Weight cyclers restrict themselves for a while and then, when they fall off the wagon, tend to indulge in treats and other unhealthy foods. This intermittent cycling between restriction and indulgence is a real problem. What happens to animals when they get junk food all the time? They overeat and get obese. But when you withhold junk food most of the time, giving it to them only every few days, something even more disturbing happens. The rats' brain chemistry changes; the brain's reward pathways start to look like the brains of people who are suffering from drug addiction. When the rats don't get their sugary, chocolaty rat junk food, they develop withdrawal symptoms, and their brains release the stress chemical CRH (short for corticotropin-releasing hormone). The CRH makes the rats feel so bad that they are driven to seek the junk food, to get relief from their stressed state of withdrawal. When the rats finally do get the chocolaty stuff, they eat it as if they will never have the chance again. They binge.[14]

Sound like anyone you know? Or like Peter eating pound cake on his way to eat a healthy salad for lunch? Studies of obese people suggest a similar compulsive aspect of overeating, with dysregulation in the brain's reward system.

Dieting can create a semiaddictive state, and it's also just plain stressful. Monitoring calories causes cognitive load, meaning that it uses up the brain's limited attention and increases how much stress you feel.[15]

Think of Peter, spending years trying to eat fewer sweets and calories. Obesity researchers have a name for this kind of long-term dieting mentality: cognitive dietary restraint. Restrainers devote a lot of their time to wishing, wanting, and trying to eat less, but their actual caloric intake is no lower than people who are unrestrained. We asked a group of women questions such as "Do you try to eat less at mealtimes than you like to eat?" and "How often do you try not to eat between meals because you are watching your weight?" The women who answered in ways that revealed a high level of dietary restraint had shorter telomeres than carefree eaters, regardless of how much they weighed.[16] It's just not healthy to spend a lifetime thinking about eating less. It's not good for your attention (a precious limited resource), it's not good for your stress levels, and it's not good for your cell aging.

Instead of dieting by restricting calories, focus on being physically active and eating nutritious foods—and in the next chapter we will help you choose the foods that are best for your telomeres and overall health.

SUGAR: NOT A SWEET STORY

When we want to spot the parties responsible for metabolic disease, we point a finger straight at highly processed, sugary foods and sweetened drinks.[17] (We're looking at you, packaged cakes, candies, cookies, and sodas.) These are the foods and drinks most associated with compulsive eating.[18] They light up the reward system in your brain. They are almost immediately absorbed into the blood, and they trick the brain into thinking we are starving and need more food. While we used to think all nutrients had similar effects on weight and metabolism— "a calorie is a calorie" —this is wrong. Simply reducing sugars, even if you eat the same number of calories, can lead to metabolic improvements.[19] Simple carbs wreak more havoc on our metabolism and control over appetite than other types of foods.

EXTREME CALORIC RESTRICTION: IS IT GOOD FOR TELOMERES?

You're in a cafeteria, standing in line with your tray. When you get to the front of the line, you notice that everyone is using pairs of tongs to select tiny morsels of food, which they carry over to a scale and weigh carefully. Once they are satisfied with the number of grams of food they have chosen, they take their trays—which bear much less food than you'd normally choose for yourself—to a table and sit down. You join them and watch them eat their meager lunches. When their plates are empty, they say, "Still a bit hungry," and smile.

Why are these people weighing out small portions of food? Why are they smiling when they're hungry? This is a thought exercise— no such cafeteria exists in the real world—but it reflects the habits of people who believe that by restricting their calories to 25 or 30 percent less than a normal healthy intake, they will live longer. People who practice caloric restriction teach themselves to have a different reaction to hunger. When they feel the pang of an empty stomach, they don't feel stressed or unhappy. Instead, they say to themselves, *Yes! I'm reaching my goal.* They are incredibly good at planning and thinking about the future. For example, a caloric restriction practitioner in one of our studies was eagerly organizing his 130th birthday, even though he was only around sixty years old at the time.[20]

If only these people were worms. Or mice. There is little doubt that extreme caloric restriction extends the longevity of various lower species. In at least some breeds of mice on restricted diets, telomeres appear to lengthen. They also have fewer senescent cells in the liver, an organ that is one of the first places senescent cells will build up.[21] Caloric restriction can improve insulin sensitivity, too, and reduce oxidative stress. But it's harder to pinpoint the effects of caloric restriction on larger animals. In one study, monkeys who ate 30 percent fewer calories than normal had a longer healthspan and longer life—but only when compared to a control group of monkeys who ate a lot of sugar and fat. In a second study, monkeys on a similarly restricted diet were compared to monkeys who ate normal portions of healthy food. Those monkeys did not have more longevity, though they stayed in the healthspan a bit longer. Adding to the uncertainty is that in both studies, the monkeys ate in solitude. Monkeys are highly social animals; in the wild, they eat together. Having them eat in circumstances that were abnormal, and quite possibly stressful, could have affected the outcome in ways we don't yet understand.

For now, it looks as if caloric restriction has no positive effect on human telomeres. Janet Tomiyama, now a psychology professor at UCLA, conducted a study during her postdoctoral fellowship at UCSF. She managed to round up a group of people from across the United States who were successful at long-term caloric restriction for an intensive study where she also examined telomeres in different blood cell types. (As you may imagine, such people are rare.) To our surprise, their telomeres weren't any longer than the normal or even the overweight control group. In fact, their telomeres tended to be slightly shorter in their peripheral mononuclear blood cells, which are types of immune cells that include the T-cells. Another study looked at rhesus monkeys who were restricted to 30 percent fewer calories than a normal rhesus monkey diet. The researchers measured telomere length in various tissues—not just blood, which is the typical source of telomere measurements, but also in fat and muscle. Once again, there were no differences in telomere length in the calorie-restricted monkeys—not in any of the cell types.

Thank goodness. Most people can't practice extreme calorie restriction, and few people want to. As one of our friends said, "I'd rather eat good dinners until I'm eighty than starve until I'm one hundred." He's got a point. You do not have to suffer to eat in a way that is good for your telomeres and good for your healthspan. To learn more, turn to the next chapter.

TELOMERE TIPS

- Telomeres tell us not to focus on weight. Instead, use your level of belly protrusion and insulin sensitivity as an index of health. (Your doctor can measure your insulin sensitivity by testing your fasting insulin and glucose.)
- Obsessing about calories is stressful and possibly bad for your telomeres.
- Eating and drinking low-sugar, low-glycemic-index food and beverages will boost your inner metabolic health, which is what really matters (more than weight).

RENEWAL LAB

SURF YOUR SUGAR CRAVINGS

Cutting back on sugar may be the single most beneficial change you can make to your diet. The American Heart Association recommends limiting added sugar to nine teaspoons a day for men and six teaspoons for women, but the average American has almost twenty teaspoons a day. A high-sugar diet is associated with more belly fat and insulin resistance, and three studies have found a link between shorter telomeres and drinking sugared beverages. (In the next chapter we'll talk about sugary drinks in more detail.)

When you get a craving for sugar (or any other food that isn't good for you), you need a tool to help you cope. Cravings are strong, and they're backed up by dopamine activity in the reward center of the brain. Fortunately, cravings are impermanent. They will pass. Psychologist Alan Marlatt has applied the idea of "Surfing the Urge" to help people with addictions resist their cravings until those cravings dissipate. Andrea Lieberstein, a mindful eating expert, has found this practice works even better for food cravings when adding a heart focus to the end, taking the edge off of the craving with feelings of compassion and kindness.

Here's how to surf your cravings:

SURF YOUR CRAVINGS

Sit comfortably and close your eyes. Picture the snack or treat you're craving: Conjure up its texture, its color, its smell. As the image becomes vivid, let yourself feel your craving. Let your attention wander throughout your body to observe the nature of this craving.

Describe this craving to yourself. What are its sensations and its qualities? What are the shapes, sensations, and any thoughts or feelings associated with it? Where is it located in your body? Does it change as you notice it, or as you exhale? Feel any discomfort. Remind yourself this is not an itch that needs to be scratched. This is a feeling that changes and will pass. Perhaps imagine it as a wave that builds, crests, and dissipates back into the ocean. Breathe into the sensation, and let it release tension, as you notice the waves falling back gently.

You might put your attention and your hand on your heart, imagining a sense of warmth and kindness flowing outward from your heart. Let this sense of warmth spread throughout your body, enveloping the sense of craving with loving-kindness. Take a moment just breathing in this feeling of compassion for yourself. Now look at the image of the food again. What has changed? What are you aware of? You can experience the craving without acting on it. Just notice it, breathe, and envelope it with a sense of loving-kindness.

You can record yourself reading this script (using, for example, the Voice Memo app on the iPhone) and listen to it whenever a craving arises. You can also download an audio version of this script from our website.

TUNE IN TO YOUR BODY'S SIGNALS OF HUNGER AND FULLNESS

By mindfully tuning in to your body's cues of hunger and fullness, you may be able to reduce overeating. When you pay attention to your level of physical hunger, you are less likely to confuse it with psychological hunger. Stress, boredom, and emotions (even happy ones) can make you feel as if you're hungry even when you're really not. In a small pilot study led by psychology researcher Jennifer Daubenmier at UCSF, we found that when women are trained to do a mindful check-in before meals, they have lower blood glucose and cortisol, particularly if they are obese. And the more they improve on mental and metabolic health, the more their telomerase increases.[22] In a larger trial, psychology researcher Ashley Mason found that the more men and women practiced mindful eating, the fewer sweets they ate, and the lower their glucose was one year later.[23] Mindful eating seems to have a small effect on weight but may be critical to breaking the craving sweets–glucose link.

Below are some mindful eating strategies that I (Elissa) and my colleagues use for our studies of weight management. They are based on Mindfulness-Based Eating Awareness Training, a program developed by Jean Kristeller, a psychologist at Indiana University. (See more resources on mindful eating.)[24]

1. Breathe. Bring your awareness to your entire body. Ask yourself: How physically hungry am I right now? What information and sensations help me answer this question?

2. Rate your physical hunger on this scale:

Not at all hungry				Moderately hungry				Very hungry	
1	2	3	4	5	6	7	8	9	10

Try to eat *before* you get to 8 so you're less likely to overeat. Definitely don't wait until you're at 10. If you're famished, it's easy to eat too much, too fast.

3. When you do eat, fully savor the taste of the food and the experience of eating.

4. Pay attention to the hunger in your stomach, to the physical sensations of fullness and distention. (We call this "listening to the stretch receptors.") After you've spent a few minutes eating, ask yourself "How physically full do I feel?" Rate your answer:

Not at all full				Moderately full				Very full	
1	2	3	4	5	6	7	8	9	10

Stop when your score is 7 or 8—in other words, when you're moderately full. Your biological signals of fullness, caused by increases in blood sugar and satiety hormones in the blood, kick in slowly, and you won't feel their full effect until twenty minutes later. Stopping before you get those signals, before you've eaten too much, is usually the hard part, but this becomes much easier once you start paying attention.

CHAPTER TEN

Food and Telomeres: Eating for Optimal Cell Health

Some foods and supplements are healthy for your telomeres, and some just aren't. We are happy to report that you do not need to give up carbs or milk products to be healthy! A whole-foods diet that features fresh vegetables, fruits, whole grains, nuts, legumes, and omega-3 fatty acids is not only good for your telomeres, it also helps reduce oxidative stress, inflammation, and insulin resistance—factors that, as we'll explain here, can shorten your healthspan.

It happens every single day: Morning arrives. I (Liz) am not a morning person, but I get out of bed and stagger to the kitchen, slowly waking up as I go. My husband, John, who is an early bird by nature, has kindly brewed me a cup of coffee.

"Milk?" he asks.

Well, that's a tough question for the predawn hours, made tougher by nutrition advice that often feels confusing. Yes, I like milk in my coffee. But should I pour it in? Milk is healthy, right? After all, it contains calcium and protein and is fortified with vitamin D. But should I reach for whole milk or skim? Or should I take a pass altogether?

Each additional breakfast item poses its own set of nutrition dilemmas:

Toast. Too many carbs, even if it's whole wheat? What about a potential reaction to gluten?

Butter. Will a little fat increase feelings of fullness, which is good, or will it clog the arteries, which is bad?

Fruit. Better to just ditch the toast idea and make a smoothie instead? Or . . . is fruit dangerously loaded with sugars?

These are a lot of questions to answer when you're still barely awake and the coffee hasn't kicked in yet. We are both scientists, trained in sifting through complicated evidence, but sometimes we still struggle to figure out what is healthiest to eat.

On mornings like this, telomeres offer a fundamental guide to the foods that are best for us. We trust telomere evidence because it looks at how the body responds to foods at the microlevel. And we take the evidence seriously, because it aligns well with the emerging knowledge in nutrition science. These findings tell us that diets don't work, and that the most empowering choice we can make is to eat fresh, whole foods instead of processed ones. As it turns out, eating for healthy telomeres is very pleasant, satisfying, and nonrestrictive.

THREE CELLULAR ENEMIES AND HOW TO STOP FEEDING THEM

You've heard us issue warnings about inflammation, insulin resistance, and oxidative stress, which create an environment that is toxic for telomeres and cells. Think of these conditions as three enemies that lurk inside each of us. You can eat foods that feed these three villains—or you can eat foods that fight them, shifting the cell environment to one that is healthier for telomere upkeep.

The First Cellular Enemy: Inflammation

Inflammation and telomere damage share a mutually destructive relationship. One makes the other worse. As we've explained, aging cells, with their short or damaged telomeres (plus any other breaks in the DNA that do not get repaired), send out proinflammatory signals that cause the body's immune system to turn on itself, damaging tissues all over the body. Inflammation can also cause immune cells to divide and replicate, which shortens telomeres even more. Thus a vicious cycle is set up.

Here's what can happen to an inflamed mouse: Researchers took a group of mice and knocked out part of a gene that protects against inflammation; without that part of the genetic code, the mice quickly developed a serious case of chronic inflammation. Their tissues accumulated short telomeres and senescent cells. The more senescent cells in their liver and intestines, the faster the mice died.[1]

One of the best ways for you to protect yourself against inflammation is to stop feeding it. The glucose absorbed from French fries or from refined carbohydrates (white bread, white rice, pasta), and from sugary candies, sodas, juices, and most baked goods, hits your bloodstream fast and hard. That uptick of blood glucose also causes an increase in cytokines, which are inflammatory messengers.

Alcohol acts as a kind of carb as well, and too much alcohol consumption appears to increase C-reactive protein (CRP), a substance that is produced in the liver and rises when there is more inflammation in the body.[2] Alcohol is also converted to a chemical (acetaldehyde, which is a carcinogen) that can damage DNA and in high doses could also harm telomeres. At least, it harms telomeres in cells in the laboratory—we have no idea if such high doses are ever achieved in humans. So far, it appears that chronic heavy alcohol use may be associated with shorter telomeres and other signs of an aged immune system, but there are no consistent relationships between light alcohol intake and telomeres.[3] It is okay to enjoy your occasional drink!

There is more good news, too, especially if you are concerned about those mice who were genetically engineered for chronic inflammation. When the mice were given an anti-inflammatory or an antioxidant drug, the telomere dysfunction was reversed. The mice's telomeres rebounded, and senescent cells stopped accumulating, so that cells could continue their dividing and renewing. This suggests that all of us can protect our telomeres from inflammation, but it's safest and smartest to do it without drugs. For a start, we can simply eat the foods that help prevent an inflammatory response from happening in the first place. And what a marvelous selection of sweet and savory plant foods we have to choose from: think of red, purple, and blue berries; red and purple grapes; apples; kale; broccoli; yellow onions; juicy red tomatoes; and green scallions. All these foods contain flavonoids and/or carotenoids, a broad class of chemicals that gives plants pigment. They are also especially high in anthocyanins and flavonols, subclasses of flavonoids that are related to lower levels of inflammation and oxidative stress.[4]

Other anti-inflammatory foods include oily fish, nuts, flaxseed, flax oil, and leafy vegetables—because all these items are rich in omega-3 fatty acids. Your body requires omega-3s to reduce inflammation and keep telomeres healthy. Omega-3s help form cell membranes throughout the body, keeping the cell structure fluid and stable. In addition, the cell can convert omega-3s into hormones that regulate inflammation and blood clotting; they help determine whether artery walls are rigid or relaxed.

It's been known for a while that people with higher blood levels of omega-3s have lower cardiovascular risk. Newer research suggests an exciting additional possibility: omega-3s may be helping to do that by keeping your telomeres from declining too quickly. Remember, telomeres shorten with age; the goal is for this shortening process to happen as slowly as possible. One study looked at the blood cells of 608 people, all of whom were middle-aged and had stable heart disease. The more omega-3s in their blood cells, the less their telomeres declined over the next five years.[5] And the less the telomeres declined, the more likely it was that these subjects, who were not so healthy to begin with, would survive the next four years.[6] Of those who had telomere shortening, 39 percent died—whereas of those who had apparent lengthening, only 12 percent died. The less your telomere length declines, the less likely *you* are to fall into the diseasespan and early death.

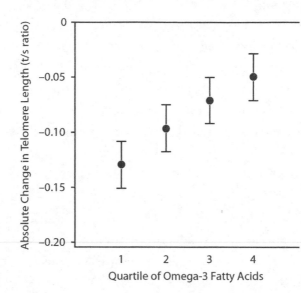

Figure 23: Omega-3 Fatty Acids and Telomere Length over Time. The higher the levels of omega-3s in the blood (EPA and DHA), the less telomere shortening over the next five years. Each one standard deviation above the average omega-3 levels predicted 32 percent lower chances of shortening. This effect was even stronger in those who started off with longer telomeres (since longer telomeres shorten more quickly).[7]

So enjoy fresh oily fish (including sushi), salmon and tuna, leafy vegetables, and flax oil and flaxseeds. (For state-by-state U.S. recommendations for fish that are caught or farmed in ways that cause less harm to the environment, consult the Monterey Bay Aquarium Seafood Watch website at http://www.seafoodwatch.org/seafood -recommendations/consumer-guides.) But should you take omega-3 supplements, otherwise known as fish-oil capsules? There has been only one randomized trial on omega-3 supplementation and telomeres, a study by psychologist Janice Kiecolt Glaser at Ohio State, and the results were suggestive. She found that people who took fish-oil supplements for four months did not have longer telomeres than people who took a placebo. However, across all the groups, the greater the increases in omega-3s in the blood relative to their levels of omega-6 fatty acids, the greater the telomere lengthening over that period.[8] The omega supplementation also reduced inflammation, and the greater decreases in inflammation were associated with increases in telomere length. (Omega-6s are polyunsaturated fats that come from sources like corn oil, soybean oil, sunflower oil, seeds, and certain nuts.) We must note, though, that those taking the supplements had other significant telomere-friendly changes: reduced levels of oxidative stress and inflammation. The results appear to depend on how well each person absorbed the levels of omega-3 polyunsaturated fats from the supplements.

Your blood levels of omega-3s, or any nutrient, are not always directly related to whether you're consuming dietary or supplement sources. All kinds of complicated and mostly unknowable factors affect this number: how well you absorb the nutrient, how well your cells use it, how fast you metabolize and lose it. (This is a good piece of information to keep in mind whenever you read recommendations for diet and supplementation.) In general, we suggest that everyone try to get their nutrients from their diet, but when that's just not possible, supplementation can be a reasonable alternative (make sure you check with your physician first). Even the most innocent-seeming supplements can have side effects or interact with medications you're already taking. They could also be contraindicated for people with certain health conditions. A general consensus seem to be a daily dosage of at least 1,000 milligrams of a mixture of EPA and DHA, which is similar to the low dose tested in the Ohio State study. For sustainability reasons, we strongly suggest the vegetarian alternative, which is made from algae. Fish have omega-3s because they eat algae. We can eat algae, too, sustainably farmed algae that contains DHA. The oceans cannot support enough fish oil to maintain the world's healthy telomeres. So far, it appears the DHA from algae promotes similar benefits to cardiovascular health as the DHA from fish.

Telomere research suggests that you should make consumption of omega-3s a priority. But you also have to keep an eye on the balance between your omega-3s and omega-6s, because the typical Western diet tilts us more toward omega-6s than omega-3s. To keep your omegas in balance, we suggest that you keep eating healthy, unprocessed foods like nuts and seeds—but dramatically reduce your consumption of fried foods, packaged crackers, cookies, chips, and snacks, which often contain oils made with high amounts of omega-6s, as well as saturated fats, which are a risk factor for cardio-vascular disease.

There's another chemical in our body worth getting to know: homocysteine, which is chemically related to cysteine, one of the amino acid building blocks of proteins. Homocysteine levels go up with aging, and correlate with inflammation, wreaking havoc on the lining of our cardiovascular system to promote heart disease. In many studies, having high homocysteine is associated with having short telomeres. But telomeres reflect the input of many factors. So it is no surprise that in one study, the relationship between telomeres and mortality appears to be due in part to both high inflammation and high homocysteine—we don't know which came first.[9] The good news here is that if you have especially high homocysteine, this is one of the cases where a vitamin pill might help—B vitamins (folate or B_{12}) appear to reduce homocysteine.[10] (Check with your doctor to see if you should be taking this supplement.)

The Second Enemy: Oxidative Stress

Human telomeres have a DNA sequence that looks like this: TTAGGG, repeated in tandem over and over, commonly over a thousand times at each chromosome end. Oxidative stress—that dangerous condition that occurs when there are too many free radicals and not enough antioxidants in your cells—damages this precious sequence, especially its GGG segments. Free radicals take aim at that big juicy row of GGG pieces, a particularly sensitive target. After free radicals have their way, the DNA strand is broken; the telomere gets shorter faster.[11] It's as if the rich meal of GGGs has been fed to the cellular enemy, oxidative stress. In cells grown in the lab, oxidative stress damages telomeres, and it also reduces the telomerase activity that can replenish shorter telomeres. It's a double whammy.[12]

But if you pump up the cells' medium (the liquid soup that supports a cell's life when it's sitting in a lab flask) with vitamin C, the telomere is protected from the free radicals.[13] Vitamin C and other antioxidants (like vitamin E) are scavengers that gobble up free radicals, preventing them from harming your telomeres and cells. People with higher blood levels of vitamins C and E have longer telomeres, but only when they also have lower levels of a molecule known as F2-isoprostane, which is an indicator of oxidative stress. The higher this ratio between blood antioxidants and F2-isoprostane, the less oxidative stress there is in the body. This is just one of the many reasons you should eat fruits and vegetables every day; they offer some of the best sources of antioxidant protection. To get sufficient antioxidants in your diet, eat plenty of produce, especially citrus, berries, apples, plums, carrots, green leafy vegetables, tomatoes, and, in smaller portions, potatoes (red or white, with the skin on). Other plant-based sources of antioxidants are beans, nuts, seeds, whole grains, and green tea.

At this point, we don't suggest getting your antioxidants from a supplement if your goal is telomere health. That's because the evidence for a connection between antioxidant supplements and healthy telomeres is still inconclusive. Some studies have found that the higher the level of certain vitamins in the blood, the longer the telomeres, and we have listed these in the table on page 238. However, while some studies find multivitamin use accompanies longer telomeres,[14] at least one study found that taking a multivitamin was related to shorter telomeres.[15] Also, high antioxidant levels even provoked laboratory-grown human cells to take on certain cancer-ous properties, a finding that, again, may warn us that too much of a good thing may be, simply, too much. In general, antioxidants from food are typically better absorbed by the body and may have more powerful effects than supplements.

OUR EARLIEST NUTRITION

Can you feed your baby's telomeres? Possibly, by making sure your baby is exclusively breast-fed in its first weeks. Janet Wojcicki, a health researcher at UCSF who has been following cohorts of pregnant women, found that children who were breast-fed only (no formula or solid foods) in the first six weeks of life have longer telomeres. Solid foods can cause inflammation and oxidative stress when they are given to infants whose guts are not yet ready.[16] Perhaps that is why introducing solid food before six weeks of age is linked to shorter telomeres.

The Third Enemy: Insulin Resistance

Nikki, a physician and administrator at her hometown hospital, has a vice: massive consumption of the sugared soda Mountain Dew. She developed the habit in residency, when she learned to rely on its sugar and caffeine to stay awake. The habit has remained with her. Early each morning, Nikki pulls a one-liter bottle of Mountain Dew from a small refrigerator in her garage, which is dedicated to warehousing her stash. She sets the bottle in the passenger seat of her car on the way to work. At each stoplight, she unscrews the bottle and takes a swig. When she arrives at work, the bottle goes into the fridge. After grand rounds: a swig. After a meeting: swig. After finishing some paperwork: swig. By the end of her long, grueling day, the bottle is empty. "I couldn't get through without it," Nikki says, with a fatalistic shrug of her shoulders.

As a doctor, Nikki knows that a daily one-liter dose of Mountain Dew is not a healthy habit. But like nearly half of all Americans, she drinks soda anyway. These folks might as well give the third enemy—insulin resistance—a straw and say, "Drink up; this stuff will help you get as big and as terrifying as you want to be."

Here's a frame-by-frame shot of what happens when you swallow sugary soda, or "liquid candy": Almost instantaneously, the pancreas releases more insulin, to help the glucose (sugar) enter cells. Within twenty minutes, glucose has built up in the bloodstream and you have high blood sugar. The liver starts to turn sugar into fat. In about sixty minutes, your blood sugar falls, and you start thinking about having more sugar to pick you back up after the "crash." When this happens often enough, you can end up with insulin resistance.

Is soda the new smoking? Maybe. Cindy Leung, a nutritional epidemiologist at UCSF and one of our collaborators, found that people who drink twenty ounces of sugary soda daily have the equivalent of 4.6 extra years of biological aging, as measured by telomere shortness.[17] That, astonishingly, is about the same level of telomere shortness caused by smoking. When people drink eight ounces of soda, their telomeres are the equivalent of two years older. You may be wondering if people who drink soda have other unhealthy habits that may affect the results—and that's a great question. In this study, which looked at around five thousand people, we did what we could to address confounding factors. We corrected for some available factors, including diet and smoking; and then we corrected for all available factors, including diet, smoking, BMI, waist circumference (to gauge belly fat), income, and age, that might have otherwise explained away this association. The association did not go away. This association between soda and telomeres exists in young children, too. Janet Wojcicki found that at three years old, children who were drinking four or more sodas a week had a greater rate of telomere shortening.[18]

Sports drinks and sweetened coffee drinks are liquid candy,

too. They contain as much sugar as a typical soda (42 grams in a 12-ounce-tall Peppermint Mocha from Starbucks) so it is wise to stay away from them, or to drink them only rarely, as a special treat.[19] Sodas and sweetened beverages are a dramatic example of sugar's harm to telomeres, because of the delivery method. It's a fast rush of sugar with no fiber to slow it down. Almost anything that's considered a dessert or a treat is a source of high sugar: cookies, candy, cakes, ice cream. Once again, refined products like white bread, white rice, pasta, and French fries are high in simple or rapidly absorbed carbohydrates and can wreak havoc on your blood sugar levels, too.

To prevent insulin spikes that can eventually lead to insulin resistance, focus on foods that are higher in fiber: Whole-wheat bread, whole-wheat pasta, brown rice, barley, seeds, vegetables, and fruits

Figure 24: Finding a Balance—as Guided by Telomeres. Choose more foods high in fiber, antioxidants, and flavenoids, like fruits and vegetables. Include foods high in omega-3 oils, like seaweed and fish. Choose less refined sugars and red meat. A healthy dietary balance, like the one pictured above, will lead to healthy shifts in your blood to high nutrients and less oxidative stress, inflammation, and insulin resistance.

are all excellent sources. (Fruits, although they contain simple carbohydrates, are healthy because of their fiber content and overall nutritional value; fruit juices, from which the fiber has been extracted, are generally not.) These foods are also filling, which helps you avoid eating excess calories. They are the same foods that help reduce the belly fat that is so closely associated with insulin resistance and metabolic disorder.

VITAMIN D AND TELOMERASE

Higher levels of vitamin D in the blood predict lower overall mortality rates.[20] Some studies find that vitamin D is related to longer telomere length, more so in women than men, and other studies do not find a relationship. So far we've found one study that tested the effects of supplements: In that small study, 2,000 IU a day of vitamin D (in the form of vitamin D_3) for four months led to increased telomerase by around 20 percent compared to a placebo group.[21] While the jury is still out on a relationship with telomeres, it's notable that vitamin D levels are often low, depending on where you live and on sunlight exposure. The best dietary sources of vitamin D are salmon, tuna, sole, flounder, fortified milk and cereals, and eggs. It can be hard to get enough vitamin D from diet and sunlight alone, depending on where you live, so this is a case when you may want to consider supplements (consult your doctor).

A HEALTHY EATING PATTERN

Platters of freshly caught fish, bowls heaped with fruits and vegetables in deep, rich hues, dishes of hearty beans, whole grains, nuts and seeds...it's a menu for a feast. It's also a recipe for supporting a healthy cellular environment. These foods reduce inflammation, oxidative stress, and insulin resistance. These foods fit into a healthy eating pattern that is great for telomeres and overall health.

Around the world—from Europe to Asia to the Americas—eating habits can be very roughly divided into two categories. There are people whose diets feature lots of refined carbohydrates, sweetened sodas, processed meat, and red meat. And then there are people who have a high intake of vegetables, fruits, whole grains, legumes, and low-fat, high-quality sources of protein, including seafood. This healthier diet is sometimes called the Mediterranean diet, but most cultures around the globe have some version of this eating pattern. Some of the details vary—some cultures eat more dairy or seaweed—but the general idea is to eat a variety of fresh, whole foods, and for most of these foods to come from a low spot on the food chain. Some researchers call this the "prudent dietary pattern." That's an accurate label, though it doesn't capture just how delicious and healthy these foods are.

People who follow this prudent pattern have longer telomeres, no matter where they live. In Southern Italy, for example, elderly people who followed the Mediterranean diet had longer telomeres. The more closely the adhered to this type of a diet, the better their overall health and the more they could fully participate in the activities of daily living.[22] And in a population study of middle-aged and older people in Korea, people who followed the local version of a prudent dietary pattern (i.e., more seaweed and fish) had longer telomeres ten years later than people who ate a diet high in red meat and refined, processed foods.[23]

We've been speaking of broad dietary patterns, but what are the best particular foods for healthy telomeres? The Korean study gives us a clue. The more that people ate legumes, nuts, seaweed, fruits, and dairy products, and the less they consumed red meat or processed meat and sweetened sodas, the longer their telomeres in their white blood cells.[24]

The benefits of eating wholesome, unprocessed foods—and not too much red meat or processed meats—hold strong across the world, through adulthood, and all the way into old age. In 2015, the World Health Organization identified red meat as a probable cause of cancer and processed meat as a cause.[25] When types of meat are examined in telomere studies, processed meat appears worse for telomeres than unprocessed red meat.[26] Processed meat refers to meat that has been altered (smoked, salted, cured), such as hot dogs, ham, sausage, or corned beef.

Of course, it is best to eat well throughout your entire life, but it is never too late to begin. The chart that follows can help guide your daily food choices. In general, though, we suggest that you worry less about any particular food item (an attitude that makes mornings easier for Liz) and focus instead on eating a variety of fresh, wholesome foods. You'll find yourself enjoying foods that fight inflammation, oxidative stress, and insulin resistance, without needing to plan carefully in advance. And you will find that you naturally follow the kind of eating plan that is healthy for your telomeres. Plus, you won't shorten your telomeres by worrying too much about all of the food choices you make every day!

BEANS ABOUT COFFEE?

Coffee's effects on health have been questioned in hundreds of studies. Those of us who love our morning cup will be happy to hear that it almost always turns up innocent. Meta-analyses show coffee reduces the risk of cognitive decline, liver disease, and melanoma, for example. Only one trial has been done on coffee and telomere length, but the news so far is good: Researchers tested whether coffee might improve the health of forty people with chronic liver disease. They were randomized to either drink four cups of coffee a day for a month or refrain (the latter were the controls). After the period of drinking coffee, the patients had significantly longer telomeres and lowered oxidative stress in their blood than the control group.[27] Further, in a sample of over four thousand women, those who drank caffeinated coffee (but not decaffeinated) were likely to have longer telomeres.[28] More reasons to enjoy the aroma of your morning coffee brewing.

We have discussed vitamin D and omega-3 supplements, which are often found to be deficient. However, outside of these, we do not make specific recommendations on supplements, because each person's needs are different, and nutrition studies' conclusions about supplements are notoriously changed by new studies. It's hard to be confident about the effects and safety of high doses of anything.

NUTRITION AND TELOMERE LENGTH**

Food, Drinks, and Telomere Length	
Associated with Shorter Telomeres	**Associated with Longer Telomeres**
Red meat, processed meat[29] White bread[30] Sweetened drinks[31] Sweetened soda[32] Saturated fat[33] Omega-6 polyunsaturated fats (linoleic acid)[34] High alcohol consumption (more than 4 drinks per day)[35]	Fiber (whole grains)[36] Vegetables[37] Nuts, legumes[38] Seaweed[39] Fruits[40] Omega-3s (e.g., salmon, arctic char, mackerel, tuna, or sardines)[41] Dietary antioxidants, including fruits, vegetables, but also beans, nuts, seeds, whole grains, and green tea[42] Coffee[43]
Vitamins	
Associated with Shorter Telomeres	**Associated with Longer Telomeres**
Iron-only supplements[44] (probably because they tend to be high doses)	Vitamin D[45] (mixed evidence) Vitamin B (folate), C, and E Multivitamin supplements (mixed evidence)[46][47]

**Note that the scientific literature here is growing and changing all the time. Check our website for updates!

TELOMERE TIPS

- Inflammation, insulin resistance, and oxidative stress are your enemies. To fight them, follow what's been called a "prudent" pattern of eating: Eat plenty of fruits, vegetables, whole grains, beans, legumes, nuts, and seeds, along with low-fat, high-quality sources of protein. This pattern is also known as the Mediterranean diet.

- Consume sources of omega-3s: salmon and tuna, leafy vegetables, and flax oil and flaxseeds. Consider supplementation with an algae-based omega-3 supplement.

- Minimize red meat (especially processed meat). You might try to go vegetarian for at least a few days each week. Eliminating meat can benefit your cells as well as the environment.

- Avoid sugary foods and drinks, and processed foods.

RENEWAL LAB

TELOMERE-FRIENDLY SNACKS

Healthy snacks are important to have on hand, because the alternative is usually *unhealthy* snacks. Typical snack foods are often processed and contain unhealthy fats, sugars, and salts. We recommend any whole-food snack high in protein and low in sugars. Here are a few ideas that also include high levels of either antioxidants or omega-3 polyunsaturated fats.

Homemade trail mix. Making your own trail mix is easy, and it's the best way to make sure it is low in sugar. (Store-bought trail mix often hides added sugars in dried fruits.) This mix is high in omega-3s and antioxidants. It's also rich in energy, so be sure to enjoy it in moderate quantities.

Combine:

- 1 cup walnuts
- ½ cup cacao nibs or dark chocolate chips
- ½ cup goji berries or other dried berries

Optional additions:

- ½ cup dried unsweetened coconut flakes
- ½ cup raw or unsalted sunflower seeds
- 1 cup raw almonds

Homemade chia pudding. Chia seeds are high in antioxidants, calcium, and fiber. These unassuming little seeds from South America also house 28 grams of omega-3s in every ounce. Chia pudding is a great snack, but it makes a tasty part of breakfast, too.

Combine:

- ¼ cup chia seeds
- 1 cup unsweetened almond or coconut milk
- ⅛ teaspoon cinnamon
- ½ teaspoon vanilla extract

After stirring the ingredients together, let the mixture sit for five minutes. Stir the pudding again and place in the fridge for 20 minutes, or until thick, or overnight.

Optional garnishes:

- dried coconut flakes
- goji berries
- cacao nibs
- sliced apple
- honey

Seaweed. Yes, seaweed. It's an easy grab, and it's telomere friendly. Seaweed snacks, such as SeaSnax, can be found in health-food stores and are made from seaweed sheets lightly roasted in olive oil with a pinch of sea salt. They come in different flavors (we especially like wasabi or onion) and are a great snack for people who crave salty or savory foods. Seaweed is also extremely rich in micronutrients, so enjoy. If you are watching your sodium, choose unsalted sheets of seaweed.

KICK A BAD FOOD HABIT: FIND YOUR MOTIVATION

Adding healthy foods into your diet is great, but it may be even more important to avoid the kind of processed, sugary, junky foods that feed your cellular enemies. Breaking an unhealthy food habit is easier said than done. When people identify their personal motivation for changing a habit, they are more likely to successfully make that change. Here are some of the questions that we ask our research volunteers to help them identify their most meaningful goals when they are trying to make changes to their diet:

- *How is your diet affecting you? Has anyone ever encouraged you to cut down on something? Why? What do you most want to change?*
- *Why exactly are you concerned about how much fast food (or junk food, sugar, or other unhealthy food) you eat? Do you have diabetes or heart disease in your family history? Do you want to lose weight? Are you worried about your telomeres?*
- *What part of you wants to change? What part of you doesn't? What are the things you care most about? How would making this change impact you and people you care about?*

When you've identified your strongest source of motivation, visualize it. If your motivation is to live a long, healthy life, create a vivid image of yourself being active and healthy at age ninety, or cheering at your grandchild's graduation. Do you want to make sure you're around to see your children grow up? Picture yourself dancing at their wedding receptions. Perhaps thinking of those tiny telomeres bravely protecting the future of your chromosomes in billions of cells throughout your body will motivate you! Whenever you're facing temptation, call that image to mind. Our colleague Professor Len Epstein of SUNY Buffalo has found that thinking vividly about the future helps people resist overeating and other impulsive behaviors.[48]

MASTER TIPS FOR RENEWAL:
Science-Based Suggestions for Making Changes That Last

Behavior change is simple, and behavior change is hard. For some people, learning about telomeres is a potent motivator. They imagine their telomeres eroding away—and they are spurred to get more exercise, for example, or adopt a challenge response to stress.

Often, though, motivation is not enough.

The science of behavior change tells us that if you want to make a change, you need to know why you're making the change—but for that change to really last, you need *more* than knowledge. When it comes to change, our minds don't work rationally. We operate largely out of automatic patterns and impulses. Thus the donut instead of the vegetable omelet, the firm resolutions that weaken when it's time to work out or meditate. As a species, we humans have much less personal control than we would like to think. Fortunately, behavioral science tells us how to make changes that stick.

First, identify a change you'd like to make. **The self-test (Telomere Trajectory Quiz) starting on page 162 can help you see where your telomeres need the most help.** Choose one area (such as exercise) and a change you'd like to make (such as starting a walking program). Before you make that change, ask yourself three questions:

1. On a scale of 1 to 10, how do you rate your readiness

to make this change? (A ranking of 1 would mean you're not at all ready; 10 means extremely ready.) If you rank your readiness at 6 or lower, go to the question below to explore what truly motivates you. Then, rate your readiness again. If your readiness score does not increase, choose a different goal.

Many of us engage in behaviors we'd like to change, but we feel stuck or ambivalent. Find one small behavior you feel ready to focus on now. One change leads to another, so focusing on one small change is the right place to put your efforts now. For tough, compulsive behaviors like excessive smoking, drinking, and overeating, you might consult a professional coach or therapist who is an expert in "motivational interviewing," a dialogue that helps people develop clear goals, get past obstacles, and meet the goals.[1]

2. What about this change is meaningful to you? Ask yourself what things are most important to you. Try to tie your goal to your deepest priorities in life, as in "I want to begin a walking program, because I want to be healthy and independent, in my own home, for as long as possible." Or "I want to be active in the lives of my child and grandchildren." The tighter the connection between your goal and your values and priorities, the more likely you are to stick with the change. Choosing intrinsic goals—those related to relationships, enjoyment, and meaning in life—works better than choosing external goals (which tend to be about wealth, fame, or how others view us). They have more lasting power for behavior change and bring us more happiness.[2]

Ask yourself the hard questions in the Renewal Lab for Chapter 10 (page 242) about finding your motivation. Then take a mental snapshot of the answer, an image that represents your motivation. This visual picture is a weapon you can use in those difficult moments when part of you is looking hard for a way out of the new behavior.

3. On a scale of 1 to 10, how confident are you that you can make this change? If you're at a 6 or lower, change your goal to make it smaller and easier to achieve. Identify any obstacles that dragged your rating down and make a realistic plan for overcoming them. Think of obstacles with that "challenge" mind-set—this is an opportunity to bring in some good stress. Another way to increase efficacy and success is to think of a past proud moment when you overcame an obstacle.[3]

Self-efficacy ratings like this are our crystal ball; they've been shown to be one of the biggest predictors of our future behavior. Confidence about whether we can carry out a specific task determines a cascade of events: whether we will even try a new behavior in the first place, whether we will persist at it once we hit obstacles.[4] Get into the self-efficacy positive loop—achieving a small part of our goal boosts our confidence, which carries us to the next step, which boosts our confidence further.

Next, consider whether you're trying to create a new habit or to break an old one. The answer will determine which strategies apply to you.

TIPS FOR CREATING NEW HABITS

Our brains are equipped for automaticity, for making the least effort possible. Make automaticity work for you, not against you. Here's how:

- **Small changes.** Slip into your new habit painlessly, in small doses. If you want to get more sleep, don't try to go to bed an hour earlier each night. That's too hard. Start by going to bed fifteen minutes earlier each night. If that's not doable, start with an even smaller goal: ten minutes, five minutes...whatever feels easy and nonthreatening. From there, you can build slowly toward your goal.
- **Staple it.** Tack your small change onto an activity that's already a routine part of your day.[5] That way, you'll have to think less about when to make the change, and eventually it will become

routine, too. For example, whenever I (Liz) wait for my computer to complete loading my e-mail, it's a trigger for me to do a micro-meditation. For other people, the lunch break is a trigger that it's time to take a walk. Hitching the behavior to an already embedded one helps you stick to your plan.

- **Mornings are green light zones.** Try to schedule your change for the morning. The earlier in the day, the less likely it is that other urgent priorities will nudge your new behavior off your schedule. You may feel stronger determination, which you can visualize as a green light that flashes "DO IT."
- **Don't decide—just do.** When it's time to go to the gym (or make any other change), don't ask yourself "Should I?" Making decisions is exhausting. And in a weak moment the answer may well be "Tomorrow." Just go. Walk there like an unthinking zombie if you have to.
- **Celebrate it.** Have a quick mini-celebration each time you practice your new habit. Consciously say to yourself, "Great!" or "I did it!" or "DONE!" and let yourself feel pride. Or put aside a dollar each time in a collection for some personal indulgence after ten times.

TIPS FOR BREAKING OLD HABITS

Trying to end an old, unwanted habit takes willpower, which is, sadly, a limited resource. Plus, plenty of unhealthy habits make us feel good, at least for a few moments. Sugary foods and drinks, for example, make your brain's reward system light up. We can become neurobiologically dependent on that sugar rush. Breaking the habit takes patience and persistence.

- **Increase your brain's ability to execute your plans.** We are best able to exert control when the brain networks that foster analytical thought are activated. When there is more activity in the prefrontal cortex, some of the more emotional areas in the amygdala are inhibited. Exercise, relaxation meditation, and foods that are high in quality protein promote this optimal mental state (and stress thwarts it).

▪ **Don't try the change when you are feeling depleted.** Sleep loss, low blood sugar, or high emotional stress can deplete you of your willpower. Wait until the conditions are in your favor.[6]

▪ **Shape your surroundings to reduce the number of times you're tempted.** Don't keep sweets, soda, or other reminders of your unwanted habit around the house, and certainly not within sight. Cookies and chips, when they do end up at home, should be out of sight in a high cupboard, not on the kitchen counter in a bowl. You may be able to resist temptation once, but saying no several times a day is exhausting. Your limited supply of willpower may run out. These tips are called stimulus control—we attempt control over our environment as much as we can so we are not surrounded by the tempting stimuli.

▪ **Follow your natural alertness rhythms.** You'll have more energy to stoke your willpower. If you're a night owl, you'll be more able to resist temptation in the evening, and more likely to succumb in the early morning. Plan accordingly. And take healthy snack breaks at your personal low points— the times of day when you tend to feel tired. This will sustain your energy for when you need to draw on your willpower.

Last, there is one strategy that helps almost everyone in every case, regardless of whether you are trying to start or stop something: social support. Ask your family and friends to help support your new goal. Tell them what would be helpful. Turn your accomplices (those who help you do exactly what you don't want to do) into supportive influences or…avoid them! You could find a partner with similar goals to share the journey with you. I (Elissa) would go running less often if I didn't have a running partner relying on me.

To help you think about ways to make small changes throughout your day, we've created the reference "Your Renewed Day" on the next page. It's a time-based table that shows you which routine behaviors can endanger your telomeres. It also suggests substitute actions that are telomere healthy.

Your Renewed Day

Each day you have an opportunity to forestall, maintain, or accelerate the aging of your cells. You can stay in balance or maybe even forestall unnecessary acceleration of biological aging by eating well, getting enough sleep for restoration, being active and maintaining or building fitness, and sustaining yourself through meaningful work, helping others, and social connection.

Or you can do the opposite—consume junk food or too many sweets, get too little sleep, and stay sedentary or decondition the fitness you have. Throw high stress into the mix of a vulnerable body, and you'd have a day of wear and tear on your cells. It's possible that you might even lose a few base pairs of telomere length. We don't *really* know how responsive telomeres are on a daily basis, but we do know that chronic behavior over time has important effects. We can all strive to have more days of renewal rather than wear and tear. Begin by making small changes. There are suggestions for telomere-healthy change throughout the book, and we've created an example of how you can build some of these behaviors into your day. Circle any you might like to try.

We've also included a blank Renewed Day schedule that you can customize with the telomere-healthy changes you'd like to make. You can copy it, or print it from our website, and stick it to your refrigerator or mirror to help remind you of easy ways to promote healthy cell renewal. Fill in several new behaviors you'd like to add to your day. What do you want to say to yourself when you wake up? Do you want to fit in a few minutes of a morning renewal

Your Renewed Day

Time	Telomere-Shortening Behavior	Telomere-Supporting Behavior
Waking up	Anticipatory stress or dread.	**Reappraise your stress response** (page 120). Wake with joy. "I am alive!"
	Mentally rehearse your to-do list. Check phone immediately.	Set an intention for the day. Look forward to any positive aspects.
Early morning	Regret that there's no time for exercise.	Perform a **cardio or interval workout** (page 186). Or do energizing **Qigong** (page 156)
Breakfast	Sausage and bagel.	Oatmeal with fruit; fruit smoothie with yogurt and nut butter; vegetable omelet.
Morning commute	Rush, hostile thoughts, maybe a little road rage.	Practice the **three-minute breathing break** (page 149).
Arrival at work	Play catch-up from the moment you arrive.	Give yourself a ten-minute window of habituation and settling before work begins.
	Anticipate, worry about the workday.	Meet situations as they arise.
Workday	Self-critical thoughts.	Notice your thoughts. Take a **self-compassion break** (page 122) or **manage your eager assistant** (page 123).
	Multitask to deal with work overload.	Focus on one task at a time. (Can you turn off your e-mail and ringer for an hour?)
Lunch	Eat fast food, deli meats.	Enjoy a lunch made from fresh, whole foods.
	Eat quickly.	Practice **mindful eating** (page 222).
		Connect with someone. Have lunch or walk with a partner; text, call, or e-mail someone you have a supportive relationship with.
Afternoon	Give in to cravings for a sugared drink, baked goods, or candy.	**Surf the urge** (page 220). Have a **telomere-friendly snack** (page 240).
		Stretch.
Evening Commute	Ruminate.	**Mentally distance yourself** (page 97).
	Negative mind wandering.	Take a **three-minute breathing break** (page 149).
Dinner	Eat processed food.	Have a whole-foods dinner (see our website for ideas).
	Look at screens.	Give the gift of focused attention to others.
Evening	Run through your evening activities and chores without a break.	**Exercise**, or try a **stress-reducing technique** (page 153).
		Ask, "Did I live my intentions today?"
	Suffer from a head buzzing with the aftereffects of a busy nonstop day.	Review your day; try a **challenge reappraisal** (page 87). Savor the things that made you happy.
		Engage in a **relaxing sleep ritual** (page 203).

My Renewed Day

Waking up	
Early morning	
Breakfast	
Morning Commute or Arrival at Work	
Workday	
Lunch	
Afternoon	
Evening Commute or Arrival at Home	
Dinner	
Evening	

mind–body activity? Think about transitions in the day when you can build in more physical activity, shift your awareness to the moment to promote stress resilience, connect with other people, and add some telomere–healthy foods to your diet.

Just remember that the path to lasting change is traveled one small step at a time.

PART IV

OUTSIDE IN:
THE SOCIAL WORLD
SHAPES YOUR
TELOMERES

CHAPTER ELEVEN

The Places and Faces That Support Our Telomeres

Like the thoughts we think and the food we eat, the factors beyond our skin—our relationships and the neighborhoods we live in—affect telomeres. Communities where people do not trust one another, and where they fear violence, are damaging to telomere health. But neighborhoods that feel safe and look beautiful—with leafy trees and green parks—are related to longer telomeres, no matter what the income and education level of their residents.

When I (Elissa) was a graduate student at Yale, I routinely worked late into the evening. By the time I walked back home from the psychology building, it was dark. I had to pass a church where someone had been murdered a few years earlier, and even though the area was usually quiet when I walked by at around 11:00 p.m., my heart would beat faster. Next, I turned down my street, where the rent was quite affordable on a student stipend. It was a long street, known for occasional muggings. As I walked, I listened carefully for steps behind me. I could feel my heart thump more powerfully. It is a good bet that my blood pressure went up and that glucose was recruited from its stores in my liver, giving me energy to run if needed. Every night, my body and mind mobilized themselves for danger. That experience lasted for just ten minutes each evening.

Imagine how stressful it would feel if the risk was much worse, the duration was longer, and you couldn't afford to move away.

Where we live affects our health. Neighborhoods shape our sense of safety and vigilance, which in turn affects levels of physiological stress, emotional state, and telomere length. Besides violence and lack of safety, there is another critical aspect that makes neighborhoods potent influences on health, and that is the level of "social cohesion"—the glue, the bond, among people who live in the same area. Are your neighbors mutually helpful? Do they trust one another? Do they get along and share values? If you were in need, could you rely on a neighbor?

Social cohesion is not necessarily a product of income or social class. We have friends in a beautiful gated neighborhood, where the houses sit on acres of rolling hills. There are positive signs of social cohesion, including Fourth of July picnics and holiday dances. But there's also mistrust and infighting, and it's not free of crime. It's a neighborhood full of doctors and lawyers, but if you live there you might wake up in the morning to the sound of a police helicopter hovering over your house, searching for an armed robbery suspect who has jumped over the gate. When you take out the trash, a neighbor who is unhappy about your plans for remodeling might accost you. Check your messages, and you could find that your neighbors are in a heated e-mail fight about whether to hire a security patrol and who will pay for it. You may not even know the person who lives next door. There are also neighborhoods that are poor but have people who know each other and have a strong sense of community and trust. While income plays a role, a neighborhood's health goes way beyond income.

People in neighborhoods with low social cohesion and who live in fear of crime have greater cellular aging in comparison to residents of communities that are the most trusting and safe.[1] And in a study in Detroit, Michigan, feeling stuck in your neighborhood—wanting to move but not having the money or opportunity to do so—is also linked to shorter telomeres.[2] In a study based in the Netherlands (known as the NESDA study), 93 percent of the sample rated their neighborhood as generally good (or higher). Despite that these neighborhoods were good environments, the more specific ratings of quality—including levels of vandalism and perception of safety—were associated with telomere length.

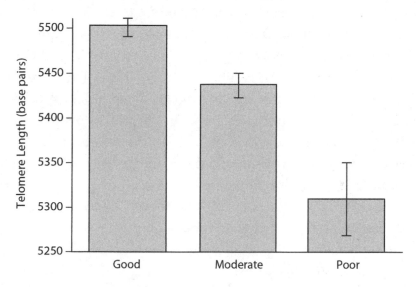

Figure 25: Telomeres and Neighborhood Quality. Here in the NESDA study, residents of neighborhoods with higher quality had significantly longer telomeres than those with moderate or poor quality.[3] This is even after adjusting for age, gender, demographic, community, clinical, and lifestyle characteristics.

Maybe the people living in lower quality neighborhoods have more depression. Was that a possibility that occurred to you? It makes sense that people who live in neighborhoods with low social cohesion would feel worse psychologically. And we know that depressed people have shorter telomeres. The NESDA researchers tested this—and found that the emotional stress of living in an unsafe neighborhood has an effect independent of how depressed or anxious its residents are.[4]

Exactly how does low social cohesion penetrate to your cells and telomeres? One answer has to do with vigilance, that sense of needing to be on high alert to maintain your safety. A group of German scientists performed a fascinating study of vigilance that pitted country folks against city dwellers. People from both groups were invited to take one of those nerve-wracking math tests that are designed to elicit a stress response, the kind where volunteers perform complex mental math while researchers give instant feedback. In this case, the participants were hooked up to a functional MRI, which allowed the researchers to watch their brain activity, and so the researchers gave their feedback over headphones, saying things like "Can you go faster?" or "Error! Please start from the beginning." When the city dwellers took the math test, they had a bigger threat response in their amygdala, a tiny brain structure that is the seat of our fear reactions, than the people who lived in the country.[5] Why the difference between the two groups? Urban living tends to be less stable, more dangerous. People in cities learn to be more vigilant; their bodies and brains are always prepared to mount a big, juicy stress response. This ultrapreparedness is adaptive but is not healthy, and it may be part of the reason that people in threatening social environments have shorter telomeres. (It's interesting, and a source of relief to us city dwellers, that the noise and crowds of urban living are *not* associated with shorter telomeres.)[6]

Some neighborhoods may shorten telomeres because they are places where it's harder to maintain good health habits. For example, people tend to get less sleep when they live in neighborhoods that are disorderly and unsafe, with low social cohesion.[7] Without adequate sleep, your telomeres suffer.

I (Liz), who also lived in New Haven for a time, experienced firsthand another way that a neighborhood can inhibit health habits. Prior to moving to New Haven, I had studied in Cambridge, England. With its flat terrain, Cambridge is a bicycling haven, and I rode my bike everywhere. When I arrived in New Haven to start postdoctoral research at Yale, I noticed that its geography was ideal for cycling. One of the first questions I asked my new lab mates was, "Where can I get a bike to ride to and from work?"

A short silence followed. Someone said, "Well, maybe biking home in the evening is not such a good idea. Bikes tend to be stolen."

Airily I replied that when that had happened in Cambridge, I simply bought a cheap, secondhand bike to replace it. Another silence, and then someone kindly explained that when their colleague said "stolen," he had meant "stolen while the person was still riding the bike." So I didn't bike in New Haven.

Other residents of low-trust, high-crime neighborhoods may draw similar conclusions. It's hard enough for many of us to fit exercise into our schedules, or to resist the call of the easy chair—and for people in unsafe neighborhoods, certain kinds of exercise may be too dangerous to even contemplate. Safety is only one barrier. Another is a lack of parks and other places to exercise. The social and "built environment" of poor neighborhoods stacks the deck against exercise. Without exercise, telomeres are shorter.

LITTERED OR LEAFY?

San Francisco is one of the world's great cities. Its citizens live within walking distance of museums, restaurants, and theaters; they can hike to spectacular views of the hillsides and bay. But as with many cities, parts of San Francisco are also quite dirty. They have a litter problem. This is not good for the residents, especially the young ones. Children who live in a neighborhood that is physically disorderly, with vacant buildings and trash in the streets, have shorter telomeres. The presence of litter or broken glass right outside the house is an especially strong predictor of telomere trouble.[8]

Have you ever been to Hong Kong? There is a stark contrast between the densely populated bustle, bright neon lights, and chaos of Kowloon, the city's downtown, and the sprawling green hills of the New Territories, which are located just outside the city. There, the citizens enjoy trees, parks, and rivers. A 2009 study looked at nine hundred elderly men; some lived in Kowloon and others lived in the lush New Territories. Guess who had shorter telomeres? The men who lived in the city. (The study controlled for social class and health behaviors.) While other factors could be responsible for the association, this study suggests that there's a role for green space in telomere health.[9]

When you're in the thick of a forest, breathing the crisp, clean air, it's not hard to believe that telomeres could benefit from exposure to nature. We're intrigued by this possibility because it's supported by what we already know about nature and a phenomenon called psychological restoration. Being in nature provides a dramatic change in context. It can inspire us with beauty and stillness. It takes us out of small thinking about small problems. It can also relieve us of the moving, blinking, wailing, shuddering, shaking, noisy urban stimuli that keep our arousal systems jacked up. Our brains get a break from registering dozens of simultaneous sensations, any of which could mean danger. Exposure to green spaces is associated with lower stress and healthier regulation of daily cortisol secretions.[10] People in England who are economically deprived have almost double (93 percent) the early mortality of the wealthiest in their country—*except* when they live in neighborhoods surrounded by greenery. Then their relative mortality dips, so that they are only 43 percent more likely to die early from any cause.[11] Nature halves their comparative risk. It's still a sad statistic about poverty, but it leads us to believe that the greenery-telomere connection deserves more exploration.

CAN MONEY BUY LONGER TELOMERES?

You don't need to be rich to have long telomeres, but having enough money for basic needs does help. One study of around two hundred African American children in New Orleans, Louisiana, found that poverty was associated with shorter telomeres.[12] Once you have basic needs met, having more money doesn't seem to help further—there are no consistent relationships between gradients of how much money you make and your telomere length. But with education, there does appear to be a dose-response gradient—the more education, the longer the telomeres.[13] Educational level is one of the most consistent predictors of early disease, so these results aren't too surprising.[14]

In a UK study, occupation mattered more than other indicators of social status: White-collar jobs (versus manual labor) were associated with longer telomere length. This was true even among twins who were raised together but as adults had different occupational status.[15]

CHEMICALS THAT ARE TOXIC TO YOUR TELOMERES

Carbon monoxide: It's odorless, flavorless, and colorless. Deep underground, in coal mines, it can build up without detection, especially after an explosion or fire. At high enough levels, it can cause a miner to asphyxiate. So in the early 1900s, miners began carrying caged canaries down into the mines with them. The miners considered them friends and would sing to the birds as they worked. If there was carbon monoxide in the mine, the canaries would show distress by swaying, reeling, or falling off their perches. The miners would know that the mine was contaminated, and they would either exit or use their breathing apparatuses.[16]

Telomeres are the canaries in our cells. Like those caged birds, telomeres are captive inside our bodies. They are vulnerable to their chemical environment, and their length is an indicator

261

of our lifelong exposure to toxins. Chemicals are like litter in our neighborhoods—they are a part of our physical surroundings. And some are silent poisons.

Let's start with pesticides. So far, seven pesticides have been linked to significantly shorter telomeres in agricultural workers who apply them to crops: alachlor; metolachlor; trifluralin; 2,4–dichloro-phenoxyacetic acid (also known as 2,4-D); permethrin; toxaphene; and DDT.[17] In one study, the greater the cumulative exposure to the pesticides, the shorter the telomeres. It wasn't possible to determine whether one type of pesticide alone was worse or better for telo-meres than any others; the study looked at an aggregate of all seven. Pesticides cause oxidative stress—and oxidative stress, when it accu-mulates, shortens telomeres. This study is supported by another finding, in which agricultural workers who are exposed to a mix-ture of pesticides while working in tobacco fields have been found to have shorter telomeres.[18]

Fortunately some of these chemicals have been banned in parts of the world. For example, there is a worldwide ban on the agricultural use of DDT (although it is still used in India). Once released, how-ever, these chemicals don't just disappear. They live on and on in the food chain ("bioaccumulation"), so any hope to live completely free of chemicals is impossible. There are probably many toxic chemicals in small amounts in each of our cells. They end up in breast milk as well, although the benefits of breast-feeding are thought to far outweigh the exposure to chemicals. Unfortunately, many compounds on the toxic list (alachlor; metolachlor; 2,4-D; permethrin) are still used in farming and gardening and are still being produced at high levels.

Another chemical, cadmium, is a heavy metal with weighty effects on our health. Cadmium is found mostly in cigarette smoke, though we all carry low but potentially toxic levels around in our bodies because of our contact with environmental contributors such as house dust, dirt, the burning of fossil fuels such as coal or oil, and the incineration of municipal waste. Cigarette smoking has been linked to shorter telomeres—no surprise there, given the other dangerous effects of smoking.[19] Some of that relationship is due to cadmium.[20] Smokers have twice the levels of cadmium in their blood compared to nonsmokers.[21] In some countries and industries, people are exposed to cadmium through factory work. In an electronic waste-recycling town in China with known high cadmium pollution, higher cadmium levels in blood were linked to shorter telomeres in placentas.[22] In a large study of U.S. adults, those with the worst cadmium exposure have up to eleven additional years of cellular aging.[23]

Lead is another heavy metal to watch out for. Lead is found in some factories, some older homes, and developing nations that do not yet regulate lead paint and still use leaded gasoline, and it is another potential culprit in telomere shortening. Although the study of the electronic waste-recycling plant found no association between lead levels and telomere length, another study of Chinese battery factory workers who were exposed to lead as part of their work environment found some striking relationships.[24] In this study of 144 workers, almost 60 percent had lead levels high enough to meet the definition of chronic lead poisoning, and they had significantly shorter immune cell telomere length than those with normal or lower lead levels. The only difference between groups was that the group with poisoning had worked at the factory longer. Fortunately, once the lead poisoning was discovered, victims were hospitalized and given treatment (lead chelation therapy). During treatment, urine was assessed for how much lead was excreted, a measure called the "total body burden" of lead. Body burden indicates long-term lead exposure. The greater the body burden of lead, the shorter the telomeres. The correlation was .70, which is very high (the highest a correlation can be is 1). This relationship was so strong that the usual relationships of telomere length with age, sex, smoking, and obesity were not detectable in those exposed to lead. Lead exposure overrode all these factors.[25]

While severe occupational hazards have the strongest effects, it is alarming that households can also carry genotoxic hazards. Older homes may still have lead paint, which can be a danger if the paint is peeling. Many cities still use lead pipes, and the lead can travel into our homes and drinking water. Consider the tragic and shameful crisis in Flint, Michigan, in which the water supply is so corrosive that lead was leached from the pipes. The water became highly contaminated—and so did the residents' blood. While this disturbing drama has unfolded publicly on our screens, the same problem is silently taking place in many other cities that use old pipes. Particularly troublesome is that children are more sensitive to lead than adults. In one study, eight-year-old children with lead exposure had telomeres that were shorter than those of children without lead exposure.[26]

One category of chemicals, **polycyclic aromatic hydrocarbons (PAHs)**, is airborne, which makes it especially hard to avoid. PAHs are by-products of combustion and can be breathed in from fumes from cigarette and tobacco smoke, coal and coal tar, gas stoves, wildfires, hazardous waste, asphalt, and traffic pollution. You can also be exposed to PAHs if you eat foods grown in affected soil or that have been cooked on a grill. Beware. Higher exposure to PAHs has been shown to be associated with shorter telomere length in several studies.[27] An investigation of PAHs offered a caution for pregnant women: the closer a pregnant mother lived to a major roadway, and the fewer trees and plants in her neighborhood (which can reduce air pollution levels), the shorter were the telomeres of her placenta, on average.[28]

CHEMICALS, CANCER, AND *LONGER* TELOMERES

Some chemicals are associated with *longer* telomeres. This may sound good, but remember that very long telomeres in some cases are associated with uncontrolled cell growth—in other words, cancer. So when genotoxic chemicals get into our bodies, we are more likely to get mutations and cancerous cells, and if the telomeres of those cells are long, they are more likely to divide and divide and divide into cancerous tumors. This is one reason we are so concerned about the widespread use and marketing of supplements and other products that claim to lengthen your telomeres.

We are concerned that chemical exposures and telomerase activating supplements may damage cells, or that they may increase telomerase and change telomeres in radical or inappropriate ways that our bodies have not learned to cope with. But when you practice naturally healthy habits such as stress management, exercise, good nutrition, and good sleep, your telomerase efficiency increases slowly, steadily, and over time. This natural process protects and maintains your telomeres. In some cases, lifestyle changes may even help your telomeres grow a bit longer, but in a way that won't trigger uncontrolled cell growth. Healthy lifestyle factors that have been correlated with longer telomeres have *never* been shown to increase cancer risk. Lifestyle changes influence telomeres through mechanisms that are different from and safer than chemical exposures or supplements.

Which chemicals might unnaturally lengthen telomeres too much? Exposure to **dioxins and furans** (toxic by-products that are released through various industrial processes and that are commonly found in animal products), **arsenic** (common in drinking water and some foods), **airborne particulate matter**, **benzene** (exposure occurs via tobacco smoke as well as gasoline and other petroleum products), and **polychlorinated biphenyls** (or PCBs, a class of banned compounds that is still found in some high-fat animal products) is associated with longer telomere length.[29] What is so interesting is that some of these chemicals have also been linked to cancer risks. Some have been linked to higher rates of cancer in animals; others have been studied in labs, where heavy doses are put into cells and create cancer-promoting molecular changes. It is possible that chemicals can both create fertile ground for mutations and cancerous cells, and create high telomerase or longer telomere length, promoting greater likelihood the cancerous cells might be replicated. We speculate that telomeres thus may be one link in the chemical–cancer relationship.

To put this into perspective, the American Association for Cancer Research Cancer Progress Report of 2014 informs us that 33 percent of the relative contribution to overall risks of developing cancer is from tobacco use alone, and about 10 percent is attributable to occupational and environmental exposures to pollutants.[30] But that low percentage is for the United States; it is not known how much higher it is in countries and regions of the world where environmental pollution and exposures at work are much less well controlled. Furthermore, a 10 percent increase in risk may seem small, but since there are over 1.6 million new cases of cancers every year in the United States alone, that 10 percent translates into 160,000 new

Telomere Toxins

Chemicals Linked to Shorter Telomeres	Chemicals Linked to Longer Telomeres (Long telomeres in these conditions indicate a possible risk of uncontrolled cell growth and some forms of cancer.)
Heavy metals, such as cadmium and lead	Dioxins and furans Arsenic Particulate matter Benzene PCBs
Agricultural pesticides and lawn products: alachlor metolachlor trifluralin 2,4-dichlorophenoxyacetic acid (also known as 2,4-D) permethrin Mostly no longer produced but still present in the environment: toxaphene DDT	
Polycyclic aromatic hydrocarbons (PAHs)	

cancer cases per year. Think about it. Every year, 160,000 additional people and their families have their lives irrevocably changed by a diagnosis of cancer. And that's just in the United States; the World Health Organization estimates that there are 14.2 million new cases of cancer around the globe each year, so we could estimate that 1.4 million new cases of cancer each year come from environmental pollution.[31]

PROTECT YOURSELF

What can you do? More research is needed to fully understand the connection between these chemicals and cell damage, but in the meantime it is reasonable to take all the precautions you can. I have always had a preference to use natural products—but only when it was convenient for me to buy them. After realizing that so many of our household cleaners and cosmetics contained genotoxic and telomere-damaging chemicals, I now actively seek out natural products.

You may also want to change the way you eat and drink. Arsenic is naturally found in wells and groundwater, so you can either have your water tested or use a filter. Avoid plastic drinking bottles and cookware. Even BPA (Bisphenol A)–free plastic bottles may not be free of other harmful chemicals. BPA substitutes may be as unsafe; they just haven't been studied to the same extent (plus, we may soon have more plastic in the ocean than fish if we don't reduce our reliance on plastic bottles). Try not to microwave plastics, even the ones that say they are microwavable. It's true that microwaveable plastics won't warp when you heat them, but there are no promises that you won't get a dose of plastics in your food.

How can you reduce your exposure to smoke, air, and traffic pollution? Avoid living near major roadways if possible. Don't smoke (yet another good reason to quit), and avoid passive smoke. Greenery—trees, green space, and even house plants—can help reduce the levels of air pollutants inside your home and in a city, including volatile organic compounds. There is no direct evidence that living with more plants leads to longer telomeres, but there are correlations to suggest that increasing your exposure to greenery can be protective. Try to walk in parks, plant trees, and support urban forestry.

For more ways to protect yourself, see the Renewal Lab on page 276.

FRIENDS AND LOVERS

Long ago, when most of humanity lived in tribes, each group would delegate a few of its members to keep watch at night. The folks on watch would remain alert for fires, enemies, and predatory animals, and everyone else could sleep soundly, knowing that they were protected. In those perilous days, belonging to a group was a way to ensure your safety. If you couldn't trust your night watchmen, you weren't going to get your much-needed sleep. Our ancestors' version of poor social capital and lack of trust!

Flash-forward to contemporary life. When you lie down in your bed at night, you probably don't worry very much about panthers dropping on you from above, or enemy warriors skulking behind the drapes. Nevertheless, the human brain hasn't changed much since tribal days. We're still wired to need someone around who "has our backs." Feeling connected to others is one of the most basic human needs. Social connection is still one of the most effective ways to soften the danger signal; its absence will amplify it. That's why it feels so good to belong to a cohesive group. It feels good to be in connection with others—to give or receive advice, borrow or lend something, work together, or share tears and feel understood. People with relationships that allow for this kind of mutual support tend to have better health, whereas people who are socially isolated are more stress reactive and depressed, and are more likely to die earlier.[32]

In animal research, even rats, who are social animals, suffer when they are caged solo. Little did we know how stressful isolation is for this social animal. Now we know that when rats are caged alone, they don't receive the safety signals from being in close proximity to others and feel more stressed out. They get three times more mammary tumors than the rats who live in a group.[33] The rats' telomeres weren't measured. But a similar experiment found that parrots caged alone have faster telomere shortening than when they are with a mate.[34]

Aside from my bicycling disappointment, I (Liz) was generally happy as a postdoctoral fellow at Yale. But as it became time to think about finding a job, I began to worry. I'd wake up in the night in a cold sweat of anxiety, wondering how on earth I would ever become employed. One of the hurdles I had to overcome was preparing a job seminar, a lecture that I'd deliver when interviewing for academic positions. Feeling insecure, I overdid it. Desperate to convince a skeptical world about the validity of my scientific conclusions, I poured every bit of my data into the text. When I practiced the talk in front of my colleagues, the reaction was...muted, to say the least. The lecture was so dense that it was unintelligible. I went back to my shared office and succumbed to despairing tears. The head of the lab, Joe Gall, came by and offered kind, encouraging words. That helped. Then Diane Juricek (later Lavett), dropped in. Diane was a visiting junior professor working in a neighboring lab, and she and I shared group meetings and lunch tables. Diane volunteered to help me work my talk into shape, taking out the excessive quantities of data description and forming it into a more coherent whole. Then she helped me rehearse the lecture in the big, old-fashioned hall near the building where we worked. This enormous generosity to a younger, less experienced colleague—Diane didn't even know me well—made a huge impression on me. I realized what an academic scientific community could be about.

At the time, I was simply grateful for Diane's help. I didn't know then that my cells were likely responding to the support. Good friends are like the trusted night watchmen; when they're around, your telomeres are more protected.[35] Your cells beam out fewer C-reactive proteins (CRPs), proinflammatory signals that are considered a risk factor for heart disease when they appear in high levels. [36]

Do you have someone in your life who is close to you but also causes unease? About half of all relationships have positive qualities with less helpful interactions, what researcher Bert Uchino calls "mixed relationships." Unfortunately, having more of these mixed-quality relationships is related to shorter telomeres.[37] (Women with mixed friendships have telomeres that are shorter; both women and men have shorter telomeres when the mixed relationship is with a parent.) That makes sense. These mixed relationships are characterized by friends who don't always know how to offer support. It's stressful when a friend misunderstands your problems or doesn't give you the kind of support you really want. (For example, a friend may decide you need a long pep talk when what you really need is a shoulder to cry on.)

Marriages come in all flavors, and the better the quality of the marriage the better the health benefits, although these are what we consider statistically small effects.[38] Put someone from a satisfying marriage into a difficult situation, and they'll likely have more resilient patterns of stress reactivity.[39] Happily married people also have a lower risk of early mortality. Marriage quality has not been examined with telomere length yet, but we do know that married people, or people living with a partner, have longer telomeres.[40] (This was a surprise finding from a genetic study of 20,000 people, and the relationship was stronger in the older couples.)[41]

Sexual intimacy in marriage may matter for telomeres, too. In one of our recent studies, we asked married couples if they had been physically intimate during the previous week. Those who answered yes tended to have longer telomeres. This finding applied to both women and men. This effect could not be explained away by the quality of the relationship or other factors relating to health. Sexual activity declines less in older couples than stereotypes would have us believe. Around half of married thirty- to forty-year-olds, and 35 percent of sixty- to seventy-year-olds, engage in sexual activity anywhere between weekly and a few times a month. Many couples remain sexually active well into their eighties.[42]

Couples in unhappy relationships, on the other hand, suffer from a high level of "permeability"—they pick up on each other's stress and negative moods. If one spouse's cortisol rises during a fight, so does the cortisol of the other spouse.[43] If one spouse wakes up in the morning with a big stress response, the other is more likely to as well.[44] Both are operating at a high level of distress, leaving no one in the relationship who can put the brakes on the tension, no one who can say, "Whoa, wait. I see you're upset. Let's take a breath here and talk about it, before things get out of control." It's easy to imagine that these relationships are wearying and depleting. Our physiological responses moment to moment are more synced with our partner's than we may realize. For example, in one study examining couples having both positive and stressful discussions in the lab, heart rate variability followed the pattern of the other partner with a slight lag.[45] We suspect the next generation of research on relationships is going to reveal many more ways that we are connected physiologically to people we are close to.

RACIAL DISCRIMINATION AND TELOMERES

One Sunday morning, thirteen-year-old Richard decided to attend a friend's church in a town a few miles outside his Midwestern city. "I guess there weren't too many black people at the church to begin with," says Richard, who is black, "and I guess the two of us were dressed differently." Richard sat quietly with his friend in the reception area, waiting for the service to begin. As a minister's son, Richard had grown up in churches; he had always known them to be places where he felt welcomed, accepted, and safe. Then a woman who ran one of the church programs walked up to them.

"What are you guys doing here?" she asked in a pointed tone. They explained that they were planning to attend the Sunday service.

"I don't think you're in the right place," she said, and told them to leave.

"I felt so uncomfortable," Richard recalls now of the incident. "She kind of convinced me that I actually didn't belong there. We ended up leaving the church and not going to the service. I almost couldn't believe it had happened, but then my dad e-mailed the minister, and he confirmed that the details were correct. The woman really had said all those things. It seems inhuman that people would go to such lengths to get me out of a church."

Discrimination is a serious form of social stress. Discriminatory acts of any kind, whether they target sexual orientation, gender, ethnicity, race, or age, are toxic. Here we're zeroing in on race, because that is where telomere research has been focused. In the United States, being black, and especially being a black man, means you are more vulnerable to encounters like the one Richard had. He says, "When I talk about racism, people think I mean something extreme. But it can be small, like when a white mother grabs her child's hand when an African American teenager walks by. It hurts."

Unfortunately, racism in its extreme form is also common. African American men are more likely to be accused of a crime and attacked by the police. Now, given dashboard cameras and iPhones, we see these painful scenes on our TVs often. Police officers are like every other human: they make automatic judgments about people from a visibly different social group. Meet someone new, and within milliseconds your brain is assessing whether the person is "same" or "other." Does the person look like me? Is he or she familiar in some way? When the answers are yes, we instinctively judge the person as being warmer, more friendly, and more trustworthy. When the person seems different from us, our brains judge them to be potentially hostile and dangerous.[46]

As we said, this is an instant, unconscious reaction. It is a reason that skin color can set off automatic judgments—but it's not an excuse for acting on those judgments. All of us have to consciously work to counter this internal bias. Tim Parrish, who was raised in a close-knit but racist community in Louisiana during the 1960s and '70s, is now an adult in his fifties. Tim, who is white, admits that sometimes racist assumptions pop into his head, even though he doesn't want them there and no longer believes them to be true. But, as Parrish explained in a opinion piece for the New York *Daily News*, "What gets injected into us as beliefs is not fully our choice. What is our choice is to be constantly vigilant, to deconstruct the assumptions we make, to combat impulses we may have that lead us in the direction of thinking we are somehow the generalized victim and the more civilized color."[47] In a relatively low-stress situation, this mental work against bias may be easier to accomplish than in fast-moving, tense situations. It is a reason that "driving while black" means you are more likely to be pulled over. If you're a black man in America and your behavior seems dangerous, or is hard to interpret, you are more likely to be shot. My (Elissa's) husband, Jack Glaser, a public policy professor at the University of California, Berkeley, works on training police officers to reduce racial bias. He is helping to adapt police procedures so that they are not so heavily influenced by automatic judgments that can lead to racial discrimination. Although he and his academic colleagues categorize this as policy work, I think of it as stress reduction at a societal level, and possibly telomere relevant!

The amount of suffering people experience when they are targets of discrimination runs very deep. African Americans tend to develop more chronic diseases of aging. For example, they have higher rates of stroke than other racial and ethnic populations in the United States. Poor health behaviors, poverty, and lack of access to good medical care may explain some of these statistics, but so does a lifetime of greater stress exposure. In a study of older adults, African Americans who experience more daily discrimination had shorter telomeres, and this relationship did not hold up for whites (who experience less discrimination in the first place).[48] But this is probably not a simple, straightforward relationship; it may depend on attitudes we are not even aware of within ourselves.

David Chae at the University of Maryland performed a fascinating study that looked at low-income, young black men living in San Francisco. He wanted to know what happens to telomeres when people internalize the common societal bias, meaning that they come to believe society's negative opinions of them at an unconscious level. Discrimination alone had a weak effect. The men who had been discriminated against *and* internalized the disparaging cultural attitudes toward blacks had shorter telomeres.[49] Internalized bias toward blacks is tested by a computer task using reaction times to see how quickly people pair the word *black* with negative words. You can test your own bias at this website: https://implicit.harvard.edu/implicit/user /agg/blindspot/indexrk.htm. Just don't berate yourself for having automatic biases; most of us do. We suspect we will see more data on discrimination and telomeres in the coming years.

Knowing how places and faces affect your telomere health can be reassuring, or it can be unsettling. It all depends on your situation— where you live, the quality of your relationships, and how much you've internalized discrimination (discrimination toward any aspect of yourself—race, sex, sexual orientation, age, disability). But *all* of us can take steps to reduce toxic exposures, improve the health of our neighborhoods, become more aware of our biases toward other groups, and create positive social connections. The Renewal Lab at the end of this chapter offers some ways to get started.

TELOMERE TIPS

- We are interconnected in ways we cannot see, and telomeres reveal these relationships.
- We are affected by the toxic stress of discrimination.
- We are affected by toxic chemicals.
- We are affected in more subtle ways, by how we feel in our neighborhood, by the abundance of green plants and trees nearby, and by the emotional and physiological states of those around us.
- When we know how we are affected by our surroundings, we can begin to create healthful, supportive environments in our homes and our neighborhoods.

RENEWAL LAB

REDUCE YOUR TOXIC EXPOSURES

We've already described some basic precautions against plastics and pollution that could shorten—or dangerously lengthen—your telomeres. Here are some more advanced moves:

- **Eat less animal and dairy fat.** The fatty parts of meat are where certain bioaccumulative compounds collect and concentrate. The same goes for the fat in large, long-lived fish, except that there is a balancing issue to weigh. Fatty fish such as salmon and tuna also contain omega-3s, which are good for your telomeres, so eat in moderation.
- **Think about the air when you turn up the heat with meat.** If you cook meat on a grill or on a gas stove, use ventilation. Try to avoid exposing the food directly to open flames, and try not to eat the charred portions, no matter how tasty they are. A good idea for any food.
- **Avoid pesticides in your produce.** Eat foods that are free of pesticides when possible; at the very least, wash your produce thoroughly before consuming. Purchase organic fruits, vegetables, and meats, or grow your own. Consider growing lettuce, basil, herbs, and tomatoes in pots on your balcony. Safe alternatives for dealing with pesky critters can be found here: http://www.pesticide.org/pests_and_alternatives.

- **Use housecleaning products containing natural ingredients.** You can make many of these products yourself. We like the housecleaning "recipes" found at http://chemical -free-living.com/chemical-free-cleaning.html.
- **Find safe personal-care products.** Carefully read the labels on personal-care products such as soap, shampoo, and makeup. You can also visit http://www.ewg.org/skindeep to identify which chemicals are in your beauty products. When in doubt, buy products that are organic or all natural.
- **Buy nontoxic house paints.** Avoid paints that contain cadmium, lead, or benzene.
- **Go greener.** Buy more house plants: two per one hundred square feet is ideal for keeping your air filtered. Good choices include philodendrons, Boston ferns, peace lilies, and English ivy.
- **Support urban forestry** with your money or your labor. Green spaces offer so many benefits to mind and body, as well as to healthy communities. **One newer idea can be considered in dense urban megacities, where one cannot plant enough trees to rid the air of toxins.** If you live in a city, consider lobbying your municipal government to install air-purifying billboards. These billboards do the work of 1,200 trees, cleaning a space of up to 100,000 cubic meters by removing pollutants such as dust particles and metals from the air.[50]
- **Stay up to date about toxic products by downloading the "Detox Me" app by Silent Spring:** http://www .silentspring.org/.

INCREASE THE HEALTH OF YOUR NEIGHBORHOOD: SMALL CHANGES ADD UP

To brighten a corner of your own neighborhood, follow the example of our San Francisco neighbors and place a few benches and tables on a bare cement sidewalk, along with a little greenery. These "parklets" attract neighbors and promote socializing and peaceful lounging. Or consider one of these:

- **Add art.** A mural or even a beautiful poster can infuse a drab area with hope, truth, faith, and positivity. Residents in a Seattle neighborhood painted boarded-up shop windows with pictures of the businesses they hoped to attract: an ice-cream parlor, a dance studio, a bookshop, and so on. The paintings helped entrepreneurs see the potential of the neighborhood. They brought their small businesses to the block, revitalizing the area and bringing economic growth to the community.[51]

- **Get greener**, especially if you are a city dweller. More green space in a neighborhood is associated with lower cortisol and lower rates of depression and anxiety.[52] Turn a vacant lot into a sustainable community fruit or vegetable patch, or plant trees and flowers in a small park space. "Greening" a vacant lot has been associated with a decrease in gun violence and vandalism and an increase in the residents' general feeling of safety.[53]

- **Warm your neighborhood tone.** Social capital is an invaluable resource that predicts good health. It's defined by the level of community engagement and positive activities and resources that exist in a neighborhood, and one of its most important ingredients is trust. So be the one to make the first move. Cook or bake a little extra and drop it off at your neighbors' house on a small plate. Share vegetables or flowers from your garden. Help out by shoveling snow, giving a ride to an elderly person, or starting a neighborhood watch. Leave a welcoming note for newcomers to the neighborhood, or plan a block party. You could also join the trend of opening a Little Free Library in front of your house by putting out a wooden cabinet where books are shared. (See https://Little FreeLibrary.org.)

▪ **Smiles matter.** Acknowledge people you pass on the street. As social animals, we are exquisitely sensitive to social cues, noticing signs of acceptance and especially rejection. Each day, we interact with strangers or acquaintances, and we can either feel separate from them—or we can connect with them in a small way that has a positive effect. Give people an "air gaze" (looking past the face, with no eye contact) and they will tend to feel more disconnected from others. Give them a smile and eye contact, and they feel more connected.[54] Plus, when people are given a smile, they are more likely to help someone else in their next moments.[55]

STRENGTHEN YOUR CLOSE RELATIONSHIPS

Then there are the people we wake up to almost every day—our family, and colleagues we work with. The quality of these relationships is important to our health. It is easy to be neutral, to take those we see all the time for granted. Investigate what it is like to really acknowledge your close ones in a significant way:

▪ Show gratitude and appreciation. Say, "Thanks for doing the dishes" or "Thanks for supporting me at the meeting."
▪ Be present. This means not looking at a screen or around the room. Give your full, sincere attention. That is a gift you can give another person, and it doesn't cost a dime.
▪ Hug or touch your loved ones more often. Touch releases oxytocin.

CHAPTER TWELVE

Pregnancy: Cellular Aging Begins in the Womb

When I (Liz) found out I was pregnant, I instantly felt protective toward my tiny unborn baby. On getting back the test results, I immediately stopped smoking. Luckily, I had been smoking only lightly, a few cigarettes a day at most. I found the transition easy to make—especially because I was so concerned about the baby's wellbeing. I have never smoked again. I also became very interested in what to eat. Listening to my obstetrician and her team, I paid attention to getting nutrients from foods (like fish, chicken, and leafy greens). I also took the micronutrient supplements for iron and vitamins they recommended.

Now, many years later, we have a much deeper understanding of how a mother's nutrition and health status affects her developing baby. We are also learning what happens to a baby's telomeres in the womb. Little did I suspect, all those years ago, that my decisions may have helped to protect my baby's telomeres. Or, more spectacular, that the choices I made—and the events that had happened to me years before the baby was born—might even have affected the starting point of my son's telomeres.

Telomeres continue to be shaped throughout adulthood. Our choices can make our telomeres healthier, or they can hasten their shortening. But long before we're old enough to make decisions about what to eat or how much to exercise, and before chronic stress starts to threaten our DNA base pairs, we begin life with an initial telomere setting. Some of us arrive in this world with shorter telomeres. Some of us are lucky to have longer ones.

As you can imagine, telomere length at birth is influenced by genetics, but that is not the whole story. We are learning astonishing things about how parents can shape their children's telomeres—before those children are even born. And this matters—the telomere length at birth and early childhood is a major predictor of what we have left as we become adults.[1] The nutrients that a pregnant mother consumes, and the level of stress she experiences, can influence her baby's telomere length. It is even possible that parents' life histories can affect telomere length in the next generation. In a sentence: Aging begins in utero.

PARENTS CAN PASS THEIR SHORTENED TELOMERES TO THEIR CHILDREN

Chloe, now age nineteen, became pregnant two years ago. Without much support or understanding from her parents, she left home and moved in with a friend. To help pay her share of the rent, she dropped out of high school and began working a minimum-wage retail job. Despite her difficult circumstances, Chloe has been determined to give her baby a good start in life. While she was pregnant, she did her best to get prenatal care. She took the prenatal vitamins she was prescribed, even though she says they made her sick. When her son was born, Chloe pledged that he would always, always feel loved.

Chloe is determined to give her child what she didn't have— better health and greater satisfaction—and to help lift him as part of the next generation. But there is shocking evidence that Chloe's low education level could have indirectly shaped her baby's telomeres— *while he was still in the womb*. Babies whose mothers never completed high school have shorter telomeres in their cord blood compared to those whose moms had a high school diploma—meaning that they have shorter telomeres from the first day of their lives.[2] Older children whose parents have lower levels of education also have shorter telomeres.[3] These findings are based on studies that controlled for other factors that could have influenced the results, such as, in the baby study, whether their babies had a low birth weight.

Let this sink in for a moment, because the implications, if borne out in subsequent studies, are revolutionary. How could a parent's education level affect the telomeres of her developing baby?

The answer is that telomeres are transgenerational. Parents can, of course, hand down *genes* that affect telomere length. But the really profound message is that parents have a second way of transmitting telomere length, known as *direct transmission*. Because of direct transmission, both parents' telomeres—at whatever length they are at the time of conception in the egg and sperm—are passed to the developing baby (a form of epigenetics).

Direct transmission of telomere length was discovered when researchers were investigating telomere syndromes. Telomere syndromes, as you'll remember, are genetic disorders that lead to hyper-accelerated aging. Their victims have extremely short telomeres. People with telomere syndromes—think of Robin in an earlier chapter—often watch their hair turn gray while still in their teens. Their bones can become fragile, or their lungs can stop working properly, or they can develop certain cancers. In other words, they make an early and tragically dramatic entrance into the diseasespan. Telomere syndromes are inherited, caused when parents pass a single mutated telomere-related gene down to their children.

But there was a mystery. Some children in these families are lucky enough not to inherit the bad gene that causes the telomere syndrome. You'd think these children would escape premature cellular aging, right? Yet some children without the bad gene *still* showed mild to moderate signs of early aging—not as severe as what you might find in a full-blown telomere syndrome, but beyond what is normal, such as very early graying. Researchers decided to measure these children's telomeres and found that their telomeres were, in fact, unusually short. The children had escaped the gene that causes inherited telomere syndrome, but somehow they had still been born with short telomeres that persisted in being short. These children had received short telomeres from their parents—but not through inheritance of a bad gene. Although the children were growing up with normal telomere maintenance genes, because their telomeres had started off so short, the telomeres simply could not be replenished fast enough to catch up and attain normal lengths.[4]

How can this be? How can children receive short telomeres from their parents, if not through genes? The answer, once you know it, is immediately obvious. It turns out that parents can directly transmit their telomere length to the child in the womb. Here's how it happens: A baby begins with a mother's egg, fertilized by the father's sperm. That egg contains chromosomes. Those chromosomes contain genetic material, of course. That's how genetic material is passed down to the baby. But the material of the chromosomes of the fertilized egg also includes the telomeres at their ends. Because the baby is made from the egg, the baby receives those telomeres directly—of whatever length they are at that time. **If the mother's telomeres are short throughout her body (including those in the egg) when she contributes the egg, the baby's telomeres will be short, too. They'll be short from the moment the baby starts developing.** That's how children without the bad gene received shorter telomeres. And this suggests that if the mother has been exposed to life factors that have shortened her telomeres, she can pass those shortened telomeres directly to her baby. On the other hand, a mother who has been able to keep her telomeres robust will pass her stable, healthy telomeres to the growing child.

What does Dad contribute? Upon fertilization of the egg, the chromosomes that come in from the dad via the sperm join the chromosomes from the mother. The sperm, like the egg, also bears its own telomeres that are directly transmitted to the developing baby. The research to date suggests a father *can* directly transmit short telomeres, but just not to the extent that a mother with very short telomeres would. In a new study of 490 newborns and parents, babies' cord blood telomeres were more related to their moms' telomere length than their dads', but they are both clearly influential.[5]

So far, there have been only a few studies that look at direct transmission of telomere length in humans. That would involve measuring both the genetics for telomeres, and the telomeres themselves, so we can separate out the effects of genetics from life experience. Those studies have all been focused on families with telomere syndromes.[6] But we and other researchers suspect that it happens in the normal population, too.[7] As you're about to see, the science of direct transmission suggests a way that poverty and disadvantage may have effects that echo through the generations.

CAN SOCIAL DISADVANTAGE BE PASSED DOWN THROUGH THE GENERATIONS?

Did your parents suffer from prolonged, extreme stress before you were born? Were they poor, or did they live in a dangerous neighborhood? You already know that the way that your parents lived before you were conceived probably affected their telomeres. It may have also affected *yours*. If your parents' telomeres were shortened by chronic stress, poverty, unsafe neighborhoods, chemical exposures, or other factors, they may have passed their shortened telomeres to you through direct transmission in the womb. There is even the possibility that you, in turn, could pass those shortened telomeres to your own children.

Direct transmission has strong and chilling implications for all of us who care about future generations. It raises a controversial idea. In our view, the evidence from telomere syndrome families suggests that it is possible for the effects of social disadvantage to accumulate over the generations. We can already see the pattern in large epidemiological studies: Social disadvantage is associated with poverty, worse health—and shorter telomeres. Parents whose telomeres are shortened by this disadvantage may directly transmit those shorter telomeres to their babies in utero. Those children will be born a step behind, or base pairs behind, with telomeres shortened by their parents' life circumstances. Now imagine that as these children grow up, they are also exposed to poverty and stress. Their telomeres, already shortened, will erode even further. In a downward spiral, each generation directly transmits its ever-shortening telomeres to the next. And each new baby could be born further and further behind, with cells that are more and more vulnerable to premature aging and an early diseasespan. This pattern is exactly what happens in the rare telomere syndrome families: With each successive generation, the progressively shorter and shorter telomeres cause earlier and worse disease impacts than in the generation before.

From the first moments of life, telomeres may be a measure of social and health inequalities. They may help explain the disparity among different postal codes in the United States. People living in certain ZIP codes that represent wealthier areas have life expectancies up to ten years longer than people in other ZIP codes that cover poorer areas. This difference has often been explained by risky behaviors or exposure to violence. But the actual biology of babies

Figure 26: Aging at Birth? "Mom, what happened to the level playing field?" Babies are born with short telomeres depending on their mothers' genes but also their mothers' biological health, level of stress, and, possibly, level of education.

born into these neighborhoods may also be different. Tragically, a neighborhood's health challenges may be compounded from generation to generation. But biology is not destiny; there are many things we can do to maintain our telomeres through our own lifetime.

NUTRITION IN PREGNANCY: FEEDING A BABY'S TELOMERES

"You're eating for two now." Pregnant women hear this advice all the time. It's true: A developing baby gets its calories and nutrition from the food the mother eats (and it's not true the mom needs to eat twice as much). Now it appears that what a pregnant woman eats can also affect her baby's telomeres. Here, we'll look at the nutrients that have been connected to telomere length in utero.

Protein

Animal research suggests that modest protein deprivation in pregnancy causes accelerated telomere shortening in the offspring in a number of tissues, including the reproductive tract, and can lead to earlier mortality.[8] When a mother rat is fed a low-protein diet during pregnancy, her daughters have shorter telomeres in their ovaries. They also have more oxidative stress and higher mitochondria copy numbers, suggesting that the cells are under high stress and to cope they are rapidly producing more mitochondria.[9]

The damage can even travel to the third generation. When the researchers looked at the granddaughter rats, they found that their ovaries had undergone accelerated tissue aging. They had more oxidative stress, higher mitochondria copy numbers, and shorter telomeres in their ovaries. The granddaughters were victims of early cellular aging, all as a result of a low-protein diet two generations earlier.[10]

Co-enzyme Q

There is strong evidence in humans and animal models that maternal malnourishment during pregnancy leads to increased risk of heart disease in the offspring. If a pregnant mother doesn't get enough to eat, or isn't adequately nourished, her child may be born at a low birth weight. Often, there's a rebound effect, with the underweight baby playing a game of catch-up that eventually leads to overeating and obesity. Babies born at a low birth weight carry an increased risk for cardiovascular disease as they get older, and babies who experience this postnatal rebound of rapid weight gain have a risk that's even higher.

As we said, this scenario links maternal malnourishment to heart disease—and one of the links in the chain may be telomere shortening. Rat pups that are born to mothers who don't get enough protein tend, like their human counterparts, to have a low birth weight. And just like human babies, they often experience a later rebound of weight gain. Susan Ozanne at the University of Cambridge has found that these rat pups have shorter telomeres in the cells of several organs, including the heart aorta. They also have lower levels of an enzyme known as CoQ (ubiquinone). CoQ is a natural antioxidant that is found mostly in our mitochondria, which play a role in energy production. A CoQ deficit has been associated with faster aging of the cardiovascular system. But when the pups' diets were supplemented with CoQ, the negative effects of protein deprivation were wiped out, including the effects on telomeres.[11] Ozanne and her colleagues concluded that "early intervention with CoQ in at-risk individuals may be a cost-effective and safe way of reducing the global burden of [cardiovascular diseases]."

Of course, it's a long leap from rat to human. What's good for one may not be good for another. Even in rats, we don't know whether the benefits are restricted to pups whose mothers were deprived of protein. CoQ should be put on the list of nutrients for further study of their potentially positive effects on telomeres. If those benefits exist, they could be harnessed for babies of mothers who had inadequate nutrition during pregnancy, or even for adults who are at risk for heart disease. Note that no studies we are aware of have used CoQ during pregnancy, or examined the safety, and thus we are not recommending it.

Folate

Folate, a B vitamin, is another crucial nutrient during pregnancy. You probably know that folate decreases the risk of spina bifida, a birth defect, but it also prevents DNA damage by shielding the regions of chromosomes known as the centromere (all the way in the middle of the chromosome) and the subtelomere (the chromosome region just inside and next to the telomere). When folate levels drop too low, the DNA becomes hypomethylated (losing its epigenetic marks), and the telomeres become too short—or, in a few cases, abnormally elongated.[12] Low folate levels also cause an unstable chemical, uracil, to be incorporated into the DNA, and possibly into the telomere itself, perhaps causing temporary elongation.

Babies of mothers who have inadequate folate during pregnancy have shorter telomeres, further pointing to folate as vital for optimal telomere maintenance.[13] And gene variants that make it harder for the body to use folate are associated with shorter telomeres in some studies.[14]

The U.S. Department of Health and Human Services recommends that pregnant women get between 400 and 800 micrograms of folate daily.[15] Just don't assume that getting even *more* folate is better. At least one study hints that a mom overdoing vitamin supplementation of folate may decrease her baby's telomere length.[16] To repeat a theme of this book: moderation and balance are essential!

BABY'S TELOMERES ARE LISTENING TO MOM'S STRESS

A mother's psychological stress may affect her developing baby's telomere length. Our colleagues Pathik Wadhwa and Sonja Entringer from the University of California, Irvine, asked if we would collaborate on a study of prenatal stress and telomeres, and we were delighted to join them and study the start of life. The study was small, but it showed that when mothers experience severe stress and anxiety during pregnancy, their babies tend to have shorter telomeres in their cord blood.[17] A baby's telomeres can suffer from prenatal stress. A recent study extended this finding by examining stressful life experiences. Researchers added up the stressful events that happened in the year before giving birth. The mothers with the highest number of stressful life events had babies with telomeres that were shorter by 1,760 base pairs at birth.[18]

Sonja and Pathik wanted to know how long the effect of prenatal stress on the baby might last. They recruited a group of adult men and women and asked if their mothers had experienced any extremely stressful events while they were pregnant. (The volunteers interviewed their mothers about major events, such as the death of a loved one or a divorce.) As adults, the volunteers who had been exposed to prenatal stress were different in a several ways—even after controlling for factors that might influence their current health. They had more insulin resistance. They were more likely to be overweight or obese. When they underwent a lab-stressor test, they released more cortisol. When their immune cells were stimulated, they responded with higher levels of proinflammatory cytokines.[19] Finally, they had shorter telomeres.[20] A pregnant mother's severe psychological stress appears to have echoes into the next generation, affecting the trajectory of telomere length for decades of the child's life.

We're speaking of very serious stress here. Almost all pregnant mothers experience some mild to moderate stress—not necessarily because they are pregnant but because they are human. At this point, there is no reason to believe that these lower levels of stress are harmful to a baby's telomeres.

The main player that has been examined in pregnancy stress is cortisol. This hormone is released from the mother's adrenal glands and can cross the placenta to affect the fetus.[21] In birds, cortisol from a stressed pregnant bird will make its way into the egg to affect the offspring. Either injecting cortisol into the egg or stressing out the mother can lead to shorter telomeres in chicks. These studies suggest

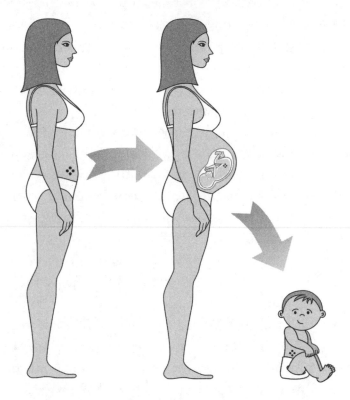

Figure 27: Telomere Transmission. There are at least three paths for telomere transmission from a parent to a grandchild. If a mother has short telomeres in her eggs, those short telomeres can be transmitted directly to the baby (this is known as germline transmission). All the baby's telomeres would then be shorter, including his or her own germline cells (sperm or eggs). During fetal development, maternal stress or poor health can lead to telomere loss in the baby, thanks to excessive cortisol exposure and other biochemical factors. Postnatally, the child's life experiences can shorten his or her telomeres. This child's short telomeres in germline can then be transmitted to his or her future offspring. Mark Haussman and Britt Heidinger have described such transmission pathways in animals and humans.[22]

the possibility that a human mother's stress could be passed on to her baby in the form of short telomeres. Again, what can happen in birds may not happen in humans—but we know enough about chronic stress and telomeres to state that pregnant women must be protected from life's harshest stressors. These include any kind of emotional or physical abuse, violence, war, chemical exposures, food insecurity, and grinding poverty. At the very least, we can support local efforts to provide services and support that buffer pregnant women from survival threats like hunger and violence from the earliest days of pregnancy.

It's clear that parents, especially mothers, influence the telomere health of their babies. And as you're about to see, telomere health is also heavily determined by the way we raise our children and teenagers.

While the health of future generations is important to all societies, it is not, in reality, often paid attention to. Our investment in our most vulnerable young citizens can now be thought of in terms of also investing in base pairs of telomeres, for our collective future of robust health and extended healthspans.

TELOMERE TIPS

- Some transmission of telomere length is out of our control. This includes genetics and direct transmission from eggs and sperm. Telomere transmission to children can happen when a parent has very short telomeres, regardless of genetics. It is a real possibility that we could be unknowingly transmitting disparities in health through this direct telomere transmission.
- Some of what we transmit is under our control. A mother's severe stress during pregnancy, smoking, and intake of certain nutrients, such as folate, are related to her baby's telomere length.
- The transmission of severe social disadvantage through telomeres can likely be blocked through policies that protect the health of women of childbearing age, and especially pregnant women, from toxic stressors and food insecurity.

RENEWAL LAB

GREENING OF THE WOMB

Pediatrician Julia Getzelman of San Francisco recommends that pregnant mothers think about "greening the womb" as well as their house. If you are pregnant, review our Renewal Lab ideas for minimizing chemical exposure in the previous chapter (page 276). Here are some key ways to think of greening the womb:

- Avoid **negative stress**, such as toxic relationships in which you know there will be conflict, unrealistic deadlines, and other situations in which you will not get enough sleep or be able to eat well for days. Life happens, including major events, while you are pregnant, but try to control what you can, and prioritize supportive relationships.
- Increase **well-being time**. Take prenatal yoga classes, or use a yoga video. Find ways to socialize with other pregnant women. Enjoy going out for walks, preferably in green areas.
- Eat a "rainbow" by consuming foods in a variety of rich, deep colors. Amp up **protective nutrients for a healthy developing baby**: Ensure adequate dietary protein, vitamin D_3 and B vitamins, including folate and B_{12}, fish or a high quality omega-3 fatty acid supplement, and probiotics.
- **Avoid pesticides and chemicals in food by eating an organic diet.** Limit your consumption of large and farmed

fish, which often contains accumulations of heavy metals and other industrial chemicals. Limit saccharin or other artificial sweeteners, as these can cross the placenta. (The newer artificial sweeteners may do the same; we expect more and more alarming findings.) Canned foods contain BPA (bisphenol A), a significant endocrine disruptor. Stick to what nature provides and consume a whole-foods diet. Avoid packaged foods with their many questionable additives.

- **Avoid chemical exposures at home** by wet-mopping frequently, using a vinegar-and-water mixture to clean most surfaces, and checking out safer cleaning products and cosmetics here: http://www.ewg.org/consumer-guides. Additionally, plastic PVC shower curtains, perfumes, and other fragrance-containing items such as scented candles may be a significant source of toxins.

CHAPTER THIRTEEN

Childhood Matters for Life: How the Early Years Shape Telomeres

Childhood exposures to stress, violence, and poor nutrition affect telomeres. But there are factors that appear to protect vulnerable children from damage—including sensitive parenting and mild "good stress."

In the year 2000, Harvard psychologist and neuroscientist Charles Nelson walked into one of Romania's notorious orphanages, a legacy of the brutal policies of the Nicolae Ceauşescu regime. The institution housed about four hundred children, all segregated by age as well as by disability. There was a ward full of children with untreated hydrocephalus, a disorder in which the skull expands to accommodate excess fluid, and spina bifida, a defect of the spinal cord and the bones along the spine. There was an infectious disease ward that housed children with HIV and children with syphilis so advanced that it had gone to the brain. On this same day, Nelson entered a ward full of supposedly healthy children who were around two or three years old. One of these children—they'd all been given similar haircuts and clothing, so it was hard to identify them by gender—stood in the middle of the floor, pants soaking wet, sobbing. Nelson asked one of the caregivers why the child was crying.

"His mother abandoned him here this morning," she said. "He's been crying all day."

With so many children under their care, the staff had no time for comforting or soothing. Leaving newly abandoned children alone was a way for the staff to quickly extinguish unwanted behaviors like crying. Babies and toddlers were left in their cribs for days at a time, with nothing to do but stare up at the ceiling. When a stranger walked by, the children would reach their arms out through the crib railings, begging to be held. Although the children were adequately fed and sheltered, they received almost no affection, no stimulation. As Nelson and his team built a lab inside the orphanage to study the effects of early childhood neglect on the developing brain, they had to establish a behavior rule of their own to avoid adding to the residents' distress: No crying in front of the children.

What Nelson and his colleague, Dr. Stacy Drury, learned from studies at the orphanage is both heartbreaking *and* hopeful. Early childhood neglect shortens telomeres—but there are interventions that can help neglected or traumatized children, if we can catch them at a young age. Although the conditions of the orphanages in Romania have improved generally, there are still around seventy thousand orphans and fewer international adoptions to rescue them.[1] Institutional care of children is an ongoing global crisis. War, along with diseases like HIV and Ebola, rob children of their parents and have left an estimated eight million children currently housed in orphanages around the world. We can't afford to turn away from this story.[2]

It is also a story that may have relevance inside our own homes. Telomere knowledge can guide our actions as parents, illuminating a path to raising our children in a way that is healthy for their telomeres. For adults who experienced trauma as children, understanding the long-lasting cellular effects of the past can offer motivation for treating telomeres with tender care now, in the present.

TELOMERES TRACK CHILDHOOD SCARS

When you were growing up, did you have a parent who drank too much? Was anyone in your family depressed? Were you often afraid that your parents would humiliate you or even hurt you?

In a study that painted a disturbing portrait of childhood in the United States, seventeen thousand people were asked to answer a list of ten questions much like the ones above. Around half the sample had experienced at least one such adverse event or situation in childhood, and 25 percent had experienced two or more. Six percent experienced at least *four*. Substance abuse in the family was most common, then sexual abuse and mental illness. Adverse childhood events happen across all levels of incomes and education. Worse, the more events that a person ticked off on the list, particularly if the person had four or more, the more likely the person was to have health problems in adulthood: obesity, asthma, heart disease, depression, and others.[3] Those with four or more adverse events were twelve times more likely to have attempted suicide.

Biological embedding is the name for the effects of childhood adversity that lodge themselves in the body. When telomeres are measured in healthy adults who were exposed to adverse childhood events, a dose-response relationship is often seen. The more traumatic events that a person experienced back then, the shorter their telomeres as an adult.[4] Shorter telomeres are one way that early adversity embeds itself in your cells.

Those short telomeres could have searing effects on a child. If you take a group of young children with shorter telomeres and peer inside their cardiovascular systems a few years later, you'll find that they are more likely to have greater thickening of the walls of their arteries. These are *kids* we're talking about here—and for them, short telomeres could mean a higher risk of early cardiovascular disease.[5]

That damage may begin at a very young age, though it can be halted or possibly reversed if children are rescued from adversity early enough. Charles Nelson and his team compared the children living in Romanian orphanages to ones who'd left the orphanages for quality care in foster homes. The more time the children had spent in the orphanage, the shorter their telomeres.[6] Many of the orphans showed low levels of brain activity during EEG scans. "Instead of a hundred-watt light bulb," Nelson has said, "it was a forty-watt light bulb."[7] Their brains were measurably smaller, and their average IQ was 74, which put them on the borderline of mental retardation. For most of the institutionalized children, their language was delayed and in some cases disordered. Their growth was stunted; they had smaller heads; and they had abnormal attachment behavior, which affects the ability to form lasting relationships. But, says Nelson, "the kids in foster care were showing dramatic recoveries." The children who'd been moved to foster care showed remarkable gains although they had not completely caught up to the children who had never been in orphanages at all; for example, although their IQ was still below that of the never-institutionalized children in the study, it was ten or more points higher than children in the institution.[8] There seemed to be a critical period of brain development: "The kids placed in foster care before age two had improvements in many domains that were better than kids placed after age two," Nelson says.[9] Drury, Nelson, and their team have continued to track these children over the years—and even now, the adolescents who lived at the orphanage as children experience telomere shortening at an accelerated rate.

What about the telomeres of children who are exposed to conditions that, while violent, are not quite so brutal? Scientists Idan Shalev, Avshalom Caspi, and Terri Moffitt, of Duke University, took cheek swabs from five-year-old British children. (Telomeres can be obtained from buccal cells, which live in cheeks.) Five years later, when the children were ten, they swabbed the children's cheeks again. During the five years, the researchers asked the children's mothers about whether their children had been bullied, hurt by someone in their household, or witnessed domestic violence between the parents. The children who had been exposed to the most violence had the greatest telomere shortening over the five years.[10] Maybe this effect on children is short-lasting, or it can change if their life circumstances improve. We hope so. But studies of adults in which people are asked to recall whether they had early adversity also show that those who did have early adversity have shorter telomeres, revealing what may be a lifelong imprint of childhood adversity inside them.[11] In a large study of adults in the Netherlands, reporting several traumatic events as a child was one of the few predictors of having a greater rate of shortening as an adult.[12] In addition, childhood trauma, particularly maltreatment, has been related to greater inflammation and a smaller prefrontal cortex.[13]

That imprint of early trauma can change the way you think, feel, and act. People who have faced early adversity aren't as flexible in their responses to life's varied experiences. They have a higher number of bad days, and their bad days feel more stressful to them. When something good happens, they also feel more joyful.[14] This pattern isn't unhealthy in itself. It just leads to a more intense and dynamic emotional experience. However, that intensity makes it harder to ride out the transitions between emotions. People with a traumatic childhood background tend to have more difficulties in relationships. They're more likely to engage in emotional eating and addictive behaviors.[15] They're not as good at taking care of themselves. These psychological reverberations of abuse may continue to shape mental and physical health all through life. In this way, early adversity may plant the seeds for a greater rate of telomere shortening, unless these resulting patterns of behavior are halted.

ADD UP YOUR ACES (ADVERSE CHILDHOOD EXPERIENCES)

Here's a version of the ACES test, used to measure the number of adverse childhood experiences. Take it now to evaluate your own adversity in childhood.[16]

When you were a child (up to eighteen years old):

1. Did a parent or other adult in the household often or very often swear at you, insult you, put you down, or humiliate you? Or act in a way that made you afraid that you might be physically hurt?
    No _____ If Yes, enter 1 _____

2. Did a parent or other adult in the household often or very often push, grab, or slap you or throw something at you? Or ever hit you so hard that you had marks or were injured?
    No _____ If Yes, enter 1 _____

3. Did an adult or a person at least five years older than you ever touch or fondle you or have you touch his or her body in a sexual way? Or attempt or actually have oral, anal, or vaginal intercourse with you?
    No _____ If Yes, enter 1 _____

4. Did you often or very often feel that no one in your family loved you or thought you were important or special? Or that your family didn't look out for each other, feel close to each other, or support each other?
    No _____ If Yes, enter 1 _____

5. Did you often or very often feel that you didn't have enough to eat, had to wear dirty clothes, and had no one to protect you?

Or that your parents were too drunk or high to take care of you or take you to the doctor if you needed it?

No _____ If Yes, enter 1 _____

6. Was a biological parent ever lost to you through divorce, abandonment, or other reasons?

No _____ If Yes, enter 1 _____

7. Was your mother or stepmother often or very often pushed, grabbed, or slapped? Or sometimes, often, or very often kicked, bitten, hit with a fist, hit with something hard, or made the target of a thrown object? Or ever repeatedly hit for at least a few minutes or threatened with a gun or knife?

No _____ If Yes, enter 1 _____

8. Did you live with anyone who was a problem drinker or an alcoholic, or who used street drugs?

No _____ If Yes, enter 1 _____

9. Was a household member depressed or mentally ill, or did a household member attempt suicide?

No _____ If Yes, enter 1 _____

10. Did a household member go to prison?

No _____ If Yes, enter 1 _____

Total score _____

Typically, having one adverse event is not related to health, whereas having three or four events may be. If you've had several adverse childhood events, and you feel lasting imprints on your current "mindstyle" or lifestyle, don't panic. Your childhood does not have to determine your future. If for example, you developed emotional eating as a coping strategy, you can shed that as an adult. It involves understanding why that pattern developed, and that it doesn't have to be your coping solution going forward. But before you can shed the behavior, it's important to discover alternative coping that works for you, and practice healthier ways to tolerate painful feelings over and over. There are so many ways to buffer residual effects of childhood trauma. If you are still bothered by thoughts about a difficult past, it may warrant seeking help from a professional in mental health. Remember: You are not powerless, and you not alone. Caring professionals can help you undo some of the damage that you were once powerless to stop. And remember there are positive attributes still with you. For example, severe adversity is related to feeling more compassion and empathy for others.[17]

DON'T STEP ON MY PAW! THE EFFECTS OF MONSTROUS MOTHERING

Dr. Frankenstein, step aside. Today's researchers know how to take perfectly nice rats and turn them into maternal monsters. In the lab, they can "build" a rat mother who mistreats her own pups. This is a hard subject for animal lovers to process, but it's helpful reading for anyone who wants to understand the physiology of childhood adversity.

One of the more stressful circumstances for a lactating mama rat is a lack of adequate bedding. Rats don't need luxurious mattresses to be comfortable, but mother lab rats do rely on things like facial tissues and strips of paper to build a little nest for their families. Another cause of high stress for rats is moving to a new place without enough time to habituate to it. By depriving mother rats of material for bedding and by moving them abruptly to a new cage, scientists can create highly stressed animals. Think of how stressful it would be to come home from the hospital with a newborn baby and *then* be greeted by a landlord who says, "Good, you're finally here! Before you put the

301

baby down, let me explain that we've moved you to a new house. Also, we took all your clothing and furniture to the dump. Bye!" You'll have an inkling of what the mother rats were feeling.

These stressed mother rats mistreated their pups. They dropped them. They stepped on them. They spent less time nursing, licking, and grooming—supportive maternal activities that calm rat pups and lead to long-term changes in calming their neural stress responses. The poor pups cried out loudly, signaling their distress. This abusive early environment shaded the contours of the pups' neural development. Compared to rats who were raised by nurturing mothers, these pups had longer telomeres in a part of their brains known as the amygdala, which governs the alarm response.[18] The alarm response had apparently been switched on so often that the telomeres there were strong and robust. Not exactly a sign of a happy upbringing.

Having a strong connection between the amygdala and the prefrontal cortex, which can dampen that response, is critical for good emotion regulation. Sadly, the mistreated rat pups had shorter telomeres in a part of the prefrontal cortex. We already know that severe stress causes the nerve cells of the amygdala to branch out, to enlarge and connect to the nerve cells in other parts of the brain. The opposite tends to happen in the neurons of the prefrontal cortex, so that the connection between the two areas becomes weaker, and the rats can't turn off the stress response as easily.[19]

LACK OF MOTHERING

Parental neglect is another condition that can harm telomeres. Steve Suomi of the National Institutes of Health in Bethesda, Maryland, has been studying parenting in rhesus monkeys for the past forty years. He has found that when they are raised in a nursery from birth, without their mother but socializing with peers, they show a range of problems. They are less playful, and more impulsive, aggressive, and stress reactive (and have lower levels of serotonin in their brains).[20] He wanted to examine whether they have greater telomere attrition as well. He and his colleagues recently had the opportunity to study this in a small group of monkeys. They randomized some to be raised by their mothers and the others to be raised in a nursery for the first seven months of life. When their telomeres were measured four years later, the monkeys raised by their mothers had dramatically longer telomeres, around 2,000 base pairs longer, than the nursery-raised monkeys.[21] While some of the shorter telomere length we see in disadvantaged children might have existed from birth, in this case the newborn monkeys were randomized at birth, so these differences were purely stemming from their early experiences. Fortunately, corrective experiences later in life, like being cared for by a grandparent, can reverse some of the problems of parentless monkeys.

NURTURING CHILDREN FOR HEALTHIER TELOMERES AND BETTER EMOTION REGULATION

It is depressing to read about the maltreatment of the rat pups, or motherless monkeys. But there is a bright side to the story: The rats who were raised by nurturing mothers had healthier telomeres. Same with the monkeys. Of course, nurturing parenting is essential for human babies and children, too. Nurturing parenting can help children develop good emotion regulation, meaning that they can experience negative feelings without getting overpowered by them.[22] Think for a second, and you'll surely have no problem producing examples of adults you know who struggle to regulate their emotions. These are the people who detonate at the slightest provocation. Road rage, anyone?

Maybe you know folks at the other extreme, who find their emotions so frightening that they'd rather end a friendship than work their way through a messy disagreement. They withdraw from anything that may stir up difficult feelings—careers, friendships, even the world outside their homes. Most of us hope that our children will learn more effective means of coping.

We can teach them. From early in life, children learn to regulate their emotions through nurturing care from their parents or caregivers. The baby cries; by showing concern, the parent acts as a kind of emotional copilot, guiding the child toward an understanding of his or her emotions. By soothing the baby and tending to its needs, the parent teaches the child that it's possible to take care of feelings and to trust others. The child learns that distressing situations will eventually pass.

Fortunately for all of us who sometimes get angry in traffic or jump under the bedcovers when emotions run high, parents don't have to have perfect emotion regulation to help their children. In the reassuring words of the great English pediatrician and researcher D. W. Winnicott, they just need to be "good enough." They need to be caring, loving, and stable, with good psychological health, but they definitely don't need to be perfect. Children raised in group homes and orphanages, however, don't get anything close to good-enough parenting; they do not get the attention they need to develop normal emotional expression and regulation. They tend to have blunted emotional expression, an effect that can last throughout their lives.

The delicious act of snuggling with a baby, offering warmth, comfort, and care, has wondrous physiological effects on the child. Scientists believe that well-nurtured children learn to use their prefrontal cortex—the brain's seat of judgment—as a brake on the amygdala and its fear response. Their cortisol levels are better regulated. Put these children on a blinking, whirling kiddie ride at the state fair, or tell them that they need to take an important test, and they'll feel a healthy amount of excitement or worry. That's what stress hormones are there for—to pump us up. When the ride comes to a stop, or when they put down their pencils, the cortisol begins its retreat. They're not constantly swimming around in a flood of stress hormones.

Nurtured children also experience the delights of oxytocin, the hormone that's released when you feel close to someone. Oxytocin is a stress-busting hormone; it reduces our blood pressure and imbues us with a glowing sense of wellbeing.[23] (Women who breast-feed their children can experience the rush of oxytocin in an intense, palpable way.) Alas, the stress-buffering effect of having one's parents nearby seems to wane once children reach adolescence.[24]

A LITTLE ADVERSITY CAN BE PROTECTIVE

There is usually no upside to serious childhood adversity, just suffering and greater risk for depression and anxiety later in life. Shorter telomeres, too. Moderate adversity in childhood, however, can be healthy. Adults who report having a few—but *just* a few— adverse experiences in their youth have healthy cardiovascular responses to stress. Their hearts pumped more blood and got them ready to face the situation; in other words, they experienced a vigorous challenge response. They felt excited, invigorated— so perhaps their early experiences had given them confidence in their ability to overcome obstacles. People with *no* adverse events actually did worse. They felt more threatened, with more vasoconstriction in their peripheral arteries. (Meanwhile, those who had experienced the most severe adversity had excessive threat reactivity.)[25] We are not prescribing a dose of adversity for any child, just pointing out that it is common. If it happens *in a moderate amount, and if the child has enough support to cope with it*, there may be a benefit. Teaching children how to cope well with stress (versus protecting them from all disappointments) is the key. As Helen Keller said, "Character cannot be developed in ease and quiet. Only through experiences of trial and suffering can the soul be strengthened, vision cleared, ambition inspired and success achieved."

THE ABCS OF PARENTING VULNERABLE CHILDREN

In children who have begun their lives under traumatic circumstances, enhanced parenting techniques may help heal some of the telomere damage from early mistreatment. Mary Dozier, of the University of Delaware, has studied children who were exposed to adversity. Some lived in inadequate housing; some were neglected or witnessed or experienced domestic violence; some had parents who abused substances or who hurt each other. Dozier and her colleagues found that these children had shorter telomeres—except when their parents interacted with them in a very sensitive, responsive way.[26] To give you a sense of what this kind of parenting looks like, here's a very short assessment:

1. Your toddler bumps his head hard on a coffee table and looks at you as if ready to cry. What do you say?

- "Oh, honey, are you okay? Do you need a hug?"
- "You're okay. Hop up."
- "You shouldn't be that close to the table. Move away from there."
- You say nothing, hoping that he'll move on to something else.

2. Your child comes home from school and says that her best friend doesn't want to be friends anymore. You say:

- "I'm so sorry, honey. Do you want to talk about it?"
- "You'll have plenty of friends over time. Don't worry."
- "What did you do that made her not want to be your friend?"
- "Why don't you get on your bike and go for a ride?"

All of these answers can sound reasonable, and under individual circumstances, any of them might be. But there is only one correct response for a child who has been through trauma, and in both

306

cases, that response is the first one. Under normal conditions, it may sometimes be appropriate to help a child learn to brush off a minor bump or scrape, for example. But children who've suffered from adversity are different. They may have a harder time regulating their emotions. They still need parents to be the emotional copilot—to reassure them that the parent has noticed their troubles and can be relied upon to help soothe them. They may need this reassurance over and over and over again. It takes time, but eventually children will learn how to respond to problems in a more adaptive way. And when they are older, they will be more likely to go to their parents with issues that they are worried about.

Dozier has developed a program known as Attachment and Biobehavioral Catch-Up, or ABC, to teach this kind of exquisite responsiveness to parents of at-risk children. One group included American parents who were adopting international children. These weren't people who lacked parenting skills. They were caring and committed. But the children they were adopting were statistically much more likely to have lived in group homes, to suffer from poor emotion regulation, to have telomere damage—the whole bushel of problems that come with childhood adversity. During this program, parents are coached to *follow their child's lead*. For example, when a child starts to play a game by banging a spoon, a parent might be tempted to say, "Spoons are for stirring pudding" or "Let's count the number of times you tap the bowl." But these responses reflect the parent's agenda, not the child's. In Dozier's program, the parent would be encouraged to join in the game, or to comment on what the child is doing: "You're making a sound with your spoon and bowl!" These smooth interactions with the parent help at-risk children learn to regulate their emotions.

It's a simple intervention, but the results are dramatic. Dozier also taught ABC to a group of parents who had been reported to Child Protective Services for allegedly neglecting their children. Before the course, the children's cortisol levels had that blunted, broken response that characterizes burnout from overuse. After the parents had taken this short course, the children had a much more normal cortisol response. Their cortisol rose in the morning (a good, healthy sign that they were ready to take on the day), and declined throughout the day. This effect wasn't just temporary. It lasted for *years*.[27]

TELOMERES AND STRESS-SENSITIVE CHILDREN

Was Rose a difficult baby? Her parents smile at the question. "Rose had colic for *three years*," they say, laughing at their exaggeration as well as the kernel of truth that is behind it. Colic, in which babies cry incessantly for more than three hours a day, three days a week, generally begins at around two weeks of age and usually peaks at about six weeks. Rose was colicky, all right. As a newborn, she would nurse, nap briefly, have about five minutes of peaceful time...and then begin wailing again. Despite her name, Rose was no demure flower. Her parents, desperate to calm their crying baby, would take her for walks and strolls through the neighborhood—only to have older ladies rush up to them, exclaiming, "Something must be wrong with your child! Healthy babies do not cry this way!"

Nothing was wrong. Rose was clean, fed, warm, and cared for. She was just very, very sensitive. She was quick to cry and slow to settle down for sleep or quiet—thus her parents' joke about her colic lasting for years. Small noises, like the running of the refrigerator motor, bothered her. When strangers held her, Rose would scream and try to wriggle out of their arms. As Rose got older, she wouldn't wear clothes with tags; they felt too itchy. When the family signed up for a professional photography session, Rose hid her eyes from the bright lights. And any change in her daily routine was upsetting.

Was Rose sensitive because of the way her parents raised her? Were they too indulgent of her demands? Should they have taught her a lesson by insisting, say, that Rose wear whatever clothes they picked out for her, itchy or not? We can begin to answer these questions by talking about temperament. Temperament, the set of personality traits we're born with, is like the deep cement foundation of a building. It can provide a stable undergirding, or it can make us tilt and shake in certain ways, especially during an "earthquake." We can recognize our temperament and learn to deal with it, but we can't really change our foundation. Temperament is biologically determined.

One aspect of temperament is stress sensitivity. Stress-sensitive children are more "permeable," which means that for good or for ill, their environment doesn't just bounce off them. It penetrates. These kids have bigger stress reactions to light, noise, and physical irritations. They are jolted by transitions, like going to back to school after the weekend (the "Monday effect"), or new situations, like staying at a grandparent's house overnight. They have a stronger, magnified response to shifts in their environment, even small shifts that other children might not notice. Some of these children may react by acting angry or aggressive; others may internalize their feelings, coming across as quiet or sullen. Telomeres tend to be shorter in children who internalize their emotions.[28] But when children have severe externalizing or acting-out disorders, such as attention deficit disorder with hyperactivity, and oppositional defiant disorder, their telomeres are shorter, too.[29]

Developmental pediatrician Tom Boyce has followed a group of kindergartners as they transition into their first year of school—a time that can be tough for stress-sensitive children. He and his colleagues hooked them up to sensors and then measured their physiological reactions to harmless but modestly stressful situations, like watching a scary video, having a few drops of lemon juice squirted onto their tongues, and (of course) performing one of those memory tasks. Most kids showed some signs of stress. But in a few kids, the stress responses were cranked up to their full force, both the hormonal responses and the autonomic nervous system. It was as if their bodies and brains thought the room was on fire. The bigger the stress responses, the shorter their telomeres tended to be.[30]

IS YOUR CHILD AN ORCHID?

It can all sound quite tragic. It may seem that people who are born with high-stress sensitivity have drawn the unlucky short straw—or, in this case, the short telomere. Actually, Boyce and others have found that certain environments allow stress-sensitive people to thrive, sometimes even more than their less sensitive peers.

In many studies Boyce has found that children who are especially stress sensitive do poorly when they are in large, crowded, chaotic classrooms or harsh family environments, but when they are in classrooms or families with warm, nurturing adults, they actually do better than the average child. They are less sick with colds and flu; they show fewer symptoms of depression or anxiety; they are even injured less often than other children.[31]

Boyce calls these stress-sensitive children "orchids." Without exquisite care and attention, an orchid won't bloom. Put it in the optimal conditions of a greenhouse, though, and it produces flowers of surpassing beauty. Around 20 percent of children have an orchid-like temperament. Again, it's not something that parents create. Those orchid seeds are planted long before birth.

A way to understand these "seeds" is to analyze the genetic signatures of orchid children. Children (and adults) with more variations in the genes for neurotransmitters that regulate mood, like dopamine and serotonin, tend to be more sensitive to stress. They're orchids. Those most stress sensitive, based on genetics, tend to benefit more from supportive interventions and will thrive.[32] To test whether this genetic signature affects how children's telomeres respond to adversity, a small and preliminary study looked at forty boys. Half were from stable homes; the other half were from harsh social environments characterized by poverty, unresponsive parenting, and family structures that kept changing. The boys exposed to harsh environments had shorter telomeres—but especially if they had the more stress-sensitive genes. That's the obvious disadvantage of being permeable to the environment—a rough situation is going to do deep damage. Then the boys revealed the flip side, the beauty of permeability: When they lived in stable environments, their telomeres weren't just okay. They were longer, healthier, than the telomeres of the boys without the genetic variations. This early study suggests that being sensitive and permeable may be a benefit when in a supportive environment.[33]

This is a fascinating story in personality research, and one of the hottest topics in the stress field. Sensitivity is neither a good nor a bad trait. It's just one of the cards we're dealt. It's best if we can clearly identify the card so that we can know how to play our hand. Orchid children benefit from warmth, gentle correction, and a consistent routine. They need assistance and patience as they make transitions to a new situation. As high-stress reactors, orchid children can benefit from learning the challenge response—and you can also teach them techniques like thought awareness and mindful breathing, which help them put some calming distance between themselves (their thoughts) and their active stress responses.

PARENTING TEENS FOR TELOMERE HEALTH

Parent: Look at what I found underneath that mess on your desk today. Am I correct in thinking that this is an assignment for a history paper?

Teen: I don't know.

Parent: It's due *tomorrow*. Have you even started it yet?

Teen: I don't know.

Parent: Answer me respectfully! Let's try again: Is this or is this not an assignment for a history paper that is due tomorrow?

Teen: I don't have to listen to this! You're just jealous because you never had fun when you were my age. You didn't know how!

Parent: You just bought yourself a grounding. You'll be staying home this Friday night.

Teen [shouting]: Go to hell!

Parent [also shouting]: AND all day Saturday!

So far we've talked about children, mostly younger ones. But what about teenagers? Parent-teen conflicts like the one above, in which an issue (like homework) is raised, fought over, but left unresolved, are common. These open-ended conflicts leave the teen with a lot of anger—and psychologists know what anger does to that cauldron of physiological responses known as stress soup. Anger heats that soup up to a rolling boil. And anger can have telomere-shortening effects, but fortunately this can be turned around through a shift in parenting style.

Gene Brody, a researcher of family studies at the University of Georgia, gives us insight into the role of parental support during the teen years, and how to bolster it. Brody tracked a group of African American teens in the impoverished rural south of the United States. It's an area where young adults leave high school only to find that there are few jobs of any kind, let alone satisfying jobs, and few resources to help them make the transition to adult life. Alcohol use in particular is high. Brody recruited a group of these teens for his Adults in the Making program, in which teens are given emotional support and job advice. The instructors also provide strategies for handling racism. The teens' parents are included in the program, too—they're taught to tell their child in clear, vigorous terms to stay away from drugs and alcohol, for example. They have six classes where parents and teens learn skills in separate groups and then practice them together at the end. Half the teens did not get the classes. Five years later, Brody measured their telomeres. First of all, having unsupportive parenting—lots of arguments and little emotional support—was associated with shorter telomere length and more substance use five years later. However, among this vulnerable group, the teens that had received the supportive intervention had longer telomeres compared to teens who hadn't. This effect is partly explained by the teens feeling less angry.[34]

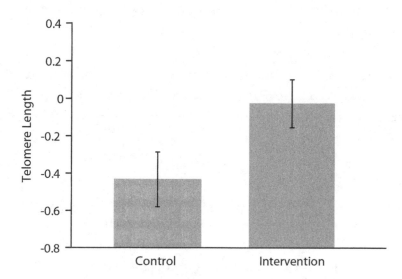

Figure 28: Family Resilience Classes and Telomeres. Among the teenagers whose parents showed very unsupportive parenting, those who were in the supportive intervention group had significantly longer telomeres five years later. (This is after adjusting for factors such as social status, stressful events, smoking, alcohol use, and body mass index.)[35]

Brody's study looked at teens in a very particular setting and at a certain income level. But his findings provide food for thought for all of us. No matter where they live, and no matter how rich or poor, *all* children's brains and bodies are undergoing tremendous changes during adolescence. It's common for teens to follow a jagged path for a while, especially because the teen brain experiences risk differently. They tend to react to threat as a thrill; when they take risks, they feel good.[36] The same behaviors are, naturally, terrifying to the more seasoned adults in their lives. Cue the parental worries, dead-of-night ruminations, and fears that explode into fights between parent and teen. A few conflicts are probably unavoidable. But when conflicts are constant, or when the tension becomes so toxic that it pollutes the air of the household, teens can become angry and rebellious. Or depressed and anxious, if they're the type to drive their feelings underground. The Renewal Lab at the end of this chapter offers a few suggestions for staying attuned to teens when they are in a difficult, hyperreactive mode.

We have been talking about how to help children heal the telomere damage caused by adversity. Early intervention, support, and emotional attunement can provide buffers for at-risk children. But you may have had prolonged, severe stress in early life yourself. If you grew up in a dangerous neighborhood, in an abusive home, or if your family had to struggle just to get food and shelter, your telomeres may have experienced some damage. Use this knowledge as motivation to take care of your telomeres now. Recognize old patterns, such as turning to food for comfort. You have more control over what happens to you now that you are an adult. And now you know how to protect the base pairs of telomeres you have left. You may especially want to take advantage of techniques that help soothe the stress response. By becoming less stress reactive, you will protect your telomeres—and there is a bonus. You will also be calmer and stronger for the children (and other loved ones) who are in your life.

TELOMERE TIPS

- Severe childhood trauma is linked to shorter telomeres. Trauma can also reverberate into adulthood in the form of poor health behaviors and relationship difficulties, which may continue to shorten telomeres. If you suffered severe childhood adversity, you can take steps now to buffer its effects on your wellbeing and telomeres.

■ Although severe childhood adversity can be damaging, moderate childhood stress may actually be healthy, provided that the child has enough support during the stressful time.

■ Parents can support their young children's telomeres by practicing warm, nurturing attunement. This responsiveness is especially important for children who have already experienced trauma or who are born with the sensitive "orchid" temperament.

RENEWAL LAB

WEAPONS OF MASS DISTRACTION

The ABC program teaches parents to avoid unresponsive behaviors, including something that almost all of us are guilty of: distraction. No matter what a child's situation or temperament, being connected to a screen means we're not connected to the child. And it's easier to be distracted than you might think. When a cell phone is present on a nearby table, people engage in conversation that is more shallow, and their attention is more divided.[37] Digital conversations limit the opportunity for full empathy and connection. No wonder the writer Pico Iyer refers to smartphones as "weapons of mass distraction."

This Renewal Lab invites you to engage with the children in your life *without* the interference of screens. *See if you can spend twenty minutes talking to a child, playing a game, or just enjoying his or her presence, without a phone or tablet computer nearby. Limit your children's screen time as well. Make it intentional—sometimes naming something gives it a lot more power and makes it more effective. Though your child may resist this screen-free drive or mealtime, he or she may also welcome it, albeit secretly. Decide on a few critical screen-free times such as meals, the car ride to and from school, and the first half hour after coming into the house after being away (when attention should be focused on reconnecting with the family). If the screen-free time is a clear rule, you won't need to get into complex negotiations every day. For tips on how to "outsmart*

smart screens" and limit your child's use, Harvard's Prevention Research Center has a free guide for parents: http://www.hsph .harvard.edu/prc/2015/01/07/outsmarting-the-smart-screens/. Your family can also participate in Screen-Free Week, a campaign hosted by the Campaign for a Commercial Free Childhood each spring (http://www.screenfree.org/).

TUNING IN TO YOUR CHILD

Vulnerable children need tremendous sensitivity and parental attunement. You can soothe some of their frustration by tuning in to their feelings. Homework, for example, is a common stressor. Kids get upset about the homework itself, and they can also get irritated with their parents when they try to help. Daniel Siegel, author of *The Whole-Brain Child* and coauthor of *Brainstorm*, offers ways to attune especially when riding waves of high emotions. He explains that parents can't help their child manage homework (or any other stressful activity) until they acknowledge and empathize with the child's feelings.

So the next time your child is under stress, try saying something that acknowledges his or her feelings, such as "You look frustrated." You can also help your child identify his or her feelings, since labeling feelings and putting together a story of what happened turns down the volume on emotions. Siegel calls this strategy "name it to tame it." You can say things like, "Wow, that seemed like a tough situation. What was that like for you? How are you feeling?" If you want to reach the child's rational thinking, you have to meet the child emotionally first, with empathy.[38] Siegel calls this "connect and redirect."

DON'T OVERREACT TO YOUR REACTIVE TEEN

Don't let your teen's emotional, thrill-seeking brain pull you into an escalating conflict. If your teen is ranting at you, you have options other than your automatic one…reacting. An argument cannot escalate if you are not part of it. Sometimes it helps to say you need a time-out for yourself, a small dose of time and space in a different place. Given the quick half-life of emotions, your child's emotions (and yours) will most likely subside, and a conversation can resume with both sides of the brain working.

In the heat of the moment, you may remind yourself that although adolescents may look like grown-ups on the outside, they are still children on the inside. They need you to be clear and steady, not to get tangled up in their drama. Remind yourself *you're* the one in the room with the adult brain—and that you have the power to remain calm and to avoid escalating the argument. Also, in calm moments, be curious. Rather than tell your teen what to do, ask him or her questions.

BE A MODEL OF LOVING ATTACHMENT

A loving relationship with your partner is not only precious but a tool for better parenting. One study tracked children's reactions to their parents' daily interactions over three months. The study examined how much emotional resonance or mirroring children had to their parents' interactions. When the parents showed affection to each other and the children felt more positive affect, the children tended to have longer telomeres. Conversely, when the parents had conflict, and the child responded with negative emotions, the child tended to have shorter telomeres.[39] So remember that emotions are permeable, and especially for sensitive children. Consider warming up your family environment and showing your affection. This is hard to do if angry emotions are running high! But by showing love for your partner, you may also be promoting wellbeing in your child (and maybe his or her telomeres, too).

Conclusion
Entwined: Our Cellular Legacy

> A human being is part of the whole, called by us "Universe,"
> a part limited in time and space. He experiences himself, his
> thoughts and feelings, as something separated from the rest—
> a kind of optical delusion of his consciousness. This delusion
> is a kind of prison for us, restricting us to our personal desires
> and to affection for a few persons nearest to us. Our task
> must be to free ourselves from this prison by widening our
> circle of compassion to embrace all living creatures and the
> whole of nature in its beauty. Nobody is able to achieve this
> completely, but the striving for such an achievement is in itself
> a part of the liberation and a foundation for inner security.
> —Albert Einstein, as quoted in the *New York Times*,
> March 29, 1972

A long life of good health and wellbeing is our hope for you. Lifestyle, mental health, and environment all contribute significantly to physical health— that is not new. What is new here is that telomeres are impacted by these factors, and thus quantify their contribution in a clear and powerful way. The fact that we see the transgenerational impacts of these influences makes the message of telomeres all the more urgent. Our genes are like computer hardware; we cannot change them. Our epigenome, of which telomeres are a part, is like software, which requires programming. We are the programmers of the epigenome. To some extent, we control the chemical signals that orchestrate the changes. Our telomeres are responsive, listening, calibrating to the current circumstances in the world. Together we can improve the programming code.

The preceding pages have been full of our best suggestions, gleaned from hundreds of studies, about how to protect your precious telomeres. You've seen how telomeres are affected by your mind. You've seen how they're shaped by your habits of movement, by the quality and length of your sleep, and by the foods you eat. Telomeres are also affected by the world beyond the mind and body—because your neighborhood and relationships foster a sense of safety that can shape your telomere health.

Unlike humans, telomeres do not make judgments. They are objective and unbiased. Their reaction to their environment is quantifiable, down to their base pairs. This makes them an ideal index for measuring the effects of our internal and external environment on our health. If we listen to what they have to tell us, telomeres provide insight into how we can prevent premature cellular aging and promote our healthspan. But as it turns out, the story of healthspan is also the story of what a beautiful life and world can look like. What is good for our own telomeres is also good for our children, our community, and people around the globe.

TELOMERES SOUND THE ALARM

Telomeres teach us that from the earliest days of life, severe stress and adversity reverberate all the way into adulthood, setting up our youngest generation for a life marred by higher likelihood of early chronic disease. In particular, we've learned that childhood exposure to stressors like violence, trauma, abuse, and socioeconomic hardship have all been linked to shorter telomeres in adulthood. The damage may begin even before the child is born; high maternal stress may be transmitted to the developing fetus in the form of shorter telomeres.

This early imprint of stress on telomeres is an alarm bell. We call for policymakers to add a new phrase, **societal stress reduction**, to the vocabulary of public health. We're not speaking here of exercise or yoga classes, though these are helpful to many people. We're talking about broad social policies that have the goal of buffering the ubiquitous socioenvironmental and economic chronic stressors faced by so many.

The worst stressors—exposure to violence, trauma, abuse, and mental illness—are shaped by a surprising factor: the level of income inequality in a region. For example, countries with the biggest gap between their richest citizens and their poorest have the worst health and the most violence. As you can see from figure 29, these countries also have the highest rates of depression, anxiety, and schizophrenia.[1]

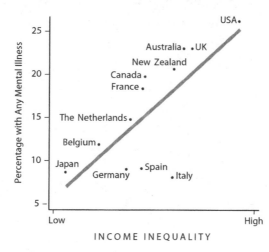

Figure 29: Income Inequality and Mental Health. A large body of research has shown that income inequality in regions and countries is associated with worse behavior (less trust, more violence, drug abuse) and worse health for all, whether it's physical or mental health. Kate Pickett and Richard Wilkinson have summarized this massive set of research findings,[2] and here they show relations with mental health. In this data set, Japan has the lowest inequality and the lowest rate of mental illness, whereas the United States has the highest of each.

A substantial number of studies have demonstrated this relationship. And it's not just the poor who suffer from the gap. Everyone in these stratified societies is at a higher risk for impaired mental and physical health—and the more unequal the society, the lower the child wellbeing. You see this effect among rich and poor states in the United States. The inequality gap has been widening such that, in the United States, the top 3 percent in the distribution owns 50 percent of the wealth[3] (no wonder the United States has the biggest gap among the rich countries). Tellingly, Sweden, which has the lowest income gap of all countries, also has the highest wellbeing, including the wellbeing of children. But it's also one of the countries with the fastest growing inequality and decreasing child wellbeing (due to a reduction of the redistributive effect of Sweden's tax and benefit system).[4]

We believe that the income gap drives the difference between the likelihood of healthy, long, stable telomeres in old age and mere stumps of short ones from aged, senescent cells. This gap represents excessive social stress, competitive stress, and the sickness in societies that leads to an early and prolonged diseasespan for both rich and poor. An essential element of societal stress reduction is the narrowing of this enormous gap. Understanding how we are interconnected is the fuel that will drive this work.

INTERCONNECTEDNESS AT ALL LEVELS

We are connected to one another and to all living things at all levels from the macro to the micro, from the societal to the cellular. The separation we all feel, as if we are each on a path alone, is an illusion. The reality is that we all share much more than we can ever comprehend, both in mind and body. We are deeply interconnected with each other and nature in phenomenal ways.

Within our body and cells, we're connected to other living organisms. Our bodies are made up of eukaryotic cells. It's thought that about a billion and a half years ago, long before humans evolved, the single eukaryotic cell swallowed up bacterial organisms that lived together interdependently as one cell. The mitochondria that live in our cells today are the legacy of these bacteria and this interdependence. We're symbiotic creatures.

Inside our bodies, we carry around a shared part of the outside world. Roughly two to three pounds of human weight is made up of other beings: microbes. Microbes live as complex communities within our gut and on our skin. Far from being our sworn enemies, they keep us balanced. Without these colonies of microbes, our immune system would be weak and underdeveloped; they send signals to our brain and can make us depressed when they are out of balance. And it works the other way, too—when we are feeling depressed or stressed, we are affecting our microbiota, impairing their balanced state, and impairing our mitochondria.[5]

Humans are increasingly interconnected to one another—from technology to financial markets, to media and social network groups. We are always embedded in a social culture, and our thoughts and feelings are shaped by our immediate social and physical surroundings.[6] Our perceptions of how supported and connected we are matter for our health. This has always been true, but now those connections are becoming both broader and tighter. A global broadband will soon encircle the world, allowing everyone on the planet to be connected via the Internet in a highly affordable way. On any random day last year, one out of seven people worldwide had logged on to Facebook.[7] This growing interconnectedness opens up further opportunities to unite around the issues that are most significant to us.

We're also sharing the same physical environment. Pollution on one side of the world can travel to the other, blowing in the wind or floating in the water. Together, we are heating up our globe, and we are all affected by it. It's another sign of how we are connected, an urgent reminder that our daily behavior matters.

Finally, we're connected from generation to generation. We now know that telomeres are transmitted through the generations. The disadvantaged unknowingly pass on that disadvantage—through economic and social problems, but also likely through shorter telomeres and other epigenetic paths. In this way, telomeres are our message to our future society. Worse, children are being exposed to toxic stress at epidemic levels, leaving them with shorter telomeres and premature cellular aging. As John F. Kennedy reminded us, "Children are the living messages we send to a time we will not see." We do not want that message to include early chronic disease. This is why it's so important to cultivate our inborn sense of compassion. We must rewrite that message.

THE LIVING MESSAGES

Telomere science has grown into a clarion call. It tells us that social stressors, especially as they affect children, will result in exponentially higher costs down the line—costs that are personal, physical, social, and economic. You can respond to that call by, first, taking good care of yourself.

The call doesn't end there. Now that you know how to protect your telomeres, we want to issue you a friendly challenge. What will you do with your many decades of brimming good health? A long healthspan makes a vital, energetic life more possible, and that vitality can ripple outward, allowing us to spend some of our time creating conditions for better health and wellbeing in other people.

We can't eliminate stress and adversity, of course, but there are ways to relieve some of the extreme pressure on the most vulnerable populations. We've told you about painful aspects of some people's lives, but that's just one aspect of their lives. Robin Huiras, the woman with an inherited telomere disorder, who has helped recruit some of the best minds in telomere science to write the first clinical handbook for treatment of telomere disorders, is helping to alleviate suffering. Peter, the medical researcher who struggles with a brain bent on overeating, travels around the world on medical missions to underserved people and has filled his life with purpose and contribution. Tim Parrish, the man who grew up in a racist community in Louisiana, writes and speaks about this painful subject, risking his own comfort to help us face our prejudices more effectively.

What is your cellular legacy? Each of us has a time-limited opportunity to leave a legacy. Just as your body is a community of individual but mutually dependent cells, we are a world of interdependent people. We all have an impact on the world, whether we realize it or not. Large changes, such as implementing policies for societal stress reduction, are vital. Small changes are important, too. How we interact with other people shapes their feelings and sense of trust. *Every day, each of us has the chance to positively influence the life of another person.*

The story of telomeres can inspire our determination to elevate our collective health. Helping to change our communities and shared environment gives us that vital sense of mission and purpose, which itself may improve our telomere maintenance.

The foundation for a new understanding of health in our society is not about "me" but "we." Redefining healthy aging is not just about accepting gray hair and focusing on inner health; it's also about our connections with others and building safe, trusting communities. Telomere science offers molecular proof of the importance of societal health to our individual wellbeing. We now have a way to index and measure the interventions we create to improve that health. Let's get started.

THE TELOMERE MANIFESTO

Your cellular health is reflected in the wellbeing of your mind, body, and community. Here are the elements of telomere maintenance that we believe to be the most crucial for a healthier world:

Mind Your Telomeres

- Evaluate sources of persistent, intense stress. What can you change?
- Transform a threat to a challenge appraisal.
- Become more self-compassionate and compassionate to others.
- Take up a restorative activity.
- *Practice thought awareness and mindful attention. Awareness opens doors to wellbeing.*

Maintain Your Telomeres

- Be active.
- Develop a sleep ritual for more restorative and longer sleep.
- Eat mindfully to reduce overeating and ride out cravings.
- Choose telomere-healthy foods—whole foods, omega-3s, skip the bacon.

Connect Your Telomeres

- Make room for connection: Disconnect from screens for part of the day.
- Cultivate a few good, close relationships.
- Provide children quality attention and the right amount of "good stress."
- Cultivate your neighborhood social capital. Help strangers.
- Seek green. Spend time in nature.
- *Mindful attention to other people allows connections to bloom. Attention is your gift to give.*

Create Telomere Health in Your Community and the World

- Improve prenatal care.
- Protect children from violence and other traumas that damage telomeres.
- Reduce inequality.
- Clean up local and global toxins.
- Improve food policies so that everyone has access to fresh, healthy, affordable food.

The future health of our society is being shaped right now, and we can measure part of that future in telomere base pairs.

Acknowledgments

We could not have written this book without drawing upon the decades of hard work of many scientists, and we thank them all for their contributions to our understanding of telomeres, human aging, and behavior, even though we couldn't reference each important contribution of our colleagues. We thank the innumerable scientific collaborators and students with whom we have worked during the past few decades; our gratitude to each of you is bottomless. Our research could not have transpired without you. We are both especially indebted to Dr. Jue Lin, PhD, who has worked tirelessly and with great talent for over ten years on all of our human telomere studies. Jue has performed tens of thousands of meticulous telomere-length and telomerase measurements for these studies, and has served as an exemplar of a translational researcher, working at all levels, from lab bench to community.

We would like to acknowledge the following people who have contributed to this book in various important ways, through enlightening discussions, providing perspectives on the book, or serving as an inspiration or support to our work. Any mistakes in the content, however, are entirely our own. We extend our deepest gratitude to: Nancy Adler, Mary Armanios, Ozlem Ayduk, Albert Bandura, James Baraz, Roger Barnett, Susan Bauer-Wu, Peter and Allison Baumann, Petra Boukamp, Gene Brody, Kelly Brownell, Judy Campisi,

Laura Carstensen, Steve Cole, Mark Coleman, David Creswell, Alexandra Croswell, Susan Czaikowski, James Doty, Mary Dozier, Rita Effros, Sharon Epel, Michael Fenech, Howard Friedman, Susan Folkman, Julia Getzelman, Roshi Joan Halifax, Rick Hecht, Jeannette Ickovics, Michael Irwin, Roger Janke, Oliver John, Jon Kabat-Zinn, Will and Teresa Kabat-Zinn, Noa Kageyama, Erik Kahn, Alan Kazdin, Lynn Kutler, Barbara Laraia, Cindy Leung, Becca Levy, Andrea Lieberstein, Robert Lustig, Frank Mars, Pamela Mars, Ashley Mason, Thea Mauro, Wendy Mendes, Bruce McEwen, Synthia Mellon, Rachel Morello-Frosch, Judy Moskowitz, Belinda Needham, Kristen Neff, Charles Nelson, Lisbeth Nielsen, Jason Ong, Dean Ornish, Bernard and Barbro Osher, Alexsis de Raadt St. James, Judith Rodin, Brenda Penninx, Ruben Perczek, Kate Pickett, Stephen Porges, Aric Prather, Eli Puterman, Robert Sapolsky, Cliff Saron, Michael Scheier, Zindel Segal, Daichi Shimbo, Dan Siegel, Felipe Sierra, the late Richard Suzman, Shanon Squires, Matthew State, Janet Tomiyama, Bert Uchino, Pathik Wadhwa, Mike Weiner, Christian Werner, Darrah Westrup, Mary Whooley, Jay Williams, Redford Williams, Janet Wojcicki, Owen Wolkowitz, Phil Zimbardo, and Ami Zota. A big thanks to Aging, Metabolism, and Emotions (AME) lab members, and especially to Alison Hartman, Amanda Gilbert, and Michael Coccia, for support on various aspects of the book. We thank Coleen Patterson of Coleen Patterson Design for her inspired illustrations, and amazing transference of images from our heads to this book.

We thank Thea Singer for covering the telomere-stress connection so beautifully in her book *Stress Less* (Hudson Street Press, 2010). We also thank the dedicated readers of our book focus group who gave us their Sunday afternoons and invaluable input: Michael Acree, Diane Ashcroft, Elizabeth Brancato, Miles Braun, Amanda Burrowes, Cheryl Church, Larry Cowan, Joanne Delmonico, Tru Dunham, Ndifreke Ekaette, Emele Faifua, Jeff Fellows, Ann Harvie, Kim Jackson, Kristina Jones, Carole Katz, Jacob Kuyser, Visa Lakshi, Larissa Lodzinski, Alisa Mallari, Chloe Martin, Heather McCausland, Marla Morgan,

Debbie Mueller, Michelle Nanton, Erica "Blissa" Nizzoli, Sharon Nolan, Lance Odland, Beth Peterson, Pamela Porter, Fernanda Raiti, Karin Sharma, Cori Smithen, Sister Rosemarie Stevens, Jennifer Taggart, Roslyn Thomas, Julie Uhernik, and Michael Worden. Thanks to Andrew Mumm of Idea Architects for his wizardry and patience connecting us across geographic and technical challenges.

We'd also like to thank the people who generously talked with us about their personal experiences, some anonymously, some named below. We weren't able to incorporate every single one of the wonderful stories we heard, but throughout the writing process the spirit of all those stories has informed and profoundly moved us. We are indebted to Cory Brundage, Robin Huiras, Sean Johnston, Lisa Louis, Siobhan Mark, Leigh Anne Naas, Chris Nagel, Siobhan O'Brien, Tim Parrish, Abby McQueeney Penamonte, Rene Hicks Schleicher, Maria Lang Slocum, Rod E. Smith, and Thulani Smith.

We extend *tremendous* thanks to Leigh Ann Hirschman, of Hirschman Literary Services, our collaborative writer. Her writing and depth of editorial experience helped make this book as readable as it is. She was a pleasure to work with: joining our immersion in the world of telomere science, ever patient with our bringing in the constant flow of new studies that entered the scientific literature as we wrote, and a balanced and guiding voice when at times we thought we would never emerge from the thickets of research.

We are also very grateful to our editor, Karen Murgolo of Grand Central Publishing, for her faith in this book and her expertise, time, and care in every decision needed throughout this process. We felt so fortunate to have benefitted from her wisdom and patience.

We have deep gratitude to Doug Abrams of Idea Architects. It was Doug who first saw the need for a book that we could not yet see. We thank him for his dedication and for his wonderful and wise curation as a developmental editor. And for making what could have been taxing to our telomeric base pairs both a delightful process and the grounding of enduring friendship.

Lastly, we are so grateful to our families (nuclear and extended) for their loving support and enthusiasm during the many seasons of the writing process, and the many more seasons that laid the scientific foundation for it.

We are also grateful for the opportunity to share this work with you, the readers, and sincerely hope that this work promotes your wellbeing and healthspan.

Information about Commercial Telomere Tests

If you'd like to estimate your telomere health, you can take the self-test on page 161. You can also take a commercial company test to determine telomere length. But should you? You don't need to have your own lungs biopsied in order to make the wise decision to stop smoking! Many of you would probably perform the same restorative activities in life whether you have a telomere test or not.

We wondered how people would react to learning the results of telomere tests. If a person learns he or she has short telomeres, for example, would that knowledge be depressing? So we tested volunteers and told them their results. Then we followed up to ask about their reactions. Most were neutral to positive, and none was very negative. But those who were short did experience some distressing thoughts about that in the ensuing months. Telomere testing is a personal decision. Only you can decide if knowing your length will benefit you. Imagine if you learn your telomeres are short— is that more motivating to you than upsetting? Learning that your telomeres are short is like seeing the "check engine" light on a dashboard; it's usually just a sign that you need to take a closer look at your health and your habits and step up your efforts.

We're often asked if we've had our own telomeres measured:

I (Liz) have, out of curiosity. My results were reassuringly good, but I always keep in mind that telomere length is a statistical indication of health, not an absolute predictor of the future.

I (Elissa) haven't had my telomeres measured yet. I would rather not know definitively if my telomeres are short. I try to engage in the life practices good for telomeres on an AMAP (as much as possible) basis, given this busy life. Telomere length trajectories over time will be more valuable than single checks. They tell us something unique about a cell's potential to replicate that no one indicator can. However, they are just one marker. It is likely that algorithms including many biomarkers and health status variables will be more beneficial for personal use once they are better developed. When the measures have more predictive value for individuals and are easier to get repeatedly, I will be more interested in getting testing done.

As of this writing, only a few commercial companies offer telomere testing.

We do not have any knowledge about—or control over—the accuracy and reliability of telomere length measurements performed by these commercial entities. Because these companies change rapidly, we list the details on our book website. At this writing, testing costs anywhere from around $100 to $500.

A few caveats: Telomere testing is an unregulated business, so there is no government agency checking whether for-profit companies are using methods and values that are accurate, or whether what they tell you about your risks is accurate. It can be interesting to learn the results of a telomere test, but we caution everyone that telomeres do not necessarily predict the future. Again, it's like smoking. Smoking does not guarantee that you'll get a lung disease, and not smoking does not guarantee that you will stay free of disease. But the statistics on smoking are in, and the message is clear: the more you smoke, the greater your chances of getting emphysema, cancer, and other serious health problems. There are plenty of good reasons to quit—or better still, not to smoke at all. In the same way, the countless studies on the relationship between telomere length and human health and disease have given us the data we need to create guidelines for keeping your telomeres (and therefore you) healthier. You may enjoy knowing your telomere length, but you don't need that information to prevent premature cellular aging.

Notes

Authors' Note: Why We Wrote This Book

1. "Oldest Person Ever," Guinness World Records, http://www.guin nessworldrecords.com/world-records/oldest-person, accessed March 3, 2016.
2. Whitney, C. R., "Jeanne Calment, World's Elder, Dies at 122," *New York Times*, August 5, 1997, http://www.nytimes.com/1997/08/05/world /jeanne-calment-world-s-elder-dies-at-122.html, accessed March 3, 2016.
3. Blackburn, E., E. Epel, and J. Lin, "Human Telomere Biology: A Contributory and Interactive Factor in Aging, Disease Risks, and Protection," *Science* 350, no. 6265 (December 4, 2015): 1193–98.

Introduction: A Tale of Two Telomeres

1. Bray, G. A. "From Farm to Fat Cell: Why Aren't We All Fat?" *Metabolism* 64, no. 3 (March 2015):349–353, doi:10.1016/j.metabol.2014.09.012, Epub 2014 Oct 22, PMID: 25554523, p. 350.
2. Christensen, K., G. Doblhammer, R. Rau, and J. W. Vaupel, "Ageing Populations: The Challenges Ahead," *Lancet* 374, no. 9696 (October 3, 2009): 1196–1208, doi:10.1016/S0140-6736(09)61460-4.
3. United Kingdom, Office for National Statistics, "One Third of Babies Born in 2013 Are Expected to Live to 100," December 11, 2013, The National Archive, http://www.ons.gov.uk/ons/rel/lifetables/historic -and-projected-data-from-the-period-and-cohort-life-tables/2012 -based/sty-babies-living-to-100.html, accessed November 30, 2015.
4. Bateson, M., "Cumulative Stress in Research Animals: Telomere Attrition as a Biomarker in a Welfare Context?" *BioEssays* 38, no. 2 (February 2016): 201–12, doi:10.1002/bies.201500127.

5. Epel, E., E. Puterman, J. Lin, E. Blackburn, A. Lazaro, and W. Mendes, "Wandering Minds and Aging Cells," *Clinical Psychological Science* 1, no. 1 (January 2013): 75–83, doi:10.1177/2167702612460234.

6. Carlson, L. E., et al., "Mindfulness-Based Cancer Recovery and Supportive-Expressive Therapy Maintain Telomere Length Relative to Controls in Distressed Breast Cancer Survivors." *Cancer* 121, no. 3 (February 1, 2015): 476–84, doi:10.1002/cncr.29063.

Chapter One: How Prematurely Aging Cells Make You Look, Feel, and Act Old

1. Epel, E. S., and G. J. Lithgow, "Stress Biology and Aging Mechanisms: Toward Understanding the Deep Connection Between Adaptation to Stress and Longevity," *Journals of Gerontology, Series A: Biological Sciences and Medical Sciences* 69 Suppl. 1 (June 2014): S10–16, doi:10.1093/gerona/glu055.

2. Baker, D. J., et al., "Clearance of p16Ink4a-positive Senescent Cells Delays Ageing-Associated Disorders," *Nature* 479, no. 7372 (November 2, 2011): 232–36, doi:10.1038/nature10600.

3. Krunic, D., et al., "Tissue Context-Activated Telomerase in Human Epidermis Correlates with Little Age-Dependent Telomere Loss," *Biochimica et Biophysica Acta* 1792, no. 4 (April 2009): 297–308, doi:10.1016/j.bbadis.2009.02.005.

4. Rinnerthaler, M., M. K. Streubel, J. Bischof, and K. Richter, "Skin Aging, Gene Expression and Calcium," *Experimental Gerontology* 68 (August 2015): 59–65, doi:10.1016/j.exger.2014.09.015.

5. Dekker, P., et al., "Stress-Induced Responses of Human Skin Fibroblasts in Vitro Reflect Human Longevity," *Aging Cell* 8, no. 5 (September 2009): 595–603, doi:10.1111/j.1474-9726.2009.00506.x; and Dekker, P., et al., "Relation between Maximum Replicative Capacity and Oxidative Stress-Induced Responses in Human Skin Fibroblasts in Vitro," *Journals of Gerontology, Series A: Biological Sciences and Medical Sciences* 66, no. 1 (January 2011): 45–50, doi:10.1093/gerona/glq159.

6. Gilchrest, B. A., M. S. Eller, and M. Yaar, "Telomere-Mediated Effects on Melanogenesis and Skin Aging," *Journal of Investigative Dermatology Symposium Proceedings* 14, no. 1 (August 2009): 25–31, doi:10.1038/jidsymp.2009.9.

7. Kassem, M., and P. J. Marie, "Senescence-Associated Intrinsic Mechanisms of Osteoblast Dysfunctions," *Aging Cell* 10, no. 2 (April 2011): 191–97, doi:10.1111/j.1474-9726.2011.00669.x.

8. Brennan, T. A., et al., "Mouse Models of Telomere Dysfunction Phenocopy Skeletal Changes Found in Human Age-Related Osteoporosis,"

Disease Models and Mechanisms 7, no. 5 (May 2014): 583–92, doi:10.1242/dmm.014928.

9. Inomata, K., et al., "Genotoxic Stress Abrogates Renewal of Melanocyte Stem Cells by Triggering Their Differentiation," *Cell* 137, no. 6 (June 12, 2009): 1088–99, doi:10.1016/j.cell.2009.03.037.

10. Jaskelioff, M., et al., "Telomerase Reactivation Reverses Tissue Degeneration in Aged Telomerase-Deficient Mice," *Nature* 469, no. 7328 (January 6, 2011): 102–6, doi:10.1038/nature09603.

11. Panhard, S., I. Lozano, and G. Loussouam, "Greying of the Human Hair: A Worldwide Survey, Revisiting the '50' Rule of Thumb," *British Journal of Dermatology* 167, no. 4 (October 2012): 865–73, doi:10.1111/j.1365-2133.2012.11095.x.

12. Christensen, K., et al., "Perceived Age as Clinically Useful Biomarker of Ageing: Cohort Study," *BMJ* 339 (December 2009): b5262.

13. Noordam, R., et al., "Cortisol Serum Levels in Familial Longevity and Perceived Age: The Leiden Longevity Study," *Psychoneuroendocrinology* 37, no. 10 (October 2012): 1669–75; Noordam, R., et al., "High Serum Glucose Levels Are Associated with a Higher Perceived Age," *Age (Dordrecht, Netherlands)* 35, no. 1 (February 2013): 189–95, doi:10.1007/s11357-011-9339-9; and Kido, M., et al., "Perceived Age of Facial Features Is a Significant Diagnosis Criterion for Age-Related Carotid Atherosclerosis in Japanese Subjects: J-SHIPP Study," *Geriatrics and Gerontology International* 12, no. 4 (October 2012): 733–40, doi:10.1111/j.1447-0594.2011.00824.x.

14. Codd, V., et al., "Identification of Seven Loci Affecting Mean Telomere Length and Their Association with Disease," *Nature Genetics* 45, no. 4 (April 2013): 422–27, doi:10.1038/ng.2528.

15. Haycock, P. C., et al., "Leucocyte Telomere Length and Risk of Cardiovascular Disease: Systematic Review and Meta-analysis," *BMJ* 349 (July 8, 2014): g4227, doi:10.1136/bmj.g4227.

16. Yaffe, K., et al., "Telomere Length and Cognitive Function in Community-Dwelling Elders: Findings from the Health ABC Study," *Neurobiology of Aging* 32, no. 11 (November 2011): 2055–60, doi:10.1016/j.neurobiolaging.2009.12.006.

17. Cohen-Manheim, I., et al., "Increased Attrition of Leukocyte Telomere Length in Young Adults Is Associated with Poorer Cognitive Function in Midlife," *European Journal of Epidemiology* 31, no. 2 (February 2016), doi:10.1007/s10654-015-0051-4.

18. King, K. S., et al., "Effect of Leukocyte Telomere Length on Total and Regional Brain Volumes in a Large Population-Based Cohort,"

JAMA Neurology 71, no. 10 (October 2014): 1247–54, doi:10.1001/jamaneurol.2014.1926.

19. Honig, L. S., et al., "Shorter Telomeres Are Associated with Mortality in Those with APOE Epsilon4 and Dementia," *Annals of Neurology* 60, no. 2 (August 2006): 181–87, doi:10.1002/ana.20894.

20. Zhan, Y., et al., "Telomere Length Shortening and Alzheimer Disease—A Mendelian Randomization Study," *JAMA Neurology* 72, no. 10 (October 2015): 1202–03, doi:10.1001/jamaneurol.2015.1513.

21. If you would like, you can contribute to studies on brain aging and disease without having to get your brain scanned, or even show up in person. Dr. Mike Weiner, a noted researcher at UCSF who leads the largest cohort study of Alzheimer's disease worldwide, developed the online Brain Health Registry. By joining the Brain Health Registry you answer questionnaires and take online cognitive tests. We are helping him study the effects of stress on brain aging. You can find the registry at http://www.brainhealthregistry.org/

22. Ward, R. A., "How Old Am I? Perceived Age in Middle and Later Life," *International Journal of Aging and Human Development* 71, no. 3 (2010): 167–84.

23. Ibid.

24. Levy, B., "Stereotype Embodiment: A Psychosocial Approach to Aging," *Current Directions in Psychological Science* 18, vol. 6 (December 1, 2009): 332–36.

25. Levy, B. R., et al., "Association Between Positive Age Stereotypes and Recovery from Disability in Older Persons," *JAMA* 308, no. 19 (November 21, 2012): 1972–73, doi:10.1001/jama.2012.14541; Levy, B. R., A. B. Zonderman, M. D. Slade, and L. Ferrucci, "Age Stereotypes Held Earlier in Life Predict Cardiovascular Events in Later Life," *Psychological Science* 20, no. 3 (March 2009): 296–98, doi:10.1111/j.1467-9280.2009.02298.x.

26. Haslam, C., et al., "'When the Age Is In, the Wit Is Out': Age-Related Self-Categorization and Deficit Expectations Reduce Performance on Clinical Tests Used in Dementia Assessment," *Psychology and Aging* 27, no. 3 (April 2012): 778784, doi:10.1037/a0027754.

27. Levy, B. R., S. V. Kasl, and T. M. Gill, "Image of Aging Scale," *Perceptual and Motor Skills* 99, no. 1 (August 2004): 208–10.

28. Ersner-Hershfield, H., J. A. Mikels, S. J. Sullivan, and L. L. Carstensen, "Poignancy: Mixed Emotional Experience in the Face of Meaningful Endings," *Journal of Personality and Social Psychology* 94, no. 1 (January 2008): 158–67.

29. Hershfield, H. E., S. Scheibe, T. L. Sims, and L. L. Carstensen, "When Feeling Bad Can Be Good: Mixed Emotions Benefit Physical Health Across Adulthood," *Social Psychological and Personality Science* 4, no.1 (January 2013): 54-61.
30. Levy, B. R., J. M. Hausdorff, R. Hencke, and J. Y. Wei, "Reducing Cardiovascular Stress with Positive Self-Stereotypes of Aging," *Journals of Gerontology, Series B: Psychological Sciences and Social Sciences* 55, no. 4 (July 2000): P205–13.
31. Levy, B. R., M. D. Slade, S. R. Kunkel, and S. V. Kasl, "Longevity Increased by Positive Self-Perceptions of Aging," *Journal of Personal and Social Psychology* 83, no. 2 (August 2002): 261–70.

Chapter Two: The Power of Long Telomeres

1. Lapham, K. et al., "Automated Assay of Telomere Length Measurement and Informatics for 100,000 Subjects in the Genetic Epidemiology Research on Adult Health and Aging (GERA) Cohort," *Genetics* 200, no. 4 (August 2015):1061–72, doi:10.1534/genetics.115.178624.
2. Rode, L., B. G. Nordestgaard, and S. E. Bojesen, "Peripheral Blood Leukocyte Telomere Length and Mortality Among 64,637 Individuals from the General Population," *Journal of the National Cancer Institute* 107, no. 6 (May 2015): djv074, doi:10.1093/jnci/djv074.
3. Ibid.
4. Lapham et al., "Automated Assay of Telomere Length Measurement and Informatics for 100,000 Subjects in the Genetic Epidemiology Research on Adult Health and Aging (GERA) Cohort." (See #1 above.)
5. Willeit, P., et al., "Leucocyte Telomere Length and Risk of Type 2 Diabetes Mellitus: New Prospective Cohort Study and Literature-Based Meta-analysis," *PLOS ONE* 9, no. 11 (2014): e112483, doi:10.1371/journal.pone.0112483; D'Mello, M. J., et al., "Association Between Shortened Leukocyte Telomere Length and Cardiometabolic Outcomes: Systematic Review and Meta-analysis," *Circulation: Cardiovascular Genetics* 8, no. 1 (February 2015): 82–90, doi:10.1161/CIRCGENETICS.113.000485; Haycock, P. C., et al., "Leucocyte Telomere Length and Risk of Cardiovascular Disease: Systematic Review and Meta-Analysis," *BMJ* 349 (2014): g4227, doi:10.1136/bmj.g4227; Zhang, C., et al., "The Association Between Telomere Length and Cancer Prognosis: Evidence from a Meta-Analysis," *PLOS ONE* 10, no. 7 (2015): e0133174, doi:10.1371/journal.pone.0133174; and Adnot, S., et al., "Telomere Dysfunction and Cell Senescence in Chronic Lung Diseases: Therapeutic

Potential," *Pharmacology & Therapeutics* 153 (September 2015): 125–34, doi:10.1016/j.pharmthera.2015.06.007.

6. Njajou, O. T., et al., "Association Between Telomere Length, Specific Causes of Death, and Years of Healthy Life in Health, Aging, and Body Composition, a Population-Based Cohort Study," *Journals of Gerontology, Series A: Biological Sciences and Medical Sciences* 64, no. 8 (August 2009): 860–64, doi:10.1093/gerona/glp061.

Chapter Three: Telomerase, the Enzyme That Replenishes Telomeres

1. Vulliamy, T., A. Marrone, F. Goldman, A. Dearlove, M. Bessler, P. J. Mason, and I. Dokal. "The RNA Component of Telomerase Is Mutated in Autosomal Dominant Dyskeratosis Congenita." *Nature* 413, no. 6854 (September 27, 2001): 432–35, doi:10.1038/35096585.

2. Epel, Elissa S., Elizabeth H. Blackburn, Jue Lin, Firdaus S. Dhabhar, Nancy E. Adler, Jason D. Morrow, and Richard M. Cawthon, "Accelerated Telomere Shortening in Response to Life Stress," *Proceedings of the National Academy of Sciences of the United States of America* 101, no. 49 (December 7, 2004): 17312–315, doi:10.1073/pnas.0407162101.

Chapter Four: Unraveling: How Stress Gets into Your Cells

1. Evercare by United Healthcare and the National Alliance for Caregiving, "Evercare Survey of the Economic Downtown and Its Impact on Family Caregiving" (March 2009), 1.

2. Epel, E. S., et al., "Cell Aging in Relation to Stress Arousal and Cardiovascular Disease Risk Factors," *Psychoneuroendocrinology* 31, no. 3 (April 2006): 277–87, doi:10.1016/j.psyneuen.2005.08.011.

3. Gotlib, I. H., et al., "Telomere Length and Cortisol Reactivity in Children of Depressed Mothers," *Molecular Psychiatry* 20, no. 5 (May 2015): 615–20, doi:10.1038/mp.2014.119.

4. Oliveira, B. S., et al., "Systematic Review of the Association between Chronic Social Stress and Telomere Length: A Life Course Perspective," *Ageing Research Reviews* 26 (March 2016): 37–52, doi:10.1016/j .arr.2015.12.006; and Price, L. H., et al., "Telomeres and Early-Life Stress: An Overview." *Biological Psychiatry* 73, no. 1 (January 2013): 15–23, doi:10.1016/j.biopsych.2012.06.025.

5. Mathur, M. B., et al., "Perceived Stress and Telomere Length: A Systematic Review, Meta-analysis, and Methodologic Considerations for Advancing the Field," *Brain, Behavior, and Immunity* 54 (May 2016): 158–69, doi:10.1016/j.bbi.2016.02.002.

6. O'Donovan, A. J., et al., "Stress Appraisals and Cellular Aging: A Key Role for Anticipatory Threat in the Relationship Between Psychological Stress and Telomere Length," *Brain, Behavior, and Immunity* 26, no. 4 (May 2012): 573–79, doi:10.1016/j.bbi.2012.01.007.

7. Ibid.

8. Jefferson, A. L., et al., "Cardiac Index Is Associated with Brain Aging: The Framingham Heart Study," *Circulation* 122, no. 7 (August 17, 2010): 690–97, doi:10.1161/CIRCULATIONAHA.109.905091; and Jefferson, A. L., et al., "Low Cardiac Index Is Associated with Incident Dementia and Alzheimer Disease: The Framingham Heart Study," *Circulation* 131, no. 15 (April 14, 2015): 1333–39, doi:10.1161/CIRCULATIONAHA.114 .012438.

9. Sarkar, M., D. Fletcher, D. J. Brown, "What doesn't kill me...: Adversity-Related Experiences Are Vital in the Development of Superior Olympic Performance," *Journal of Science in Medicine and Sport* 18, no. 4 (July 2015): 475–79. doi:10.1016/j.jsams.2014.06.010.

10. Epel, E., et al., "Can Meditation Slow Rate of Cellular Aging? Cognitive Stress, Mindfulness, and Telomeres," *Annals of the New York Academy of Sciences* 1172 (August 2009): 34–53, doi:10.1111/j.1749-6632.2009.04414.x.

11. McLaughlin, K. A., M. A. Sheridan, S. Alves, and W. B. Mendes, "Child Maltreatment and Autonomic Nervous System Reactivity: Identifying Dysregulated Stress Reactivity Patterns by Using the Biopsychosocial Model of Challenge and Threat," *Psychosomatic Medicine* 76, no. 7 (September 2014): 538–46, doi:10.1097/PSY.0000000000000098.

12. O'Donovan et al., "Stress Appraisals and Cellular Aging: A Key Role for Anticipatory Threat in the Relationship Between Psychological Stress and Telomere Length." (See #6 above.)

13. Barrett, L., *How Emotions Are Made* (New York: Houghton Mifflin Harcourt, in press).

14. Ibid.

15. Jamieson, J. P., W. B. Mendes, E. Blackstock, and T. Schmader, "Turning the Knots in Your Stomach into Bows: Reappraising Arousal Improves Performance on the GRE," *Journal of Experimental Social Psychology* 46, no. 1 (January 2010): 208–12.

16. Beltzer, M. L, M. K. Nock, B. J. Peters, and J. P. Jamieson, "Rethinking Butterflies: The Affective, Physiological, and Performance Effects of Reappraising Arousal During Social Evaluation," *Emotion* 14, no. 4 (August 2014): 761–68, doi:10.1037/a0036326.

17. Waugh, C. E., S. Panage, W. B. Mendes, and I. H. Gotlib, "Cardiovascular and Affective Recovery from Anticipatory Threat," *Biological*

Psychology 84, no. 2 (May 2010): 169–175, doi:10.1016/j.biopsycho .2010.01.010; and Lutz, A., et al., "Altered Anterior Insula Activation During Anticipation and Experience of Painful Stimuli in Expert Meditators," *NeuroImage* 64 (January 1, 2013): 538–46, doi:10.1016/j.neuroimage.2012.09.030.

18. Herborn, K.A., et al., "Stress Exposure in Early Post-Natal Life Reduces Telomere Length: An Experimental Demonstration in a Long-Lived Seabird," *Proceedings of the Royal Society B: Biological Sciences* 281, no. 1782 (March 19, 2014): 20133151, doi:10.1098/rspb.2013.3151.

19. Aydinonat, D., et al., "Social Isolation Shortens Telomeres in African Grey Parrots (*Psittacus erithacus erithacus*)," *PLOS ONE* 9, no. 4 (2014): e93839, doi:10.1371/journal.pone.0093839.

20. Gouin, J. P., L. Hantsoo, and J. K. Kiecolt-Glaser, "Immune Dysregulation and Chronic Stress Among Older Adults: A Review," *Neuroimmunomodulation* 15, nos. 4–6 (2008): 251–59, doi:10.1159/000156468.

21. Cao, W., et al., "Premature Aging of T-Cells Is Associated with Faster HIV-1 Disease Progression," *Journal of Acquired Immune Deficiency Syndromes (1999)* 50, no. 2 (February 1, 2009): 137–47, doi:10.1097/QAI.0b013e3181926c28.

22. Cohen, S., et al., "Association Between Telomere Length and Experimentally Induced Upper Respiratory Viral Infection in Healthy Adults," *JAMA* 309, no. 7 (February 20, 2013): 699–705, doi:10.1001/jama.2013.613.

23. Choi, J., S. R. Fauce, and R. B. Effros, "Reduced Telomerase Activity in Human T Lymphocytes Exposed to Cortisol," *Brain, Behavior, and Immunity* 22, no. 4 (May 2008): 600–605, doi:10.1016/j.bbi.2007.12.004.

24. Cohen, G. L., and D. K. Sherman, "The Psychology of Change: Self-Affirmation and Social Psychological Intervention," *Annual Review of Psychology* 65 (2014): 333–71, doi:10.1146/annurev-psych-010213-115137.

25. Miyake, A., et al., "Reducing the Gender Achievement Gap in College Science: A Classroom Study of Values Affirmation," *Science* 330, no. 6008 (November 26, 2010): 1234–37, doi:10.1126/science.1195996.

26. Dutcher, J. M., et al., "Self-Affirmation Activates the Ventral Striatum: A Possible Reward-Related Mechanism for Self-Affirmation," *Psychological Science* 27, no. 4 (April 2016): 455–66, doi:10.1177/0956797615625989.

27. Kross, E., et al., "Self-Talk as a Regulatory Mechanism: How You Do It Matters," *Journal of Personality and Social Psychology* 106, no. 2 (February 2014): 304–24, doi:10.1037/a0035173; and Bruehlman-Senecal, E., and O. Ayduk, "This Too Shall Pass: Temporal Distance and the Regulation

of Emotional Distress," *Journal of Personality and Social Psychology* 108, no. 2 (February 2015): 356–75, doi:10.1037/a0038324.

28. Lebois, L. A. M., et al., "A Shift in Perspective: Decentering Through Mindful Attention to Imagined Stressful Events," *Neuropsychologia* 75 (August 2015): 505–24, doi:10.1016/j.neuropsychologia.2015.05.030.

29. Kross, E., et al., "'Asking Why' from a Distance: Its Cognitive and Emotional Consequences for People with Major Depressive Disorder," *Journal of Abnormal Psychology* 121, no. 3 (August 2012): 559–69, doi:10.1037/a0028808.

Chapter Five: Mind Your Telomeres: Negative Thinking, Resilient Thinking

1. Meyer Friedman and Ray H. Roseman, *Type A Behavior and Your Heart* (New York: Knopf, 1974).

2. Chida, Y., and A. Steptoe, "The Association of Anger and Hostility with Future Coronary Heart Disease: A Meta-analytic Review of Prospective Evidence," *Journal of the American College of Cardiology* 53, no. 11 (March 17, 2009): 936–46, doi:10.1016/j.jacc.2008.11.044.

3. Miller, T. Q, et al., "A Meta-analytic Review of Research on Hostility and Physical Health," *Psychological Bulletin* 119, no. 2 (March 1996): 322–48.

4. Brydon, L., et al., "Hostility and Cellular Aging in Men from the Whitehall II Cohort," *Biological Psychiatry* 71, no. 9 (May 2012): 767–73, doi:10.1016/j.biopsych.2011.08.020.

5. Zalli, A., et al., "Shorter Telomeres with High Telomerase Activity Are Associated with Raised Allostatic Load and Impoverished Psychosocial Resources," *Proceedings of the National Academy of Sciences of the United States of America* 111, no. 12 (March 25, 2014): 4519–24, doi:10.1073/pnas.1322145111.

6. Low, C. A., R. C. Thurston, and K. A. Matthews, "Psychosocial Factors in the Development of Heart Disease in Women: Current Research and Future Directions," *Psychosomatic Medicine* 72, no. 9 (November 2010): 842–54, doi:10.1097/PSY.0b013e3181f6934f.

7. O'Donovan, A., et al., "Pessimism Correlates with Leukocyte Telomere Shortness and Elevated Interleukin-6 in Post-menopausal Women," *Brain, Behavior, and Immunity* 23, no. 4 (May 2009):446–49, doi:10.1016/j.bbi.2008.11.006.

8. Ikeda, A., et al., "Pessimistic Orientation in Relation to Telomere Length in Older Men: The VA Normative Aging Study," *Psychoneuroendocrinology* 42 (April 2014): 68–76, doi:10.1016/j.psyneuen.2014.01.001;

and Schutte, N. S., K. A. Suresh, and J. R. McFarlane, "The Relationship Between Optimism and Longer Telomeres," 2016, under review.

9. Killingsworth, M. A., and D. T. Gilbert, "A Wandering Mind Is an Unhappy Mind," *Science* 330, no. 6006 (November 12, 2010): 932, doi:10.1126/science.1192439.

10. Epel, E. S., et al., "Wandering Minds and Aging Cells," *Clinical Psychological Science* 1, no. 1 (January 2013): 75–83.

11. Kabat-Zinn, J., *Wherever You Go, There You Are: Mindfulness Meditation in Everyday Life* (New York: Hyperion, 1995), p. 15.

12. Engert, V., J. Smallwood, and T. Singer, "Mind Your Thoughts: Associations Between Self-Generated Thoughts and Stress-Induced and Baseline Levels of Cortisol and Alpha-Amylase," *Biological Psychology* 103 (December 2014): 283–91, doi:10.1016/j.biopsycho.2014.10.004.

13. Nolen-Hoeksema, S., "The Role of Rumination in Depressive Disorders and Mixed Anxiety/Depressive Symptoms," *Journal of Abnormal Psychology* 109, no. 3 (August 2000): 504–11.

14. Lea Winerman, "Suppressing the 'White Bears,'" *Monitor on Psychology* 42, no. 9 (October 2011): 44.

15. Alda, M., et al., "Zen Meditation, Length of Telomeres, and the Role of Experiential Avoidance and Compassion," *Mindfulness* 7, no. 3 (June 2016): 651–59.

16. Querstret, D., and M. Cropley, "Assessing Treatments Used to Reduce Rumination and/or Worry: A Systematic Review," *Clinical Psychology Review* 33, no. 8 (December 2013): 996–1009, doi:10.1016/j.cpr.2013.08.004.

17. Wallace, B. Alan, *The Attention Revolution: Unlocking the Power of the Focused Mind* (Boston: Wisdom, 2006).

18. Saron, Clifford, "Training the Mind: The Shamatha Project," in *The Healing Power of Meditation: Leading Experts on Buddhism, Psychology, and Medicine Explore the Health Benefits of Contemplative Practice*, ed. Andy Fraser (Boston: Shambhala, 2013), 45–65.

19. Sahdra, B. K., et al., "Enhanced Response Inhibition During Intensive Meditation Training Predicts Improvements in Self-Reported Adaptive Socioemotional Functioning," *Emotion* 11, no. 2 (April 2011): 299–312, doi:10.1037/a0022764.

20. Schaefer, S. M., et al., "Purpose in Life Predicts Better Emotional Recovery from Negative Stimuli," *PLOS ONE* 8, no. 11 (2013): e80329, doi:10.1371/journal.pone.0080329.

21. Kim, E. S., et al., "Purpose in Life and Reduced Incidence of Stroke in Older Adults: The Health and Retirement Study," *Journal of Psychosomatic*

Research 74, no. 5 (May 2013): 427–32, doi:10.1016/j.jpsychores.2013 .01.013.

22. Boylan, J.M., and C. D. Ryff, "Psychological Wellbeing and Metabolic Syndrome: Findings from the Midlife in the United States National Sample," *Psychosomatic Medicine* 77, no. 5 (June 2015): 548–58, doi:10.1097 /PSY.0000000000000192.

23. Kim, E. S., V. J. Strecher, and C. D. Ryff, "Purpose in Life and Use of Preventive Health Care Services," *Proceedings of the National Academy of Sciences of the United States of America* 111, no. 46 (November 18, 2014): 16331–36, doi:10.1073/pnas.1414826111.

24. Jacobs, T.L., et al., "Intensive Meditation Training, Immune Cell Telomerase Activity, and Psychological Mediators," *Psychoneuroendocrinology* 36, no. 5 (June 2011): 664–81, doi:10.1016/j.psyneuen.2010.09.010.

25. Varma, V. R., et al., "Experience Corps Baltimore: Exploring the Stressors and Rewards of High-Intensity Civic Engagement," *Gerontologist* 55, no. 6 (December 2015): 1038–49, doi:10.1093/geront/gnu011.

26. Gruenewald, T. L., et al., "The Baltimore Experience Corps Trial: Enhancing Generativity via Intergenerational Activity Engagement in Later Life," *Journals of Gerontology, Series B: Psychological Sciences and Social Sciences*, February 25, 2015, doi:10.1093/geronb/gbv005.

27. Carlson, M. C., et al., "Impact of the Baltimore Experience Corps Trial on Cortical and Hippocampal Volumes," *Alzheimer's & Dementia: The Journal of the Alzheimer's Association* 11, no. 11 (November 2015): 1340–48, doi:10.1016/j.jalz.2014.12.005.

28. Sadahiro, R., et al., "Relationship Between Leukocyte Telomere Length and Personality Traits in Healthy Subjects," *European Psychiatry: The Journal of the Association of European Psychiatrists* 30, no. 2 (February 2015): 291–95, doi:10.1016/j.eurpsy.2014.03.003.

29. Edmonds, G. W., H. C. Côté, and S. E. Hampson, "Childhood Conscientiousness and Leukocyte Telomere Length 40 Years Later in Adult Women—Preliminary Findings of a Prospective Association," *PLOS ONE* 10, no. 7 (2015): e0134077, doi:10.1371/journal.pone.0134077.

30. Friedman, H. S., and M. L. Kern, "Personality, Wellbeing, and Health," *Annual Review of Psychology* 65 (2014): 719–42.

31. Costa, D. de S., et al., "Telomere Length Is Highly Inherited and Associated with Hyperactivity-Impulsivity in Children with Attention Deficit/ Hyperactivity Disorder," *Frontiers in Molecular Neuroscience* 8 (2015): 28, doi:10.3389/fnmol.2015.00028; and Yim, O. S., et al., "Delay Discounting, Genetic Sensitivity, and Leukocyte Telomere Length," *Proceedings of*

the National Academy of Sciences of the United States of America 113, no. 10 (March 8, 2016): 2780–85, doi:10.1073/pnas.1514351113.

32. Martin, L.R., H. S. Friedman, and J. E. Schwartz, "Personality and Mortality Risk Across the Life Span: The Importance of Conscientiousness as a Biopsychosocial Attribute," *Health Psychology* 26, no. 4 (July 2007): 428–36; and Costa, P. T., Jr., et al., "Personality Facets and All-Cause Mortality Among Medicare Patients Aged 66 to 102 Years: A Follow-On Study of Weiss and Costa (2005)," *Psychosomatic Medicine* 76, no. 5 (June 2014): 370–78, doi:10.1097/PSY.0000000000000070.

33. Shanahan, M. J., et al., "Conscientiousness, Health, and Aging: The Life Course of Personality Model," *Developmental Psychology* 50, no. 5 (May 2014): 1407–25, doi:10.1037/a0031130.

34. Raes, F., E. Pommier, K. D. Neff, and D. Van Gucht, "Construction and Factorial Validation of a Short Form of the Self-Compassion Scale," *Clinical Psychology & Psychotherapy* 18, no. 3 (May–June 2011): 250–55, doi:10.1002/cpp.702.

35. Breines, J. G., et al., "Self-Compassionate Young Adults Show Lower Salivary Alpha-Amylase Responses to Repeated Psychosocial Stress," *Self Identity* 14, no. 4 (October 1, 2015): 390–402.

36. Finlay-Jones, A. L., C. S. Rees, and R. T. Kane, "Self-Compassion, Emotion Regulation and Stress Among Australian Psychologists: Testing an Emotion Regulation Model of Self-Compassion Using Structural Equation Modeling," *PLOS ONE* 10, no. 7 (2015): e0133481, doi:10.1371/journal.pone.0133481.

37. Alda et al., "Zen Meditation, Length of Telomeres, and the Role of Experiential Avoidance and Compassion." (See #15 above.)

38. Hoge, E. A., et al., "Loving-Kindness Meditation Practice Associated with Longer Telomeres in Women," *Brain, Behavior, and Immunity* 32 (August 2013): 159–63, doi:10.1016/j.bbi.2013.04.005.

39. Smeets, E., K. Neff, H. Alberts, and M. Peters, "Meeting Suffering with Kindness: Effects of a Brief Self-Compassion Intervention for Female College Students," *Journal of Clinical Psychology* 70, no. 9 (September 2014): 794–807, doi:10.1002/jclp.22076; and Neff, K. D., and C. K. Germer, "A Pilot Study and Randomized Controlled Trial of the Mindful Self-Compassion Program," *Journal Of Clinical Psychology* 69, no. 1 (January 2013): 28–44, doi:10.1002/jclp.21923.

40. This exercise is adapted from Dr. Neff's website: http://self-compassion .org/exercise-2-self-compassion-break/. For more information on developing self-compassion, see K. Neff, *Self-Compassion: The Proven Power of Being Kind to Yourself* (New York: HarperCollins, 2011).

41. Valenzuela, M., and P. Sachdev, "Can cognitive exercise prevent the onset of dementia? Systematic review of randomized clinical trials with longitudinal follow-up." *Am J Geriatr Psychiatry*, 2009. 17(3): p. 179–87.

Assessment: How Does Your Personality Influence Your Stress Responses?

1. Scheier, M. F., C. S. Carver, and M. W. Bridges, "Distinguishing Optimism from Neuroticism (and Trait Anxiety, Self-Mastery, and Self-Esteem): A Reevaluation of the Life Orientation Test," *Journal of Personality and Social Psychology* 67, no. 6 (December 1994): 1063–78.
2. Marshall, Grant N., et al. "Distinguishing Optimism from Pessimism: Relations to Fundamental Dimensions of Mood and Personality," *Journal of Personality and Social Psychology* 62.6 (1992): 1067.
3. O'Donovan et al., "Pessimism Correlates with Leukocyte Telomere Shortness and Elevated Interleukin-6 in Post-Menopausal Women" (see #7 above); and Ikeda et al., "Pessimistic Orientation in Relation to Telomere Length in Older Men: The VA Normative Aging Study" (see #8 above).
4. Glaesmer, H., et al., "Psychometric Properties and Population-Based Norms of the Life Orientation Test Revised (LOT-R)," *British Journal of Health Psychology* 17, no. 2 (May 2012): 432–45, doi:10.1111 /j.2044-8287.2011.02046.x.
5. Eckhardt, Christopher, Bradley Norlander, and Jerry Deffenbacher, "The Assessment of Anger and Hostility: A Critical Review," *Aggression and Violent Behavior* 9, no. 1 (January 2004): 17–43, doi:10.1016 /S1359-1789(02)00116-7.
6. Brydon et al., "Hostility and Cellular Aging in Men from the Whitehall II Cohort." (See #4 above.)
7. Trapnell, P. D., and J. D. Campbell, "Private Self-Consciousness and the Five-Factor Model of Personality: Distinguishing Rumination from Reflection," *Journal of Personality and Social Psychology* 76, no. 2 (February 1999) 284–304.
8. Ibid; and Trapnell, P.D., "Rumination-Reflection Questionnaire (RRQ) Shortforms," unpublished data, University of British Columbia (1997).
9. Ibid.
10. John, O. P., E. M. Donahue, and R. L. Kentle, *The Big Five Inventory— Versions 4a and 54* (Berkeley: University of California, Berkeley, Institute of Personality and Social Research, 1991). We thank Dr. Oliver John of UC Berkeley for permission to use this scale. John, O. P., and S.

Srivastava, "The Big-Five Trait Taxonomy: History, Measurement, and Theoretical Perspectives," in *Handbook of Personality: Theory and Research*, ed. L. A. Pervin and O. P. John, 2nd ed. (New York: Guilford Press, 1999): 102–38.

11. Sadahiro, R., et al., "Relationship Between Leukocyte Telomere Length and Personality Traits in Healthy Subjects," *European Psychiatry* 30, no. 2 (February 2015): 291–95, doi:10.1016/j.eurpsy.2014.03.003, pmid: 24768472.

12. Srivastava, S., et al., "Development of Personality in Early and Middle Adulthood: Set Like Plaster or Persistent Change?" *Journal of Personality and Social Psychology* 84, no. 5 (May 2003): 1041–53, doi:10.1037/0022-3514.84.5.1041.

13. Ryff, C. D., and C. L. Keyes, "The Structure of Psychological Wellbeing Revisited," *Journal of Personality and Social Psychology* 69, no. 4 (October 1995): 719–27.

14. Scheier, M. F., et al., "The Life Engagement Test: Assessing Purpose in Life," *Journal of Behavioral Medicine* 29, no. 3 (June 2006): 291–98, doi:10.1007/s10865-005-9044-1.

15. Pearson, E. L., et al., "Normative Data and Longitudinal Invariance of the Life Engagement Test (LET) in a Community Sample of Older Adults," *Quality of Life Research* 22, no. 2 (March 2013): 327–31, doi:10.1007/s11136-012-0146-2.

Chapter Six: When Blue Turns to Gray: Depression and Anxiety

1. Whiteford, H. A., et al., "Global Burden of Disease Attributable to Mental and Substance Use Disorders: Findings from the Global Burden of Disease Study 2010," *Lancet* 382, no. 9904 (November 9, 2013): 1575–86, doi:10.1016/S0140-6736(13)61611-6.

2. Verhoeven, J. E., et al., "Anxiety Disorders and Accelerated Cellular Ageing," *British Journal of Psychiatry* 206, no. 5 (May 2015): 371–78.

3. Cai, N., et al., "Molecular Signatures of Major Depression," *Current Biology* 25, no. 9 (May 4, 2015): 1146–56, doi:10.1016/j.cub.2015.03.008.

4. Verhoeven, J. E., et al., "Major Depressive Disorder and Accelerated Cellular Aging: Results from a Large Psychiatric Cohort Study," *Molecular Psychiatry* 19, no. 8 (August 2014): 895–901, doi:10.1038/mp .2013.151.

5. Mamdani, F., et al., "Variable Telomere Length Across Post-Mortem Human Brain Regions and Specific Reduction in the Hippocampus of Major Depressive Disorder," *Translational Psychiatry* 5 (September 15, 2015): e636, doi:10.1038/tp.2015.134.

6. Zhou, Q. G., et al., "Hippocampal Telomerase Is Involved in the Modulation of Depressive Behaviors," *Journal of Neuroscience* 31, no. 34 (August 24, 2011): 12258–69, doi:10.1523/JNEUROSCI.0805-11.2011.

7. Wolkowitz, O. M., et al., "PBMC Telomerase Activity, but Not Leukocyte Telomere Length, Correlates with Hippocampal Volume in Major Depression," *Psychiatry Research* 232, no. 1 (April 30, 2015): 58–64, doi:10.1016/j.pscychresns.2015.01.007.

8. Darrow, S. M., et al., "The Association between Psychiatric Disorders and Telomere Length: A Meta-analysis Involving 14,827 Persons," *Psychosomatic Medicine* 78, no. 7 (September 2016): 776–87, doi:10.1097/PSY.0000000000000356.

9. Cai et al., "Molecular Signatures of Major Depression." (See #3 above.)

10. Verhoeven, J. E., et al., "The Association of Early and Recent Psychosocial Life Stress with Leukocyte Telomere Length," *Psychosomatic Medicine* 77, no. 8 (October 2015): 882–91, doi:10.1097/PSY.0000000000000226.

11. Verhoeven, J. E., et al., "Major Depressive Disorder and Accelerated Cellular Aging: Results from a Large Psychiatric Cohort Study," *Molecular Psychiatry* 19, no. 8 (August 2014): 895–901, doi:10.1038/mp.2013.151.

12. Ibid.

13. Cai et al., "Molecular Signatures of Major Depression." (See #3 above.)

14. Eisendrath, S. J., et al., "A Preliminary Study: Efficacy of Mindfulness-Based Cognitive Therapy Versus Sertraline as First-Line Treatments for Major Depressive Disorder," *Mindfulness* 6, no. 3 (June 1, 2015): 475–82, doi:10.1007/s12671-014-0280-8; and Kuyken, W., et al., "The Effectiveness and Cost-Effectiveness of Mindfulness-Based Cognitive Therapy Compared with Maintenance Antidepressant Treatment in the Prevention of Depressive Relapse/Recurrence: Results of a Randomised Controlled Trial (the PREVENT Study)," *Health Technology Assessment* 19, no. 73 (September 2015): 1–124, doi:10.3310/hta19730.

15. Teasdale, J. D., et al., "Prevention of Relapse/Recurrence in Major Depression by Mindfulness-Based Cognitive Therapy," *Journal of Consulting and Clinical Psychology* 68, no. 4 (August 2000): 615–23.

16. Teasdale, J., M. Williams, and Z. Segal, *The Mindful Way Workbook: An 8-Week Program to Free Yourself from Depression and Emotional Distress* (New York: Guilford Press, 2014).

17. Wolfson, W., and Epel, E. (2006), "Stress, Post-traumatic Growth, and Leukocyte Aging," poster presentation at the American Psychosomatic Society 64th Annual Meeting, Denver, Colorado, Abstract 1476.

18. Segal, Z., J. M. G. Williams, and J. Teasdale, *Mindfulness-Based Cognitive Therapy for Depression*, 2nd ed. (New York: Guilford Press, 2013), pp. 74–75.

(The three-minute breathing space is part of the MBCT program. Our breathing break is a modified version).

19. Bai, Z., et al., "Investigating the Effect of Transcendental Meditation on Blood Pressure: A Systematic Review and Meta-analysis," *Journal of Human Hypertension* 29, no. 11 (November 2015): 653–62. doi:10.1038/jhh.2015.6; and Cernes, R., and R. Zimlichman, "RESPeRATE: The Role of Paced Breathing in Hypertension Treatment," *Journal of the American Society of Hypertension* 9, no. 1 (January 2015): 38–47, doi:10.1016/j.jash.2014.10.002.

Master Tips for Renewal: Stress-Reducing Techniques Shown to Boost Telomere Maintenance

1. Morgan, N., M. R. Irwin, M. Chung, and C. Wang, "The Effects of Mind-Body Therapies on the Immune System: Meta-analysis," *PLOS ONE* 9, no. 7 (2014): e100903, doi:10.1371/journal.pone.0100903.

2. Conklin, Q., et al., "Telomere Lengthening After Three Weeks of an Intensive Insight Meditation Retreat," *Psychoneuroendocrinology* 61 (November 2015): 26–27, doi:10.1016/j.psyneuen.2015.07.462.

3. Epel, E., et al. "Meditation and Vacation Effects Impact Disease-Associated Molecular Phenotypes," *Translational Psychiatry* (August 2016): 6, e880, doi: 10.1038/tp.2016.164.

4. Kabat-Zinn, J., *Full Catastrophe Living: Using the Wisdom of Your Body and Mind to Face Stress, Pain, and Illness*, rev. ed. (New York: Bantam Books, 2013).

5. Lengacher, C. A., et al., "Influence of Mindfulness-Based Stress Reduction (MBSR) on Telomerase Activity in Women with Breast Cancer (BC)," *Biological Research for Nursing* 16, no. 4 (October 2014): 438–47, doi:10.1177/1099800413519495.

6. Carlson, L. E., et al., "Mindfulness-Based Cancer Recovery and Supportive-Expressive Therapy Maintain Telomere Length Relative to Controls in Distressed Breast Cancer Survivors," *Cancer* 121, no. 3 (February 1, 2015): 476–84, doi:10.1002/cncr.29063.

7. Black, D. S., et al., "Yogic Meditation Reverses NF-κB- and IRF-Related Transcriptome Dynamics in Leukocytes of Family Dementia Caregivers in a Randomized Controlled Trial," *Psychoneuroendocrinology* 38, no. 3 (March 2013): 348–55, doi:10.1016/j.psyneuen.2012.06.011.

8. Lavretsky, H., et al.,"A Pilot Study of Yogic Meditation for Family Dementia Caregivers with Depressive Symptoms: Effects on Mental Health, Cognition, and Telomerase Activity," *International Journal of Geriatric Psychiatry* 28, no. 1 (January 2013): 57–65, doi:10.1002/gps.3790.

9. Desveaux, L., A. Lee, R. Goldstein, and D. Brooks, "Yoga in the Management of Chronic Disease: A Systematic Review and

Meta-analysis," *Medical Care* 53, no. 7 (July 2015): 653–61, doi:10.1097/MLR.0000000000000372.

10. Hartley, L., et al., "Yoga for the Primary Prevention of Cardiovascular Disease," *Cochrane Database of Systematic Reviews* 5 (May 13, 2014): CD010072, doi:10.1002/14651858.CD010072.pub2.

11. Lu, Y. H., B. Rosner, G. Chang, and L. M. Fishman, "Twelve-Minute Daily Yoga Regimen Reverses Osteoporotic Bone Loss," *Topics in Geriatric Rehabilitation* 32, no. 2 (April 2016): 81–87.

12. Liu, X., et al., "A Systematic Review and Meta-analysis of the Effects of Qigong and Tai Chi for Depressive Symptoms," *Complementary Therapies in Medicine* 23, no. 4 (August 2015): 516–34, doi:10.1016/j.ctim.2015.05.001.

13. Freire, M. D., and C. Alves, "Therapeutic Chinese Exercises (Qigong) in the Treatment of Type 2 Diabetes Mellitus: A Systematic Review," *Diabetes & Metabolic Syndrome: Clinical Research & Reviews* 7, no. 1 (March 2013): 56–59, doi:10.1016/j.dsx.2013.02.009.

14. Ho, R. T. H., et al., "A Randomized Controlled Trial of Qigong Exercise on Fatigue Symptoms, Functioning, and Telomerase Activity in Persons with Chronic Fatigue or Chronic Fatigue Syndrome," *Annals of Behavioral Medicine* 44, no. 2 (October 2012): 160–70, doi:10.1007/s12160-012-9381-6.

15. Ornish D., et al., "Effect of Comprehensive Lifestyle Changes on Telomerase Activity and Telomere Length in Men with Biopsy-Proven Low-Risk Prostate Cancer: 5-Year Follow-Up of a Descriptive Pilot Study," *Lancet Oncology* 14, no. 11 (October 2013): 1112–20, doi: 10.1016/S1470-2045(13)70366-8.

Assessment: What's Your Telomere Trajectory? Protective and Risky Factors

1. Ahola, K., et al., "Work-Related Exhaustion and Telomere Length: A Population-Based Study," *PLOS ONE* 7, no. 7 (2012): e40186, doi:10.1371/journal.pone.0040186.

2. Damjanovic, A. K., et al., "Accelerated Telomere Erosion Is Associated with a Declining Immune Function of Caregivers of Alzheimer's Disease Patients," *Journal of Immunology* 179, no. 6 (September 15, 2007): 4249–54.

3. Geronimus, A. T., et al., "Race-Ethnicity, Poverty, Urban Stressors, and Telomere Length in a Detroit Community-Based Sample," *Journal of Health and Social Behavior* 56, no. 2 (June 2015): 199–224, doi:10.1177/0022146515582100.

4. Darrow, S. M., et al., "The Association between Psychiatric Disorders and Telomere Length: A Meta-analysis Involving 14,827 Persons,"

Psychosomatic Medicine 78, no. 7 (September 2016): 776–87, doi:10.1097 /PSY.0000000000000356; and Lindqvist et al, "Psychiatric Disorders and Leukocyte Telomere Length: Underlying Mechanisms Linking Mental Illness with Cellular Aging," *Neuroscience & Biobehavioral Reviews* 55 (August 2015): 333–64, doi:10.1016/j.neubiorev.2015.05.007.

5. Mitchell, P. H., et al., "A Short Social Support Measure for Patients Recovering from Myocardial Infarction: The ENRICHD Social Support Inventory," *Journal of Cardiopulmonary Rehabilitation* 23, no. 6 (November–December 2003): 398–403.

6. Zalli, A., et al., "Shorter Telomeres with High Telomerase Activity Are Associated with Raised Allostatic Load and Impoverished Psychosocial Resources," *Proceedings of the National Academy of Sciences of the United States of America* 111, no. 12 (March 25, 2014): 4519–24, doi:10.1073 /pnas.1322145111; and Carroll, J. E., A. V. Diez Roux, A. L. Fitzpatrick, and T. Seeman, "Low Social Support Is Associated with Shorter Leukocyte Telomere Length in Late Life: Multi-Ethnic Study of Atherosclerosis," *Psychosomatic Medicine* 75, no. 2 (February 2013): 171–77, doi:10.1097/PSY.0b013e31828233bf.

7. Carroll et al., "Low Social Support Is Associated with Shorter Leukocyte Telomere Length in Late Life: Multi-ethnic Study of Atherosclerosis." (See #6 above.)

8. Kiernan, M., et al., "The Stanford Leisure-Time Activity Categorical Item (L-Cat): A Single Categorical Item Sensitive to Physical Activity Changes in Overweight/Obese Women," *International Journal of Obesity (2005)* 37, no. 12 (December 2013): 1597–1602, doi:10.1038/ijo.2013.36.

9. Puterman, E., et al., "The Power of Exercise: Buffering the Effect of Chronic Stress on Telomere Length," *PLOS ONE* 5, no. 5 (2010): e10837, doi:10.1371/journal.pone.0010837; and Puterman, E., et al., "Determinants of Telomere Attrition over One Year in Healthy Older Women: Stress and Health Behaviors Matter," *Molecular Psychiatry* 20, no. 4 (April 2015): 529–35, doi:10.1038/mp.2014.70.

10. Werner, C., A. Hecksteden, J. Zundler, M. Boehm, T. Meyer, and U. Laufs. "Differential Effects of Aerobic Endurance, Interval and Strength Endurance Training on Telomerase Activity and Senescence Marker Expression in Circulating Mononuclear Cells." *European Heart Journal* 36 (2015) (Abstract Supplement): P2370. Manuscript in progress.

11. Buysse D. J., et al., "The Pittsburgh Sleep Quality Index: A New Instrument for Psychiatric Practice and Research," *Psychiatry Research* 28, no. 2 (May 1989): 193–213.

12. Prather, A. A., et al., "Tired Telomeres: Poor Global Sleep Quality, Perceived Stress, and Telomere Length in Immune Cell Subsets in Obese Men and Women," *Brain, Behavior, and Immunity* 47 (July 2015): 155–162, doi:10.1016/j.bbi.2014.12.011.

13. Farzaneh-Far, R., et al., "Association of Marine Omega-3 Fatty Acid Levels with Telomeric Aging in Patients with Coronary Heart Disease," *JAMA* 303, no. 3 (January 20, 2010): 250–57, doi:10.1001/jama.2009.2008.

14. Lee, J. Y., et al., "Association Between Dietary Patterns in the Remote Past and Telomere Length," *European Journal of Clinical Nutrition* 69, no. 9 (September 2015): 1048–52, doi:10.1038/ejcn.2015.58.

15. Kiecolt-Glaser, J. K., et al., "Omega-3 Fatty Acids, Oxidative Stress, and Leukocyte Telomere Length: A Randomized Controlled Trial," *Brain, Behavior, and Immunity* 28 (February 2013): 16–24, doi:10.1016/j.bbi.2012.09.004.

16. Lee, "Association between Dietary Patterns in the Remote Past and Telomere Length" (see #14 above); Leung, C. W., et al., "Soda and Cell Aging: Associations Between Sugar-Sweetened Beverage Consumption and Leukocyte Telomere Length in Healthy Adults from the National Health and Nutrition Examination Surveys," *American Journal of Public Health* 104, no. 12 (December 2014): 2425–31, doi:10.2105/AJPH.2014.302151; and Leung, C., et al., "Sugary Beverage and Food Consumption and Leukocyte Telomere Length Maintenance in Pregnant Women," *European Journal of Clinical Nutrition* (June 2016): doi:10.1038/ejcn.2016.v93.

17. Nettleton, J. A., et al., "Dietary Patterns, Food Groups, and Telomere Length in the Multi-ethnic Study of Atherosclerosis (MESA)," *American Journal of Clinical Nutrition* 88, no. 5 (November 2008): 1405–12.

18. Valdes, A. M., et al., "Obesity, Cigarette Smoking, and Telomere Length in Women," *Lancet* 366, no. 9486 (August 20–26, 2005): 662–664; and McGrath, M., et al., "Telomere Length, Cigarette Smoking, and Bladder Cancer Risk in Men and Women," *Cancer Epidemiology, Biomarkers, and Prevention* 16, no. 4 (April 2007): 815–19.

19. Kahl, V. F., et al., "Telomere Measurement in Individuals Occupationally Exposed to Pesticide Mixtures in Tobacco Fields," *Environmental and Molecular Mutagenesis* 57, no. 1 (January 2016): 74–84, doi:10.1002/em.21984.

20. Pavanello, S., et al., "Shorter Telomere Length in Peripheral Blood Lymphocytes of Workers Exposed to Polycyclic Aromatic Hydrocarbons,"

Carcinogenesis 31, no. 2 (February 2010): 216–21, doi:10.1093/carcin /bgp278.

21. Hou, L., et al., "Air Pollution Exposure and Telomere Length in Highly Exposed Subjects in Beijing, China: A Repeated-Measure Study," *Environment International* 48 (November 1, 2012): 71–77, doi:10.1016 /j.envint.2012.06.020; and Hoxha, M., et al., "Association between Leukocyte Telomere Shortening and Exposure to Traffic Pollution: A Cross-Sectional Study on Traffic Officers and Indoor Office Workers," *Environmental Health* 8 (September 21, 2009): 41, doi:10.1186/1476- 069X-8-41.

22. Wu, Y., et al., "High Lead Exposure Is Associated with Telomere Length Shortening in Chinese Battery Manufacturing Plant Workers," *Occupational and Environmental Medicine* 69, no. 8 (August 2012): 557–63, doi:10.1136/oemed-2011-100478.

23. Pavanello et al., "Shorter Telomere Length in Peripheral Blood Lymphocytes of Workers Exposed to Polycyclic Aromatic Hydrocarbons" (see #20 above); and Bin, P., et al., "Association Between Telomere Length and Occupational Polycyclic Aromatic Hydrocarbons Exposure," *Zhonghua Yu Fang Yi Xue Za Zhi* 44, no. 6 (June 2010): 535–38. (The article is in Chinese.)

Chapter Seven: Training Your Telomeres: How Much Exercise Is Enough?

1. Najarro, K., et al., "Telomere Length as an Indicator of the Robustness of B- and T-Cell Response to Influenza in Older Adults," *Journal of Infectious Diseases* 212, no. 8 (October 15, 2015): 1261–69, doi:10.1093 /infdis/jiv202.

2. Simpson, R. J., et al., "Exercise and the Aging Immune System," *Ageing Research Reviews* 11, no. 3 (July 2012): 404–20, doi:10.1016/j .arr.2012.03.003.

3. Cherkas, L. F., et al., "The Association between Physical Activity in Leisure Time and Leukocyte Telomere Length," *Archives of Internal Medicine* 168, no. 2 (January 28, 2008): 154–58, doi:10.1001/archinternmed .2007.39.

4. Loprinzi, P. D., "Leisure-Time Screen-Based Sedentary Behavior and Leukocyte Telomere Length: Implications for a New Leisure-Time Screen-Based Sedentary Behavior Mechanism," *Mayo Clinic Proceedings* 90, no. 6 (June 2015): 786–90, doi:10.1016/j.mayocp.2015.02.018; and Sjögren, P., et al., "Stand Up for Health—Avoiding Sedentary Behaviour Might Lengthen Your Telomeres: Secondary Outcomes from a Physical

Activity RCT in Older People," *British Journal of Sports Medicine* 48, no 19 (October 2014): 1407–09, doi:10.1136/bjsports-2013-093342.

5. Werner, C., et al., "Differential Effects of Aerobic Endurance, Interval and Strength Endurance Training on Telomerase Activity and Senescence Marker Expression in Circulating Mononuclear Cells," *European Heart Journal* 36 (abstract supplement) (August 2015): P2370, http://eur heartj.oxfordjournals.org/content/ehj/36/suppl_1/163.full.pdf.

6. Loprinzi, P. D., J. P. Loenneke, and E. H. Blackburn, "Movement-Based Behaviors and Leukocyte Telomere Length among US Adults," *Medicine and Science in Sports and Exercise* 47, no. 11 (November 2015): 2347–52, doi:10.1249/MSS.0000000000000695.

7. Chilton, W. L., et al., "Acute Exercise Leads to Regulation of Telomere-Associated Genes and MicroRNA Expression in Immune Cells," *PLOS ONE* 9, no. 4 (2014): e92088, doi:10.1371/journal.pone.0092088.

8. Denham, J., et al., "Increased Expression of Telomere-Regulating Genes in Endurance Athletes with Long Leukocyte Telomeres," *Journal of Applied Physiology (1985)* 120, no. 2 (January 15, 2016): 148–58, doi:10.1152/japplphysiol.00587.2015.

9. Rana, K. S., et al., "Plasma Irisin Levels Predict Telomere Length in Healthy Adults," *Age* 36, no. 2 (April 2014): 995–1001, doi:10.1007 /s11357-014-9620-9.

10. Mooren, F. C., and K. Krüger, "Exercise, Autophagy, and Apoptosis," *Progress in Molecular Biology and Translational Science* 135 (2015): 407–22, doi:10.1016/bs.pmbts.2015.07.023.

11. Hood, D. A., et al., "Exercise and the Regulation of Mitochondrial Turnover," *Progress in Molecular Biology and Translational Science* 135 (2015): 99–127, doi:10.1016/bs.pmbts.2015.07.007.

12. Loprinzi, P. D., "Cardiorespiratory Capacity and Leukocyte Telomere Length Among Adults in the United States," *American Journal of Epidemiology* 182, no. 3 (August 1, 2015): 198–201, doi:10.1093/aje/kwv056.

13. Krauss, J., et al., "Physical Fitness and Telomere Length in Patients with Coronary Heart Disease: Findings from the Heart and Soul Study," *PLOS ONE* 6, no. 11 (2011): e26983, doi:10.1371/journal.pone .0026983.

14. Denham, J., et al., "Longer Leukocyte Telomeres Are Associated with Ultra-Endurance Exercise Independent of Cardiovascular Risk Factors," *PLOS ONE* 8, no. 7 (2013): e69377, doi:10.1371/journal.pone .0069377.

15. Denham et al., "Increased Expression of Telomere-Regulating Genes in Endurance Athletes with Long Leukocyte Telomeres." (See #8 above.)

16. Laine, M. K., et al., "Effect of Intensive Exercise in Early Adult Life on Telomere Length in Later Life in Men," *Journal of Sports Science and Medicine* 14, no. 2 (June 2015): 239–45.

17. Werner, C., et al., "Physical Exercise Prevents Cellular Senescence in Circulating Leukocytes and in the Vessel Wall," *Circulation* 120, no. 24 (December 15, 2009): 2438–47, doi:10.1161/CIRCULATIONAHA .109.861005.

18. Saßenroth, D., et al., "Sports and Exercise at Different Ages and Leukocyte Telomere Length in Later Life—Data from the Berlin Aging Study II (BASE-II)," *PLOS ONE* 10, no. 12 (2015): e0142131, doi:10.1371 /journal.pone.0142131.

19. Collins, M., et al., "Athletes with Exercise-Associated Fatigue Have Abnormally Short Muscle DNA Telomeres," *Medicine and Science in Sports and Exercise* 35, no. 9 (September 2003): 1524–28.

20. Wichers, M., et al., "A Time-Lagged Momentary Assessment Study on Daily Life Physical Activity and Affect," *Health Psychology* 31, no. 2 (March 2012): 135–144, doi:10.1037/a0025688.

21. Von Haaren, B., et al., "Does a 20-Week Aerobic Exercise Training Programme Increase Our Capabilities to Buffer Real-Life Stressors? A Randomized, Controlled Trial Using Ambulatory Assessment," *European Journal of Applied Physiology* 116, no. 2 (February 2016): 383–94, doi:10.1007/s00421-015-3284-8.

22. Puterman, E., et al., "The Power of Exercise: Buffering the Effect of Chronic Stress on Telomere Length," *PLOS ONE* 5, no. 5 (2010): e10837, doi:10.1371/journal.pone.0010837.

23. Puterman, E., et al., "Multisystem Resiliency Moderates the Major Depression–Telomere Length Association: Findings from the Heart and Soul Study," *Brain, Behavior, and Immunity* 33 (October 2013): 65–73, doi:10.1016/j.bbi.2013.05.008.

24. Werner et al., "Differential Effects of Aerobic Endurance, Interval and Strength Endurance Training on Telomerase Activity and Senescence Marker Expression in Circulating Mononuclear Cells." (See #5 above.)

25. Masuki, S., et al., "The Factors Affecting Adherence to a Long-Term Interval Walking Training Program in Middle-Aged and Older People," *Journal of Applied Physiology (1985)* 118, no. 5 (March 1, 2015): 595–603, doi:10.1152/japplphysiol.00819.2014.

26. Loprinzi, "Leisure-Time Screen-Based Sedentary Behavior and Leukocyte Telomere Length." (See #4 above.)

Chapter Eight: Tired Telomeres: From Exhaustion to Restoration

1. "Lack of Sleep Is Affecting Americans, Finds the National Sleep Foundation," National Sleep Foundation, https://sleepfoundation.org /media-center/press-release/lack-sleep-affecting-americans-finds-the -national-sleep-foundation, accessed September 29, 2015.

2. Carroll, J. E., et al., "Insomnia and Telomere Length in Older Adults," *Sleep* 39, no. 3 (March 1, 2016): 559–64, doi:10.5665/sleep.5526.

3. Micic, G., et al., "The Etiology of Delayed Sleep Phase Disorder," *Sleep Medicine Reviews* 27 (June 2016): 29–38, doi:10.1016/j.smrv.2015.06.004.

4. Sachdeva, U. M., and C. B. Thompson, "Diurnal Rhythms of Autophagy: Implications for Cell Biology and Human Disease," *Autophagy* 4, no. 5 (July 2008): 581–89.

5. Gonnissen, H. K. J., T. Hulshof, and M. S. Westerterp-Plantenga, "Chronobiology, Endocrinology, and Energy-and-Food-Reward Homeostasis," *Obesity Reviews* 14, no. 5 (May 2013): 405–16, doi:10.1111 /obr.12019.

6. Van der Helm, E., and M. P. Walker, "Sleep and Emotional Memory Processing," *Journal of Clinical Sleep Medicine* 6, no. 1 (March 2011): 31–43.

7. Meerlo, P., A. Sgoifo, and D. Suchecki, "Restricted and Disrupted Sleep: Effects on Autonomic Function, Neuroendocrine Stress Systems and Stress Responsivity," *Sleep Medicine Reviews* 12, no. 3 (June 2008): 197–210, doi:10.1016/j.smrv.2007.07.007.

8. Walker, M. P., "Sleep, Memory, and Emotion," *Progress in Brain Research* 185 (2010): 49–68, doi:10.1016/B978-0-444-53702-7.00004-X.

9. Lee, K. A., et al., "Telomere Length Is Associated with Sleep Duration but Not Sleep Quality in Adults with Human Immunodeficiency Virus," *Sleep* 37, no. 1 (January 1, 2014): 157–66, doi:10.5665/sleep.3328; and Cribbet, M. R., et al., "Cellular Aging and Restorative Processes: Subjective Sleep Quality and Duration Moderate the Association between Age and Telomere Length in a Sample of Middle-Aged and Older Adults," *Sleep* 37, no. 1 (January 1, 2014): 65–70, doi:10.5665/sleep.3308.

10. Jackowska, M., et. al., "Short Sleep Duration Is Associated with Shorter Telomere Length in Healthy Men: Findings from the Whitehall II Cohort Study," *PLOS ONE* 7, no. 10 (2012): e47292, doi:10.1371/journal .pone.0047292.

11. Cribbet et al., "Cellular Aging and Restorative Processes." (See #9 above.)

12. Ibid.

13. Prather, A. A., et al., "Tired Telomeres: Poor Global Sleep Quality, Perceived Stress, and Telomere Length in Immune Cell Subsets in Obese Men and Women," *Brain, Behavior, and Immunity* 47 (July 2015): 155–62, doi:10.1016/j.bbi.2014.12.011.

14. Chen, W. D., et al., "The Circadian Rhythm Controls Telomeres and Telomerase Activity," *Biochemical and Biophysical Research Communications* 451, no. 3 (August 29, 2014): 408–14, doi:10.1016/j.bbrc.2014.07.138.

15. Ong, J., and D. Sholtes, "A Mindfulness-Based Approach to the Treatment of Insomnia," *Journal of Clinical Psychology* 66, no. 11 (November 2010): 1175–84, doi:10.1002/jclp.20736.

16. Ong, J. C., et al., "A Randomized Controlled Trial of Mindfulness Meditation for Chronic Insomnia," *Sleep* 37, no. 9 (September 1, 2014): 1553–63B, doi:10.5665/sleep.4010.

17. Chang, A. M., D. Aeschbach, J. F. Duffy, and C. A. Czeisler, "Evening Use of Light-Emitting eReaders Negatively Affects Sleep, Circadian Timing, and Next-Morning Alertness," *Proceedings of the National Academy of Sciences of the United States of America* 112, no. 4 (January 2015): 1232–37, doi:10.1073/pnas.1418490112.

18. Dang-Vu, T. T., et al., "Spontaneous Brain Rhythms Predict Sleep Stability in the Face of Noise," *Current Biology* 20, no. 15 (August 10, 2010): R626–27, doi:10.1016/j.cub.2010.06.032.

19. Griefhan, B., P. Bröde, A. Marks, and M. Basner, "Autonomic Arousals Related to Traffic Noise During Sleep," *Sleep* 31, no. 4 (April 2008): 569–77.

20. Savolainen, K., et al., "The History of Sleep Apnea Is Associated with Shorter Leukocyte Telomere Length: The Helsinki Birth Cohort Study," *Sleep Medicine* 15, no. 2 (February 2014): 209–12, doi:10.1016/j.sleep.2013.11.779.

21. Salihu, H. M., et al., "Association Between Maternal Symptoms of Sleep Disordered Breathing and Fetal Telomere Length," *Sleep* 38, no. 4 (April 1, 2015): 559–66, doi:10.5665/sleep.4570.

22. Shin, C., C. H. Yun, D. W. Yoon, and I. Baik, "Association Between Snoring and Leukocyte Telomere Length," *Sleep* 39, no. 4 (April 1, 2016): 767–72, doi:10.5665/sleep.5624.

Chapter Nine: Telomeres Weigh In: A Healthy Metabolism

1. Mundstock, E., et al., "Effect of Obesity on Telomere Length: Systematic Review and Meta-analysis," *Obesity (Silver Spring)* 23, no. 11 (November 2015): 2165–74, doi:10.1002/oby.21183.

2. Bosello, O., M. P. Donataccio, and M. Cuzzolaro, "Obesity or Obesities? Controversies on the Association Between Body Mass Index and

Premature Mortality," *Eating and Weight Disorders* 21, no. 2 (June 2016): 165–74, doi:10.1007/s40519-016-0278-4.

3. Farzaneh-Far, R., et al., "Telomere Length Trajectory and Its Determinants in Persons with Coronary Artery Disease: Longitudinal Findings from the Heart and Soul Study," *PLOS ONE* 5, no. 1 (January 2010): e8612, doi:10.1371/journal.pone.0008612.

4. "IDF Diabetes Atlas, Sixth Edition," *International Diabetes Federation*, http://www.idf.org/atlasmap/atlasmap?indicator=i1&date=2014, accessed September 16, 2015.

5. Farzaneh-Far et al., "Telomere Length Trajectory and Its Determinants in Persons with Coronary Artery Disease." (See #3 above.)

6. Verhulst, S., et al., "A Short Leucocyte Telomere Length Is Associated with Development of Insulin Resistance," *Diabetologia* 59, no. 6 (June 2016): 1258–65, doi:10.1007/s00125-016-3915-6.

7. Zhao, J., et al., "Short Leukocyte Telomere Length Predicts Risk of Diabetes in American Indians: The Strong Heart Family Study," *Diabetes* 63, no. 1 (January 2014): 354–62, doi:10.2337/db13-0744.

8. Willeit, P., et al., "Leucocyte Telomere Length and Risk of Type 2 Diabetes Mellitus: New Prospective Cohort Study and Literature-Based Meta-analysis," *PLOS ONE* 9, no. 11 (2014): e112483, doi:10.1371/journal.pone.0112483.

9. Guo, N., et al., "Short Telomeres Compromise β-Cell Signaling and Survival," *PLOS ONE* 6, no. 3 (2011): e17858, doi:10.1371/journal.pone.0017858.

10. Formichi, C., et al., "Weight Loss Associated with Bariatric Surgery Does Not Restore Short Telomere Length of Severe Obese Patients after 1 Year," *Obesity Surgery* 24, no. 12 (December 2014): 2089–93, doi:10.1007/s11695-014-1300-4.

11. Gardner, J. P., et al., "Rise in Insulin Resistance is Associated with Escalated Telomere Attrition," *Circulation* 111, no. 17 (May 3, 2005): 2171–77.

12. Fothergill, Erin, Juen Guo, Lilian Howard, Jennifer C. Kerns, Nicolas D. Knuth, Robert Brychta, Kong Y. Chen, et al. "Persistent Metabolic Adaptation Six Years after *The Biggest Loser* Competition," *Obesity* (Silver Spring, Md.), May 2, 2016, doi:10.1002/oby.21538.

13. Kim, S., et al., "Obesity and Weight Gain in Adulthood and Telomere Length," *Cancer Epidemiology, Biomarkers & Prevention* 18, no. 3 (March 2009): 816–20, doi:10.1158/1055-9965.EPI-08-0935.

14. Cottone, P., et al., "CRF System Recruitment Mediates Dark Side of Compulsive Eating," *Proceedings of the National Academy of Sciences of*

the United States of America 106, no. 47 (November 2009): 20016–20, doi:0.1073/pnas.0908789106.

15. Tomiyama, A. J., et al., "Low Calorie Dieting Increases Cortisol," *Psychosomatic Medicine* 72, no. 4 (May 2010): 357–64, doi:10.1097 /PSY.0b013e3181d9523c.

16. Kiefer, A., J. Lin, E. Blackburn, and E. Epel, "Dietary Restraint and Telomere Length in Pre- and Post-Menopausal Women," *Psychosomatic Medicine* 70, no. 8 (October 2008): 845–49, doi:10.1097/PSY.0b013 e318187d05e.

17. Hu, F. B., "Resolved: There Is Sufficient Scientific Evidence That Decreasing Sugar-Sweetened Beverage Consumption Will Reduce the Prevalence of Obesity and Obesity-Related Diseases," *Obesity Reviews* 14, no. 8 (August 2013): 606–19, doi:10.1111/obr.12040; and Yang, Q., et al., "Added Sugar Intake and Cardiovascular Diseases Mortality Among U.S. Adults," *JAMA Internal Medicine* 174, no. 4 (April 2014): 516–24, doi:10.1001/jamainternmed.2013.13563.

18. Schulte, E. M., N. M. Avena, and A. N. Gearhardt, "Which Foods May Be Addictive? The Roles of Processing, Fat Content, and Glycemic Load," *PLOS ONE* 10, no. 2 (February 18, 2015): e0117959, doi:10.1371 /journal.pone.0117959.

19. Lustig, R. H., et al., "Isocaloric Fructose Restriction and Metabolic Improvement in Children with Obesity and Metabolic Syndrome," *Obesity* 2 (February 24, 2016): 453–60, doi:10.1002/oby.21371, epub October 26, 2015.

20. Incollingo Belsky, A. C., E. S. Epel, and A. J. Tomiyama, "Clues to Maintaining Calorie Restriction? Psychosocial Profiles of Successful Long-Term Restrictors," *Appetite* 79 (August 2014): 106–12, doi:10.1016 /j.appet.2014.04.006.

21. Wang, C., et al., "Adult-Onset, Short-Term Dietary Restriction Reduces Cell Senescence in Mice," *Aging* 2, no. 9 (September 2010): 555–66.

22. Daubenmier, J., et al., "Changes in Stress, Eating, and Metabolic Factors Are Related to Changes in Telomerase Activity in a Randomized Mindfulness Intervention Pilot Study," *Psychoneuroendocrinology* 37, no. 7 (July 2012): 917–28, doi:10.1016/j.psyneuen.2011.10.008.

23. Mason, A. E., et al., "Effects of a Mindfulness-Based Intervention on Mindful Eating, Sweets Consumption, and Fasting Glucose Levels in Obese Adults: Data from the SHINE Randomized Controlled Trial," *Journal of Behavioral Medicine* 39, no. 2 (April 2016): 201–13, doi:10.1007/s10865-015-9692-8.

24. Kristeller, J., with A. Bowman, *The Joy of Half a Cookie: Using Mindfulness to Lose Weight and End the Struggle with Food* (New York: Perigee, 2015). Also see www.mindfuleatingtraining.com and www.mb-eat.com.

Chapter Ten: Food and Telomeres: Eating for Optimal Cell Health

1. Jurk, D., et al., "Chronic Inflammation Induces Telomere Dysfunction and Accelerates Ageing in Mice," *Nature Communications* 2 (June 24, 2104): 4172, doi:10.1038/ncomms5172.

2. "What You Eat Can Fuel or Cool Inflammation, A Key Driver of Heart Disease, Diabetes, and Other Chronic Conditions," Harvard Medical School, Harvard Health Publications, http://www.health.harvard.edu/family_health_guide/what-you-eat-can-fuel-or-cool-inflammation-a-key-driver-of-heart-disease-diabetes-and-other-chronic-conditions, accessed November 27, 2015.

3. Weischer, M., S. E. Bojesen, and B. G. Nordestgaard, "Telomere Shortening Unrelated to Smoking, Body Weight, Physical Activity, and Alcohol Intake: 4,576 General Population Individuals with Repeat Measurements 10 Years Apart," *PLOS Genetics* 10, no. 3 (March 13, 2014): e1004191, doi:10.1371/journal.pgen.1004191; and Pavanello, S., et al., "Shortened Telomeres in Individuals with Abuse in Alcohol Consumption," *International Journal of Cancer* 129, no. 4 (August 15, 2011): 983–92. doi:10.1002/ijc.25999.

4. Cassidy, A., et al., "Higher Dietary Anthocyanin and Flavonol Intakes Are Associated with Anti-inflammatory Effects in a Population of U.S. Adults," *American Journal of Clinical Nutrition* 102, no. 1 (July 2015): 172–81, doi:10.3945/ajcn.115.108555.

5. Farzaneh-Far, R., et al., "Association of Marine Omega-3 Fatty Acid Levels with Telomeric Aging in Patients with Coronary Heart Disease," *JAMA* 303, no. 3 (January 20, 2010): 250–57, doi:10.1001/jama.2009.2008.

6. Goglin, S., et al., "Leukocyte Telomere Shortening and Mortality in Patients with Stable Coronary Heart Disease from the Heart and Soul Study," *PLOS ONE* (2016), in press.

7. Farzaneh-Far et al., "Association of Marine Omega-3 Fatty Acid Levels with Telomeric Aging in Patients with Coronary Heart Disease." (See #5 above.)

8. Kiecolt-Glaser, J. K., et. al., "Omega-3 Fatty Acids, Oxidative Stress, and Leukocyte Telomere Length: A Randomized Controlled Trial," *Brain, Behavior, and Immunity* 28 (February 2013): 16–24, doi:10.1016/j.bbi.2012.09.004.

9. Glei, D. A., et al., "Shorter Ends, Faster End? Leukocyte Telomere Length and Mortality Among Older Taiwanese," *Journals of Gerontology, Series A: Biological Sciences and Medical Sciences* 70, no. 12 (December 2015): 1490–98, doi:10.1093/gerona/glu191.

10. Debreceni, B., and L. Debreceni, "The Role of Homocysteine-Lowering B-Vitamins in the Primary Prevention of Cardiovascular Disease," *Cardiovascular Therapeutics* 32, no. 3 (June 2014): 130–38, doi:10.1111/1755 -5922.12064.

11. Kawanishi, S., and S. Oikawa, "Mechanism of Telomere Shortening by Oxidative Stress," *Annals of the New York Academy of Sciences* 1019 (June 2004): 278–84.

12. Haendeler, J., et al., "Hydrogen Peroxide Triggers Nuclear Export of Telomerase Reverse Transcriptase via Src Kinase Familiy-Dependent Phosphorylation of Tyrosine 707," *Molecular and Cellular Biology* 23, no. 13 (July 2003): 4598–610.

13. Adelfalk, C., et al., "Accelerated Telomere Shortening in Fanconi Anemia Fibroblasts—a Longitudinal Study," *FEBS Letters* 506, no. 1 (September 28, 2001): 22–26.

14. Xu, Q., et al., "Multivitamin Use and Telomere Length in Women," *American Journal of Clinical Nutrition* 89, no. 6 (June 2009): 1857–63, doi:10.3945/ajcn.2008.26986, epub March 11, 2009.

15. Paul, L., et al., "High Plasma Folate Is Negatively Associated with Leukocyte Telomere Length in Framingham Offspring Cohort," *European Journal of Nutrition* 54, no. 2 (March 2015): 235–41, doi:10.1007 /s00394-014-0704-1.

16. Wojcicki, J., et al., "Early Exclusive Breastfeeding Is Associated with Longer Telomeres in Latino Preschool Children," *American Journal of Clinical Nutrition* (July 20, 2016), doi:10.3945/ajcn.115.115428.

17. Leung, C. W., et al., "Soda and Cell Aging: Associations between Sugar-Sweetened Beverage Consumption and Leukocyte Telomere Length in Healthy Adults from the National Health and Nutrition Examination Surveys," *American Journal of Public Health* 104, no. 12 (December 2014): 2425–31, doi:10.2105/AJPH.2014.302151.

18. Wojcicki, et al "Early Exclusive Breastfeeding Is Associated with Longer Telomeres in Latino Preschool Children." (See #16 above.)

19. "Peppermint Mocha," Starbucks, http://www.starbucks.com/menu/drinks /espresso/peppermint-mocha#size=179560&milk=63&whip=125,accessed September 29, 2015.

20. Pilz, Stefan, Martin Grübler, Martin Gaksch, Verena Schwetz, Christian Trummer, Bríain Ó Hartaigh, Nicolas Verheyen, Andreas Tomaschitz,

and Winfried März. "Vitamin D and Mortality." *Anticancer Research* 36, no. 3 (March 2016): 1379–87.

21. Zhu et al., "Increased Telomerase Activity and Vitamin D Supplementation in Overweight African Americans," *International Journal of Obesity* (June 2012): 805–09, doi:10.1038/ijo.2011.197.

22. Boccardi, V., et al., "Mediterranean Diet, Telomere Maintenance and Health Status Among Elderly," *PLOS ONE* 8, no.4 (April 30, 2013): e62781, doi:10.1371/journal.pone.0062781.

23. Lee, J. Y., et al., "Association Between Dietary Patterns in the Remote Past and Telomere Length," *European Journal of Clinical Nutrition* 69, no. 9 (September 2015): 1048–52, doi:10.1038/ejcn.2015.58.

24. Ibid.

25. "IARC Monographs Evaluate Consumption of Red Meat and Processed Meat," World Health Organization, International Agency for Research on Cancer, press release, October 26, 2015, https://www.iarc.fr/en /media-centre/pr/2015/pdfs/pr240_E.pdf.

26. Nettleton, J. A., et al., "Dietary Patterns, Food Groups, and Telomere Length in the Multi-Ethnic Study of Atherosclerosis (MESA)," *American Journal of Clinical Nutrition* 88, no. 5 (November 2008): 1405–12.

27. Cardin, R., et al., "Effects of Coffee Consumption in Chronic Hepatitis C: A Randomized Controlled Trial," *Digestive and Liver Disease* 45, no. 6 (June 2013): 499–504, doi:10.1016/j.dld.2012.10.021.

28. Liu, J. J., M. Crous-Bou, E. Giovannucci, and I. De Vivo, "Coffee Consumption Is Positively Associated with Longer Leukocyte Telomere Length" in the Nurses' Health Study. *Journal of Nutrition* 146, no. 7 (July 2016): 1373–78, doi:10.3945/jn.116.230490, epub June 8, 2016.

29. Lee, J. Y., et al., "Association Between Dietary Patterns in the Remote Past and Telomere Length" (see #23 above); and Nettleton et al., "Dietary Patterns, Food Groups, and Telomere Length in the Multi-Ethnic Study of Atherosclerosis (MESA)" (see #26 above).

30. García-Calzón, S., et al., "Telomere Length as a Biomarker for Adiposity Changes after a Multidisciplinary Intervention in Overweight/Obese Adolescents: The EVASYON Study," *PLOS ONE* 9, no. 2 (February 24, 2014): e89828, doi:10.1371/journal.pone.0089828.

31. Lee et al., "Association Between Dietary Patterns in the Remote Past and Telomere Length." (See #23 above.)

32. Leung et al., "Soda and Cell Aging." (See #17 above.)

33. Tiainen, A. M., et al., "Leukocyte Telomere Length and Its Relation to Food and Nutrient Intake in an Elderly Population," *European Journal of Clinical Nutrition* 66, no. 12 (December 2012):1290–94, doi:10.1038/ejcn.2012.143.

34. Cassidy, A., et al., "Associations Between Diet, Lifestyle Factors, and Telomere Length in Women," *American Journal of Clinical Nutrition* 91, no. 5 (May 2010): 1273–80, doi:10.3945/ajcn.2009.28947.

35. Pavanello, et al., "Shortened Telomeres in Individuals with Abuse in Alcohol Consumption." (See #3 above.)

36. Cassidy et al., "Associations Between Diet, Lifestyle Factors, and Telomere Length in Women." (See #34 above.)

37. Tiainen et al., "Leukocyte Telomere Length and Its Relation to Food and Nutrient Intake in an Elderly Population." (See #33 above.)

38. Lee et al., "Association Between Dietary Patterns in the Remote Past and Telomere Length." (See #23 above.)

39. Ibid.

40. Ibid.

41. Farzaneh-Far et al., "Association of Marine Omega-3 Fatty Acid Levels With Telomeric Aging in Patients with Coronary Heart Disease." (See #5 above.)

42. García-Calzón et al., "Telomere Length as a Biomarker for Adiposity Changes after a Multidisciplinary Intervention in Overweight/Obese Adolescents: The EVASYON Study." (See #30 above.)

43. Liu et al., "Coffee Consumption Is Positively Associated with Longer Leukocyte Telomere Length" in the Nurses' Health Study. (See #28 above.)

44. Paul, L., "Diet, Nutrition and Telomere Length," *Journal of Nutritional Biochemistry* 22, no. 10 (October 2011): 895–901, doi:10.1016/j.jnutbio.2010.12.001.

45. Richards, J. B., et al., "Higher Serum Vitamin D Concentrations Are Associated with Longer Leukocyte Telomere Length in Women," *American Journal of Clinical Nutrition* 86, no. 5 (November 2007): 1420–25;

46. Xu et al., "Multivitamin Use and Telomere Length in Women" (see #14 above).

47. Paul et al., "High Plasma Folate Is Negatively Associated with Leukocyte Telomere Length in Framingham Offspring Cohort." (This study also found vitamin use was associated with shorter telomeres.) (See #15 above.)

48. O'Neill, J., T. O. Daniel, and L. H. Epstein, "Episodic Future Thinking Reduces Eating in a Food Court," *Eating Behaviors* 20 (January 2016): 9–13, doi:10.1016/j.eatbeh.2015.10.002.

Master Tips for Renewal: Science-Based Suggestions for Making Changes That Last

1. Vasilaki, E. I., S. G. Hosier, and W. M. Cox, "The Efficacy of Motivational Interviewing as a Brief Intervention for Excessive Drinking: A Meta-analytic

Review," *Alcohol and Alcoholism* 41, no. 3 (May 2006): 328–35, doi:10.1093 /alcalc/agl016; and Lindson-Hawley, N., T. P. Thompson, and R. Begh, "Motivational Interviewing for Smoking Cessation," *Cochrane Database of Systematic Reviews* 3 (March 2, 2015): CD006936, doi:10.1002/14651858 .CD006936.pub3.

2. Sheldon, K. M., A. Gunz, C. P. Nichols, and Y. Ferguson, "Extrinsic Value Orientation and Affective Forecasting: Overestimating the Rewards, Under-estimating the Costs," *Journal of Personality* 78, no. 1 (February 2010): 149–78, doi:10.1111/j.1467-6494.2009.00612.x; Kasser, T., and R. M. Ryan, "Further Examining the American Dream: Differential Correlates of Intrinsic and Extrinsic Goals," *Personality and Social Psychology Bulletin* 22, no. 3 (March 1996): 280–87, doi:10.1177/0146167296223006; and Ng, J. Y., et al., "Self-Determination Theory Applied to Health Con-texts: A Meta-analysis," *Perspectives on Psychological Science: A Journal of the Association for Psychological Science* 7, no. 4 (July 2012): 325–40, doi:10.1177/1745691612447309.

3. Ogedegbe, G. O., et al., "A Randomized Controlled Trial of Positive-Affect Intervention and Medication Adherence in Hypertensive African Americans," *Archives of Internal Medicine* 172, no. 4 (February 27, 2012): 322–26, doi:10.1001/archinternmed.2011.1307.

4. Bandura, A., "Self-Efficacy: Toward a Unifying Theory of Behavioral Change." *Psychological Review* 84, no. 2 (March 1977): 191–215.

5. B. J. Fogg illustrates his suggestion of making tiny changes attached to daily trigger events: "Forget Big Change, Start with a Tiny Habit: BJ Fogg at TEDxFremont," YouTube, https://www.youtube.com/watch?v=AdKU Jxjn-R8.

6. Baumeister, R. F., "Self-Regulation, Ego Depletion, and Inhibition," *Neuropsychologia* 65 (December 2014): 313–19, doi:10.1016/j.neuropsycho logia.2014.08.012.

Chapter Eleven: The Places and Faces That Support Our Telomeres

1. Needham, B. L., et al., "Neighborhood Characteristics and Leukocyte Telomere Length: The Multi-ethnic Study of Atherosclerosis," *Health & Place* 28 (July 2014): 167–72, doi:10.1016/j.healthplace.2014.04.009.

2. Geronimus, A. T., et al., "Race-Ethnicity, Poverty, Urban Stress-ors, and Telomere Length in a Detroit Community-Based Sample," *Journal of Health and Social Behavior* 56, no. 2 (June 2015): 199–224, doi:10.1177/0022146515582100.

3. Park, M., et al., "Where You Live May Make You Old: The Associa-tion Between Perceived Poor Neighborhood Quality and Leukocyte

Telomere Length," *PLOS ONE* 10, no. 6 (June 17, 2015): e0128460, doi:10.1371/journal.pone.0128460.

4. Ibid.

5. Lederbogen, F., et al., "City Living and Urban Upbringing Affect Neural Social Stress Processing in Humans," *Nature* 474, no. 7352 (June 22, 2011): 498–501, doi:10.1038/nature10190.

6. Park et al., "Where You Live May Make You Old." (See #3 above.)

7. DeSantis, A. S., et al., "Associations of Neighborhood Characteristics with Sleep Timing and Quality: The Multi-ethnic Study of Atherosclerosis," *Sleep* 36, no. 10 (October 1, 2013): 1543–51, doi:10.5665/sleep.3054.

8. Theall, K. P., et al., "Neighborhood Disorder and Telomeres: Connecting Children's Exposure to Community Level Stress and Cellular Response," *Social Science & Medicine (1982)* 85 (May 2013): 50–58, doi:10.1016/j.socscimed.2013.02.030.

9. Woo, J., et al., "Green Space, Psychological Restoration, and Telomere Length," *Lancet* 373, no. 9660 (January 24, 2009): 299–300, doi:10.1016/S0140-6736(09)60094-5.

10. Roe, J. J., et al., "Green Space and Stress: Evidence from Cortisol Measures in Deprived Urban Communities," *International Journal of Environmental Research and Public Health* 10, no. 9 (September 2013): 4086–103, doi:10.3390/ijerph10094086.

11. Mitchell, R., and F. Popham, "Effect of Exposure to Natural Environment on Health Inequalities: An Observational Population Study," *Lancet* 372, no. 9650 (November 8, 2008): 1655–60, doi:10.1016 /S0140-6736(08)61689-X.

12. Theall et al., "Neighborhood Disorder and Telomeres." (See #8 above.)

13. Robertson, T., et al., "Is Socioeconomic Status Associated with Biological Aging as Measured by Telomere Length?" *Epidemiologic Reviews* 35 (2013): 98–111, doi:10.1093/epirev/mxs001.

14. Adler, N. E., et al., "Socioeconomic Status and Health: The Challenge of the Gradient," *American Psychologist* 49, no. 1 (January 1994): 15–24.

15. Cherkas, L. F., et al., "The Effects of Social Status on Biological Aging as Measured by White-Blood-Cell Telomere Length," *Aging Cell* 5, no. 5 (October 2006): 361–65, doi:10.1111/j.1474-9726.2006.00222.x.

16. "Canary Used for Testing for Carbon Monoxide," Center for Construction Research and Training, Electronic Library of Construction Occupational Safety & Health, http://elcosh.org/video/3801/a000096 /canary-used-for-testing-for-carbon-monoxide.html.

17. Hou, L., et al., "Lifetime Pesticide Use and Telomere Shortening Among Male Pesticide Applicators in the Agricultural Health Study," *Environ-*

mental Health Perspectives 121, no. 8 (August 2013): 919–24, doi:10.1289 /ehp.1206432.

18. Kahl, V. F., et al., "Telomere Measurement in Individuals Occupationally Exposed to Pesticide Mixtures in Tobacco Fields," *Environmental and Molecular Mutagenesis* 57, no. 1 (January 2016), doi:10.1002/em.21984.

19. Ibid.

20. Zota A. R., et al., "Associations of Cadmium and Lead Exposure with Leukocyte Telomere Length: Findings from National Health and Nutrition Examination Survey, 1999–2002," *American Journal of Epidemiology* 181, no. 2 (January 15, 2015): 127–136, doi:10.1093/aje/kwu293.

21. "Toxicological Profile for Cadmium," U.S. Department of Health and Human Services, Public Health Service, Agency for Toxic Substances and Disease Registry (Atlanta, Ga., September 2012), http://www.atsdr .cdc.gov/toxprofiles/tp5.pdf.

22. Lin, S., et al., "Short Placental Telomere Was Associated with Cadmium Pollution in an Electronic Waste Recycling Town in China," *PLOS ONE* 8, no. 4 (2013): e60815, doi:10.1371/journal.pone.0060815.

23. Zota et al., "Associations of Cadmium and Lead Exposure with Leukocyte Telomere Length." (See #20 above.)

24. Wu, Y., et al., "High Lead Exposure Is Associated with Telomere Length Shortening in Chinese Battery Manufacturing Plant Workers," *Occupational and Environmental Medicine* 69, no. 8 (August 2012): 557–63, doi:10.1136/oemed-2011-100478.

25. Ibid.

26. Pawlas, N., et al., "Telomere Length in Children Environmentally Exposed to Low-to-Moderate Levels of Lead," *Toxicology and Applied Pharmacology* 287, no. 2 (September 1, 2015): 111–18, doi:10.1016/j .taap.2015.05.005.

27. Hoxha, M., et al., "Association Between Leukocyte Telomere Shortening and Exposure to Traffic Pollution: A Cross-Sectional Study on Traffic Officers and Indoor Office Workers," *Environmental Health* 8 (2009): 41, doi:10.1186/1476-069X-8-41; Zhang, X., S. Lin, W. E. Funk, and L. Hou, "Environmental and Occupational Exposure to Chemicals and Telomere Length in Human Studies," *Postgraduate Medical Journal* 89, no. 1058 (December 2013): 722–28, doi:10.1136/postgradmedj -2012-101350rep; and Mitro, S. D., L. S. Birnbaum, B. L. Needham, and A. R. Zota, "Cross-Sectional Associations Between Exposure to Persistent Organic Pollutants and Leukocyte Telomere Length Among U.S. Adults in NHANES, 2001–2002," *Environmental Health Perspectives* 124, no. 5 (May 2016): 651–58, doi:10.1289/ehp.1510187.

28. Bijnens, E., et al., "Lower Placental Telomere Length May Be Attributed to Maternal Residental Traffic Exposure; A Twin Study," *Environment International* 79 (June 2015): 1–7, doi:0.1016/j.envint.2015.02.008.

29. Ferrario, D., et al., "Arsenic Induces Telomerase Expression and Maintains Telomere Length in Human Cord Blood Cells," *Toxicology* 260, nos. 1–3 (June 16, 2009): 132–41, doi:10.1016/j.tox.2009.03.019; Hou, L., et al., "Air Pollution Exposure and Telomere Length in Highly Exposed Subjects in Beijing, China: A Repeated-Measure Study," *Environment International* 48 (November 1, 2012): 71–77, doi:10.1016/j .envint.2012.06.020; Zhang et al., "Environmental and Occupational Exposure to Chemicals and Telomere Length in Human Studies"; Bassig, B. A., et al., "Alterations in Leukocyte Telomere Length in Workers Occupationally Exposed to Benzene," *Environmental and Molecular Mutagenesis* 55, no. 8 (2014): 673–78, doi:10.1002/em.21880; and Li, H., K. Engström, M. Vahter, and K. Broberg, "Arsenic Exposure Through Drinking Water Is Associated with Longer Telomeres in Peripheral Blood," *Chemical Research in Toxicology* 25, no. 11 (November 19, 2012): 2333–39, doi:10.1021/tx300222t.

30. American Association for Cancer Research, *AACR Cancer Progress Report 2014: Transforming Lives Through Cancer Research*, 2014, http://cancer-progressreport.org/2014/Documents/AACR_CPR_2014.pdf, accessed October 21, 2015.

31. "Cancer Fact Sheet No. 297," World Health Organization, updated February 2015,: http://www.who.int/mediacentre/factsheets/fs297/en/, accessed October 21, 2015.

32. House, J. S., K. R. Landis, and D. Umberson, "Social Relationships and Health," *Science* 241, no. 4865 (July 29, 1988): 540–45; Berkman, L. F., and S. L. Syme, "Social Networks, Host Resistance, and Mortality: A Nine-Year Follow-up Study of Alameda County Residents," *American Journal of Epidemiology* 109, no. 2 (February 1979): 186–204; and Holt-Lunstad, J., T. B. Smith, M. B. Baker, T. Harris, and D. Stephenson, "Loneliness and Social Isolation as Risk Factors for Mortality: A Meta-analytic Review," *Perspectives on Psychological Science: A Journal of the Association for Psychological Science* 10, no. 2 (March 2015): 227–37, doi:10.1177/1745691614568352.

33. Hermes, G. L., et al., "Social Isolation Dysregulates Endocrine and Behavioral Stress While Increasing Malignant Burden of Spontaneous Mammary Tumors," *Proceedings of the National Academy of Sciences of the United States of America* 106, no. 52 (December 29, 2009): 22393–98, doi:10.1073/pnas.0910753106.

34. Aydinonat, D., et al., "Social Isolation Shortens Telomeres in African Grey Parrots (*Psittacus erithacus erithacus*)," *PLOS ONE* 9, no. 4 (2014): e93839, doi:10.1371/journal.pone.0093839.

35. Carroll, J. E., A. V. Diez Roux, A. L. Fitzpatrick, and T. Seeman, "Low Social Support Is Associated with Shorter Leukocyte Telomere Length in Late Life: Multi-ethnic Study of Atherosclerosis," *Psychosomatic Medicine* 75, no. 2 (February 2013): 171–77, doi:10.1097/PSY.0b013e31828233bf.

36. Uchino, B. N., et al., "The Strength of Family Ties: Perceptions of Network Relationship Quality and Levels of C-Reactive Proteins in the North Texas Heart Study," *Annals of Behavioral Medicine* 49, no. 5 (October 2015): 776–81, doi:10.1007/s12160-015-9699-y.

37. Uchino, B. N., et al., "Social Relationships and Health: Is Feeling Positive, Negative, or Both (Ambivalent) About Your Social Ties Related to Telomeres?" *Health Psychology* 31, no. 6 (November 2012): 789–96, doi:10.1037/a0026836.

38. Robles, T. F., R. B. Slatcher, J. M. Trombello, and M. M. McGinn, "Marital Quality and Health: A Meta-analytic Review," *Psychological Bulletin* 140, no. 1 (January 2014): 140–87, doi:10.1037/a0031859.

39. Ibid.

40. Mainous, A. G., et al., "Leukocyte Telomere Length and Marital Status among Middle-Aged Adults," *Age and Ageing* 40, no. 1 (January 2011): 73–78, doi:10.1093/ageing/afq118; and Yen, Y., and F. Lung, "Older Adults with Higher Income or Marriage Have Longer Telomeres," *Age and Ageing* 42, no. 2 (March 2013): 234–39, doi:10.1093/ageing/afs122.

41. Broer, L., V. Codd, D. R. Nyholt, et al, "Meta-Analysis of Telomere Length in 19,713 Subjects Reveals High Heritability, Stronger Maternal Inheritance and a Paternal Age Effect," *European Journal of Human Genetics: EJHG* 21, no. 10 (October 2013): 1163–68, doi:10.1038/ejhg.2012.303.

42. Herbenick, D., et al., "Sexual Behavior in the United States: Results from a National Probability Sample of Men and Women Ages 14–94," *Journal of Sexual Medicine* 7, Suppl. 5 (October 7, 2010): 255–65, doi:10.1111/j.1743-6109.2010.02012.x.

43. Saxbe, D. E., et al., "Cortisol Covariation within Parents of Young Children: Moderation by Relationship Aggression," *Psychoneuroendocrinology* 62 (December 2015): 121–28, doi:10.1016/j.psyneuen.2015.08.006.

44. Liu, S., M. J. Rovine, L. C. Klein, and D. M. Almeida, "Synchrony of Diurnal Cortisol Pattern in Couples," *Journal of Family Psychology* 27, no. 4 (August 2013): 579–88, doi:10.1037/a0033735.

45. Helm, J. L., D. A. Sbarra, and E. Ferrer, "Coregulation of Respiratory

Sinus Arrhythmia in Adult Romantic Partners," *Emotion* 14, no. 3 (June 2014): 522–31, doi:10.1037/a0035960.

46. Hack, T., S. A. Goodwin, and S. T. Fiske, "Warmth Trumps Competence in Evaluations of Both Ingroup and Outgroup," *International Journal of Science, Commerce and Humanities* 1, no. 6 (September 2013): 99–105.

47. Parrish, T., "How Hate Took Hold of Me," *Daily News*, June 21, 2015, http://www.nydailynews.com/opinion/tim-parrish-hate-hold-article-1 .2264643, accessed October 23, 2015.

48. Lui, S. Y., and Kawachi, I. "Discrimination and Telomere Length Among Older Adults in the US: Does the Association Vary by Race and Type of Discrimination?" under review, Public Health Reports.

49. Chae, D. H., et al., "Discrimination, Racial Bias, and Telomere Length in African American Men," *American Journal of Preventive Medicine* 46, no. 2 (February 2014): 103–11, doi:10.1016/j.amepre.2013.10.020.

50. Peckham, M., "This Billboard Sucks Pollution from the Sky and Returns Purified Air," *Time*, May 1, 2014, http://time.com/84013/this -billboard-sucks-pollution-from-the-sky-and-returns-purified-air/, accessed November 24, 2015.

51. Diers, J., *Neighbor Power: Building Community the Seattle Way* (Seattle: University of Washington Press, 2004).

52. Beyer, K. M. M., et al., "Exposure to Neighborhood Green Space and Mental Health: Evidence from the Survey of the Health of Wisconsin," *International Journal of Environmental Research and Public Health* 11, no. 3 (March 2014): 3453–72, doi:10.3390/ijerph110303453; and Roe et al., "Green Space and Stress" (see #10 above).

53. Branas, C. C., et al., "A Difference-in-Differences Analysis of Health, Safety, and Greening Vacant Urban Space," *American Journal of Epidemiology* 174, no. 11 (December 1, 2011): 1296–1306, doi:10.1093/aje /kwr273.

54. Wesselmann, E. D., F. D. Cardoso, S. Slater, and K. D. Williams, "To Be Looked At as Though Air: Civil Attention Matters," *Psychological Science* 23, no. 2 (February 2012): 166–168, doi:10.1177/0956797611427921.

55. Guéguen, N., and M-A De Gail, "The Effect of Smiling on Helping Behavior: Smiling and Good Samaritan Behavior," *Communication Reports*, 16, no. 2 (2003): 133–40, doi: 10.1080/08934210309384496.

Chapter Twelve: Pregnancy: Cellular Aging Begins in the Womb

1. Hjelmborg, J. B., et al., "The Heritability of Leucocyte Telomere Length Dynamics," *Journal of Medical Genetics* 52, no. 5 (May 2015): 297–302, doi:10.1136/jmedgenet-2014-102736.

2. Wojcicki, J. M., et al., "Cord Blood Telomere Length in Latino Infants: Relation with Maternal Education and Infant Sex," *Journal of Perinatology: Official Journal of the California Perinatal Association* 36, no. 3 (March 2016): 235–41, doi:10.1038/jp.2015.178.

3. Needham, B. L., et al., "Socioeconomic Status and Cell Aging in Children," *Social Science and Medicine (1982)* 74, no. 12 (June 2012): 1948–51, doi:10.1016/j.socscimed.2012.02.019.

4. Collopy, L. C., et al., "Triallelic and Epigenetic-like Inheritance in Human Disorders of Telomerase," *Blood* 126, no. 2 (July 9, 2015): 176–84, doi:10.1182/blood-2015-03-633388.

5. Factor-Litvak, P., et al., "Leukocyte Telomere Length in Newborns: Implications for the Role of Telomeres in Human Disease," *Pediatrics* 137, no. 4 (April 2016): e20153927, doi:10.1542/peds.2015-3927.

6. De Meyer, T., et al., "A Non-Genetic, Epigenetic-like Mechanism of Telomere Length Inheritance?" *European Journal of Human Genetics* 22, no. 1 (January 2014): 10–11, doi:10.1038/ejhg.2013.255.

7. Collopy et al., "Triallelic and Epigenetic-like Inheritance in Human Disorders of Telomerase." (See #4 above.)

8. Tarry-Adkins, J. L., et al., "Maternal Diet Influences DNA Damage, Aortic Telomere Length, Oxidative Stress, and Antioxidant Defense Capacity in Rats," *FASEB Journal: Official Publication of the Federation of American Societies for Experimental Biology* 22, no. 6 (June 2008): 2037–44, doi:10.1096/fj.07-099523.

9. Aiken, C. E., J. L. Tarry-Adkins, and S. E. Ozanne, "Suboptimal Nutrition in Utero Causes DNA Damage and Accelerated Aging of the Female Reproductive Tract," *FASEB Journal: Official Publication of the Federation of American Societies for Experimental Biology* 27, no. 10 (October 2013): 3959–65, doi:10.1096/fj.13-234484.

10. Aiken, C. E., J. L. Tarry-Adkins, and S. E. Ozanne. "Transgenerational Developmental Programming of Ovarian Reserve," *Scientific Reports* 5 (2015): 16175, doi:10.1038/srep16175.

11. Tarry-Adkins, J. L., et al., "Nutritional Programming of Coenzyme Q: Potential for Prevention and Intervention?" *FASEB Journal: Official Publication of the Federation of American Societies for Experimental Biology* 28, no. 12 (December 2014): 5398–405, doi:10.1096/fj.14-259473.

12. Bull, C., H. Christensen, and M. Fenech, "Cortisol Is Not Associated with Telomere Shortening or Chromosomal Instability in Human Lymphocytes Cultured Under Low and High Folate Conditions," *PLOS ONE* 10, no. 3 (March 6, 2015): e0119367, doi:10.1371/journal.pone.0119367; and Bull, C., et al., "Folate Deficiency Induces Dysfunctional Long and Short Telomeres;

Both States Are Associated with Hypomethylation and DNA Damage in Human WIL2-NS Cells," *Cancer Prevention Research (Philadelphia, Pa.)* 7, no. 1 (January 2014): 128–38, doi:10.1158/1940-6207.CAPR-13-0264.

13. Entringer, S., et al., "Maternal Folate Concentration in Early Pregnancy and Newborn Telomere Length," *Annals of Nutrition and Metabolism* 66, no. 4 (2015): 202–08, doi:10.1159/000381925.

14. Cerne, J. Z., et al., "Functional Variants in CYP1B1, KRAS and MTHFR Genes Are Associated with Shorter Telomere Length in Postmenopausal Women," *Mechanisms of Ageing and Development* 149 (July 2015): 1–7, doi:10.1016/j.mad.2015.05.003.

15. "Folic Acid Fact Sheet," Womenshealth.gov, http://womenshealth.gov /publications/our-publications/fact-sheet/folic-acid.html, accessed November 27, 2015.

16. Paul, L., et al., "High Plasma Folate Is Negatively Associated with Leukocyte Telomere Length in Framingham Offspring Cohort," *European Journal of Nutrition* 54, no. 2 (March 2015): 235–41, doi:10.1007/s00394-014-0704-1.

17. Entringer, S., et al., "Maternal Psychosocial Stress During Pregnancy Is Associated with Newborn Leukocyte Telomere Length," *American Journal of Obstetrics and Gynecology* 208, no. 2 (February 2013): 134.e1–7, doi:10.1016/j.ajog.2012.11.033.

18. Marchetto, N. M., et al., "Prenatal Stress and Newborn Telomere Length," *American Journal of Obstetrics and Gynecology*, January 30, 2016, doi:10.1016/j.ajog.2016.01.177.

19. Entringer, S., et al., "Influence of Prenatal Psychosocial Stress on Cytokine Production in Adult Women," *Developmental Psychobiology* 50, no. 6 (September 2008): 579–87, doi:10.1002/dev.20316.

20. Entringer, S., et al., "Stress Exposure in Intrauterine Life Is Associated with Shorter Telomere Length in Young Adulthood," *Proceedings of the National Academy of Sciences of the United States of America* 108, no. 33 (August 16, 2011): E513–18, doi:10.1073/pnas.1107759108.

21. Haussman, M., and B. Heidinger, "Telomere Dynamics May Link Stress Exposure and Ageing across Generations," *Biology Letters* 11, no. 11 (November 2015), doi:10.1098/rsbl.2015.0396.

22. Ibid.

Chapter Thirteen: Childhood Matters for Life: How the Early Years Shape Telomeres

1. Sullivan, M. C.," For Romania's Orphans, Adoption Is Still a Rarity," National Public Radio, August 19, 2012, http://www.npr.org/2012/08/19 /158924764/for-romanias-orphans-adoption-is-still-a-rarity.

2. Ahern, L., "Orphanages Are No Place for Children," *Washington Post*, August 9, 2013, https://www.washingtonpost.com/opinions/orphanages -are-no-place-for-children/2013/08/09/6d502fb0-fadd-11e2-a369-d1954 abcb7e3_story.html, accessed October 14, 2015.

3. Felitti, V. J., et al., "Relationship of Childhood Abuse and Household Dysfunction to Many of the Leading Causes of Death in Adults: The Adverse Childhood Experiences (ACE) Study," *American Journal of Preventive Medicine* 14, no. 4 (May 1998): 245–58.

4. Chen, S. H., et al., "Adverse Childhood Experiences and Leukocyte Telomere Maintenance in Depressed and Healthy Adults," *Journal of Affective Disorders* 169 (December 2014): 86–90, doi:10.1016/j.jad.2014 .07.035.

5. Skilton, M. R., et al., "Telomere Length in Early Childhood: Early Life Risk Factors and Association with Carotid Intima-Media Thickness in Later Childhood," *European Journal of Preventive Cardiology* 23, no. 10 (July 2016), 1086–92, doi:10.1177/2047487315607075.

6. Drury, S. S., et al., "Telomere Length and Early Severe Social Deprivation: Linking Early Adversity and Cellular Aging," *Molecular Psychiatry* 17, no. 7 (July 2012): 719–27, doi:10.1038/mp.2011.53.

7. Hamilton, J., "Orphans' Lonely Beginnings Reveal How Parents Shape a Child's Brain," National Public Radio, February 24, 2014, http://www .npr.org/sections/health-shots/2014/02/20/280237833/orphans -lonely-beginnings-reveal-how-parents-shape-a-childs-brain, accessed October 15, 2015.

8. Powell, A., "Breathtakingly Awful," *Harvard Gazette*, October 5, 2010, http://news.harvard.edu/gazette/story/2010/10/breathtakingly-awful/, accessed October 26, 2015.

9. Authors' interview with Charles Nelson, September 18, 2015.

10. Shalev, I., et al., "Exposure to Violence During Childhood Is Associated with Telomere Erosion from 5 to 10 Years of Age: A Longitudinal Study," *Molecular Psychiatry* 18, no. 5 (May 2013): 576–81, doi:10.1038 /mp.2012.32.

11. Price, L. H., et al., "Telomeres and Early-Life Stress: An Overview," *Biological Psychiatry* 73, no. 1 (January 1, 2013): 15–23, doi:10.1016/j .biopsych.2012.06.025.

12. Révész, D., Y. Milaneschi, E. M. Terpstra, and B. W. J. H. Penninx, "Baseline Biopsychosocial Determinants of Telomere Length and 6-Year Attrition Rate," *Psychoneuroendocrinology* 67 (May 2016): 153–62, doi:10 .1016/j.psyneuen.2016.02.007.

13. Danese, A., and B. S. McEwen, "Adverse Childhood Experiences,

Allostasis, Allostatic Load, and Age-Related Disease," *Physiology & Behavior* 106, no. 1 (April 12, 2012): 29–39, doi:10.1016/j.physbeh.2011.08.019.

14. Infurna, F. J., C. T. Rivers, J. Reich, and A. J. Zautra, "Childhood Trauma and Personal Mastery: Their Influence on Emotional Reactivity to Everyday Events in a Community Sample of Middle-Aged Adults," *PLOS ONE* 10, no. 4 (2015): e0121840, doi:10.1371/journal.pone.0121840.

15. Schrepf, A., K. Markon, and S. K. Lutgendorf, "From Childhood Trauma to Elevated C-Reactive Protein in Adulthood: The Role of Anxiety and Emotional Eating," *Psychosomatic Medicine* 76, no. 5 (June 2014): 327–36, doi:10.1097/PSY.0000000000000072.

16. Felitti, V. J., et al., "Relationship of Childhood Abuse and Household Dysfunction to Many of the Leading Causes of Death in Adults. The Adverse Childhood Experiences (ACE) Study," *American Journal of Preventive Medicine* 14, no. 4 (May 1998): 245–58, doi.org/10.1016/S0749-3797(98)00017-8.

17. Lim, D., and D. DeSteno, "Suffering and Compassion: The Links Among Adverse Life Experiences, Empathy, Compassion, and Prosoial Behavior," *Emotion* 16, no. 2 (March 2016): 175–82, doi:10.1037/emo0000144.

18. Asok, A., et al., "Infant-Caregiver Experiences Alter Telomere Length in the Brain," *PLOS ONE* 9, no. 7 (2014): e101437, doi:10.1371/journal.pone.0101437.

19. McEwen, B. S., C. N. Nasca, and J. D. Gray, "Stress Effects on Neuronal Structure: Hippocampus, Amygdala, and Prefrontal Cortex," *Neuropsychopharmacology: Official Publication of the American College of Neuropsychopharmacology* 41, no. 1 (January 2016): 3–23, doi:10.1038/npp.2015.171; and Arnsten, A. F. T., "Stress Signalling Pathways That Impair Prefrontal Cortex Structure and Function," *Nature Reviews Neuroscience* 10, no. 6 (June 2009): 410–22, doi:10.1038/nrn2648.

20. Suomi, S., "Attachment in Rhesus Monkeys," in *Handbook of Attachment: Theory, Research, and Clinical Applications*, ed. J. Cassidy and P. R. Shaver, 3rd ed. (New York: Guilford Press, 2016).

21. Schneper, L., Jeanne Brooks-Gunn, Daniel Notterman, and Stephen, Suomi, "Early Life Experiences and Telomere Length in Adult Rhesus Monkeys: An Exploratory Study." *Psychosomatic Medicine*, in press (n.d.).

22. Gunnar, M. R., et al., "Parental Buffering of Fear and Stress Neurobiology: Reviewing Parallels Across Rodent, Monkey, and Human Models," *Social Neuroscience* 10, no. 5 (2015): 474–78, doi:10.1080/17470919.2015.1070198.

23. Hostinar, C. E., R. M. Sullivan, and M. R. Gunnar, "Psychobiological Mechanisms Underlying the Social Buffering of the Hypothalamic-Pituitary-Adrenocortical Axis: A Review of Animal Models and Human

Studies Across Development," *Psychological Bulletin* 140, no. 1 (January 2014): 256–82, doi:10.1037/a0032671.

24. Doom, J. R., C. E. Hostinar, A. A. VanZomeren-Dohm, and M. R. Gunnar, "The Roles of Puberty and Age in Explaining the Diminished Effectiveness of Parental Buffering of HPA Reactivity and Recovery in Adolescence," *Psychoneuroendocrinology* 59 (September 2015): 102–11, doi:10.1016/j.psyneuen.2015.04.024.

25. Seery, M. D., et al., "An Upside to Adversity?: Moderate Cumulative Lifetime Adversity Is Associated with Resilient Responses in the Face of Controlled Stressors," *Psychological Science* 24, no. 7 (July 1, 2013): 1181–89, doi:10.1177/0956797612469210.

26. Asok, A., et al., "Parental Responsiveness Moderates the Association Between Early-Life Stress and Reduced Telomere Length," *Development and Psychopathology* 25, no. 3 (August 2013): 577–85, doi:10.1017/S0954579413000011.

27. Bernard, K., C. E. Hostinar, and M. Dozier, "Intervention Effects on Diurnal Cortisol Rhythms of Child Protective Services–Referred Infants in Early Childhood: Preschool Follow-Up Results of a Randomized Clinical Trial," *JAMA Pediatrics* 169, no. 2 (February 2015): 112–19, doi:10.1001/jamapediatrics.2014.2369.

28. Kroenke, C. H., et al., "Autonomic and Adrenocortical Reactivity and Buccal Cell Telomere Length in Kindergarten Children," *Psychosomatic Medicine* 73, no. 7 (September 2011): 533–40, doi:10.1097/PSY.0b013e318229acfc.

29. Wojcicki, J. M., et al., "Telomere Length Is Associated with Oppositional Defiant Behavior and Maternal Clinical Depression in Latino Preschool Children," *Translational Psychiatry* 5 (June 2015): e581, doi:10.1038/tp.2015.71; and Costa, D. S., et al., "Telomere Length Is Highly Inherited and Associated with Hyperactivity-Impulsivity in Children with Attention Deficit/Hyperactivity Disorder," *Frontiers in Molecular Neuroscience* 8 (July 2015): 28, doi:10.3389/fnmol.2015.00028.

30. Kroenke et al., "Autonomic and Adrenocortical Reactivity and Buccal Cell Telomere Length in Kindergarten Children." (See #27 above.)

31. Boyce, W. T., and B. J. Ellis, "Biological Sensitivity to Context: I. An Evolutionary-Developmental Theory of the Origins and Functions of Stress Reactivity," *Development and Psychopathology* 17, no. 2 (spring 2005): 271–301.

32. Van Ijzendoorn, M. H., and M. J. Bakermans-Kranenburg, "Genetic Differential Susceptibility on Trial: Meta-analytic Support from Randomized Controlled Experiments," *Development and Psychopathology* 27, no. 1 (February 2015): 151–62, doi:10.1017/S0954579414001369.

33. Colter, M., et al., "Social Disadvantage, Genetic Sensitivity, and Children's Telomere Length," *Proceedings of the National Academy of Sciences of the United States of America* 111, no. 16 (April 22, 2014): 5944–49, doi:10.1073/pnas.1404293111.

34. Brody, G. H., T. Yu, S. R. H. Beach, and R. A. Philibert, "Prevention Effects Ameliorate the Prospective Association Between Nonsupportive Parenting and Diminished Telomere Length," *Prevention Science: The Official Journal of the Society for Prevention Research* 16, no. 2 (February 2015): 171–80, doi:10.1007/s11121-014-0474-2; Beach, S. R. H., et al., "Nonsupportive Parenting Affects Telomere Length in Young Adulthood Among African Americans: Mediation through Substance Use," *Journal of Family Psychology: JFP: Journal of the Division of Family Psychology of the American Psychological Association (Division 43)* 28, no. 6 (December 2014): 967–72, doi:10.1037/fam0000039; and Brody, G. H., et al., "The Adults in the Making Program: Long-Term Protective Stabilizing Effects on Alcohol Use and Substance Use Problems for Rural African American Emerging Adults," *Journal of Consulting and Clinical Psychology* 80, no. 1 (February 2012): 17–28. doi:10.1037/a0026592.

35. Brody et al., "Prevention Effects Ameliorate the Prospective Association Between Nonsupportive Parenting and Diminished Telomere Length"; and Beach et al., "Nonsupportive Parenting Affects Telomere Length in Young Adulthood among African Americans: Mediation through Substance Use." (See #33 above.)

36. Spielberg, J. M., T. M. Olino, E. E. Forbes, and R. E. Dahl, "Exciting Fear in Adolescence: Does Pubertal Development Alter Threat Processing?" *Developmental Cognitive Neuroscience* 8 (April 2014): 86–95, doi:10.1016/j.dcn.2014.01.004; and Peper, J. S., and R. E. Dahl, "Surging Hormones: Brain-Behavior Interactions During Puberty," *Current Directions in Psychological Science* 22, no. 2 (April 2013): 134–39, doi:10.1177/0963721412473755.

37. Turkle, S., *Reclaiming Conversation: The Power of Talk in a Digital Age* (New York: Penguin Press, 2015).

38. Siegel, D., and T. P. Bryson, *The Whole-Brain Child: 12 Revolutionary Strategies to Nurture Your Child's Developing Mind* (New York: Delacorte Press, 2011).

39. Robles, T. F., et al., "Emotions and Family Interactions in Childhood: Associations with Leukocyte Telomere Length Emotions, Family Interactions, and Telomere Length," *Psychoneuroendocrinology* 63 (January 2016): 343–50, doi:10.1016/j.psyneuen.2015.10.018.

Conclusion: Entwined: Our Cellular Legacy

1. Pickett, K. E., and R. G. Wilkinson, "Inequality: An Underacknowledged Source of Mental Illness and Distress," *British Journal of Psychiatry: The Journal of Mental Science* 197, no. 6 (December 2010): 426–28, doi:10.1192/bjp.bp.109.072066.
2. Ibid; and Wilkerson, R. G., and K. Pickett, *The Spirit Level: Why More Equal Societies Almost Always Do Better* (London: Allen Lane, 2009).
3. Stone, C., D. Trisi, A. Sherman, and B. Debot, "A Guide to Statistics on Historical Trends in Income Inequality," Center on Budget and Policy Priorities, updated October 26, 2015, http://www.cbpp.org/research/poverty-and-inequality/a-guide-to-statistics-on-historical-trends-in-income-inequality.
4. Pickett, K. E., and R. G. Wilkinson, "The Ethical and Policy Implications of Research on Income Inequality and Child Wellbeing," *Pediatrics* 135, Suppl. 2 (March 2015): S39–47, doi:10.1542/peds.2014-3549E.
5. Mayer, E. A., et al., "Gut Microbes and the Brain: Paradigm Shift in Neuroscience," *Journal of Neuroscience: The Official Journal of the Society for Neuroscience* 34, no. 46 (November 12, 2014): 15490–96, doi:10.1523/JNEUROSCI.3299-14.2014; Picard, M., R. P. Juster, and B. S. McEwen, "Mitochondrial Allostatic Load Puts the 'Gluc' Back in Glucocorticoids," *Nature Reviews Endocrinology* 10, no. 5 (May 2014): 303–10, doi:10.1038/nrendo.2014.22; and Picard, M., et al., "Chronic Stress and Mitochondria Function in Humans," under review.
6. Varela, F. J., E. Thompson, and E. Rosch, *The Embodied Mind* (Cambridge, MA: MIT Press, 1991).
7. "Zuckerberg: One in Seven People on the Planet Used Facebook on Monday," *Guardian*, August 28, 2015, http://www.theguardian.com/technology/2015/aug/27/facebook-1bn-users-day-mark-zuckerberg, accessed October 26, 2015; and "Number of Monthly Active Facebook Users Worldwide as of 1st Quarter 2016 (in Millions)," Statista, http://www.statista.com/statistics/264810/number-of-monthly-active-facebook-users-worldwide/.

We thank the many authors and organizations that allowed us permissions to reprint scales and figures.

For figures, this includes:

Blackburn, Elizabeth H., Elissa S. Epel, and Jue Lin. "Human Telomere Biology: A Contributory and Interactive Factor in Aging, Disease Risks, and Protection." *Science* (New York, N.Y.) 350, no. 6265 (December 4, 2015): 1193–98. **Reprinted with permission from AAAS.**

Epel, Elissa S., Elizabeth H. Blackburn, Jue Lin, Firdaus S. Dhabhar, Nancy E. Adler, Jason D. Morrow, and Richard M. Cawthon. "Accelerated Telomere Shortening in Response to Life Stress." *Proceedings of the National Academy of Sciences of the United States of America* 101, no. 49 (December 7, 2004): 17312–15. **Permissions granted by the National Academy of Sciences, U.S.A. Copyright (2004) National Academy of Sciences, U.S.A.**

Cribbet, M. R., M. Carlisle, R. M. Cawthon, B. N. Uchino, P. G. Williams, T. W. Smith, and K. C. Light. "Cellular Aging and Restorative Processes: Subjective Sleep Quality and Duration Moderate the Association between Age and Telomere Length in a Sample of Middle-Aged and Older Adults." *SLEEP* 37, no. 1: 65–70. **Republished with permission of the American Academy of Sleep Medicine; permission conveyed through Copyright Clearance Center, Inc.**

Carroll J. E., S. Esquivel, A. Goldberg, T. E. Seeman, R. B. Effros, J. Dock, R. Olmstead, E. C. Breen, and M. R. Irwin. "Insomnia and Telomere Length in Older Adults." *SLEEP* 39, no 3 (2016): 559–64. **Republished with permission of the American Academy of Sleep Medicine; permission conveyed through Copyright Clearance Center, Inc.**

Farzaneh-Far R, J. Lin, E. S. Epel, W. S. Harris, E. H. Blackburn, and M. A. Whooley. "Association of Marine Omega-3 Fatty Acid Levels with Telomeric Aging in Patients with Coronary Heart Disease." *JAMA* 303, no 3 (2010): 250–57. **Permissions granted by the American Medical Association.**

Park, M., J. E. Verhoeven, P. Cuijpers, C. F. Reynolds III, and B. W. J. H. Penninx. "Where You Live May Make You Old: The Association between Perceived Poor Neighborhood Quality and Leukocyte Telomere Length." *PLoS ONE* 10, no.6 (2015), e0128460. http://doi. org/10.1371/journal.pone.0128460. **Permissions granted by Park et al. via the Creative Commons Attribution License. Copyright © 2015 Park et al.**

Brody, G. H., T. Yu, S. R. H. Beach, and R. A. Philibert. "Prevention Effects Ameliorate the Prospective Association between Nonsupportive Parenting and Diminished Telomere Length." *Prevention Science: The Official*

Journal of the Society for Prevention Research 16, no. 2 (February 2015): 171–80. **With permission of Springer.**

Pickett, Kate E., and Richard G. Wilkinson. "Inequality: An Underacknowledged Source of Mental Illness and Distress." *The British Journal of Psychiatry: The Journal of Mental Science* 197, no. 6 (December 2010): 426–28. **Permissions granted by the Royal College of Psychiatrists. Copyright, the Royal College of Psychiatrists.**

For scales, this includes:

Kiernan, M., D. E. Schoffman, K. Lee, S. D. Brown, J. M. Fair, M. G. Perri, and W. L. Haskell. "The Stanford Leisure-Time Activity Categorical Item (L-Cat): A Single Categorical Item Sensitive to Physical Activity Changes in Overweight/Obese Women." *International Journal of Obesity* 37 (2013): 1597–602. **Permissions granted by Nature Publishing Group and Dr. Michaela Kiernan, Stanford University School of Medicine. Copyright 2013. Reprinted by permission from Macmillan Publishers Ltd.**

The ENRICHD Investigators. "Enhancing Recovery in Coronary Heart Disease (ENRICHD): Baseline Characteristics." *The American Journal of Cardiology* 88, no. 3, (August 1, 2001): 316–22. **Permissions granted by Elsevier science and technology journals and Dr. Pamela Mitchell, University of Washington. Permission conveyed through Copyright Clearance Center, Inc. Republished with permission of Elsevier Science and Technology Journals.**

Buysse, Daniel J., Charles F. Reynolds III, Timothy H. Monk, Susan R. Berman, and David J. Kupfer. "The Pittsburgh Sleep Quality Index: A New Instrument for Psychiatric Practice and Research." *Psychiatry Research* 28, no. 2 (May 1989): 193–213. **Copyright © 1989 and 2010, University of Pittsburgh. All rights reserved. Permissions granted by Dr. Daniel Buysse and the University of Pittsburgh.**

Scheier, M. F., and C. S. Carver. "Optimism, Coping, and Health: Assessment and Implications of Generalized Outcome Expectancies." *Health Psychology* 4, no. 3 (1985): 219–47. **Permissions granted by Dr. Michael Scheier, Carnegie Mellon University, and the American Psychological Association.**

Trapnell, P. D., J. D. Campbell. "Private Self-Consciousness and the Five-Factor Model of Personality: Distinguishing Rumination from Reflection." *Journal of Personality and Social Psychology* 76 (1999): 284–330. **Permissions granted by Dr. Paul Trapnell, University of Winnipeg, and the American Psychological Association.**

John, O. P., E. M. Donahue, and R. L. Kentle. Conscientiousness: "The Big Five Inventory—Versions 4a and 54." Berkeley: University of California, Berkeley, Institute of Personality and Social Research, 1991. **Permissions granted by Dr. Oliver John, University of California, Berkeley.**

Scheier, M. F., C. Wrosch, A. Baum, S. Cohen, L. M. Martire, K. A. Matthews, R. Schulz, and B. Zdaniuk. "The Life Engagement Test: Assessing Purpose in Life." *Journal of Behavioral Medicine* 29 (2006): 291–98. **With permission of Springer. Permissions granted by Springer Publishing and Dr. Michael Scheier, Carnegie Mellon University.**

The Adverse Childhood Experiences Scale (ACES) was reprinted with permission from Dr. Vincent Felitti, MD, Co-PI, Adverse Childhood Experiences Study, University of California, San Diego.

Index

Page numbers in italics refer to illustrations, charts, and graphs in the text.

About the Authors

Dr. Elizabeth Blackburn, PhD, received the Nobel Prize in Physiology or Medicine in 2009 alongside two colleagues for the discovery of the molecular nature of telomeres, the ends of chromosomes that serve as protective caps, and for discovering telomerase, the enzyme that maintains telomeres. She is currently the president of the Salk Institute and professor emeritus at UCSF. Blackburn is a past president of the American Association for Cancer Research and the American Society for Cell Biology and is a recipient of nearly every major medical award, including the Albert Lasker Basic Medical Research Award. She was named one of *TIME* magazine's 100 most influential people. She is a member of the U.S. National Academies of Sciences and Medicine and the Royal Society of London. Blackburn has helped guide public science policy and served on the President's Council on Bioethics, an advisory committee to the president of the United States.

Blackburn was born in Tasmania, Australia. She received her bachelor of science degree from the University of Melbourne and her PhD in molecular biology from the University of Cambridge and conducted her postdoctoral fellowship at Yale University. She and her husband currently live in La Jolla, California, and part-time in San Francisco.

Dr. Elissa Epel, PhD, is a leading health psychologist who studies stress, aging, and obesity. She is a professor in the Department of Psychiatry at UCSF, the director of UCSF's Aging, Metabolism, and Emotions (AME) Center, director of COAST, a UCSF obesity research center, and associate director of UCSF's Center for Health and Community. She is a member of the National Academy of Medicine and serves on scientific advisory committees for National Institute of Health initiatives (such as the Science of Behavior Change program), the Mind & Life Institute, and the European Society of Preventive Medicine. She has received many research awards, including awards from Stanford University, the Society of Behavioral Medicine, the Academy of Behavioral Medicine Research, and the American Psychological Association.

Epel was born in Carmel, California. She received her bachelor's degree from Stanford University and her PhD in clinical and health psychology from Yale University. She completed her clinical internship at the Veterans Administration Palo Alto Healthcare System and conducted her postdoctoral fellowship at UCSF. She lives in San Francisco with her husband and son.

MAKING
GLOBALIZATION
WORK

ALSO BY JOSEPH E. STIGLITZ

The Roaring Nineties
Globalization and Its Discontents

MAKING GLOBALIZATION WORK

JOSEPH E. STIGLITZ

W. W. NORTON & COMPANY

NEW YORK   LONDON

For information about permission to reproduce selections from this book, write to
Permissions, W. W. Norton & Company, Inc., 500 Fifth Avenue, New York, NY 10110

Manufacturing by The Maple-Vail Book Manufacturing Group
Book design by Chris Welch
Production manager: Amanda Morrison

Library of Congress Cataloging-in-Publication Data

Stiglitz, Joseph E.
Making globalization work / Joseph E. Stiglitz. — 1st ed.
p. cm.
Includes bibliographical references and index.
ISBN-13: 978-0-393-06122-2 (hardcover)
ISBN-10: 0-393-06122-1 (hardcover)
1. Globalization—Economic aspects. I. Title.
HF1359.S753 2006
337—dc22
2006020633

W. W. Norton & Company, Inc., 500 Fifth Avenue, New York, N.Y. 10110
www.wwnorton.com

W. W. Norton & Company Ltd., Castle House, 75/76 Wells Street, London W1T 3QT

1 2 3 4 5 6 7 8 9 0

For Anya, forever

CONTENTS

PREFACE

My book *Globalization and Its Discontents* was written just after I left the World Bank, where I served as senior vice president and chief economist from 1997 to 2000. That book chronicled much of what I had seen during the time I was at the Bank and in the White House, where I served from 1993 to 1997 as a member and then chairman of the Council of Economic Advisers under President William Jefferson Clinton. Those were tumultuous years; the 1997–98 East Asian financial crisis pushed some of the most successful of the developing countries into unprecedented recessions and depressions. In the former Soviet Union, the transition from communism to the market, which was supposed to bring new prosperity, instead brought a drop in income and living standards by as much as 70 percent. The world, in the best of circumstances, marked by intense competition, uncertainty, and instability, is not an easy place, and the developing countries were not always doing the most they could to advance their own well-being. But I became convinced that the advanced industrial countries, through international organizations like the International Monetary Fund (IMF), the World Trade Organization (WTO), and the World Bank, were not only not doing all that they could to help these countries but were sometimes making their life more difficult. IMF programs had clearly worsened the East Asian crisis, and the "shock therapy" they had pushed in the former Soviet Union and its satellites played an important role in the failures of the transition.

I covered many of these topics in *Globalization and Its Discontents*. I felt I had a unique perspective to bring to the debate, having seen policies being formulated from inside the White House, and from inside the World Bank, where we worked alongside developing countries to help develop strategies to enhance growth and reduce poverty. Equally important, as an economic theorist, I spent almost forty years working to understand the strengths, and limitations, of the market economy. My research had not only cast doubt on the validity of general claims about market efficiency but also on some of the fundamental beliefs underlying globalization, such as the notion that free trade is necessarily welfare enhancing.

In my earlier book, I described some of the failures of the international financial system and its institutions, and showed why globalization has not benefited as many people as it could and should have. And I sketched out some of what needs to be done to make globalization work—especially for the poor and developing countries. The book included some proposals for reforming the world financial system and the international financial institutions that govern it, but space did not allow me to flesh out these proposals.

Just as my time in the White House and at the World Bank put me in a unique position to understand globalization's problems, so too has it provided me with the basis for this sequel. During my years in Washington, I traveled the world and met many government leaders and officials, as I studied the successes and failures of globalization. After I left Washington to return to academia, I remained involved in the globalization debate. In 2001, I received the Nobel Prize for my earlier theoretical work on the economics of information. Since then, I have visited dozens of developing countries, continued my discussions with academics and businesspeople, with prime ministers, presidents, and parliamentarians on every continent, and been involved in fora debating development and globalization involving every segment of our global society.

When I was about to leave the White House for the World Bank, President Clinton asked me to stay on as the chairman of his Council of Economic Advisers and as a member of his cabinet. I declined, because I thought that the task of designing policies and programs that would do something about the abject poverty which plagued the less developed world was a far more important challenge. It seemed terribly unfair that in a world of richness and plenty, so many should live in such poverty. The problems were obviously difficult, but I felt confident something could be done. I accepted the World Bank's offer, not only because it would give me new opportunities to study the problems but because it would provide me a platform from which I could support the interests of the developing countries.

In my years at the World Bank, I came to understand why there was such discontent with the way globalization was proceeding. Though development was possible, it was clear that it was not inevitable. I had seen countries where poverty was increasing rather than decreasing, and I had seen what that meant—not just in statistics but in the lives of the people. There are, of course, no magic solutions. But there are a multitude of changes to be made—in policies, in economic institutions, in the rules of the game, and in mindsets—that hold out the promise of helping make globalization work better, especially for the developing countries. Some changes will occur inevitably—China's entry into the global scene as a dominant manufacturing economy and India's success in outsourcing, for instance, are already forcing changes in policies and thinking. The instability that has marked global financial markets during the past decade—from the global financial crisis of 1997–98 to the Latin American crises of the early years of the new millennium, to the falling dollar beginning in 2003—has forced us to rethink the global financial system. Sooner, or later, the world will have to make some of the changes I suggest in the following chapters; the question is not so much *whether* these or similar changes will occur, but when—and, more important, whether they will occur before another set of global disasters or after. Haphazard changes that are done quickly in the wake of a crisis may not be the best way to reform the global economic system.

The end of the Cold War has opened up new opportunities and removed old constraints. The importance of a market economy has now been recognized and the death of communism means that governments can now turn away from ideological battles and toward fixing the problems of capitalism. The world would have benefited had the United States used the opportunity to help build an international economic and political system based on values and principles, such as a trade agreement designed to promote development in poor countries. Instead, unchecked by competition to "win the hearts and minds" of those in the Third World, the advanced industrial countries actually created a global trade regime that helped their special corporate and financial interests, and hurt the poorest countries of the world.

Development is complex. Indeed, one of the main criticisms leveled against the IMF and other international economic institutions is that their one-size-fits-all solutions do not—can not—capture these complexities. Yet, out of the myriad of global economic narratives, some general principles do arise. Many of the successful developing countries have some policies in common, which each adapted to its own situation. One of this book's objectives is to explain these points in common.

I should say a word about the relationship between my earlier research, especially that connected with the work that led to the Nobel Prize, my policy positions during my years in Washington, and my subsequent writings, especially in *Globalization and Its Discontents* and in *The Roaring Nineties*.[1]

My earlier academic work on the consequences of imperfect and limited information and imperfect competition led me to an awareness of the limitations of markets. Over the years I, and others, have extended that work into macro-economics. My work in the economics of the public sector had emphasized the need for balance between the government and the market—perspectives close to those of the Clinton administration, and which I helped articulate in the annual *Economic Report of the President* in the years I served on the Council of Economic Advisers. When I came to the World Bank, I was troubled by what I saw: the Bank—and, even more, the IMF—pushing conservative economic policies (such as the privatization of social security) that were exactly the opposite of those for which I had fought so hard when I was at the White House. Worse, they were using models that my theoretical work had done so much to discredit. (I was, of course, even more troubled to learn that Clinton's own Treasury was pushing these policies.)

My economic research had shown the deep underlying flaws in IMF economics, in "market fundamentalism," the belief that markets by themselves lead to economic efficiency. *Intellectual* consistency—consistency with my earlier academic work—impelled me to voice my concerns that the policies which they were pushing in, for instance, East Asia, might only make matters worse. To do any less would have been a dereliction of my responsibilities.

What we had fought for while I was in the Clinton administration was relevant, not just to Americans but to the rest of the world as well. As I moved from the Clinton administration to the World Bank, I continued to push for the right balance between the private and public sectors and to advance policies promoting equality and full employment. The issues I raised during my tenure at the World Bank—which received a warm reception by many of the economists there—are the same ones I raised in *Globalization and Its Discontents.*

The passions evoked by the global financial crises and the difficult transitions from communism to a market economy have now faded. Today, these matters can be looked at more calmly and, as I describe in chapter 1, on many of the pivotal issues there is an emerging consensus that resembles the ideas put forth in *Globalization and Its Discontents.* That book helped change the debate about how globalization should be reshaped. A number of these ideas are widely accepted now, and even the IMF has come round to my point of view that allowing unfettered flows of speculative capital is extremely risky. Of course, as the continuing clashes between the Left and Right in the United States and elsewhere remind us, there remain large areas of disagreement about both economics and basic values. Indeed, one of my main criticisms of the international economic institutions is that, regardless of the circumstances, they have supported one particular economic perspective—one which I think, in many ways, is misguided.

This book reflects my faith in democratic processes; my belief that an informed citizenry is more likely to provide some checks against the abuses of the special corporate and financial interests that have so dominated the globalization process; that ordinary citizens of the advanced industrial countries, as well as of the developing world, share a common interest in making globalization work. I hope that this book, like its predecessor, will help transform the globalization debate—and, ultimately, the political processes which shape globalization.

Globalization is the field on which some of our major societal conflicts—including those over basic values—play out. Among the most important of those conflicts is that over the role of government and markets.

It used to be that conservatives could appeal to Adam Smith's "invisible hand"—the notion that markets and the pursuit of self-interest would lead, as if by an invisible hand, to economic efficiency. Even if they could admit that markets, by themselves, might not engender a socially acceptable distribution of income, they argued that issues of efficiency and equity should be separated.

In this conservative view, economics is about efficiency, and issues of equity (which, like beauty, so often lies in the eyes of the beholder) should be left to politics. Today, the intellectual defense of market fundamentalism has largely disappeared.[2] My research on the economics of information showed that whenever information is imperfect, in particular when there are information asymmetries—where some individuals know something that others do not (in other words, *always*)—the reason that the invisible hand seems invisible is that it is not there.[3] Without appropriate government regulation and intervention, markets do not lead to economic efficiency.[4]

In recent years we have seen dramatic illustrations of these theoretical insights. As I described in my book *The Roaring Nineties*,[5] the pursuit of self-interest by CEOs, accountants, and investment banks did not lead to economic efficiency, but rather to a bubble accompanied by massive misallocations of investment. And the bubble, when it burst, led, as they almost always do, to recession.

Today, by and large, there is (at least among economists, if not among politicians) an understanding of the limitations of markets. The scandals of the nineties in America and elsewhere brought down "Finance and Capitalism American Style" from the pedestal on which they stood for too long. More broadly, Wall Street's perspective, which is often shortsighted, is being recognized as antithetical to development, which requires long-term thinking and planning.

There is also a growing recognition that there is not just one form of capitalism, not just one "right" way of running the economy. There are, for instance, other forms of market economies—such as that of Sweden, which has sustained robust growth—that have led to quite different societies, marked with better health care and education and less inequality. While Sweden's version may not work as well elsewhere, or may not be appropriate for a particular developing country, its success demonstrates that there are alternative forms of effective market economies. And when there are alternatives and choices, democratic political processes should be at the center of the decision making—not technocrats. One of my criticisms of the international economic institutions is that they tried to pretend that there were not trade-offs—a single set of policies made everyone better off—while the essence of economics is choice, that there are alternatives, some of which benefit some groups (such as foreign capitalists) at the expense of others, some of which impose risks on some groups (such as workers) to the advantage of others.

Among the central choices facing all societies is the role of government. Economic success requires getting the balance right between the government and the market. What services should the government provide? Should there be public pension programs? Should government encourage particular sectors with incentives? What regulations, if any, should it adopt to protect workers, consumers, and the environment? This balance obviously changes over time, and will differ from country to country. But I shall argue that globalization, as it has been pushed, has often made it more difficult to obtain the requisite balance.

I also hope to show that while globalization's critics are correct in saying it has been used to push a particular set of values, this need not be so. Globalization does not have to be bad for the environment, increase inequality, weaken cultural diversity, and advance corporate interests at the expense of the well-being of ordinary citizens. In *Making Globalization Work,* I attempt to show how globalization, properly managed, as it was in the successful development of much of East Asia, can do a great deal to benefit both the developing and the developed countries of the world.

Attitudes toward globalization, and the failures and inequities associated with the way it has been managed, provide a Rorschach test for both countries and their people, revealing their fundamental beliefs and attitudes, their perspectives on the role of government and the market, the importance they attach to social justice, and the weight they put on noneconomic values.

Economists who place less importance on reducing income inequality are more prone to think that the actions governments might take to reduce that inequality are too costly, and may even be counterproductive. These "free market" economists are also more inclined to believe that markets, by themselves, without government intervention, are efficient, and that the best way to help the poor is simply to let the economy grow—and, somehow, the benefits will trickle down to the poor. (Interestingly, such beliefs have persisted, even as economic research has undermined their intellectual foundations.)

On the other hand, those who, like me, think that markets often fail to produce efficient outcomes (producing too much pollution and too little basic research, for instance) and are disturbed by income inequalities and high levels of poverty, also believe that reducing that inequality can cost less than the conservative economists predict. Those who worry about inequality and poverty also see the enormous costs of not dealing with the problem: the social consequences, including alienation, violence, and social conflict. They are also more sanguine about the possibilities for government interventions; while governments sometimes, or even often, are less efficient than one might have hoped, there are notable instances of success, several of which I discuss in the pages that follow. All human institutions are imperfect, and the challenge for each is to learn from the successes and failures.

These perspectives on the importance of dealing with inequality and poverty are mirrored in differences in views about their origins. By and large, those who are concerned about inequality see much of it as arising out of luck—the luck of being born with good genes or with rich parents (the "sperm lottery"),[6] or the luck of buying a piece of real estate in the right place at the right time (just before oil is struck, or before a local real estate bubble develops).[7] Those who are less concerned feel that wealth is a reward for hard work. In this view, redistribution of income not only takes away incentives for work and savings but is almost immoral, for it deprives individuals of their just rewards.

Paralleling these positions are stances on a host of other issues. Those who are less concerned about inequality and more concerned about economic efficiency tend to be less concerned with noneconomic values such as social justice, the environment, cultural diversity, universal access to health care, and consumer protection. (There are many exceptions, of course—conservatives, for instance, who worry about the environment.)

I emphasize these connections between economic and cultural attitudes to emphasize how much it matters to whom we entrust key aspects of economic decision making. If one delegates decision making to "conservatives," almost inevitably one will get economic policies and outcomes that reflect their political interests and cultural values.[8] This book obviously reflects my own judgments and values; at least, I hope to be transparent, and present both sides of the ongoing economic debates.

SAVING GLOBALIZATION FROM ITS ADVOCATES

Some seventy years ago, during the Great Depression, the British economist, John Maynard Keynes, formulated his theory of unemployment, which detailed how government action could help restore the economy to full employment and growth. Keynes was vilified by conservatives, who saw his prescription as increasing the role of government. They seized on the budget deficits that inevitably accompany a downturn as an occasion to cut back on government programs. But Keynes actually did more to save the capitalist system than all the pro-market financiers put together. Had the advice of the conservatives been followed, the Great Depression would have been even worse; it would have been longer and deeper, and, the demand for an alternative to capitalism would have grown. By the same token, I believe that unless we recognize and deal with the problems of globalization, it will be difficult to sustain its current momentum.

Globalization, like development, is not inevitable—even though there are strong underlying political and economic forces behind it. By most measures, between World War I and World War II, both the pace and extent of globalization slowed, and even reversed. For example, measures of trade, as a percentage of GDP, actually declined.[9] If globalization leads to lower standards of living for many or most of the citizens of a country and if it compromises fundamental cultural values, then there will be political demands to slow or stop it.

The path of globalization will, of course, be changed not only by the force of ideas and experiences (ideas about whether trade or capital market liberalization will improve growth and the actual experiences with these reforms, for example) but also by global events. In recent years, 9/11 and the war on terrorism, the war in Iraq, and the emergence of China and India have all redefined the globalization debate in ways that I will discuss.

This book is as much about how politics has been used to shape the economic system as it is about economics itself. Economists believe that incentives matter. There are strong incentives—and enormous opportunities—to shape political processes and the economic system in ways that generate profits for some at the expense of the many.

Open, democratic processes can circumscribe the power of special interest groups. We can bring ethics back into business. Corporate governance can recognize the rights not only of shareholders but of others who are touched by the actions of the corporations.[10] An engaged and educated citizenry can understand how to make globalization work, or at least work better, and can demand that their political leaders shape globalization accordingly. I hope this book will help make this vision a reality.

ACKNOWLEDGMENTS

My list of those to whom I am indebted for my understanding of globalization has grown much longer over the past four years since writing *Globalization and Its Discontents*. In addition to those at the international economic institutions, and especially the World Bank, that I noted in that book, I now need to add Nick Stern and François Bourguignon, who succeeded me as chief economists at the World Bank and with whom I have continued to engage in discussions about the development process. I'd like to thank Supachai Panitchpakdi, former head of the World Trade Organization, with whom I have had innumerable discussions concerning the direction of the development round; Leif Pagrotsky, Sweden's education minister, who was at the forefront of arguing for a fairer trade regime when he served as Sweden's trade minister; Pascal Lamy, formerly EU commissioner for trade (now head of the WTO), especially for discussions on the Everything But Arms initiative; Kemal Dervis, with whom I worked closely at the World Bank, and who has now become head of the UNDP; and Juan Somavia, head of the ILO, who convened the World Commission on the Social Dimension of Globalization, whose report represents an important landmark in the changing perspectives on globalization.

In preparing to write this book, I revisited many of the countries that I had visited, studied, and written about earlier—including

Argentina, Ethiopia, Thailand, Korea, China, Russia, Colombia, Philippines, Indonesia, Mexico, Vietnam, Ecuador, India, Turkey, and Brazil—to see how things had changed. I also went back to a few countries I had visited only briefly before, such as Bangladesh and Nigeria, as well as some that I had not had a chance to see, including Bolivia, Madagascar, Venezuela, and Azerbaijan. I owe a great debt to the numerous government officials (from the prime minister or president to their finance ministers and economic advisers on down), to the academics and businesspeople, and to those in the donor community and in civil society (NGOs) who gave so generously of their time. Many will see their ideas reflected in the discussions here.

Various versions of some of the ideas presented here have been discussed and presented at seminars through the world. I particularly want to thank George Papandreou, former foreign minister of Greece, who convenes an annual seminar of academics and political leaders (the Symi Symposium) in which globalization issues have often come to the fore; the Vatican Academy of Social Sciences, at which some of the ideas concerning debt were discussed; the Commonwealth, which asked me to undertake a study with Andrew Charlton of the London School of Economics on what a *true* development round of trade negotiations might look like, and which helped finance that study. I want to thank the president of the UN General Assembly, the Commonwealth finance ministers, the WTO, the Center for Global Development, and the World Bank for inviting me to present the findings of that study. I have similarly benefited from the airings that the ideas in each of the chapters have received in seminars and international meetings around the world. The ideas on reforming the global financial system were presented before the UN Committee on Economic and Social Affairs, the American Economic Association meetings in Boston in January 2006, at a seminar in Sweden (with George Soros) in December 2001, and at the annual meeting of the Spanish Association of Economists in La Coruña in September 2005. The problems in the intellectual property regime and the proposed reforms were discussed at a ministerial meeting for the least developed countries held by the World Intellectual Property Organization in Seoul in October 2004 and at an international conference sponsored by the Initiative for Policy Dialogue (IDP) at Columbia University in June 2005. Some of the ideas in chapter 1 were presented and discussed in the Tanner Lectures delivered at Oxford University in the spring of 2004. I want to thank the CIDOB Foundation in Barcelona, which in the fall of 2004 co-sponsored a conference on the "post–Washington Consensus consensus," at which many of these ideas were further developed.

This book covers a wide range of ideas. Many of the topics are areas on which I have been engaged in research for more than three decades, and in that time have accumulated a huge reservoir of intellectual debts. The discussions of trade policy in chapter 3 owe a great deal to Peter Orszag, with whom I worked closely on the problems of dumping while I was at the Council, and to Alan Winters, Michael Finger, and Bernard Hoeckman at the World Bank. For the discussions of intellectual property in chapter 4, I am particularly in debt to Jamie Love, Michael Cragg, Paul David, Giovanni Dosi, Mario Cimoli, Richard Nelson, Ha Joon Chang, and all the participants at the IPD intellectual property task force meetings, as well as my several co-authors in the general theory of innovation, Richard Gilbert, Carl Shapiro, David Newbery, and Partha Dasgupta. Several of these also worked with me on problems on natural resources and the environment. Kevin Conrad and Geoff Heal at Columbia and Prime Minister Somares of Papau New Guinea and Environment Minister Rodriguez have been at the center of the Rainforest Coalition described in chapter 6. Michael Toman, Alan Krupnick, and Ray Squitieri worked closely with me at the Council of Economic Advisers on the problems of global warming; and Ruth Bell was good enough to read an earlier version of chapter 6. On the issue of debt, David Hale, Barry Herman, Kunibert Raffer, Michael Dooley, Roberto Frenkel, Jurgen Kaiser, and Susan George deserve thanks, as well as all the participants at the IPD sovereign debt task force meetings. The global reserves proposal, the Chang Mai Initiative, and the failed attempt to create an Asian Monetary Fund have been discussed at meetings, seminars, and colloquia in Stockholm, Washington, and elsewhere, and I wish to thank the participants of those seminars, especially George Soros, who put forward a similar proposal of his own. I have also benefited from discussions with Andrew Sheng and Eusake Sakikabara on these topics.

I learned firsthand about the problems of oil company practices in work I did for the states of Alaska, Texas, Louisiana, and California, and I am greatly indebted to my collaborator in that work, Jeff Leitzinger. The challenge of maintaining market competition has been an abiding concern, both in my theoretical and policy work; Steven Salop, Jason Furman, Barry Nalebuff, and Jon Orszag are among my many collaborators who have influenced my thinking.

There is another set of rather special debts that I wish to acknowledge. The debate about globalization and the question of the limits of the market economy, of which the globalization debate has become a central part, has now been going on for a long time. There are some (should I say many?) who disagree with the views presented here; I have tried to listen to their arguments carefully, to appraise the evidence, to understand the models, to ascertain the underlying source of disagreement. I spent years at the Hoover Institution, one of the more conservative think tanks in the world, with such luminaries as Nobel Prize winners Milton Friedman, George Stigler, and Gary Becker. I wish to acknowledge my gratitude to all of them for their patience and tolerance. I fear that they sometimes found it trying to have someone question what they considered obvious or proven beyond a reasonable shadow of doubt. Too often, I fear, the combatants in these debates slide past each other, each simply asserting their positions. They are more engaged in rallying their troops than in winning converts. I suspect that I may not win many converts, but I have, I think, made an effort to engage on the issues, to uncover the differences in underlying assumptions and values.

The public debate about globalization has been especially lively within the last half decade, with important contributions by Martin Wolf (*Why Globalization Works*), Jagdish Bhagwati (*In Defense of Globalization*), Bill Easterly (*The Elusive Quest for Growth*), Jeff Sachs (*The End of Poverty*), and Thomas Friedman (*The World Is Flat*). Onstage and offstage, we have continued these debates with each other, and I believe we have all benefited—even if we have not been able to convince each other of the merits of our positions. Our democracies have given us the opportunity—I would say the responsibility—to engage in these debates, which hopefully will play a role in shaping public policy in this vital area.

There are four particular debts that I wish to highlight. The first is to my colleagues in the Initiative for Policy Dialogue, a network of economists from the developed and developing world dedicated to exploring alternative approaches to development and globalization and to working to ensure that these alternatives get a hearing in public debates. While I hesitate to single out any individuals, I would be remiss if I did not mention José Antonio Ocampo, former head of CEPAL (the UN Commission for Latin America) and now undersecretary of the UN for economics; K. S. Jomo, now assistant secretary of the UN for economics; Deepak Nayyar, professor of economics at Jawaharlal Nehru University; Dani Rodrik, of Harvard University; Eric Berglof, chief economist, European Bank for Reconstruction and Development (EBRD); Patrick Bolton, professor of business and economics at Columbia University; Ha-Joon Chang, of Cambridge University; Ricardo Ffrench-Davis, also of CEPAL; Akbar Noman, with whom I worked closely on the problems of Africa; and especially Shari Spiegel, director of IPD. IPD has received financial support from the Ford, Charles Stewart Mott, John D. and Catherine T. MacArthur, and Rockefeller Foundations; the Canadian International Development Agency (CIDA); the Commonwealth Secretariat; the Open Society Institute; the Rockefeller Brothers Fund; the Swedish International Development Agency (SIDA); and the United Nations Development Programme. To all of these I am deeply grateful.

The second debt is to Bruce Greenwald, my colleague at Columbia, with whom I did much of my pioneering work in the economics of information, and with whom I have taught, for the past four years, a course on Globalization and Markets. As always, Bruce has challenged my ideas, enriched my thinking, and brought original perspectives on every aspect of globalization. His influence is evident throughout, but especially in chapter 9.

The third is to Andrew Charlton, co-author of our report to the Commonwealth, with whom I have worked closely, especially in the areas of trade.

The fourth is to Columbia University and its president, Lee Bollinger, who has helped the university focus so much of its attention on the issues of globalization and help make it a global center of knowledge and learning, creating a university-wide Committee on Global Thought, which I chair. Colleagues like Jeff Sachs and Merit Janow (now serving on the appellate panel of the WTO) have combined a commitment to academia with a deep involvement in the surrounding world. The diversity of students and faculty at Columbia is a microcosm of the world. Like any great center of learning, it has provided an environment that encourages a flourishing debate among competing ideas. I, and this book, have benefited enormously from the challenges to my ideas.

I am indebted to my assistants at Columbia University, including Jill Blacksford—and especially to Maria Papadakis for going above and beyond on so many occasions. I would like to thank my research assistants for their hard work on the book, including Hamid Rashid, Anton Korinek, Dan Choate, Josh Goodman, Megan Torau, Jayant Ray, and Stephan Litschig. Sharon Cleary assisted with the research and editing, and managed the enormous task of bringing everything together at the end. Alan Brown, Sheridan Prasso, and Gen Watanbe helped edit the final drafts of the book.

At Norton, my longtime editor Drake McFeely grasped the importance of this book from the start and worked tirelessly on the first two drafts of the manuscript. The support team—capably led by Nancy Palmquist and Amanda Morrison, and including Allegra Huston, Brendan Curry, and Don Rifkin—performed miracles under tight deadlines. And my very special thanks go to Stuart Proffitt of Penguin, with whom the idea for this book was hatched over lunch and whose close comments on the manuscript were also invaluable.

My wife, Anya Stiglitz, was involved from the beginning. A reporter who had spent years in the developing world, Anya's trained eye helped me see more clearly what was going on in these countries, to see, as we traveled together for months on end, how globalization affected the everyday lives of the people—to see beyond the narrow boundaries to which academic disciplines inevitably draw one. Her curiosity about why things were the way they were forced me to try harder to explain the underlying forces. She is, if anything, even more committed to the idea that another world is possible, one in which globalization might live up more closely to its potential of enhancing the well-being of the poor—and she challenged me to go beyond diagnosing the problems to showing how that world might be created. But she was willing to go from these lofty aspirations to the mundane, hard, and often tedious work of shaping this manuscript, as she read through each draft and had the patience to edit and reedit them.

MAKING GLOBALIZATION WORK

CHAPTER I

Another World Is Possible

In a vast field on the outskirts of Mumbai, activists from around the world gathered for the World Social Forum in January 2004. The first Forum to be held in Asia, this meeting had a very different feel from those held in Porto Alegre, Brazil, in the four previous years. Over 100,000 people attended the week-long event, and the scene was, like India itself, a colorful crush of humanity. Fair trade organizations staffed rows of stalls selling handmade jewelry, colorful textiles, and housewares. Banners strung along the streets proclaimed, "HANDLOOM IS A BIGGEST EMPLOYMENT SOURCE IN INDIA." Columns of demonstrators banged drums and chanted slogans as they wended their way through the crowds. Loincloth-clad groups of *dalit* activists (members of the castes that used to be known as untouchables), representatives of workers' rights organizations and women's groups, the UN and nongovernmental organizations (NGOs) all rubbed shoulders. Thousands gathered in temporary meeting halls the size of aircraft hangars to hear a program of speakers that included former Irish president Mary Robinson (former UN High Commissioner for Human Rights, 1997–2002) and Nobel Peace Prize winner Shirin Ebadi. It was hot and humid and there were crowds everywhere.

Many conversations took place at the World Social Forum. There was debate about how to restructure the institutions that run the world

3

and how to rein in the power of the United States. But there was one overriding concern: globalization. There was a consensus that change is necessary, summed up in the motto of the conference: "Another world is possible." The activists at the meeting had heard the promises of globalization—that it would make everyone better off; but they had seen the reality: while some were in fact doing very well, others were worse off. In their eyes, globalization was a big part of the problem.

Globalization encompasses many things: the international flow of ideas and knowledge, the sharing of cultures, global civil society, and the global environmental movement. This book, however, is mostly about economic globalization, which entails the closer economic integration of the countries of the world through the increased flow of goods and services, capital, and even labor. The great hope of globalization is that it will raise living standards throughout the world: give poor countries access to overseas markets so that they can sell their goods, allow in foreign investment that will make new products at cheaper prices, and open borders so that people can travel abroad to be educated, work, and send home earnings to help their families and fund new businesses.

I believe that globalization has the potential to bring enormous benefits to those in both the developing and the developed world. But the evidence is overwhelming that it has failed to live up to this potential. This book will show that the problem is not with globalization itself but in the way globalization has been managed. Economics has been driving globalization, especially through the lowering of communication and transportation costs. But politics has shaped it. The rules of the game have been largely set by the advanced industrial countries— and particularly by special interests within those countries—and, not surprisingly, they have shaped globalization to further their own interests. They have not sought to create a fair set of rules, let alone a set of rules that would promote the well-being of those in the poorest countries of the world.

After speaking at the World Social Forum, Mary Robinson, Delhi University chancellor Deepak Nayaar, International Labour Organization president Juan Somavia, and I were among the few who went on to the World Economic Forum in Davos, the Swiss ski resort where the global elite gather to mull over the state of the world. Here, in this snowy mountain town, the world's captains of industry and finance had very different views about globalization from those we heard in Mumbai.

The World Social Forum had been an open meeting, bringing together vast numbers from all over the world who wanted to discuss social change and how to make their slogan, "Another world is possible," a reality. It was chaotic, unfocused, and wonderfully lively—a chance for people to see each other, make their voices heard, and to network with their fellow activists. Networking is also one of the main reasons that the movers and shakers of the world attend the invitation-only event at Davos. The Davos meetings have always been a good place to take the pulse of the world's economic leaders. Though largely a gathering of white businessmen, supplemented by a roster of government officials and senior journalists, in recent years the invitation list has been expanded to include a number of artists, intellectuals, and NGO representatives.

In Davos there was relief, and a bit of complacency. The global economy, which had been weak since the bursting of the dot-com bubble in America, was finally recovering, and the "war on terror" seemed to be under control. The 2003 gathering had been marked by enormous tension between the United States and the rest of the world over the war in Iraq, and still earlier meetings had seen disagreement over the direction which globalization was taking. The 2004 meeting was marked with relief that these tensions had at least been modulated. Still there was worry about American unilateralism, about the world's most powerful country imposing its will on others while preaching democracy, self-determination, and human rights. People in the developing world had long been worried about how global decisions—decisions about economics and politics that affected their lives—were made. Now, it seemed, the rest of the world was worried also.

I have been going to the annual meetings at Davos for many years and had always heard globalization spoken of with great enthusiasm. What was fascinating about the 2004 meeting was the speed with which views had shifted. More of the participants were questioning whether globalization really was bringing the promised benefits—at least to many in the poorer countries. They had been chastened by the economic instability that marked the end of the twentieth century, and they worried about whether developing countries could cope with the consequences. This change is emblematic of the massive change in thinking about globalization that has taken place in the last five years all around the world. In the 1990s, the discussion at Davos had been about the virtues of opening international markets. By the early years of the millennium, it centered on poverty reduction, human rights, and the need for fairer trade arrangements.

At a Davos panel on trade, the contrast in views between the developed and developing countries was especially marked. A former World Trade Organization official said that if trade liberalization—the lowering of tariffs and other trade barriers—had not fully delivered on its promise of enhanced growth and reduced poverty, it was the fault of the developing countries, which needed to open their markets more to free trade and globalize faster. But an Indian running a micro-credit bank stressed the downside of free trade for India. He spoke of peanut farmers who could not compete with imports of Malaysian palm oil. He said it was increasingly difficult for small and medium-sized businesses to get loans from banks. This was not surprising. Around the world, countries that have opened up their banking sectors to large international banks have found that those banks prefer to deal with other multinationals like Coca-Cola, IBM, and Microsoft. While in the competition between large international banks and local banks the local banks often appeared to be the losers, the real losers were the local small businesses that depended on them. The puzzlement of some listeners, convinced that the presence of international banks would unambiguously be better for everyone, showed that these businessmen had paid little attention to similar complaints from Argentina and Mexico, which saw lending to local companies dry up after many of their banks were taken over by foreign banks in the 1990s.

At both Mumbai and Davos, there was discussion of reform. At Mumbai, the international community was asked to create a fairer form of globalization. At Davos, the developing countries were enjoined to rid themselves of their corruption, to liberalize their markets, and to open up to the multinational businesses so well represented at the meeting. But at both events there was an understanding that something had to be done. At Davos the responsibility was placed squarely on the developing countries; at Mumbai, it was on the entire international community.

THE TWO FACES OF GLOBALIZATION

In the early 1990s, globalization was greeted with euphoria. Capital flows to developing countries had increased sixfold in six years, from 1990 to 1996. The establishment of the World Trade Organization in 1995—a goal that had been sought for half a century—was to bring the semblance of a rule of law to international commerce. Everyone was supposed to be a winner—those in both the developed and the developing world. Globalization was to bring unprecedented prosperity *to all*.

No wonder then that the first major modern protest against globalization—which took place in Seattle in December 1999, at what was supposed to be the start of a new round of trade negotiations, leading to further liberalization—came as a surprise to the advocates of open markets. Globalization had succeeded in unifying people from around the world—against globalization. Factory workers in the United States saw their jobs being threatened by competition from China. Farmers in developing countries saw their jobs being threatened by the highly subsidized corn and other crops from the United States. Workers in Europe saw hard-fought-for job protections being assailed in the name of globalization. AIDS activists saw new trade agreements raising the prices of drugs to levels that were unaffordable in much of the world. Environmentalists felt that globalization undermined their decades-long struggle to establish regulations to preserve our natural heritage. Those who wanted to protect and develop their own cultural heritage saw too the intrusions of globalization. These protestors did not accept the argument that, economically at least, globalization would ultimately make everybody better off.

There have been many reports and commissions devoted to the topic of globalization. I was involved in the World Commission on the Social Dimensions of Globalization, which was established in 2001 by the International Labour Organization (created in 1919 in Geneva to bring together government, business, and labor). Co-chaired by President Benjamin W. Mkapa of Tanzania and President Tarja Kaarina Halonen of Finland, our commission issued a highly skeptical report in 2004. A few lines go a long way to understanding how much of the world feels about globalization:

> The current process of globalization is generating unbalanced outcomes, both between and within countries. Wealth is being created, but too many countries and people are not sharing in its benefits. They also have little or no voice in shaping the process. Seen through the eyes of the vast majority of women and men, globalization has not met their simple and legitimate aspirations for decent jobs and a better future for their children. Many of them live in the limbo of the informal economy without formal rights and in a swathe of poor countries that subsist precariously on the margins of the global economy. Even in economically successful countries some workers and communities have been adversely affected by globalization. Meanwhile the revolution in global communications heightens awareness of these disparities . . . these global imbalances are morally unacceptable and politically unsustainable.[1]

The commission surveyed seventy-three countries around the world. Its conclusions were startling. In every region of the world except South Asia, the United States, and the European Union (EU), unemployment rates increased between 1990 and 2002. By the time the report was issued, global unemployment had reached a new high of 185.9 million. The commission also found that 59 percent of the world's people were living in countries with growing inequality, with only 5 percent in countries with declining inequality.[2] Even in most of the developed countries, the rich were getting richer while the poor were often not even holding their own.

In short, globalization may have helped some countries—their GDP, the sum total of the goods and services produced, may have increased—but it had not helped most of the people even in these countries. The worry was that globalization might be creating rich countries with poor people.

Of course, those who are discontented with economic globalization generally do not object to the greater access to global markets or to the spread of global knowledge, which allows the developing world to take advantage of the discoveries and innovations made in developed countries. Rather, they raise five concerns:

- The rules of the game that govern globalization are unfair, specifically designed to benefit the advanced industrial countries. In fact, some recent changes are so unfair that they have made some of the poorest countries actually worse off.
- Globalization advances material values over other values, such as a concern for the environment or for life itself.
- The way globalization has been managed has taken away much of the developing countries' sovereignty, and their ability to make decisions themselves in key areas that affect their citizens' well-being. In this sense, it has undermined democracy.
- While the advocates of globalization have claimed that everyone will benefit economically, there is plenty of evidence from both developing and developed countries that there are many losers in both.
- Perhaps most important, the economic system that has been pressed upon the developing countries—in some cases essentially forced upon them—is inappropriate and often grossly damaging. Globalization should not mean the Americanization of either economic policy or culture, but often it does—and that has caused resentment.

The last is a topic that touches both those in developed and developing countries. There are many forms of a market economy—the American model differs from that of the Nordic countries, from the Japanese model, and from the European social model. Even those in developed countries worry that globalization has been used to advance the "Anglo-American liberal model" over these alternatives—and even if the American model has done well as measured by GDP, it has not done well in many other dimensions, such as the length (and, some would argue, the quality) of life, the eradication of poverty, or even the maintenance of the well-being of those in the middle. Real wages in the United States, especially of those at the bottom, have stagnated for more than a quarter century, and incomes are as high as they are partly because Americans work far longer hours than their European counterparts. If globalization is being used to advance the American model of a market economy, many elsewhere are not sure they want it. Those in the developing world have an even stronger complaint—that globalization has been used to advance a version of market economics that is more extreme, and more reflective of corporate interests, than can be found even in the United States.

Globalization and poverty

Critics of globalization point to the growing numbers of people living in poverty. The world is in a race between economic growth and population growth, and so far population growth is winning. Even as the percentages of people living in poverty are falling, the absolute number is rising. The World Bank defines poverty as living on less than $2 a day, absolute or extreme poverty as living on less than $1 a day.

Think for a minute what it means to live on one or two dollars a day.[3] Life for people this poor is brutal. Childhood malnutrition is endemic, life expectancy is often below fifty years, and medical care is scarce. Hours are spent each day searching for fuel and drinkable water and eking out a miserable livelihood, planting cotton on a semi-arid plot of land and hoping that this year the rains will not fail, or in the backbreaking toil of growing rice in a meager half acre, knowing that no matter how hard one works there will be barely enough to feed one's family.

Globalization has played a part both in the biggest successes—and in some of the failures. China's economic growth, which was based on exports, has lifted several hundred million people out of poverty. But China managed globalization carefully: it was slow to open up its own markets for imports, and even today does not allow the entry of hot speculative money—money that seeks high returns in the short run and rushes into a country in a wave of optimism only to rush out again at the first hint of trouble. China's government realized that while the rush in might bring a short-lived boom, the recessions and depressions that could be expected to follow would bring long-lasting damage, more than offsetting the short-run gain. China avoided the boom-and-bust that marked other countries in East Asia and Latin America (as we will see in chapter 2), maintaining growth in excess of 7 percent every year.

The sad truth, however, is that outside of China, poverty in the developing world has increased over the past two decades. Some 40 percent of the world's 6.5 billion people live in poverty (a number that is up 36 percent from 1981), a sixth—877 million—live in extreme poverty (3 percent more than in 1981). The worst failure is Africa, where the percentage of the population living in extreme poverty has increased from 41.6 percent in 1981 to 46.9 percent in 2001. Given its increasing population, this means that the number of people living in extreme poverty has almost doubled, from 164 million to 316 million.[4]

Historically, Africa is the region most exploited by globalization: during the years of colonialism the world took its resources but gave back little in return. In recent years, Latin America and Russia have also been disappointed by globalization. They opened up their markets, but globalization did not deliver on its promises, especially to the poor.

Income and higher living standards are important, but the deprivations of poverty go beyond a lack of money. When I was chief economist of the World Bank, we published a study called *Voices of the Poor*. A team of economists and researchers interviewed some 60,000 poor men and women from sixty countries in order to find out how they felt about their situation.[5] Unsurprisingly, they stressed not just their inadequate income but their feelings of insecurity and powerlessness. Those without jobs, especially, felt marginalized, shunted aside by their societies.

For those who have a job, much of this insecurity arises from the risk of being thrown out of it or of wages plummeting—seen so dramatically in the crises in Latin America, Russia, and East Asia at the end of the 1990s. Globalization has exposed developing countries to more risks, but markets to insure against these risks are notably absent. In more advanced countries, governments fill in the gap by providing pensions for senior citizens, disability payments, health insurance, welfare, and unemployment insurance. But in developing countries, governments are typically too poor to implement social insurance programs. What little money they have is more likely to be spent on basic education and health, and on building infrastructure. The poor are left to fend for themselves and so are vulnerable when the economy slows down or jobs are lost due to competition from foreign countries. The wealthy have a buffer of savings to protect them, but the poor do not.

Insecurity was one of the major concerns of the poor; a sense of powerlessness was another. The poor have few opportunities to speak out. When they speak, no one listens; when someone does listen, the reply is that nothing can be done; when they are told something can be done, nothing is ever done. A remark in the World Bank report, from a young woman in Jamaica, captures this sense of powerlessness: "Poverty is like living in jail, living under bondage, waiting to be free."

What is true for poor people is too often true for poor countries. While the idea of democracy has spread and more countries have free elections than, say, thirty years ago,[6] developing countries find their ability to act eroded both by new constraints imposed from outside and by the weakening of their existing institutions and arrangements to which globalization has contributed. Consider, for instance, the demands imposed on developing countries as a condition for aid. Some might make sense (though not all, as we will see in chapter 2). But that is not the point. Conditionality undermines domestic political institutions. The electorate sees its government bending before foreigners or giving into international institutions that it believes to be run by the United States. Democracy is undermined; the electorate feels betrayed. Thus, although globalization has helped spread the idea of democracy, it has, paradoxically, been managed in a way that undermines democratic processes within countries.

Moreover, it is perceived—quite rightly, I think—that the way globalization is currently managed is not consistent with democratic principles. Little weight is given, for instance, to the voices and concerns of the developing countries. At the International Monetary Fund, the international institution charged with oversight of the global financial system, a single country—the United States—has effective veto. It is not a question of one man one vote, or one country one vote: dollars vote. The countries with the largest economies have the most votes—and it is not even today's dollars that count. Votes are determined largely on the basis of economic power at the time the IMF was estab-

lished sixty years ago (with some adjustments since). China, with its bur-
geoning economy, is underrepresented. As another example, the head of
the World Bank, the international organization charged with promoting
development, has always been appointed by the president of the United
States (without even having to consult his own Congress). American politics, not qualifications, are what matters: experience in development, or
even experience in banking, is not required. In two instances—the
appointments of Paul Wolfowitz and Robert MacNamara—the background was defense, and both these former secretaries of defense were
associated with discredited wars (Iraq and Vietnam).

REFORMING GLOBALIZATION

The globalization debate has gone from a general recognition that all
was not well with globalization and that there was a real basis for at
least some of the discontent to a deeper analysis that links specific policies with specific failures. Experts and policymakers now agree on the
areas where change has to take place. This book is concerned with the
hardest question of all: What changes, large and small, will enable
globalization to live up to its promise, or at least more nearly do so?
How do we make globalization work?

Making globalization work will not be easy. Those who benefit from
the current system will resist change, and they are very powerful. But
forces for change have already been set in motion. There will be
reforms, even if they are piecemeal ones. I hope that this book will help
lead to reforms based on a broader vision of what is currently wrong.
It also provides a number of specific suggestions for how to make globalization work better. Some of these are small, and should meet little
resistance; others are big, and may not be implemented for years.

There are many things that must be done. Six areas where the international community has recognized that all is not well illustrate both
the progress that has been made and the distance yet to go.

The pervasiveness of poverty
Poverty has, at last, become a global concern. The United Nations and
multinational institutions such as the World Bank have all begun

focusing more on poverty reduction. In September 2000, some 150 heads of state or government attended the Millennium Summit at the United Nations in New York and signed the Millennium Development Goals, pledging to cut poverty in half by 2015.[7] They recognized the many dimensions to poverty—not just inadequate income, but also, for instance, inadequate health care and access to water.

Until recently, IMF perspectives have been paramount in economic policy discussions, and the IMF traditionally focused on inflation rather than on wages, unemployment, or poverty. Its view was that poverty reduction was the mandate of the World Bank, while its own mandate was global economic stability. But focusing on inflation and ignoring employment led to the obvious result: higher unemployment and more poverty. The good news is that, at least officially, the IMF has now made poverty reduction a priority.

By now it has become clear that opening up markets (taking down trade barriers, opening up to capital flows) *by itself* will not "solve" the problem of poverty; it may even make it worse. What is needed is both more assistance and a fairer trade regime.

The need for foreign assistance and debt relief
At Monterrey, Mexico, in March 2002 at the International Conference on Financing for Development, which was attended by 50 heads of state or government and 200 government ministers, among others, the advanced industrial countries committed themselves to substantial increases in assistance—to 0.7 percent of their GDP (though so far few countries have lived up to those commitments, and some—especially the United States—are a far way off).[8] In tandem with the recognition that aid should be increased has come a broad agreement that more assistance should be given in the form of grants and less in loans—not surprising given the constant problems in repaying the loans.

Most telling of all, however, is the altered approach to conditionality. Countries seeking foreign aid are typically asked to meet a large number of conditions; for instance, a country may be told that it must quickly pass a piece of legislation or reform social security, bankruptcy, or other financial systems if it is to receive aid. The enormous number of conditions often distracted governments from more vital tasks.

Excessive conditionality was one of the major complaints against the IMF and the World Bank. Both institutions now admit that they went overboard, and in the last five years they have actually greatly reduced conditionality.

Many developing countries face a huge burden of debt. In some, half or more of their governmental spending or foreign exchange earnings from exports has to be used to service this debt—taking away money that could be used for schools, roads, or health clinics. Development is difficult as it is; with this debt burden, it becomes virtually impossible.

Once a year, the leaders of the major industrial countries (called the G-8) get together to discuss major global problems. At the 2005 G-8 summit, held in Gleneagles, Scotland, the leaders of the advanced industrial countries agreed to write off completely the debt owed to the IMF and the World Bank by the poorest eighteen countries of the world, fourteen of which are in Africa.[9] Even after two previous attempts at debt reduction, many developing countries still have an enormous debt overhang. As I write this, the world's developing countries owe roughly $1.5 trillion to creditors including international banks, the IMF, and the World Bank. Approximately one-third of that is owed by low-income countries.[10] And in spite of debt forgiveness, the level of indebtedness by low-income countries has continued to increase.

Debt and how the world deals with countries that cannot fulfill their debt obligations is unfortunately not just a problem for low-income countries. Russia's default threatened, at least for a moment, to precipitate a global financial crisis. Argentina's default at the end of 2001— the largest in history—prompted even the IMF to weigh the advantages of some regular restructuring mechanism, the analogue to bankruptcy proceedings for private debt. This was a major step forward.

The aspiration to make trade fair

Trade liberalization—opening up markets to the free flow of goods and services—was supposed to lead to growth. The evidence is at best mixed.[11] Part of the reason that international trade agreements have been so unsuccessful in promoting growth in poor countries is that they were often unbalanced: the advanced industrial countries were allowed to levy tariffs on goods produced by developing countries that were, on average, four times higher than those on goods produced by other advanced industrial countries.[12] While developing countries were forced to abandon subsidies designed to help their nascent industries, advanced industrial countries were allowed to continue their own enormous agricultural subsidies, forcing down agricultural prices and undermining living standards in developing countries.

In the aftermath of the Seattle riots, as a closer look was taken at past trade agreements, it became clear that at least some of the discontent was justified. The poorest countries had actually been made worse off by the last trade agreement. And the world responded: at Doha, in November 2001, there was an agreement that the next round of trade negotiations should focus on the needs of the developing countries. (Regretfully, as we shall see in chapter 3, in the subsequent years Europe and the United States largely reneged on the promises that had been made at Doha.)

The limitations of liberalization

In the 1990s, when the policies of liberalization failed to produce the promised results, the focus was on what the developing countries had failed to do. If trade liberalization did not produce growth, it was because the countries had not liberalized enough, or because corruption created an unfavorable climate for business. Today, even among many of the advocates of globalization, there is more awareness of shared blame.

The most hotly contested policy issue of the 1990s was capital market liberalization, opening up markets to the free flow of short-term, hot, speculative money. The IMF even tried to change its charter at its annual meeting in 1997, held in Hong Kong, to enable it to push countries to liberalize. By 2003, even the IMF had conceded that, at least for many developing countries, capital market liberalization had led not to more growth, just to more instability.[13]

Trade and capital market liberalization were two key components of a broader policy framework, known as the Washington Consensus—a consensus forged between the IMF (located on 19th Street), the World Bank (on 18th Street), and the U.S. Treasury (on 15th Street)—on what constituted the set of policies that would best promote development.[14] It emphasized downscaling of government, deregulation, and rapid liberalization and privatization. By the early years of the millennium, confidence in the Washington Consensus was fraying, and a post–Washington Consensus consensus was emerging. The Washington Consensus had, for instance, paid too little attention to issues of equity, employment, and competition, to pacing and sequencing of reforms, or to how privatizations were conducted. There is by now also a consensus that it focused too much on just an increase in GDP, not on other things that affect living standards, and focused too little on sustainability—on whether growth could be sustained economically, socially, politically, or environmentally. The fact that countries like Argentina—which got an A+ rating from the IMF for following the Washington Consensus precepts—did well for a few short years only to later face calamity has helped to reinforce the new emphasis on sustainability.

Protecting the environment
A failure of environmental stability poses an even greater danger for the world in the long run. A decade ago, concern about the environment and globalization was limited mostly to environmental advocacy groups and experts. Today, it is almost universal. Unless we lessen environmental damage, conserve on our use of energy and other natural resources, and attempt to slow global warming, disaster lies ahead. Global warming has become a true challenge of globalization. The successes of development, especially in India and China, have provided those countries the economic wherewithal to increase energy usage, but the world's environment simply cannot sustain such an onslaught. There will be grave problems ahead if everybody emits greenhouse gases at the rate at which Americans have been doing so. The good news is that this is, by now, almost universally recognized, except in some quarters in Washington; but the adjustments in lifestyles will not be easy.

A flawed system of global governance
There is now also a consensus, at least outside the United States, that something is wrong with the way decisions are made at the global level;

there is a consensus, in particular, on the dangers of unilateralism and on the "democratic deficit" in the international economic institutions. Both by structure and process, voices that ought to be heard are not. Colonialism is dead, yet the developing countries do not have the representation that they should.

World War I made clear our growing global interdependence, and when it was over several international institutions were created. The most important, the League of Nations, failed in its mission to preserve the peace. As World War II was coming to an end, there was a resolve to do better. The United Nations was created to prevent the wars that had proven such a scourge during the first half of the twentieth century. With memories of the Great Depression of the 1930s still fresh, two new economic institutions were established: the International Monetary Fund and the World Bank. At the time, much of the developing world was still colonized; these institutions were clubs of the rich countries, and their governance reflected this. They quickly established "old boy" rules to enhance their control: the United States agreed that Europe could appoint the head of the IMF, with an American in the number two position; and Europe agreed that the U.S. president could appoint the head of the World Bank. If these institutions had been more successful in ameliorating the problems they were supposed to address—if, for instance, the IMF had succeeded in ensuring the stability of the world's economy—these anachronisms in governance might have been forgiven. But the IMF failed in its major mission of ensuring global financial stability—as evidenced so starkly in the global crises at the end of the 1990s, which affected every major emerging market economy that had followed the IMF's advice. As the IMF crafted policies to respond to the crises, it seemed more often to focus on saving the Western creditors than on helping the countries in crisis and their people. There was money to bail out Western banks but not for minimal food subsidies for those on the brink of starvation. Countries that had turned to the IMF for guidance failed in sustained growth, while countries like China, which followed its own counsel, had enormous success. Deeper analyses exposed the role that particular IMF policies such as capital market liberalization had played in the failures. While the IMF complained about problems of governance and lack of transparency in developing countries, it seemed that the IMF itself was beset by these same problems. It lacked some of the basic rules of democratic institutions: namely, transparency, so that citizens could see what issues were on the table and have time to react, and also so they could see how officials had voted, so that they could be held accountable. In addition, there was a need for regulations restricting officials from moving quickly to private firms as they departed their public service to the IMF; such restrictions are standard fare in modern democracies, to reduce the appearance—or reality—of conflicts of interests, the incentive of servants rewarding potential future employers through favorable procurement or regulation.

There is a growing consensus both that there is a problem of governance in the international public institutions like the IMF that shape globalization and that these problems contribute to their failures. At the very least, the democratic deficit in their governance has contributed to their lack of legitimacy, which has undermined their efficacy—especially when they speak on issues of democratic governance.

The Nation-State and Globalization

Some 150 years ago, the lowering of communication and transportation costs gave rise to what may be viewed as the earlier precursor of globalization. Until then, most trade had been local; it was the changes of the nineteenth century that led to the formation of national economies and helped to strengthen the nation-state. New demands were put on government: markets might be producing growth, but they were accompanied by new social, and in some cases even economic, problems. Governments took on new roles in preventing monopolies, in laying the foundations of modern social security systems, in regulating banks and other financial institutions. There was mutual reinforcement: success in these endeavours helped shape and strengthen the process of nation building, and the increased capabilities of the nation-state led to greater success in strengthening the economy and enhancing individual well-being.

The conventional wisdom that the United States' development was the result of unfettered capitalism is wrong. Even today, the U.S. government, for instance, plays a central role in finance. It provides, or provides guarantees for, a significant fraction of all credit, with programs for mortgages, student loans, exports and imports, cooperatives, and small businesses. Government not only regulates banking and insures depositors but also tries to ensure that credit flows to underserved groups and, at least until recently, to all regions in the country—not just the big money centers.

Historically, the U.S. government played an even larger role in the economy in promoting development, including the development of technology and infrastructure. In the nineteenth century, when agriculture was at the center of the economy, government created the whole system of agricultural universities and "extension" services. Huge land grants spurred the development of the western railroads. In the nineteenth century, the U.S. government funded the first telegraph line; in the twentieth, it funded the research that led to the Internet.

The United States was successful partly because of the role that its government played in promoting development, in regulating markets, and in providing basic social services. The question facing developing countries today is, will their governments be able to play a comparable role? While the process of globalization has put new demands on nation-states to address the increasing inequality and insecurity that it can cause and to respond to the competitive challenges that it presents, globalization has, in many ways, limited their capacity to respond. For instance, globalization has unleashed market forces that by themselves are so strong that governments, especially in the developing world, often cannot control them. Governments that attempt to control capital flows may find themselves powerless to do so, as individuals find ways of circumventing the regulations. A country may want to raise the minimum wage but discovers it can't, because foreign companies operating there will decide to move to a country with lower wages.

Increasingly, a government's inability to control the actions of individuals or companies is also limited by international agreements that impinge on the right of sovereign states to make decisions. A government that wants to ensure that banks lend a certain fraction of their portfolio to underserved areas, or to ensure that accounting frameworks accurately reflect a company's true status, may find it is unable to pass the appropriate laws. Signing on to international trade agreements can prevent governments from regulating the influx and outflow of hot, speculative money, even though capital market liberalization can lead to economic crises.

The nation-state, which has been the center of political and (to a large extent) economic power for the past century and a half is being squeezed today—on one side, by the forces of global economics, and on the other side, by political demands for devolution of power. Globalization—the closer integration of the countries of the world—has resulted in the need for more collective action, for people and countries to act together to solve their common problems. There are too many problems—trade, capital, the environment—that can be dealt with only at the global level. But while the nation-state has been weakened, there has yet to be created at the international level the kinds of democratic global institutions that can deal effectively with the problems globalization has created.

In effect, economic globalization has outpaced political globalization. We have a chaotic, uncoordinated system of global governance without global government, an array of institutions and agreements dealing with a series of problems, from global warming to international trade and capital flows. Finance ministers discuss global finance matters at the IMF, paying little heed to how their decisions affect the environment or global health. Environment ministers may call for something to be done about global warming, but they lack the resources to back up those calls.

There is a clear need for strong international institutions to deal with the challenges posed by economic globalization; yet today confidence in existing institutions is weak. The fact that the institutions which make the decisions suffer, as we have noted, from a democratic deficit is clearly a problem. It results in decisions that are too often not in the interests of those in the developing world. Making matters even worse is the fact that those in the advanced industrial countries, whose governments dictate the direction of economic globalization, have not yet developed the underlying sympathies which are necessary to make the global community work. Of course, when we see earthquakes in Turkey, or a famine in Ethiopia, or a tsunami in Indonesia—images that globalization has enabled us to bring into every person's living room—we feel enormous sympathy for the victims, and there is an outpouring of help. But more than that is required.

As the nation-state developed, individuals felt connected to others within the nation—not as closely as to those in their own local community, but far more closely than to those outside the nation-state. The problem is that, as globalization has proceeded, loyalties have changed little. War shows these differences in attitude most dramatically: Americans keep accurate count of the number of U.S. soldiers lost, but when estimates of Iraqi deaths, up to fifty times as high, were released, it hardly caused a stir. Torture of Americans would have generated outrage; torture by Americans seemed mainly to concern those in the anti-war movement; it was even defended by many as necessary to protect the United States. These asymmetries have their parallel in the economic sphere. Americans bemoan the loss of jobs at home, and do not celebrate a larger gain in jobs by those who are far poorer abroad.

Most of us will always live locally—in our own communities, states, countries. But globalization has meant that we are, at the same time, part of a global community. Europeans are, sometimes with difficulty, learning how to think of themselves both as German or Italian or British *and* as European. Closer economic integration has helped. So too at the global level: we may live locally, but increasingly we will have to think globally, think of ourselves as part of a global community. This will entail more than just treating others with respect. It will entail thinking about what is fair: what, for instance, would a fair trade regime look like? It will entail putting ourselves in others' shoes: what would we think is fair or right if we were in their position?[15] And it will entail thinking carefully about when we need to impose rules and regulations to make the global system work, and when we should respect national sovereignty, allowing each to make the decisions appropriate for themselves.

A change in mindset will be essential if we are to change the way globalization is managed. Such a change is already under way. This chapter has highlighted the enormous changes in attitudes toward globalization that have occurred in the last decade alone. The debate is, to a large extent, no longer "anti-" or "pro-" globalization. We have realized the positive potential of globalization: almost half of humanity—

Asia, including China and India—is being integrated into the global economy; 2.4 billion people whose countries have suffered colonialism and exploitation, wars and internal disarray, have seen unprecedented rates of growth for a quarter of a century or more. This is an event of historic proportions, and it too has to be put into historical context. Even in the most successful years of the West, during the Industrial Revolution or the boom that followed World War II, growth seldom exceeded 3 percent. China's average growth over the past three decades has been triple that. These successes are partly due to globalization. But we have also seen the darker side of globalization: the recessions and depressions that global instability has brought with it; the degradation of the environment as global growth proceeds without global rules; a continent, Africa, stripped of its assets, its natural resources, and left with a debt burden beyond its ability to pay. Even the advanced industrial countries are beginning to question globalization, as it brings with it economic insecurity and inequality; as economic materialism trumps other values; as countries realize that their well-being, even their survival, depends on others that they may not trust, such as the unstable oil regimes in the Middle East and elsewhere. There may be growth, but most of the people may be worse off. Trickle-down economics, which holds that so long as the economy as a whole grows everyone benefits, has been repeatedly shown to be wrong.

Some say globalization is inevitable, that one has to simply accept it with its flaws. But as most of the world has come to live in democracies, if globalization does not benefit most of the people they will eventually react. They can be fooled for a while—they can, for a while, believe stories that, while the pain is here today, the gain is around the corner—but after a quarter century or more, such stories lose their credibility. There have been reversals in globalization before—the degree of global economic integration, by most measures, fell after World War I;[16] and it can happen again. Already the world has seen the beginnings of a backlash against globalization, even in the countries that have been its greatest beneficiaries, as attempts by Indian, Chinese, and Dubai firms to buy companies in the developed world have met with resistance.

Some of the problems with globalization are inevitable, and we have to learn to cope with them: long-standing economic theories, explained in later chapters, argue that globalization will lead to increasing inequality in the advanced industrial countries as wages, especially of unskilled workers, are depressed. Downward pressure on wages can be resisted, but then unemployment will increase. Even the most powerful politicians cannot repeal these laws of economics, try as they might. But they can help our societies adjust to this great transformation of our global society, as the nation-state helped the transition to industrialization more than a century ago.[17]

Today, there is an understanding that many of the problems with globalization are of our own making—are a result of the way globalization has been managed. I am heartened as I see mass movements, especially in Europe, calling for debt relief, and as I see the leaders of most of the advanced industrial countries calling for a fairer trade regime, doing something about global warming, and committing themselves to cutting poverty in half by 2015. But there is a gap between the rhetoric and the reality—and many of these leaders are ahead of the people in their democracies, who may be fully committed to these lofty goals, but only so long as it does not cost them anything.

I hope that this book will help to change mindsets—as those in the developed world see more clearly some of the consequences of the policies that their governments have undertaken. I hope it will convince many, in all countries, that "another world is possible." Even more: that "another world is necessary and inevitable." We cannot carry on along the course we have been on. The forces of democracy are too strong: voters will not allow the continuation of the way that globalization has been managed. We are already beginning to see manifestations of this in elections in Latin America and elsewhere. The good news is that economics is not zero-sum. We can restructure globalization so that those in both the developed and the developing world, the current generations and future generations, can all benefit—though there are some special interests who will lose out, and they will resist these changes. We can have stronger economies *and* societies that put more weight on values, like culture, the environment, and life itself.

CHAPTER 2

The Promise of Development

The back roads of Karnataka, in southern India, are filled with potholes, and even short distances can take hours by car. Women labor on the roads breaking stones by hand. The landscape is dotted with lone men plowing the dusty fields with oxen. At roadside stalls, shopkeepers sell biscuits and tea. It's a typical scene in India, where much of the population is still illiterate and the median income is just $2.70 a day.

Just a few miles away, in the city of Bangalore, a revolution is taking place. The gleaming global headquarters of the giant Indian high-tech and consulting firm Infosys Technologies has become a symbol of a controversial outsourcing movement, in which American companies hire Indian workers to do work that was previously done in the United States and Europe. Although companies have been sending manufacturing work to low-wage countries for decades, India's success at attracting high-skill jobs such as computer programming and customer service has caused a lot of worry in the United States.

Infosys, which generates some $1.5 billion a year in revenues, has been a boon to the local economy. Its employees spend money on cars, housing, and clothes, and at the new restaurants and bars that have sprung up in Bangalore. Any visitor to Bangalore can feel the rising prosperity. But the enthusiasm for this new world is not universally shared. In the 2004 national election, the ruling Bharatiya Janata Party (BJP) ran on a platform of "India shining"—and on the lives of some 250 million people India was indeed shining, as their standard of living had improved immensely over the previous two decades. But just ten miles outside Bangalore, and even in parts of the city, poverty can be seen everywhere; for the other 800 million people of India, the economy has not shone brightly at all.

About 80 percent of the world's population lives in developing countries, marked by low incomes and high poverty, high unemployment and low education. For those countries, globalization presents both unprecedented risks and opportunities. Making globalization work in ways that enrich the whole world requires making it work for the people in those countries.

We will see in this chapter that there are no magic solutions or simple prescriptions. The history of development economics is marked by the quixotic quest to find "the answer," disappointment in the failure of one strategy leading to the hope that the next will work.[1] For instance, education is important—but if there are no jobs for those who are educated, there will not be development. It is important for developed countries to open up their markets to poorer countries—but if the developing countries have no roads or ports with which to bring their goods to market, what good does it do? If productivity in agriculture is so low that farmers have little to sell, then ports and roads will make little difference. Development is a process that involves every aspect of society, engaging the efforts of everyone: markets, governments, NGOs, cooperatives, not-for-profit institutions.

A developing country that simply opens itself up to the outside world does not necessarily reap the fruits of globalization. Even if its GDP increases, the growth may not be sustainable, or sustained. And even if growth is sustained, most of its people may find themselves worse off.

The debate about economic globalization is mixed with debates about economic theory and values. A quarter century ago, three major schools of economic thought competed with each other—free market capitalism, communism, and the managed market economy. With the fall of the Berlin Wall in 1989, however, the three were reduced to two, and the argument today is largely between those who push free market ideology and those who see an important role for both government and the private sector. Of course, these positions overlap. Even free market advocates recognize that one of the problems in Africa is the lack of government. And even critics of unfettered capitalism respect the importance of the market.

Still, there is a huge gap between the different perspectives, and we should not let ourselves be fooled into thinking there are no differences. In the last chapter, we described the Washington Consensus strategy for development. These policies focused on minimizing the role of government, emphasizing privatization (selling off government enterprises to the private sector), trade and capital market liberalization (eliminating trade barriers and impediments to the free flow of capital), and deregulation (eliminating regulations on the conduct of business). Government had a role in maintaining macro-stability, but the attention was on price stability rather than on output stability, employment, or growth. There was a large set of dos and don'ts: do privatize everything, from factories to social security; don't have the government involved in promoting particular industries; do strengthen property rights; don't be corrupt. Minimizing government meant lowering taxes—but keeping budgets in balance.

In practice, the Washington Consensus put little emphasis on equity. Some of its advocates believed in trickle-down economics, that somehow all would benefit—though there was little evidence to support such a conclusion. Others believed that equity was the province of politics, not economics: economists should focus on efficiency, and the Washington Consensus policies, they believed, would deliver on that.

The alternative view, which I hold, sees government having a more active role, in both promoting development and protecting the poor.[2] Economic theory and historical experience provide guidance on what government needs to do. While markets are at the center of any successful economy, government has to create a climate that allows business to thrive and create jobs. It has to construct physical and institutional infrastructure—laws ensuring, for instance, a sound banking system and securities markets in which investors can have confidence that they are not being cheated. Poorly developed markets are marked by monopolies and oligopolies; high prices in a vital area like telecommunications hinder development, so governments must have strong competition policies. There are many other areas in which markets, by themselves, do not work well. There will be too much of some things, like pollution and environmental degradation, and too little of others, like research. What separates developed from less developed countries is not just a gap in resources but a gap in knowledge, which is why investments in education and technology—largely from government—are so important.

In practice, the advocates of this alternative view also put more emphasis on employment, social justice, and nonmaterialistic values such as the preservation of the environment than do those who advocate a minimalist role for government. Unemployment, for instance, is seen not just as a waste of resources; it also undermines the individual's sense of self-worth, and it has a host of undesirable social consequences— including violence. Proponents of this view often argue for political reforms as well, to give citizens more voice in decision making; they point out that conditionality and economic institutions like independent central banks that are not politically accountable undermine democracy. By contrast, advocates of the Washington Consensus express a lack of confidence in democratic processes, arguing, for instance, that the independence of central banks is essential for ensuring good monetary policy.

How is it, one might ask, that economists—all trained with years of schooling, culminating in advanced degrees—cannot agree on what will lead to development? What is the prime minister of a country to do, as he is visited by an adviser from the IMF and told to follow the IMF prescriptions, and then visited by an academic adviser who recommends the contrary? Both begin with an appeal to economic theory, to the universal laws of economics, the laws of supply and demand. But economic theory is not monolithic. The Washington Consensus prescription is based on a theory of the market economy that assumes perfect information, perfect competition, and perfect risk markets—an idealization of reality which is of little relevance to developing countries in particular. The results of any theory depend on its assumptions—

and if the assumptions depart too far from reality, policies based on that model are likely to go far awry.

Advances in economic theory in the 1970s and 1980s illuminated the limits of markets; they showed that unfettered markets do not lead to economic efficiency whenever information is imperfect or markets are missing (for instance, good insurance markets to cover the key risks confronting individuals). And information is always imperfect and markets are always incomplete.[3] Nor do markets, by themselves, necessarily lead to economic efficiency when the task of a country is to absorb new technology, to close the "knowledge gap": a central feature of development. Today, most academic economists agree that markets, by themselves, do not lead to efficiency; the question is whether government can improve matters.

While it is difficult for economists to perform experiments to test their theories, as a chemist or a physicist might, the world provides a vast array of natural experiments as dozens of countries try different strategies. Unfortunately, because each country differs in its history and circumstances and in the myriad of details in the policies—and details do matter—it is often difficult to get a clear interpretation. What is clear, however, is that there have been marked differences in performance, that the most successful countries have been those in Asia, and that in most of the Asian countries, government played a very active role. As we look more carefully at the effects of particular policies, these conclusions are reinforced: there is a remarkable congruence between what economic theory says government should do and what the East Asian governments actually did. By the same token, the economic theories based on imperfect information and incomplete risk markets that predicted that the free flow of short-term capital—a key feature of market fundamentalist policies—would produce not growth but instability have also been borne out.

Twenty-five years ago, it was understandable that there could be a debate about market fundamentalism and the Washington Consensus policies. They had not really been tried. (Of course, the theoretical objections and historical experiences provided a strong word of caution.) Today, as we see the successes and failures, it is hard to understand the continuation of that debate—apart from the role of ideology and the interests that are served by the Washington Consensus policies. (Even when the economy does not grow, there are some that may do well from those policies.)

The task of less developed countries today is in some ways easier than that which faced Europe and the United States as they industrialized in the nineteenth century: they simply have to catch up, rather than forge into unknown territory. Nevertheless, the task has proven insurmountable almost everywhere outside of Asia—the most successful example of economic development the world has ever seen. Their success has been so strong—and they have been successful for so long—that it is easy to take it for granted. But Asia's growth would have surprised many experts of the 1950s and 1960s, such as the Nobel Prize–winning economist Gunnar Myrdal, who assessed Asia's prospects as truly bleak.[4] Conventional wisdom then was that countries such as Korea should stick to what they were best at: growing rice. The East Asian miracle shows that rapid development—and growth with equity, in which the poor and the rich both benefit—is possible, even though no particular preconditions were in place. Failures elsewhere show that development is not inevitable.

The differences in performance across regions are startling. While East Asia averaged 5.9 percent growth over the past thirty years (6.5 percent during the past fifteen years), Latin America and Africa have been in a race for the lowest overall growth rate, with sub-Saharan Africa's per capita income actually dropping an average 0.2 percent each year over the past thirty years.[5] But both have been outdistanced by Russia. Russia has seen its income decline since the beginning of its transition from communism to a market economy by a total of 15 percent; per capita income actually dropped by 40 percent in the first decade, but the Russian economy has finally begun growing again in the last five years.

East Asia

Globalization—in the form of export-led growth—helped pull the East Asian countries out of poverty. Globalization made this possible, providing access to international markets as well as access to technology that enabled vast increases in productivity. But these countries managed globalization: it was their ability to take advantage of globalization, without being taken advantage of by globalization, that accounts for much of their success.

These countries simultaneously achieved growth and stability: some had not a single year of negative growth over a span of almost a quarter century, others had one bad year; in this respect, their performance was better than that of any of the advanced industrial countries. Even during the downturn of 1997–98, China and Vietnam continued to grow. China followed standard expansionary macro-policies (not the policies recommended by the IMF elsewhere in East Asia) and saw its growth slow to a respectable 7 percent before resuming the higher levels of 8 percent and 9 percent. (Some think these numbers underestimate true growth.) If the provinces of China were treated as separate countries—and with populations sometimes in excess of 50 million, they are far larger than most countries around the world—then most of the fastest-growing countries in the world would be in China.[6]

Importantly, these governments made sure that the benefits of growth did not go just to a few, but were widely shared.[7] They focused not only on price stability but on real stability, ensuring that new jobs were created in pace with new entrants to the labor force. Poverty fell dramatically—in Indonesia, for example, the poverty rate (at the $1-a-day standard) fell from 28 percent to 8 percent between 1987 and 2002[8]—while health and life expectancy improved and literacy became close to universal. In 1960, Malaysia's per capita income was $784 (in 2000 U.S. dollars), slightly lower than that of Haiti at the time. Today, it is over $4,000. The average level of education in South Korea in 1960 was less than four years; today, South Korea leads in high-tech industries such as chip production, and its income has increased sixteenfold in the past forty years.[9] China began its journey later, but its achievements have in some ways been even more remarkable. Incomes have increased more than eightfold since 1978; poverty at the $1-a-day standard has fallen by three-quarters.[10]

But while these "market" economy countries were deeply engaged in globalization, their own markets were far from unfettered. Globalization was measured and paced, and government intervened carefully, but pervasively, in the economy. Of course, they did all the usual things that are expected of government. They expanded primary education and higher education simultaneously, recognizing that success required both universal literacy and a cadre of highly skilled individuals capable of absorbing advanced technology. They invested heavily in infrastructure such as ports, roads, and bridges, all of which made it easier to transport goods and so drove down the cost of doing business and of shipping goods out of the country.

They also went beyond the usual list of what governments typically do. Governments in East Asia played a large role in planning and in advancing technology, choosing which sectors their countries would develop rather than leaving it up to only the market to decide. From the 1960s onward, these countries made great efforts to develop local industries. Investment in the high-tech sector helped Taiwan, Korea, and Malaysia become major producers of electronics, computers, and computer chips. In addition, they became among the most efficient producers in the world of traditional products like steel and plastics.

The intent of government was not to "outsmart" the market—picking winners better than the market would do. But they realized that there were often enormous spillovers: technological advances in one area could help stimulate growth in another. They realized that markets often failed to coordinate new activities well: firms using plastics would not develop without a local supplier of plastics, but it was an enormous risk for a firm to produce plastics without an assured demand for its output. They realized too that banks often were less interested in lending to new industries than in providing finance for speculative real estate or (as is so often the case in developing countries) just lending to the government.

Economists had long talked about the importance of saving and investment for growth, but before East Asia took on the task, policymakers simply left it to the market. Economists might have bemoaned the low level of savings, but they thought there was little that government could do. The East Asian governments showed that this was not true. The money to make their investments came from their own people, as the governments encouraged saving; and so these countries did not have to depend on volatile capital flows from abroad. Nearly all the countries of the region saved 25 percent or more of GDP; today, China has a national savings rate in excess of 40 percent of GDP, in contrast to 14 percent in the United States. In Singapore, 42 percent of wage income was compulsorily placed in a "provident fund." In other countries, such as Japan, government-created savings institutions, which reached deep into rural areas, provided a safe and convenient way for people to save.

All these countries believed in the importance of markets, but they realized that markets had to be created and governed, and that sometimes private firms might not do what needs to be done. If private banks are not setting up branches in rural areas to garner savings, government must step in. If private banks are not providing long-term credit, government must step in. If private firms are not providing the basic inputs for production—like steel and plastic—government should step in if it can do so efficiently. Korea and Taiwan showed that it could; the Korean government proceeded cautiously, but, after determining that it could invest profitably, went ahead and created, in 1968, one of the most efficient steel companies in the world. Earlier, in 1954, Taiwan's government helped establish the enormously successful Formosa Plastics Corporation.

While most of the region liberalized—opening up markets and scaling down government regulations—it did so slowly, at a pace consistent with the economies' capacity to cope. While Asian governments focused on export-led growth, especially in the early days of their development, they limited imports that would undercut local manufacturing and agriculture.

Some countries, such as China, Malaysia, and Singapore, invited in foreign investment; others, namely South Korea and Japan, felt more comfortable without it and grew just as well. Even those that invited in foreigners made sure that the guest firms transferred technology and trained local workers, so that they were contributing to the nation's development effort. Malaysia did not just turn over its oil to foreign oil companies, but had them help it develop its resources, learning all the while; today its government-owned oil company, Petronas, is providing training for other developing countries. By managing its own oil company, it was able to ensure that more of the value of the resource stayed in Malaysia, rather than being sent abroad as profits.

The debate about capital market liberalization was more tendentious. Even as they have opened up their markets for long-term investment, the two Asian giants—India and China—have restricted short-term capital flows. They recognized that you cannot build factories and create jobs with money that can move in and out overnight. They had seen the record of instability that had accompanied these flows, risk that came without evident reward.

With their high savings rate, the countries of East Asia were hardly in need of additional capital. But during the 1980s, many of these countries—perhaps succumbing to pressure from the IMF and the U.S. Treasury—opened up their markets to the free flow of capital. For a while capital flowed in, but then sentiment changed and it fled out. The result was a crisis that spread across the region and beyond. In 1997 speculators attacked the Thai baht, causing the currency to go into freefall beginning in early July. Foreign banks called in their loans to Korea. Indonesia faced problems from both the banks and the speculators. Central banks around the region spent billions of dollars trying to prop up their currencies. When they ran out of funds they turned to the IMF, but it provided money only with a long list of conditions, including government spending cuts, tax increases, and higher interest rates. As central banks raised interest rates, local companies found they were unable to meet their interest obligations. There were massive bankruptcies, and the currency crisis turned into a banking crisis.

It was a terrible time: there were riots and social unrest in Indonesia, unemployed businessmen wandering the parks in Seoul because they were too ashamed to tell their wives that they had no office to go to anymore, people selling their clothes and housewares on the streets of Bangkok. Many people went back to the countryside to live with their families because they could not find work in the capital. Koreans lined up to turn in their gold jewelry so the government could melt it down and use it to repay off part of the national debt.

The IMF policies failed to stabilize the currencies; they only succeeded in making the economic downturns far worse than they otherwise would have been—just as standard economic theory had predicted. Critics of the IMF argue that the policies were not really designed to protect the countries from a recession, but to protect the lenders; their intent was to quickly rebuild reserves so that international creditors could be repaid. The countries did, in fact, quickly restore their reserves, and even managed to repay the IMF the money owed within a few years.

Much of Asia has recovered now, but the crisis was damaging and unnecessary. East Asia had learned that while globalization, well managed, had brought them enormous prosperity, globalization—when it meant opening themselves up to destabilizing speculative flows—had also brought economic devastation. As officials there reflect on the lessons of that brutal experience, they have come to reject even more firmly the Washington Consensus market fundamentalism which opened their countries to the ravages of the speculators. And they have put more emphasis on equity and on policies to help the poor. Growth has recovered, but these students of the "class of '97" have not forgotten the lessons.

Latin America

East Asia demonstrated the success of a course markedly different from the Washington Consensus, with a role for government far larger than the minimalist role allowed by market fundamentalism. Meanwhile, Latin America embraced the Washington Consensus policies more wholeheartedly than any other region (indeed, the term was first coined with reference to policies advocated for that region). Together, the failures of Latin America and the successes of East Asia provide the strongest case against the Washington Consensus.

In earlier decades, Latin America had had notable success with strong government interventionist policies that were neither as refined as those employed in East Asia nor as subtle, being focused more on restriction of imports than on expansion of exports. High tariffs were placed on certain imports, to encourage the development of local industries—a strategy often referred to as import substitution. While its success did not match that of East Asia, Latin America's per capita income still grew at an average of more than 2.8 percent annually from

1950 to 1980 (at 2.2 percent for 1930 to 1980).[11] Brazil, whose government intervened most aggressively in the economy, grew at an average of 5.7 percent for the half century that began in 1930.

In 1980, fighting its own problem of inflation, the United States initiated interest rate increases that climbed to over 20 percent. These rates spilled over to loans to Latin America, triggering the Latin American debt crisis of the early 1980s, when Mexico, Argentina, Brazil, Costa Rica, and a host of other countries defaulted on their debt. As a result of the debt crisis, the region suffered three years of decline and ten years of stagnation, a performance so poor that it came to be called the lost decade.

It was during this period that Latin American economic policies changed dramatically, with most countries adopting Washington Consensus policies. As high inflation broke out in many of the countries, the Washington Consensus's focus on fighting inflation made sense. Their governments had not been working well for them, and the appeal of the Washington Consensus—minimizing the role of government—was understandable. As countries like Argentina adopted the Washington Consensus policies, praise was heaped upon them. When price stability was restored and growth resumed, the World Bank and the IMF claimed credit for the success; the case for the Washington Consensus had been made. But, as it turned out, the growth was not sustainable. It was based on heavy borrowing from abroad and on privatizations which sold off national assets to foreigners—the proceeds from which were not invested. There was a consumption boom. GDP was increasing, but national wealth was diminishing. Growth was to last a short seven years, and was to be followed by recession and stagnation. Growth for the decade of the 1990s was only half what it had been in the decades prior to 1980, and what growth there was went disproportionately to the rich.

While East Asia saw enormous reductions in poverty, progress in Latin America was minimal. At this writing, it is fair to say there is widespread disillusionment in Latin America with the Washington Consensus: a growing consensus against the Washington Consensus reflected in the election of leftist governments in Brazil, Venezuela, and Bolivia. These governments have often been castigated for being populist, because they promise to bring education and health benefits to the poor, and to strive for economic policies that not only bring higher growth but also ensure that the fruits of that growth are more widely shared. In a democracy, it seems natural—not wrong—for politicians to strive to enhance the well-being of the average citizen; and it is clear that earlier policies failed to meet the legitimate needs of the average citizen, even as those at the top of the income distribution were doing very well. It is too soon to tell whether they will succeed in those promises. Venezuelan president Hugo Chavez seems to have succeeded in bringing education and health services to the barrios of Caracas, which previously had seen little of the benefits of that country's rich endowment of oil. If these leaders fail to deliver on their promises, it is hard to predict how the currents of unrest will play out.

Countries in transition from communism
Just as the successes of East Asia are far greater than even the impressive GDP statistics suggest, the failures of Russia and most of the other countries making the transition from communism to capitalism were far deeper than GDP statistics alone show. Decreases in life expectancy—in Russia it fell by a stunning four years between 1990 and 2000—confirmed the impression of increasing destitution.[12] (Elsewhere in the world, life expectancy was rising.) Crime and lawlessness were rampant.

After the Berlin Wall fell, there was hope of democracy and economic prosperity throughout the former Soviet Union and its satellite states. Advisers from the West rushed to Eastern Europe to guide those countries through their transitions. Many believed, mistakenly, that "shock therapy" was needed—that the transition to Western-style capitalism should take place overnight through rapid privatization and liberalization. Instantaneous price liberalization brought with it— predictably—hyperinflation. Prices in Ukraine at one point increased at the rate of 3,300 percent a year. Tight monetary policy (high interest rates with little credit available) and fiscal austerity (tight budgets) were used to bring down the hyperinflation; they also brought down the economies, which slid into deep recessions and depressions. Meanwhile, rapid privatizations were giving away hundreds of billions of dollars of the countries' most valuable assets, creating a new class of oligarchs who took money out of the country far faster than the inflow of billions that the IMF was pouring in as assistance. Capital markets were liberalized in the mistaken belief that money would be induced to come in. Instead, there was massive capital flight, including the famous purchase of the Chelsea football club and numerous country estates in the U.K. by one of the oligarchs, Roman Abramovich. Ordinary Russians, naturally, found it hard to see how this helped Russia's growth. It was as if the advisers believed that opening a birdcage would encourage birds to fly into the cage, rather than encouraging the birds in the cage to fly out.

When I was chief economist of the World Bank, we had an intense debate about those privatizations. I was among those who worried that rapid privatizations not only generated lower revenue for governments desperately in need of money but undermined confidence in the market economy. Without appropriate laws concerning corporate governance, there might be massive theft of corporate assets by managers; there would be incentives to strip assets rather than to build wealth. I worried too about the huge inequality to which these privatization could give rise. The other side said: Don't worry, just privatize as rapidly as possible; the new owners will make sure that resources are well used and the economy will grow. Unfortunately, what happened in Russia and elsewhere was even worse than I had feared. Though the Russian government had been told repeatedly by its advisers from the IMF, the U.S. Treasury, and elsewhere that privatization would lead to growth and investment, the outcome was disappointing: output fell by one-third.

The rapid and corrupt privatizations in Russia set in motion a vicious circle. The meager amounts received by the government led to questioning the legitimacy of the transfer of public resources to the private sector. Investors—those who had acquired the assets—then felt, quite rightly, that their property rights were not secure, that a new government might, under popular pressure, reverse the privatization. As a result, they limited their investment and took as much of their profits out of the country as they could—leading to further disillusionment with the privatization process, making property rights still less secure.

The capital market liberalization pushed by the IMF made matters worse because it made it easier for the oligarchs who had stripped assets from the corporations they controlled to take their money offshore, to places where secure property rights were already well established. They enjoyed benefits of weak legal frameworks at home and strong property protections abroad.

To some who visited Moscow in those early days of the transition, it seemed a success. The stores were filled with goods, the roads with cars. But the goods were imported luxury goods for the newly established wealthy who had managed to get the vast assets of the state into their private hands; while a few were driving Mercedes and enjoying the New Russia, millions more were seeing their meager pensions being eroded below even the level of subsistence.

It is now widely agreed that the speed of the reforms in the former Soviet bloc countries was a mistake. The privatizations were done before sound regulations and strong tax laws were put into place. As government revenue dropped, spending on health and infrastructure collapsed. One of the legacies of Russia's past was a high-quality education system, but this quickly deteriorated as budgets were slashed. At the same time, the old social safety nets were being cast aside. The results were grim: poverty in the former Soviet bloc countries increased from 1987 (shortly before the beginning of transition) to 2001 by a factor of ten. The contrast between the claims of free market advocates, who predicted an unleashing of forces that would bring record prosperity, and the unprecedented increases in poverty that actually occurred could not have been greater.

Some countries, like Poland and Slovenia, managed the transition better, partly because they did not embrace shock therapy as strongly.[13] The countries of Eastern Europe as a whole did well largely, I think, because of the prospect of joining the EU. It forced them to adopt quickly a sound legal framework, and that reassured investors. As they joined the EU, they obtained access to a huge market—and their low wages combined with their highly educated labor forces gave them a distinct advantage.

The Soviet bloc countries were not the only ones transitioning from communism. China and Vietnam, while retaining a Communist political regime, also began to move to a market economy, and the contrast was striking. As incomes in Russia plummeted—falling by a third from 1990 to 2000—incomes in those countries soared, increasing by 135 percent in China and 75 percent in Vietnam. They rejected shock therapy in favor of a slower and more gentle transition to a market economy. Today the vibrancy of their economies suggests that the tortoise has outpaced the hare.

The performance gap between China and Russia has put the advocates of shock therapy—rapid change, with little sensitivity to social costs and little concern for the prerequisites which make a market economy work—on the defensive.[14] They say that China's task was easier because it was a less developed, mainly agrarian country. But development is itself difficult—success stories outside of East Asia are rare—and the defenders of shock therapy have never adequately explained why compounding two difficult problems, development and transition, should have made the task easier. Many of the less developed countries of the former Soviet bloc that followed their shock therapy advice fared as badly as Russia itself; the mainly agrarian economies of Mongolia and Moldova showed even greater decline. Those that fared better, like Kazakhstan, did so because of oil.

Africa

I was in East Africa during the early days of independence, in the late 1960s. There was a sense of euphoria, although the countries knew that colonialism had left them ill-prepared for development and democracy. They didn't have even a modicum of experience in self-government—there were few trained individuals, and the countries lacked the institutional infrastructure necessary for democracy and the physical infrastructure necessary for growth. In Uganda, the British had promoted Idi Amin within the military and so groomed him to be one of the leaders of the future. But Britain's legacy stood bright and shining when compared to the bloody history of Belgium's activities in the Congo.

It was hardly surprising that by the 1980s many African countries had fallen on hard times. Each country has its own story: corrupt and often ruthless dictators in Uganda, Congo, Kenya, and Nigeria; well-intentioned and mostly honest, but highly flawed, policies of "African socialism" in Tanzania; misguided macro-economic policies in Ivory Coast. In the 1980s, many turned to the World Bank and the IMF for help. They were provided with assistance—typically loans rather than grants—accompanied by conditions designed to assist their "structural adjustment." Too often, though, the conditions were misguided, the projects for which the money was lent misconceived. The borrowing countries were required to adapt the structure of their economy to the IMF's market fundamentalism, to Washington Consensus policies. Liberalization opened up African markets to goods from foreign countries, but the African countries had little to sell abroad. Opening up capital markets did not bring an inrush of capital; investors were more interested in taking out Africa's bountiful natural resources. Often, the IMF requirements brought fiscal austerity; while all countries have to learn to live within their means, the IMF went much further than necessary. It imposed constraints that prevented the borrowing country even from making good use of the limited amount of foreign assistance it received. In Ethiopia, for instance, the Fund went so far as to demand that the country ignore foreign assistance in assessing whether its budget was balanced; in effect, foreign assistance went to increase reserves, not to build hospitals or schools or roads. Not surprisingly, the policies failed to bring growth. But the burden of debt remained.

In the 1990s, many of the African countries, including Nigeria, Kenya, Tanzania, Uganda, Ethiopia, and Ghana, found themselves with new leadership, and the new leaders seemed more committed than the old to pursuing good economic policies. Deficits and inflation were brought under control. Some, such as Olusejun Obasanjo in Nigeria, Yoweri Museveni in Uganda, Benjamin Mkapa in Tanzania, and Meles Zenawi in Ethiopia, took strong stands against corruption; even if it was not eliminated entirely, remarkable progress was made. Uganda and Ethiopia had periods of growth: Ethiopia grew at more than 6 percent annually between 1993 and 1997, when war broke out with neighboring Eritrea; Uganda grew, on average, more than 4 percent annually from 1993 to 2000. Several countries made major strides in improving literacy, and were it not for the AIDS epidemic there would have been great advances in health and life expectancy. But even these successful countries failed to attract much foreign investment. The vast markets of Asia, with their more highly educated labor force, their better infrastructure, their fast-growing economies, were simply more attractive for most of the multinationals.

While Africa's economies did not grow, its population did. Africa had been a continent with an abundance of land; land maintained its productivity by being left fallow for long periods of time. But with the new population pressures, this was no longer possible. Agricultural productivity declined, and poverty grew. Again, globalization bypassed Africa. Just as even the countries with good macro-economic policies failed to attract investment, the Green Revolution, which increased agricultural productivity enormously in Asia, bypassed Africa. Today, agricultural productivity is a third of that of Asia. And as if that were not enough, the AIDS epidemic hit it with devastating force. Even countries like Botswana that managed their economy well and husbanded their resources—growing at 9 percent annually for the almost four decades since independence from Britain—have seen reversals in life expectancy. As a result of these forces, by the early years of the twenty-first century, as we have seen, the number of people in poverty in Africa had doubled from the levels two decades earlier.

South Asia

For the past two decades, with the exception of an economic crisis in the early 1990s, India—a country of some 1.1 billion people—has been growing at 5 percent a year or more. In 2006, it is expected to grow at 8 percent.

For decades after independence, socialist doctrines prevailed and the economy stagnated. But even in this era, the government was sowing the seeds of future success. It created a number of institutes of technology and science, investments in education and research that were eventually to pay off in the new millennium. The emergence of Bangalore as the capital of India's information technology sector can be traced back to the founding of the Indian Institute of Science there in 1909, on land donated by the Maharajah of Mysore and endowed by the industrial baron J. N. Tata.

The Green Revolution of the 1970s, which promoted the use of better farming techniques and new seeds, increased yields enormously. Growth did not really take off, however, until the early 1980s, when the government ended its open hostility to business and removed many of the restrictions that had stymied the private sector.[15] The liberalizations of the early 1990s were critical in continuing the momentum of the earlier reforms, but even as the government opened up the country to foreign direct investment, it continued to restrict short-term capital flows. Only in 2006, fifteen years after the liberalization reforms began, have discussions begun on adjusting—not eliminating—those restrictions.

The advent of the Internet proved to be the most important turning point. New technology meant that at last India could reap the benefits of its long-term investments in education, and inadequacies in infrastructure were less of a hindrance. Opportunities created by America's bubble economy of the 1990s helped too, in an indirect way.[16] While technology brought down the costs of communication, massive overinvestment in telecommunications brought it down even further, as excess capacity in the cables that line the floor of the Pacific and satellites drove down the communication costs further. Typically, firms thinking about investing in a developing country have to weigh a long list of advantages and disadvantages: wages may be low compared to developed countries, but a lack of infrastructure frequently means higher transportation costs, as well as an unreliable and expensive supply of electricity and communications services. What was different in the case of India's new high-tech sector was that these infrastructure problems were either irrelevant (the costs of transportation simply didn't matter) or could be sidestepped. Companies built their own generators in order to bypass the erratic local electricity supply. Satellites, which could in a nanosecond link India's firms with those in Silicon Valley or elsewhere in Europe and the United States, meant that calls could be made around the world without depending on India's unreliable phone system.

India's success, in fact, has much in common with that of China. In both, there is emerging a middle class of several hundred million that is beginning to enjoy the bountiful life that those in the West have had for so long, and in both countries there are still huge gaps between rich and poor. India did far less well than China in reducing poverty—but it has done far better in preventing the rise of inequality, the disparities across regions and between the very top and the rest. Still, both China and India, even as they reach new peaks of success, have recognized that they cannot continue as they have done until now. Both governments have committed themselves to focusing on helping the lagging rural sector; both are worried about creating new jobs for the new entrants to the labor force (India has actually created a guaranteed employment scheme for rural areas). Both recognize the importance of technology and education in the competitive global marketplace, and know that this will require strengthening their already huge investments in education—Asia today graduates more than three times the number of engineers and scientists that the United States does. The challenge is to improve the quality as they increase the quantity.

Today, developing countries around the world are looking to Asia, to the examples of success, to see what they can learn. It is not surprising that global support for the Washington Consensus has waned. Its failures can be seen around the world, in Africa, Latin America, and the economies in transition. The clearest test was in the transition from communism to a market economy; those that followed the Washington Consensus failed, almost to a country. At best, they achieved meager growth; at worst, they are suffering growing inequality and instability. Even democracy looks less secure.

A VISION OF DEVELOPMENT

In the array of statistics and anecdotes describing developing countries—some totally depressing, some conveying enormous hope—it's important to remember the big picture: success means sustainable, equitable, and democratic development that focuses on increasing living standards, not just on measured GDP. Income is, of course, an important part of living standards, but so too is health (measured, for instance, by life expectancy and infant mortality) and education.[17] The king of Bhutan has spoken of GNH, gross national happiness, as he sought growth strategies that improved education, health, and the quality of life in rural areas as well as in the towns, all the while maintaining traditional values.

GDP is a handy measure of economic growth, but it is not the be-all and end-all of development. Growth must be sustainable. Everyone knows that by cramming for an exam you get your grade up, but what you learn is soon forgotten. You can get GDP up by despoiling the environment, by depleting scarce natural resources, by borrowing from abroad—but this kind of growth is not sustainable. Papua New Guinea is cutting down its tropical rainforests, home to an immense range of species; the sales improve its GDP today, but in twenty years there will be nothing more to cut.[18]

Still, because GDP is relatively easy to measure, it has become a fixation of economists. The trouble with this is that what we measure is what we strive for. Sometimes, increases in GDP are associated with poverty reduction, as was the case in East Asia. But that was not an accident: governments designed policies to make sure that the poor shared in the benefits. Elsewhere, growth has often been accompanied by increased poverty and sometimes even lower income for individuals in the middle. This is what has been happening in the United States: between 1999 and 2004, average disposable income went up by 11 percent in real terms, but median household income—the income of the family at the center, the true middle middle-class family—fell by some $1,500, adjusting for inflation, or around 3 percent. In Latin America, from 1981 to 1993, while GDP went up by 25 percent, the portion of the population living on under $2.15 a day increased from 26.9 percent to 29.5 percent. If economic growth is not shared throughout society, then development has failed.

The East Asian governments realized that success requires social and political stability, and that social and political stability in turn require both high levels of employment and limited inequality. Not only was conspicuous consumption discouraged, but so too were large wage disparities. In China, at least in the earlier stages of development, senior management typically received no more than three times the income of an ordinary worker; in Japan, ten times. (By contrast, in recent years the compensation of senior managers in the United States has been hundreds of times that of ordinary workers.)[19]

I believe that it is important for countries to focus on equity, on ensuring that the fruits of growth are widely shared. There is a compelling moral case for equity; but it is also necessary if there is to be sustained growth. A country's most important resource is its people, and if a large fraction of its people do not live up to their potential—as a result of lack of access to education or because they suffer the life-long effects of childhood malnutrition—the country will not be able to live up to its potential. Countries that don't invest widely in education find it hard to attract foreign investment in businesses that depend on a skilled labor force—and today, more and more businesses depend in part on skilled labor. At the other extreme, high levels of inequality, especially as a result of unemployment, can result in social unrest; crime is likely to increase, creating a climate that is unattractive to businesses.

It is not just income—even the income of the average individual—that matters but overall standards of living. There can be a discrepancy between the two. Development is typically accompanied by urbanization, and many cities in developing countries are squalid, marred by noise, congestion, poor sanitation, and dirty air. In March 1991, air pollution got so bad in Mexico City that President Carlos Salinas de Gortari ordered a major oil refinery to be shut down. In the nineteenth-century transformation that marked the Industrial Revolution in Europe and the United States, environmental problems were so serious that health deteriorated and life spans were shortened.[20] In Britain, the first country to enter the Industrial Revolution, average height—a measure of physical well-being—declined from the late eighteenth century to the middle of the nineteenth.[21] Fortunately, improvements in medicine and nutrition have managed partially to overcome environmental factors, so that in most developing countries, other than those devastated by AIDS, life spans are increasing.

Today, there is more concern in the development community about the importance of health and the environment. There is also more concern about economic security—reflecting the importance that ordinary workers place on this, as we saw in chapter 1.[22]

The Role of Markets

Recent decades have seen marked changes in thinking, not only about what successful development means but also how to go about it.[23] During the 1960s and 1970s it was thought that what separated less developed from more developed countries was the former's lack of capital. Emphasis was placed on savings and investment. That is one of the reasons the World Bank was created in 1944, to help provide more capital to developing countries. When it turned out that foreign aid and easier access to capital did not lead to the hoped-for results, many in the development community pushed the idea that markets were the solution—although they had failed to produce development in the years before the end of colonialism.[24] When the question "Why hadn't markets already delivered?" was posed, there was an easy answer: governments were in the way. All that was required for development, then, was to get government out of the way, privatizing and liberalizing—stripping away regulation, cutting government expenditure, and tightening restraints on borrowing.

The emphasis on the importance of markets, which had begun in the 1980s when Thatcher and Reagan were in office, was strengthened after the fall of communism—a natural reaction to the failure of the planned economy in the former Communist states. By the last decade of the twentieth century, the examples of Russia and Latin America had shown that the strategy of just getting government out of the way also had failed. At that point, the search for alternatives began in earnest. Some economists turned to small variations on the theme, various forms of "market plus" (or Washington Consensus plus)—adding in, for instance, the importance of human capital and especially female education. When these policies too failed, it became clear that what was needed was a deeper change in strategy, a more comprehensive approach to development—with emphases differing from country to country and from time to time. These strategies, however, were not really new: they were variants of the strategies that had worked so well and for so long in East Asia and elsewhere, but which had long been ignored by the believers in the Washington Consensus and market fundamentalism.

A Comprehensive Approach to Development

The World Bank endorsed this "comprehensive" approach to development while I was its chief economist and Jim Wolfensohn was its president.[25] It was criticized for lacking focus, but that claim is simply wrong. At any moment, there may be several areas on which attention is focused—bottlenecks, for instance, in the economy. The comprehensive approach recognized, though, the dangers of the kind of single-minded focus that had characterized development policies of the past: schools without jobs would not lead to development, nor would trade liberalization without roads and ports lead to more trade. China has been adept at changing the focus of its attention as it has moved along in its three-decades-long development. Its eleventh five-year plan, adopted in March 2006, shifted from exports to increasing domestic demand, in recognition of growing protectionist pressures around the world. For China, with a savings rate of over 40 percent of GDP, capital with which to invest was no longer the problem; the current necessity is to stimulate consumption. At one stage, the focus was on attracting foreign investors; when that proved enormously successful, focus shifted to developing domestic entrepreneurs.

Providing more resources and strengthening markets—the key elements of the development strategy of the World Bank in early decades—are still important elements in successful development. Countries cannot grow without capital. Markets are essential; markets help allocate resources, ensuring that they are well deployed, which is especially important where resources are scarce. The comprehensive approach has involved strengthening markets, but equally important has been strengthening government and figuring out, for each country as it reaches each stage of development, what the right mix of government and market might be.

The successes in Asia echoed those of the United States and other countries in the industrialized world: government has a large role to play. The right mix of government and markets will differ between countries and over time. In China, for instance, where there was already plenty of government, the challenge was to develop the market. This is what took place in the period after the Cultural Revolution, the 1980s, when China's economy began the astounding takeoff that continues today.[26] What matters, of course, is not just the size of government but what government does. A central component of China's rapid growth was township and village enterprises established by the local communes. The government got out of agriculture and gave families the right to control the land, and agricultural productivity soared. At the same time, the central government moved away from micromanaging every detail of the economy to managing the overall economic framework, including ensuring a supply of finance for the development of infrastructure. As China's transition evolved, the government realized that continued success would require stronger laws concerning corporate governance. It realized too that, in the zeal to strengthen the market, areas such as rural education and health had been left behind. The 2006 five-year plan sets out to redress these imbalances.

The list of potential arenas for government action is large. Today, nearly everyone agrees that government needs to be involved in providing basic education, legal frameworks, infrastructure, and some elements of a social safety net, and in regulating competition, banks, and environmental impacts. The East Asian countries believed, as we have seen, that government should do more. East Asian nations feel that it is their responsibility to maintain full employment and actively promote growth, and their governments remain concerned about inequality and social stability. In Malaysia, the role of government has extended in yet another direction. For decades, the Malaysian government has carried out an aggressive affirmative action program to help the ethnic Malays. This was an important part of nation building; the view that all groups would benefit from a more stable and equitable society was widely accepted, even though some members of Malaysia's ethnic Chinese community may have lost opportunities as a result. However, because the government made sure that all shared in the fruits of development, ethnic conflict has largely been avoided.

People are at the core of development

Development is about transforming the lives of people, not just transforming economies. Policies for education or employment need to be looked at through this double lens: how they promote growth and how they affect individuals directly. Economists talk about education as human capital: investment in people yields a return, just as investment in machinery does. But education does more. It opens up minds to the notion that change is possible, that there are other ways of organizing production, as it teaches the basic principles of modern science and the elements of analytic reasoning and enhances the capability to learn. The Nobel laureate Amartya Sen has emphasized the enhanced capabilities that education brings, and the resulting freedom that development brings to individuals.[27]

Just as the focus on GDP results in too narrow a focus for development strategies, so too a focus on the number of years of schooling may lead to too narrow a focus for education policies. The number of years of schooling is an important indicator of how well a country is doing in advancing education, but just as important is what schools teach. Education needs to be compatible with the work that people will do once they leave school. In Ethiopia, the government of Meles Zenawi realized that even if its most ambitious development programs succeed, most of the people going to rural schools today will still be farmers when they grow up, so it has been working to redirect curriculum in order to make them better farmers. Education had been viewed as a way out, an opportunity to get a better job in the cities. Now it is also being viewed as a way up, enhancing income even for those who remain in the rural sector. Education can be used to promote health and the environment as well as to impart technical skills. Students can learn in school the dangers of locating latrines uphill from their source of drinking water, or the dangers of indoor air pollution—the choking smoke in huts without ventilation—and what can be done about it.

With education, a broad approach is important. Too often, international development institutions such as the World Bank focused narrowly on primary education. This was understandable: the returns are high, and many countries were spending a disproportionate part of their education budgets on university education for children of the elite. Moreover, having a strong base of primary education is essential if one wants to identify the most able for advanced training. Still, if the knowledge gap between the developed and less developed countries is to be narrowed, there also has to be a strong secondary school and university system.[28]

Of course, it does little good to have highly educated individuals without jobs for them. Without appropriate jobs, developing countries will lose this much-needed intellectual capital, their brightest children, in whom they have invested enormously through elementary and secondary education and sometimes even through college, to developed countries. This is often referred to as the "brain drain," another way in which developing countries wind up subsidizing the developed.[29] Former Malaysian prime minister Mahathir bin Mohamad referred to this loss, in his usual colorful language, as stealing the developing countries' intellectual property. In defense of intellectual property protections, as we will see in chapter 4, developed countries point out that drug prices are high in order to pay for the failures, the research that does not lead to the blockbuster drug. Mahathir points out that the same logic applies to education: the country provides education to all its youth, only to find that sometimes the best move to the West—and the developing countries receive nothing in compensation.

The importance of community

Markets, government, and individuals are three of the pillars of successful development strategy. A fourth pillar is communities, people working together, often with help from government and nongovernmental organizations. In many developing countries, much important collective action is at the local level. In Bali, as in much of Asia, irrigation for agriculture is provided by a network of canals. These are maintained by the community which ensures that the water is shared fairly among the villages and villagers.

The story of the Grameen micro-credit bank in rural Bangladesh, which gives small loans to poor rural women—who have a far better repayment rate than the rich urban borrowers—is well known. These schemes have been so successful because they entail groups of women who take responsibility for one another, helping one another out and ensuring that each pays what is due.[30] Its sister organization BRAC (originally the Bangladesh Rural Advancement Committee), also a nongovernmental organization, is even larger than Grameen, and both have branched out into a wide variety of activities. Today they build schools and even run a university, and provide cell phones, mortgage financing, and health and legal services. Seeing their work is an extraordinary experience: groups of women sitting in rows on the ground proudly discussing what they have done with the small loans they were given, children in the most basic rural schools chanting the day's lessons, and signs throughout Bangladesh advertising the cell phone programs that have connected thousands of poor people and helped them join the world after centuries of isolation.

In August 2003, I visited a chicken-feed factory run by BRAC. One of the first things women had done with the loans they got from BRAC was to buy newborn chicks, so they could raise chickens for meat and eggs. It soon turned out that many of the baby chicks died, because raising chicks in the first few days of life required skills and attention that the women could not provide. Instead of shutting the project down, the BRAC workers set up a program to take care of the baby chicks and pass them on to the women when the chicks were old enough to survive. They found that higher-quality chicken feed was necessary, so they opened an animal feed company and sold the feed to the women raising the chicks. Thus BRAC created wealth and jobs throughout the supply chain: from eggs to baby chicks, to processing nutritious feed for those chicks.

Were it not for BRAC and Grameen, the Bangladeshi farmers would be even poorer than they are now. Health is better and birth rates are lower as a result of their efforts and those of similar organizations. Life expectancy is up 12 percent in twelve years, to sixty-two years in 2002, and population growth rates are down to 1.7 percent from 2.4 percent in 1990. The micro-finance model used by BRAC and Grameen has been copied all over the world. What makes their programs so successful is that they come out of the communities that they service and address the needs of the people in those communities.

Grameen Bank and BRAC knew, for instance, that success wasn't just a matter of raising chicks. It was about changing the power structure within the community by giving more economic resources to the poorest of the poor, especially to the women, who had for so long been treated as second-class citizens. The community was strengthened by the health, legal aid, and education programs they established. I was taken to an elementary class in family law set up by BRAC, which taught women their basic legal rights, including the rudiments of divorce law, so that they knew what protection they had from physical abuse and abandonment by their husbands. Many had not known that Bangladesh law does not allow quick Islamic divorce. BRAC's classes empowered them, not only by teaching them about their rights but by helping them realize those rights. Grameen's lending programs reinforced this: by only providing mortgages on houses that were put in a woman's name, they provided an economic incentive for men to stay with their wives.

World Bank studies have highlighted the importance of community involvement, finding that local participation in the choice and design of projects leads to a higher likelihood of success.[31] The World Bank now has a program that allocates $25,000 grants to communities to spend as they please. Thailand is one of several countries imitating the program and putting decision making into the hands of local communities. There is a compelling argument for these programs: the people in the village know better than anyone else what will make a difference to their lives; they know how the money is spent, and any corruption hurts them directly. Having invested in the planning and execution of a project, they are more likely to feel ownership, a commitment to see it through to success, and therefore more likely to see it receive the funds required to maintain it. For example, in India and in many other developing countries, women spent vast amounts of time trudging back and forth to the water supply, bringing water for cooking and washing. People in the community know best where a new well should be put, and that is why Indian water projects with local participation have done so much better than programs designed outside the community. Of course, there have been failures, such as in East Timor where some of the local grants were misspent, but overall it is clear that development will happen best with community commitment.

The Challenges of Implementation

Successful development requires not just a vision and a strategy; ideas have to be converted into projects and policies. When I was at the World Bank, it would often be said in the face of obvious failure that our strategy was correct, it just wasn't implemented well. The fault would be put on bureaucrats—especially in the developing countries, though sometimes at the World Bank or the IMF—for failing to pay attention to certain details. But policies have to be designed to be implemented by ordinary mortals, and if they seemingly cannot be, if time after time there are implementation problems, then something is fundamentally wrong.

Managing change is extraordinarily difficult. It is clear that rushing into major reforms does not work. Shock therapy failed in Russia. China's Great Leap Forward in the 1960s was a catastrophe. What matters, of course, is not just the pace of change but the sequencing of reforms. Privatization was done in Russia before adequate systems of collecting taxes and regulating newly privatized enterprises were put in place. Liberalizing the free flow of foreign exchange before the banking system was strengthened turned out to be a disaster in Indonesia and Thailand. Educating people but not having jobs for them is a recipe for disaffection and instability, not for growth. Balance is also important: allowing urban–rural income differences to grow is another prescription for trouble. Many of the development strategies that were not well implemented failed because they were based on a flawed vision of development. Successful countries have a broader vision of what development entails and a more comprehensive strategy for bringing it about. Sensitive to concerns such as those just described, they were better at implementing change.

Governance

Much of the debate about development centers on how the advanced industrial countries can best provide more resources—through aid, debt relief, and direct investment—and how they can best provide more opportunities, through reforming global trade arrangements. But even if globalization succeeds in increasing resources to developing countries and opening up new opportunities, development is not assured. Countries must be able to use the resources well, and take advantage of new opportunities. This is the responsibility of each country. A major factor determining how well a country will do is the "quality" of the public and private institutions, which in turn is related to how decisions get made and in whose interest, a subject broadly referred to as "governance."

Today, throughout the developing world, there is enormous focus on one vital aspect of governance: corruption. I believe it is having its effect. Of course, there will continue to be stories of corruption. No country is immune from corruption, and it takes different forms in different countries. The corruption of campaign contributions by major corporations in advanced industrial countries, which we will discuss in chapter 7, is greater in magnitude and, in some ways, more insidious to democratic processes than the petty but more pervasive corruption involving small bribes to government officials. When government officials are eking out a living on a minimal wage, it is understandable, though not forgivable, that they demand bribes before they will do the job they were hired for. At least these ill-gotten gains are used to pay for food or education for their children.

Singapore showed that with strong punishment and high government salaries, this kind of corruption could be quickly stamped out. More remarkable has been the progress made by countries that could not afford to do what Singapore did. In Ethiopia, the government is so adamant about fighting corruption that the business community complains about excessive zealousness. In Uganda, the government has been publicizing all checks sent to the local level, so that villagers know what they should be receiving—and can make sure that those between Kampala and the village do not take their cut. In Nigeria, the government has promised to publish what it receives from the oil companies, so that citizens can see that money is not being stolen. In Thailand, the new constitution includes the notion that citizens have a basic right to know what their government is doing—a version of the Freedom of Information Act. Similar bills are being enacted throughout the developing world. These successes are striking moves in the right direction—but too often they have made only a small dent in prevailing cultures of corruption.

There are two things that those in the West can do to help developing countries strengthen democratic governance. The first is simple: don't undermine democracy. (Though many of the more successful countries have political systems that are far from democratic, the continued success of the East Asian countries after democratization, and the success of India, suggest that economic success is fully consistent with democracy.) In country after country, people are told about the importance of democracy, but no sooner have they grasped the message than they are told that what they care most about—the overall performance of their economy, which determines the pace of job creation and inflation—is too important to be left to democratic political processes. IMF conditionality undermines democracy, as, arguably, do demands that monetary policy be taken out of the hands of democratic political processes and turned over to "experts." Many international trade agreements—especially bilateral trade agreements, which we will discuss in the next chapter—by circumscribing the legitimate activities of democratically elected governments, do that too.

The second is equally important and will be discussed at greater length in chapter 5: the developed countries should do more to reduce opportunities for corruption, by limiting bank secrecy, increasing transparency, and enforcing anti-bribery measures. Every bribe requires both a briber and a bribee—and too often the briber comes from a developed country. Corruption would occur even if there were not safe havens to which the money can go, and in which the corrupt can sustain their lifestyle after their wrongdoing has been discovered; but secret bank accounts make it easier.

MAKING GLOBALIZATION WORK—FOR MORE PEOPLE

In his 2005 book, *The World Is Flat*, Thomas L. Friedman says that globalization and technology have flattened the world, creating a level playing field in which developed and less developed countries can compete on equal terms.[32] He is right that there have been dramatic changes in the global economy, in the global landscape; in *some* directions, the world is much flatter than it has ever been, with those in various parts of the world being more connected than they have ever been, but the world is not flat.[33]

Countries that want to participate in the new world of high-tech globalization need new technologies, computers, and other equipment in order to connect with the rest of the world. Individuals who want to compete in this global economy have to have the skills and resources to do so. Parts of India, such as Bangalore, have both the technology and the people with skills to use it, but Africa does not. As globalization and new technology reduce the gap between parts of India and China and the advanced industrial countries, the gap between Africa and the rest of the world is actually increasing. Within countries, too, the gap between the rich and the poor is increasing—and, with it, the gap between those who can effectively compete globally and those who can't.

High technology is a high-stakes game, in which large investments (by governments and countries) are required. The advanced industrial countries and their large firms have the resources; many others do not. What is remarkable is how well India and China have done, given their handicaps.

Not only is the world not flat: in many ways it has been getting less flat. The countries of East Asia made globalization work for them; their success is the best argument for the good that globalization can do for other developing countries. But for some of the poorest countries of the world, dependent as they are on aid from the World Bank, the IMF, or donors in Europe, America, and Japan, conditions imposed in order to receive that aid—though less onerous than in the past—may still be precluding them from following economic policies of their own choosing, including policies of the kind that proved so successful in East Asia. And recent trade agreements have made those policies—promoting technology, closing the knowledge gap, using financial markets as catalysts for growth—more difficult, if not impossible, to pursue.

It is bad enough that the developing countries are at a natural disadvantage—but the rules of the game are tilted against them, and in some ways increasingly so. The global trade and financial regimes give the advanced industrial countries a marked advantage. In later chapters I will describe in detail how they benefit the advanced industrial countries at the expense of the poor.

Equally worrying, in some respects, is how new technologies (reinforced by new trade rules) are enhancing the market power of incumbent, dominant firms, such as Microsoft, which are all from the developed world; for the first time, in a key global industry, there is a near-global monopolist, so powerful that even highly innovative firms in the United States like Netscape, the developer of the first major browser, get easily squashed. What chance, then, do much less capitalized, innovative firms from the developing countries have? At most, they can pick up the crumbs—occupy niches too small for the giants to bother with. So much market power does Microsoft have that it brazenly threatened to withdraw from Korea if Korea pursued its antitrust action against the firm—in a sense, it confirmed the allegations of overweening market power, for if that were not the case its threat to withdraw would have been meaningless.

The following chapters will detail these failures of globalization, including how trade agreements, rather than creating the opportunities that were promised, have sometimes created an even more unlevel playing field—a playing field so increasingly unlevel that recent trade agreements have actually made the poorest countries worse off. These agreements have also condemned to death thousands in the developing world suffering from diseases like AIDS, for which there are already medicines that work wonders. We will see how corporations strip countries of their natural resources, leaving behind a trail of environmental devastation—and how commonly accepted legal frameworks allow them to get away with it. We will see how the richest country in the world refuses to do anything about the world's greatest environmental problem—global warming—whose devastating effects will be especially felt in some of the world's poorest countries. We will see how Western governments have sometimes let stand global monopolies and cartels, to the detriment of those in the developing world.

Of course, if the developing countries had solved all of their own problems better, if they had had more honest governments, less influential special interests, more efficient firms, better educated workers—if, in fact, they did not suffer from all the afflictions of being poor—then they could have managed this unfair and dysfunctional globalization better. But development is hard enough in any case.

There are few success stories—our brief tour of the world has shown us a world replete with failures. The rest of the world cannot solve the problems of the developing world. They will have to do that for themselves. But we can at least create a more level playing field. It would be even better if we tilted it to favor the developing countries. There is a compelling moral case for doing this. I think there is also a compelling case that it is in our self-interest. Their growth will enhance our growth. Greater stability and security in the developing world will contribute to stability and security in the developed world.

CHAPTER 3

Making Trade Fair

I f any trade agreement were to be a success, it should have been the
one among Mexico, the United States, and Canada. Enacted in
1994, the North American Free Trade Agreement (NAFTA) created what was at the time the largest free trade area in the world, with
376 million people and a GDP of nearly $9 trillion.[1] The pact opened
up the world's richest country, the United States, to Mexico. These two
countries had a shared—though not always pleasant—history. Mexican immigration to the United States has been large; vast parts of the
United States are Spanish-speaking; and the United States relies on
Mexican labor in areas such as agriculture, manufacturing, and unskilled
services. Some 10 million Mexicans—a tenth of Mexico's population—
are living, legally or illegally, in the United States.[2] As Mexicans come
to the United States to work, many stay, marry American citizens, raise
their children, and now even dominate communities in states like California, Texas, and Arizona. Even before NAFTA, Mexico and Canada
were America's biggest trading partners, as well as the countries most
visited by U.S. citizens.

The ties between the two countries, combined with the disparity in
economic and political power, bring tensions. As the Mexican saying
goes, "Mexico—so far from God, so close to the United States." America's per capita income is six times that of Mexico. The corresponding

sixfold wage difference, together with Mexico's high unemployment rates, exerts an enormous pull across the border, with thousands risking their lives to enter illegally. It is not in the United States' interests to have a poor, unstable country on its southern border, and NAFTA supporters hoped the pact would bring Mexico's economy forward and help this country, rich with art and history and culture, prosper. Instead, more than ten years later, it is clear that NAFTA has not succeeded. While it has not been the disaster that its critics predicted, neither has it brought all the benefits that were claimed by its advocates.

Advocates of trade liberalization believe it will bring unprecedented prosperity. They want developed countries to open themselves up to exports from developing countries, liberalize their markets, take away man-made barriers to the flows of goods and services, and let globalization work its wonders. But trade liberalization is also among the most controversial aspects of globalization; many see the alleged costs—lower wages, growing unemployment, loss of national soveignty—as outweighing the purported benefits of greater efficiency and increased growth.

In part, free trade has not worked because we have not tried it: trade agreements of the past have been neither free nor fair. They have been asymmetric, opening up markets in the developing countries to goods from the advanced industrial countries without full reciprocation. A host of subtle but effective trade barriers has been kept in place. This asymmetric globalization has put developing countries at a disadvantage. It has left them worse off than they would be with a truly free and fair trade regime.

But even if trade agreements had been truly free and fair, not all countries would have benefited—or at least benefited much—and not all people, even in the countries that did benefit, would share in the gains. Even if trade barriers are brought down symmetrically, not everyone is equally in a position to take advantage of the new opportunities. It is easy for those in the advanced industrial countries to seize the opportunities that the opening up of markets in the developing countries affords—and they do so quickly. But there are many impediments facing those in the developing world. There is often a lack of infrastructure to bring their goods to market, and it may take years for the goods they produce to meet the standards demanded by the advanced industrial countries. These are among the reasons that when, in February 2001, Europe unilaterally opened up its markets to the poorest countries of the world, almost no new trade followed. To fulfill the promise that more trade will follow from trade liberalization, much else is required, as we shall see.

Moreover, trade liberalization exposes countries to more risk, and developing countries (and their workers) are less prepared to bear that risk. Workers in the United States and Europe worry about being thrown out of their jobs as a result of a surge in imports. But workers in these countries have a strong safety net to fall back on: they have the education that makes it easier to move from one job to another; they often have bank accounts and receive severance pay to buffer their transition between jobs. Workers in developing countries have none of these.

Finally, even if trade does follow, not everyone is a winner. The theory of trade liberalization (under the assumption of perfect markets, and under the hypothesis that the liberalization is fair) only promises that the country as a whole will benefit. Theory predicts that there will be losers. In principle, the winners could compensate the losers; in practice, this almost never happens. If all the benefits go to a few at the top, then trade liberalization leads to rich countries with poor people, and even those in the middle may suffer. Thus, if liberalization is not managed well, the majority of citizens may be worse off—and see no reason to support it. It is not a matter of special interests opposing liberalization, but of citizens correctly perceiving the world as it is.

But this is not the world as it has to be. Trade liberalization can, when done fairly, when accompanied by the right measures and the right policies, help development. As we saw in chapters 1 and 2, the most successful developing countries in the world have achieved their success through trade—through exports. The question is: can the benefits that they enjoy be sustained, and be brought to all of the people of the world? I believe they can be; but if that is to be the case, trade liberalization will have to be managed in a way very different from that of the past.

The North American Free Trade Area

Understanding why NAFTA failed to live up to its promise can help us to understand the disappointments of trade liberalization. One of the main arguments for NAFTA was that it would help close the gap in income between Mexico and the United States, and thus reduce the pressure of illegal migration.[3] Yet the disparity in income between the two countries actually grew in NAFTA's first decade—by more than 10 percent. Nor did NAFTA result in a rapid growth in Mexico's economy. Growth during that first decade was a bleak 1.8 percent on a real per capita basis, better than in much of the rest of Latin America but far worse than earlier in the century (in the quarter century from 1948 to 1973, Mexico grew at an average annual rate per capita of 3.2 percent).[4] President Fox promised 7 percent growth when he took office in 2000; in fact, in real terms, growth during his term of office averaged only 1.6 percent per annum—and real growth per capita has been negligible. In fact, NAFTA made Mexico more dependent on the United States, which meant that when the U.S. economy did poorly, so did Mexico's.

Not only did NAFTA not lead to robust growth; it can even be argued that in some ways it contributed to Mexico's poverty. Poor Mexican corn farmers now have to compete in their own country with highly subsidized American corn (though the relatively better-off Mexican city dwellers benefit from lower corn prices). A fairer trade agreement would have eliminated America's agricultural subsidies and its restrictions on imports of agricultural goods, like sugar, into the United States. Even if the United States did not eliminate all its subsidies, Mexico should have been given the right to countervail—that is, to impose duties on US imports to offset the subsidies. But NAFTA does not allow that.

While NAFTA eliminated tariffs, it allowed a whole set of nontariff barriers to stand. After NAFTA was signed, the United States continued to use nontariff barriers to bar Mexican products that had begun to make inroads in its markets, including avocadoes, brooms, and tomatoes. When, for instance, Mexican tomato exports to the United States began to increase in 1996, Florida tomato growers pressured

Congress and the Clinton administration to take action. If Mexico could be shown to be selling tomatoes below cost, it could be charged with dumping, and anti-dumping duties could be imposed. But Mexico was not dumping tomatoes. The reason that Mexico could be charged with selling below cost was because prices were measured in a deliberately lopsided fashion (I will discuss this more fully later in the chapter). Mexico did not want to risk a trial, so agreed to raise its price. American consumers and Mexican tomato growers were hurt, but Florida tomato producers got what they wanted—less competition from Mexican tomatoes.

The one part of Mexico's economy that was successful, at least in the years immediately after NAFTA, was the area just south of the border. So-called maquiladora factories sprang up, supplying American manufacturers like General Motors and General Electric with low-cost parts. Employment grew 110 percent over NAFTA's first six years, compared with 78 percent over the previous six years.[5] (Elsewhere, employment stagnated.)[6] Advocates of NAFTA are quick to take credit for these successes, while arguing that the failures are not NAFTA's fault and that matters would have been far worse without the agreement. There is, of course, no easy answer to this sort of counterfactual argument, which supposes an imaginary alternative. But careful studies do shed some light. One can ask whether, given the expansion of the U.S. economy and the dramatic fall of real wages in Mexico after 1994 in comparison both to the United States and to its competitors in Asia, one would have expected an increase in Mexican exports to the United States comparable to what was observed. The answer, based on standard economic models, is yes. NAFTA seems to have added little, if anything.[7]

Equally telling is what happened after the first flush of NAFTA. After the early years of growth in the maquiladora region, employment there too actually started to decline, with some 200,000 jobs lost in the first two years of the new millennium.[8] Some of the factors that had led to growth, like the strong U.S. economy, had waned. But there was a more fundamental problem. Not only was the United States growing faster than Mexico in the years after NAFTA, but so was China.[9] Trade liberalization is important for growth, but not as important as NAFTA supporters had hoped. NAFTA gave Mexico a slight advantage over other U.S. trading partners; but Mexico, with its low investment in education and technology, has had a hard time competing with China, which invests twice as much (as a percentage of GDP) in research. Countries often hope that trade agreements will boost foreign investment and create jobs. But when companies make investment decisions they look at many factors, including the quality of the workforce, infrastructure, location, and political and social stability.

Tariffs play only a limited role, as China's success makes clear. By focusing on tariffs, NAFTA diverted attention from other things that needed to be done to make Mexico competitive. Indeed, reduced tariffs have created their own problems. Prior to NAFTA, tariffs made up 7 percent of Mexico's tax revenue; after NAFTA, the figure dropped to 4 percent. Mexico's public expenditures of around 19 percent of GDP—more than a third financed by oil revenues—are markedly lower than those of Brazil or the United States, and are insufficient to finance needed public investment in education, research, and infrastructure.

TRADE LIBERALIZATION: THEORY AND PRACTICE

The British economist Adam Smith, the founder of modern economics, was a strong champion of both free markets and free trade, and his arguments are compelling: free trade allows countries to take advantage of their comparative advantage, with all nations benefiting as each one specializes in the areas in which it excels. Large trading areas allow firms and individuals to specialize further and become even better at what they do. Imagine a small village with only one baker, then consider that a larger village might have two or three. A bigger town would support a larger number of bakers, some of whom will make only bread and others who will make only cakes. An even bigger city will have not only bread makers and cake makers; its bakers will have so many customers that they can specialize even further, making a wide variety of very good cakes and gourmet breads. Bigger markets enhance the efficiency of each producer and the choice available to consumers.

Without free trade, capital and labor will earn different returns in different countries (assuming capital and labor cannot move freely—which is a fair assumption, especially in the short run). In a country that lacks capital, such as machinery and technology, labor will be less productive and wages will be lower. If labor moves from a country where productivity and wages are low to one where they are high, the increase in output can be enormous, and the world's economy grows. Free trade is a substitute for people actually having to move. We can sit at home in the developed world and buy inexpensive goods from China, a country where labor is cheap. Conversely, the Chinese can stay in China and get high-tech goods from the United States, a country with more advanced technology, highly skilled labor, and large capital investment. In theory, this will mean that as the demand for Chinese goods increases, the demand for their unskilled labor increases, and eventually unskilled wages in China will be higher.[10]

The Fear of Job Loss

The downside to this rosy scenario is the possibility that jobs will be lost as they move from one country to another—for example, as people in the United States buy cheap goods made in China instead of in the United States. Free trade advocates say that although jobs are lost, new opportunities are created. High-productivity/high-wage jobs replace low-productivity/low-wage jobs. The argument is persuasive, except for one detail: in many countries, unemployment rates are high and those who lose their jobs do not move on to higher-wage alternatives but onto the unemployment rolls. This has happened especially in many developing countries around the world when they liberalized so fast that the private sector did not have time to respond and create new jobs, or when interest rates were so high that the private sector could not afford to make the investments necessary to create new jobs.

It even happens in developed countries, though there, if monetary and fiscal policies are working well, jobs should be created in tandem with jobs that are lost. But too often, that does not happen. Unemployment in Europe has remained stubbornly high. People who lose their jobs do not automatically get new jobs. Especially when the unemployment rate is high, there may be an extended period of unemployment as workers search for a new employer. Middle-aged workers often fail to find any job at all—they simply retire earlier. Low-skilled workers are particularly likely to suffer. That is why people in the advanced industrial countries worry about losing manufacturing jobs to China or service sector jobs (like back offices of financial companies) to India.

When the result of rapid trade liberalization is that unemployment goes up, then the promised benefits of liberalization are likely not to be realized.[11] When workers move from low-productivity, protected jobs into unemployment, it is poverty, not growth, that is likely to increase.[12]

Even if they do not actually lose their jobs, unskilled workers in advanced industrial countries see their wages decrease. They are told that unless they agree to lower wages, the reduction of benefits, and the weakening of job protections, competition will force the firm to move the jobs overseas. Young workers in France have been mystified by how the removal of long-fought-for job protections and the lowering of wages—necessary, it is alleged, to compete in the global marketplace—will make them better off. They are told to be patient, that in the long run they will see that they are better off; but, given the number of cases in which those promises have failed to be fulfilled ten or twenty years after liberalization, their skepticism is understandable. John Maynard Keynes, the great economist of the mid-twentieth century, had responded to those who urged patience in the midst of the Great Depression as markets would in the long run restore the economy to full employment, by saying yes, but "In the long run, we are all dead."[13]

Politicians and economists who promise that trade liberalization will make everyone better off are being disingenuous. Economic theory (and historical experience) suggests the contrary: even if trade liberalization may make the country as a whole better off, it results in some groups being worse off.[14] And it suggests that, at least in the advanced industrial countries, it is those at the bottom—unskilled workers—who will be hurt the most.[15]

The world of Adam Smith and the free trade advocates, in which free trade will make everyone better off, is not only a mythical world of perfectly working markets with no unemployment; it is also a world in which risk doesn't matter because there are perfect insurance markets to which risk can be shifted, and where competition is always perfect, with no Microsofts or Intels dominating the field. In such a world, workers wouldn't worry about losing their jobs because of trade liber-alization; they would move seamlessly into other jobs. Even if there was some glitch, workers could buy insurance against the risk of being tem-porarily unemployed, or against the risk that the new job paid less than the old. Even in the best-functioning market economies, this kind of insurance can't be bought; while in developed countries the govern-ment provides some unemployment insurance, in most developing countries workers are left to fend for themselves.

That is why trade liberalization requires more than just onetime assis-tance to move from the old industries to the new. More open economies may be subject to all manner of shocks—domestic firms, for instance, may find it hard to compete with an onslaught of imports that sud-denly become cheaper when a foreign country devalues its currency, as in a crisis. When Korea's currency was devalued, Korean steel exports to the United States increased, and American steelworkers complained. When Brazil has a good orange crop, Florida orange growers cry for help, and sometimes get it through one of the nontariff protectionist mechanisms described below.[16] Everyone feels the insecurity.

It is not just those who lose their jobs, and their families, who are affected. Almost everyone is at risk. For example, when local industries shut down because of competition from imports, their suppliers are adversely affected. Increased insecurity is one of the reasons that oppo-sition to trade liberalization is so widespread.

But while globalization has led to more insecurity and contributed to the growing inequality in both developed and less developed coun-tries, it has limited the ability of governments to respond. Not only does liberalization require removing tariffs, which are an important source of public revenue for less developed countries, but to compete a country may have to lower other taxes as well.[17] As taxes are lowered, so are pub-lic revenues, forcing cuts in education and infrastructure and expendi-tures on safety nets such as unemployment insurance at a time when they are more important than ever, in order both to respond to the competi-tion and to help people cope with the consequences of liberalization.

While developing countries may suffer from trade liberalization, they are not always in a position to reap its benefits through increased exports. There are several reasons for this: One already noted is that they often lack the infrastructure (ports and roads) needed to move their products. The other is they may not have anything to export. Capital markets are highly imperfect, with interest rates in developing countries at a much higher level than those with which even the best of entrepreneurs in the developed world could cope; even if someone sees a new export opportunity he cannot get the necessary finance, at least at reasonable terms. These supply-side constraints are a big problem in many of the poorest countries of the world, such as in Africa. By now, there are numerous instances in which advanced industrial countries have opened up their markets, but the gains in exports have been limited. These countries will need some form of assistance—aid for trade—to help them take advantage of the new opportunities. Some used to argue that trade was more important than aid; trade helps a country to stand on its own. But it is better to see aid and trade as complements: both are needed for successful development.[18]

Infant Industries and Infant Economies

Countries often need time to develop, in order to compete with foreign companies; to get this time, they may have to protect their nascent industries temporarily. The standard argument for free trade is based on efficiency. More goods can be produced with given resources if each country focuses on its own comparative advantage. But even more important in determining the pace of growth in developing countries is how fast they acquire the knowledge and technology of the advanced industrial countries. We saw in the last chapter that developing countries not only lag in resources but also in technology; for achieving sustained growth, closing the knowledge gap is more vital than improving efficiency or increasing available capital. The question is: how best to learn? Some argue that the best way—probably the only way—to learn how to produce steel is to produce steel, as Korea did when it started a steel industry. At the time, its comparative advantage was growing rice. But even if Korean farmers became the most efficient rice producers in the world, their incomes would still be limited. The Korean government realized that if it was to succeed in becoming developed, it had to transform its economy from agriculture to industry.

If developing countries are to enter into such industries, those industries have to be protected until they are strong enough to compete with established international giants. Tariffs result in higher prices—high enough that the new industries can cover costs, invest in research, and make the other investments that they need in order to be able eventually to stand on their own feet. This is called the "infant industry argument" for protection.[19] It was a popular idea in Japan in the 1960s—and in the United States and Europe in the nineteenth century. Most successful countries did in fact develop behind protectionist barriers; critics of globalization accuse countries like Japan and the United States, which have climbed the ladder of development, of wanting to kick the ladder away so that others can't follow.

Advocates of free trade respond with two main criticisms of the infant industry argument. First, they say, the appropriate response is not protection; if in the long run the firm will be profitable, it can obtain a loan to tide it over the hard times. In the real world, however, new firms have a difficult time getting capital. The United States government has only partially overcome this problem by having a Small Business Administration (SBA) that provides loans for small businesses. (The U.S. shipping and logistics giant FedEx began with an SBA loan.) In developing countries, these problems are even more acute.

Second, critics argue that, too often, protected infants never grow up, and demand to be permanently insulated from outside competition.

More generally, special interests grab hold of any argument, including the infant industry argument, to push protectionist measures in pursuit of higher profits—which impose enormous costs on the rest of the economy.[20] In Bangladesh, protection of textile producers puts apparel makers in jeopardy by raising the cost of raw materials. These experiences are a warning for any country contemplating using protection as a basis for encouraging new industries.

But the politics of different countries differ, and there is nothing inevitable in such a political failure. East Asia did manage to wean its infants; the question is whether others have political systems capable of doing the same.

One of the responses to the last criticism of the infant industry argument is to focus on broad-based protection, a uniform tariff on, say, manufactured goods. This is the approach of the infant economy (as opposed to the infant industry) argument for protection.[21] Without protection, a country whose static comparative advantage lies in, say, agriculture risks stagnation; its comparative advantage will remain in agriculture, with limited growth prospects. Broad-based industrial protection can lead to an increase in the size of the industrial sector, which is, almost everywhere, the source of innovation; many of these advances spill over into the rest of the economy, as do the benefits from the development of institutions, like financial markets, that accompany the growth of an industrial sector. Moreover, a large and growing industrial sector (and the tariffs on manufactured goods) provides revenues with which the government can fund education, infrastructure, and other ingredients necessary for broad-based growth. In chapter 4, we will see that advocates of strong intellectual property protections argue for exactly the same trade-off: they claim that the short-run inefficiencies (in that case, arising from monopoly; in this case, arising from tariff protection) are more than offset by long-run dynamic gains. In each case, it is a question of getting the balance right: almost surely, some intellectual property protection is desirable; and almost surely, some trade protection is desirable. While the economic rationale behind the infant economy argument is similar to that behind the infant industry argument, the political argument is far stronger: broad-based protection reduces the scope for special interest.

If advocates of the infant industry argument have sometimes been excessively optimistic about the virtues of protection, advocates of liberalization sometimes seem even more to live in a dreamland, believing that almost any trade agreement, especially with the United States or European Union, no matter how unfair, will magically bring investment and create jobs. They cite statistical studies claiming that trade liberalization enhances growth. But a careful look at the evidence shows something quite different. It shows that countries, like those in East Asia, that have become more integrated into the global economy have grown faster. It is exports—not the removal of trade barriers—that is the driving force of growth. Studies that focus directly on the removal of trade barriers show little relationship between liberalization and growth. The advocates of quick liberalization tried an intellectual sleight of hand, hoping that the broad-brush discussion of the benefits of globalization would suffice to make their case.[22]

Fair Trade versus Free Trade

Economists focus on how trade liberalization affects efficiency and growth. But popular discussions focus more on *fairness*. When people in the developed world talk of unfair trade, what they often have in mind is developing countries' huge advantage of low wages. But these countries have offsetting disadvantages as well, including a high cost of capital, poor infrastructure, lower skill levels, and overall low productivity. Those in the developing world complain equally vociferously of the difficulties of competing with the advanced industrial countries. Economists emphasize that these different strengths and weaknesses mean that each country has a comparative advantage, the things at which it is relatively good, and they should determine what it exports. It is not unfair to be poor and have low wages; it is unfortunate.

Too often, in political discourse, there is almost a presumption that if some country or firm is undercutting an American firm, it must be because that firm is playing unfairly. After all, American firms must be more efficient than those anywhere else; on a level playing field they would win. The dumping laws (often dubbed "fair trade laws"), described in greater detail later in this chapter, are almost based on this presumption: since American firms are more efficient, their costs must be lower; if foreign firms are outcompeting American firms, it must be because they are cheating—selling below cost. But this ignores the basic principle of trade: trade is based not on the absolute strengths of a country but on its relative strengths, on its *comparative* advantage; and even if America were more efficient in every industry (which it is not), industries in which it was *relatively* less efficient would find themselves losing to competition.

What, then, should one mean by fair trade? There is a natural benchmark: the trade regime that would emerge if all subsidies and trade restrictions were eliminated.[23] The world, of course, is nowhere near such a regime. Asymmetries in liberalization can benefit some groups at the expense of others. For instance, trade agreements now forbid most subsidies—except for agricultural goods. This depresses incomes of those farmers in the developing world who do not get subsidies. And since 70 percent of those in the developing world depend directly or indirectly on agriculture, this means that incomes of the developing countries are depressed. But by whatever standard one uses, today's international trading regime is unfair to developing countries.[24]

Even with an unfair trading system, China, India, and a few other developing countries have been growing enormously, and their growth is based in no small part on trade. But others have not been so fortunate. The unlevel playing field means that there will be more countries as a whole that lose, and more people even in successful countries who will lose. China, by most accounts one of the true winners in the global trade competition, faces a problem of growing inequality; its farmers are suffering because of American and European agricultural subsidies, which drive down prices. China and other developing countries face a cruel dilemma—they can spend scarce resources to subsidize their farmers in order to offset the developed world's largesse to theirs, but that will mean less to spend on development and therefore slower growth for the country as a whole.

THE HISTORY OF TRADE AGREEMENTS

Economists have been arguing for free trade for two centuries, but it was the Great Depression of the 1930s, more than abstract arguments, that was responsible for the wave of liberalization that began sixty years ago. Successive increases in tariffs in the late 1920s and early 1930s were thought to have played an important role in deepening the Great Depression. Each country saw its economy shrinking and so tightened restrictions on imports. These restrictions hurt other countries, which responded by tightening their own restrictions; as they did so, a vicious circle emerged. It was natural that after World War II, when global leaders sought to create a new, more prosperous international economic order, they not only sought to enhance financial stability through the creation of the International Monetary Fund but also attempted to establish an International Trade Organization (ITO) to regulate trade. This did not happen. The United States rejected the proposal for the

ITO in 1950 because of concerns on the part of some conservatives and corporations that it would lead to an infringement of national sovereignty and excessive regulation. It was not until forty-five years later that the World Trade Organization (WTO) came into being.

In the interim, trade negotiations led by the advanced industrial countries under the auspices of GATT, the General Agreement on Tariffs and Trade, greatly reduced tariffs on manufactured goods and created the foundations of the modern trade regime. The GATT system was built on the principle of nondiscrimination: countries would not discriminate against other members of GATT. This meant that each country would treat all others the same—all would be the most favored, hence the name: the most favored nation principle, the bedrock of the multilateral system. Alongside this went the principle of national treatment: foreign producers would be treated the same, and be subject to the same regulations, as domestic producers.

Trade negotiations occur in a series of rounds, in which many issues are put on the table, with complex bargaining among the countries. Each country agrees to lower tariffs and to open up markets if others reciprocate. By having enough issues on the table, it is hoped that negotiators can find a set of trade concessions that will make every country feel better off. GATT focused on liberalization of trade in manufactured goods, the comparative advantage of the advanced industrial countries. There was limited trade liberalization in the areas important for developing countries, such as agriculture and textiles. Textiles remained subject to strong limits (quotas) on a country-by-country, product-by-product basis;[25] likewise, agriculture remained highly protected and subsidized.

The Uruguay Round, the round of trade negotiations that began in Punta del Este, Uruguay, in September 1986, ended with an agreement signed in Marrakech on April 15, 1994. Under this agreement GATT, which had 128 member countries, was replaced by the World Trade Organization, which today has 149 member countries. Ministers from these countries meet at least every two years. The WTO was designed to provide a faster expansion of trade agreements, reaching into new areas like services and intellectual property rights, than had occurred under GATT.

Most important, for the first time there was an effective—if limited—enforcement mechanism. The WTO did not itself punish violators, but it authorized countries that had suffered injury as a result of a violation to retaliate by imposing trade restrictions on the offending country. The EU has become quite sophisticated in using this instrument against the United States. It draws up a long list of potential candidates for retaliation, targeting areas in which tariffs will be particularly painful, or goods produced in the districts of congressmen whom they are trying to sway. The threats have worked remarkably well.

The first step toward a rule of law in international trade was the great achievement of the Uruguay Round. Without a rule of law, brute power wins. The WTO's international law is an imperfect rule of law; the rules are derived from bargaining, including bargaining between the rich and the poor countries, and in that bargaining it is the rich and powerful that typically prevail. Enforcement is asymmetric—a threat of trade restriction by the United States against a small country like Antigua will elicit a response, but the United States does not pay much attention if Antigua threatens a trade restriction. Only when the practice affects a large number of countries—such as in the case of the cotton subsidies that the United States doles out to its farmers—is the threat of retaliation even credible.[26] Even so, an imperfect rule of law is better than none.

From Seattle to Cancún

Half a decade after the completion of the Uruguay Round, on November 30, 1999, the WTO convened in Seattle, Washington, for what was supposed to be the launch of a new round of trade negotiations, intended to be the crowning achievement of the Clinton administration's efforts at trade liberalization, which included the creation of NAFTA in 1994 and the World Trade Organization in 1995.[27] Instead, the meeting was a disaster. The negotiations were quickly overshadowed by massive street protests. Beginning at 5 a.m. on the first day of the conference, hundreds of activists began to take control of street intersections near the convention center. By the end of the day, the mayor had declared a state of civil emergency and imposed curfews, and the governor had called up the National Guard. The scale of the demonstrations dwarfed any previous protest associated with globalization.

While the protestors represented a melange of views and did not offer any coherent alternatives, there was much to complain about (though the WTO itself should not have borne the brunt of the complaints; it simply provides a forum in which trade negotiations occur). The Uruguay Round had been based on what became known as the "Grand Bargain," in which the developed countries promised to liberalize trade in agriculture and textiles (that is, labor-intensive goods of interest to exporters in developing countries) and, in return, developing countries agreed to reduce tariffs and accept a range of new rules and obligations on intellectual property rights, investments, and services. Afterward, many developing countries felt that they had been misled into agreeing to the Grand Bargain: the developed countries did not keep their side of the deal. Textile quotas would remain in place for a decade, and no end to agricultural subsidies was in sight.

For forty years, trade liberalization had focused on opening up markets for manufactured goods—at the time, the comparative advantage of the United States and Europe. But I emphasized earlier the dynamic nature of comparative advantage: today it is China and other developing countries that have a comparative advantage in many areas of manufacturing. Unknowingly, for four decades, trade negotiators had been working to open up markets for China! With manufacturing in the developed world shrinking—today it represents only 11 percent of American employment and output—American and European trade negotiators would have to deliver something in services (which are now over 70 percent of America's economy, and nearly that in Europe and Japan) and in intellectual property to satisfy their constituents. They succeeded.

The list of complaints against the Uruguay Round trade agreement was long:

- It was so asymmetric that the poorest countries were actually worse off; sub-Saharan Africa, the poorest region with an average income of just over $500 per capita per year, lost some $1.2 billion a year.[28]

- Seventy percent of the gains went to the developed countries—some $350 billion annually. Although the developing world has 85 percent of the world's population and almost half of total global income, it received only 30 percent of the benefits—and these benefits went mostly to middle-income countries like Brazil.[29]
- The Uruguay Round made an unlevel playing field less level. Developed countries impose far higher—on average four times higher—tariffs against developing countries than against developed ones. A poor country like Angola pays as much in tariffs to the United States as does rich Belgium; Guatemala pays as much as New Zealand.[30] And this discrimination exists even after the developed countries have granted so-called preferences to developing countries. Rich countries have cost poor countries three times more in trade restrictions than they give in total development aid.[31]
- The focus was on liberalization of capital flows (which developed countries wanted) and investment rather than on liberalization of labor flows (which would have benefited the developing countries), even though the latter would have led to a far greater increase in global output.
- By the same token, liberalization of unskilled labor services would have led to a far greater increase in global efficiency than liberalization of skilled labor services (like financial services), the comparative advantage of the advanced industrial countries. Yet negotiators focused on liberalizing skill-intensive services.
- The strengthening of intellectual property rights largely benefited the developed countries, and only later did the costs to developing countries become apparent, as lifesaving generic medicines were taken off the market and developed-world companies began to patent traditional and indigenous knowledge. (We will discuss this more fully in chapter 4.)

The United States and Europe have perfected the art of arguing for free trade while simultaneously working for trade agreements that protect themselves against imports from developing countries. Much of the success of the advanced industrial countries has to do with shaping the agenda—they set the agenda so that markets were opened up for the goods and services that represented their comparative advantage.

Western negotiators almost take it for granted that they can control what gets discussed, and determine the outcomes. As the United States and the EU push for opening up markets for services, they do not think (as they logically should): by and large, services are labor intensive; by and large, it is the developing countries that have an abundance of labor; and therefore, by and large, a fair service sector liberalization will be of especial benefit to developing countries. They think: we can liberalize the high-skilled services which represent our comparative advantage now, and we can make sure, one way or the other, not to liberalize services that are intensive in unskilled labor. From the very beginning of the discussion, they had in mind an unbalanced agreement.

Special interests are largely to blame—not special interests in the developing countries resisting trade liberalization, as proponents of trade liberalization complain, but special interests in the developed world shaping the agenda to benefit themselves, while leaving even the average citizen in their own countries worse off. The negotiators, in representing their immediate "clients"—the corporations that lobby them heavily and constantly, partly directly, partly through lobbying Congress and the administration—often lose sight of the big picture, confusing the interests of these companies with America's national interests or, even worse, with what is good for the global trading system. And the story is much the same in other industrial countries. Within each country export-corporation interests pressure negotiators to get agreements that provide more access for their goods, while import industries press for protection. The negotiators strive not for intellectual consistency, not for an agreement based on principles, but only to balance the competing interests.

The Seattle protests sent an important message of discontent to the trade ministers, but the advanced industrial countries were not yet ready to give up on their push for further liberalization. The trade ministers met next at Doha in Qatar, a small country off the Persian Gulf, in November 2001—a far-flung location well chosen for those not wanting to be bothered by demonstrators questioning what was going on behind closed doors. The developed countries promised to make the talks a "development round"; in other words, they committed themselves to creating a trade regime that would actively enhance development prospects and redress the imbalances of previous rounds.[32] The developing countries were hesitant to go along; they were afraid that another unfair trade agreement would be foisted on them, one which, like the last, would leave some of them actually worse off; they worried that, once the negotiations began, their arms would be twisted in one way or another and they would be forced to sign on to a new agreement against their best interests. They were skeptical about the promises being made at Doha; and, as the negotiations evolved over succeeding years, their skepticism seems to be have been justified.

The negotiations stalled over the refusal of the developed world to cut back on agricultural subsidies—in fact, in 2002 the United States enacted a new farm bill that nearly doubled its subsidies. In September 2003 the trade ministers met again at Cancún, which, in the local Mayan language, means "snake pit"—and so it proved for the negotiators. The ministers were supposed to appraise the progress that had been made and give directions to their negotiators for concluding the "development round." Despite still refusing to make concessions in agriculture or any other major issue of concern to the developing world—in effect, reneging on their promise—the developed countries insisted on pushing their own agenda of reduced tariffs and opening access for the goods and services the EU and the United States wanted to export. They even wanted to impose new demands on the developing countries. While the advanced industrial countries still talked about a development round, it was mere rhetoric: there was a real risk that this new round, rather than undoing the imbalances of the past, would make them worse. The talks collapsed on the fourth day of the meeting. Never before had trade negotiations ended in such disarray.

The next global meeting of trade ministers in Hong Kong in December 2005—originally intended to wrap up the development round—did not end in disaster, but neither could it be called a success: Pascal Lamy, the head of the WTO, had managed to lower expectations so far that any agreement, even one which would have little effect on global trade, would be viewed as the best that could be expected in the circumstances. More effort was put into managing the press than into making meaningful offers. The United States, which because of its huge cotton subsidies is the world's largest cotton exporter, to much fanfare offered to open its markets to African cotton producers—an offer worth little since it would not be importing much cotton (because of its huge cotton subsidies, America is a cotton exporter, not a major importer).

The era of multilateral trade liberalization seems to be nearing an end (at least for a while), as well-founded disillusionment in the developing countries combines with growing protectionist sentiment in the developed world. Whatever emerges from the so-called development round—if anything—will not be deserving of the epithet. It will do little either to create a trade regime that is fair to the developing countries or that will promote their development: tariffs imposed by developed countries against developing countries will still be far higher than those imposed against other developed countries, and developed countries will still be providing massive agricultural subsidies, doing enormous harm to the developing countries.

The real danger today is not that something will or will not be agreed to at the conclusion of the development round which will harm the developing countries significantly: the scale of reforms is so low that it is likely to matter little. Any eventual agreement will do only limited damage, or be of only limited benefit. The real danger is that the world will think that it has accomplished what was set out in Doha, so that, going forward, there is no need for a development round. Trade negotiators will then return to business as usual—another round of trade negotiations in which hard bargaining results in the lion's share of the gains going to the developed countries.

MAKING GLOBALIZATION WORK

Doha failed.[33] While it may be difficult to define precisely what is a fair global trade regime, it is clear that the current arrangements are not fair, and it is clear that the development round will do little to make the trade regime fairer or more pro-development.[34] I believe, however, that it is possible to design a global trade regime that promotes the well-being of the poorest countries and that is, at the same time, good for the advanced industrial countries as a whole—though, of course, some special corporate interests might well suffer. This was, of course, the promise of Doha. The reforms would cost the developed countries little—in most cases nothing at all, as taxpayers would save billions from subsidies and consumers would save billions from lower prices— and developing countries would benefit enormously.

While Doha has failed to deliver on its promise, sometime in the future the challenge of creating a fair trade regime—and a trade regime that will give the poor countries of the world the opportunity to develop through trade—remains. There is a full agenda of reforms, going well beyond the agricultural issues on which so much of the discussion has focused: reforms that are both pro-poor and pro-development. These reforms are what a *true* development round would look like.

Developing Countries Should Be Treated Differently

Developing countries are different from more developed countries— some of these differences explain why they are so much poorer. The idea that developing countries should, as a result, receive "special and differential treatment" is now widely accepted and has been included in many trade agreements.[35] Developed countries are allowed, for instance, to deviate from the most favored nation principle by allowing lower tariffs on imports from developing countries—though even with this so-called preferential treatment, developed country tariffs against imports from developing countries are, as we have seen, four times higher than tariffs against goods produced by other developed countries.

The current system, however, makes preferential treatment completely voluntary, provided by each of the advanced industrial countries on its own whim. Preferences can be taken away if the developing country does not do what the granting country wants. Preferential treatment has become a political instrument, a tool for getting developing countries to toe the line.

Free trade for the poor: an extended market access proposal

One single reform would simultaneously simplify negotiations, promote development, and address the inequities of the current regime. Rich countries should simply open up their markets to poorer ones, without reciprocity and without economic or political conditionality. Middle-income countries should open up their markets to the least developed countries, and should be allowed to extend preferences to one another without extending them to the rich countries, so that they need not fear that imports from those countries might kill their nascent industries. Even the advanced industrial countries would benefit, because they could proceed more rapidly with liberalization among themselves—which their economies are capable of withstanding—without having to satisfy the worries of the developing world. This reform replaces the principle of "reciprocity for and among all countries—regardless of circumstances" with the principle of reciprocity among equals, but differentiation between those in markedly different circumstances.[36]

The European Union recognized the wisdom of this basic approach when in 2001 it unilaterally opened up its markets to the poorest countries of the world, taking away (almost) all tariffs and trade restrictions without demanding political or economic concessions.[37] The rationale was that European consumers would benefit from lower prices and more product diversity; while it would cost European producers a negligible amount, it could be of enormous benefit to the poorest countries; and it was a strong demonstration of goodwill. The European initiative should be extended to all advanced industrial countries, and markets should be opened up not just to the poorest but to all developing countries. (In one of the high points of hypocrisy and cynicism in the Hong Kong meeting in December 2005, the United States offered to open itself up to 97 percent of the goods produced by the least developed countries, a number carefully calibrated to exclude most of the products, such as Bangladeshi textiles and apparel, that it wanted to keep out. Bangladesh would be free, of course, to export jet engines and all manner of other products which are beyond its capacity to produce.)[38]

Broadening developing countries' development agenda

Development is hard enough: we should not restrict what developing countries can do to help themselves grow. But that is what the Uruguay Round has done, as it restricts their ability to use a variety of instruments to encourage industrialization.

There is a difference between the effects on the global economy of agricultural subsidies given by the United States and Europe, which are allowed, and the subsidies that developing countries might want to give to help start new industries, or even to protect their industries and farmers against subsidized competition, which are prohibited. When the United States subsidizes cotton, global prices are affected; farmers in the developing world are hurt because of U.S. generosity to its farmers. (Economists call this an "externality.") But if Jamaica protects its milk producers, global prices are unaffected. Moreover, developing countries have limited tools to deal with the consequences of liberalization: the Jamaican dairy farmers who are put out of business as a result of America's highly subsidized milk industry have few viable alternatives. There are few jobs in the cities, and turning to some lower-paying alternative crop may make the subsistence farmer even poorer. The government has a tough choice to make: supplement the income of the individual farmers or spend government funds on an investment that the whole country needs. There is not enough money to do both. Protection against America's subsidized milk may be the only sensible alternative, at least in the short run.

If the extended market access proposal is adopted, then countries will have the scope to pursue their pro-development strategies and policies aimed at protecting their very poor citizens. But if it is not, then there must be exceptions that allow developing countries more leeway, especially to utilize uniform revenue-raising tariffs (the effect on imports being little different from that of a change in the exchange rate) and temporary industrial subsidies. As Europe has rightly pointed out, the United States often uses its defense expenditures to subsidize a range of industries. Boeing has benefited from military expenditures in aircraft design, and the software industry has benefited enormously from a whole range of government expenditures that helped develop the Internet and even the browser. Indeed, commercial benefits are often put forward as one of the justifications for the huge level of defense expenditures. The United States is wealthy enough to afford an inefficient industrial policy hidden within its military; developing countries are not—and they should be free, if they choose, to have one appropriate to their circumstances.

Agriculture

A decade after the Uruguay Round, more than two-thirds of farm income in Norway and Switzerland came from subsidies, more than half in Japan, and one-third in the EU. For some crops, like sugar and rice, the subsidies amounted to as much as 80 percent of farm income.[39] The aggregate agricultural subsidies of the United States, EU, and Japan (including hidden subsidies, such as on water), if they do not actually exceed the total income of sub-Saharan Africa, amount to at least 75 percent of that region's income, making it almost impossible for African farmers to compete in world markets.[40] The average European cow gets a subsidy of $2 a day (the World Bank measure of poverty); more than half of the people in the developing world live on less than that. It appears that it is better to be a cow in Europe than to be a poor person in a developing country.

The Burkina Faso cotton farmer lives in a country with an average annual income of just over $250.[41] He ekes out a living on small plots of semi-arid land; there is no irrigation, and he is too poor to afford fertilizer, a tractor, or high-quality seeds. Meanwhile, a cotton farmer in California farms a huge tract of hundreds of acres, using all the technology of modern farming: tractors, high-grade seeds, fertilizers, herbicides, insecticides. The most striking difference is irrigation—and the water he uses to irrigate the land is in effect highly subsidized. He pays far less for it than he would in a competitive market. But even with the water subsidy, even with all of his other advantages, the California farmer simply couldn't compete in a fair global marketplace were it not for further direct government subsidies that provide half or more of his income. Without these subsidies, it would not pay for the United States to produce cotton; with them, the United States is, as we have noted, the world's largest cotton exporter. Some 25,000 very rich American cotton farmers get to divide $3 billion to $4 billion in subsidies among themselves, which encourages them to produce even more. The increased supply naturally depresses global prices, hurting some 10 million farmers in Burkina Faso and elsewhere in Africa.[42]

In globally integrated markets, international prices affect domestic prices. As global agricultural prices are depressed by the huge American and EU subsidies, domestic agricultural prices fall too, so that even those farmers who do not export—who only sell at home—are hurt. And lower incomes for farmers translate into lower incomes for those who sell goods to the farmers: the tailors and butchers, storekeepers and barbers. Everyone in the country suffers. The subsidies may not have been intended to do so much harm to so many, but this is the *foreseen* consequence.

The most often-heard reason for continuing these subsidies in the United States is that subsidies are essential to maintaining the small family farmer and traditional ways of life. But the vast bulk of the money goes to large farms, often corporate ones. These subsidies have become simply another form of corporate welfare. Looking across all crops, some 30,000 farms (1 percent of the total) receive almost 25 percent of the total amount spent, with an average of more than $1 million per farm. Eighty-seven percent of the money goes to the top 20 percent of the farmers, each of whom receives on average almost $200,000. By contrast, the 2,440,184 small farmers at the bottom—the true family farmers—get 13 percent of the total, less than $7,000 each.[43] The huge subsidies—including the allegedly non–trade-distorting ones—actually drive out the small farmer. When farming becomes more lucrative because of the subsidies, the demand for land is increased, driving up the price. With the price of land so high, farming has to become capital-intensive. It has to make heavy use of fertilizers and herbicides, which are as bad for the environment as the increased output is for farmers in the developing world. As a result, small farmers, who don't have the resources for this kind of capital-intensive farming, find it attractive to sell out to large farmers and cash in the capital gain. As land increasingly moves to the large farms, with their heavy use of fertilizers, herbicides, and technology, output increases further, and those in the developing world are hurt once again.[44]

If the developed countries believe they need a transition period for the abolition of subsidies, it should be done by eliminating all subsidies to farmers making in excess of, say, $100,000, and capping subsidies to any one farmer at, say, $100,000.

Since the vast majority of those living in developing countries depend directly or indirectly on agriculture for their livelihood, eliminating subsidies and opening agricultural markets would, by raising prices, be of enormous benefit. Not all developing countries, however, would benefit. Importers of agricultural goods would suffer as prices rise. Among and within the developing countries, there would be losers and winners: farmers would be better off, while urban workers would face higher food prices. The way to solve this transitional problem would be for industrial countries to provide assistance to help the developing countries through the adjustment period—even a fraction of what they now spend on agricultural subsidies would do.

Cotton is an exception. If cotton subsidies were removed, the effect on producers would be significant but the effect on consumers would be negligible. Since the cost of the raw material represents such a small fraction of the value of a finished garment, a substantial increase in the price of cotton would hardly be reflected in the prices paid for textiles and apparel. This is one of the reasons that there is currently such a strong demand by developing countries for the elimination of cotton subsidies.

Escalating Tariffs

While reducing agricultural tariffs and subsidies has received enormous attention, that is not enough to create fairness. Tariff structures themselves need to be made pro-development. One would think that agricultural countries could can the fruits and vegetables they grow, and so earn more than they make from exporting raw produce. It would be easy to do and would create jobs. But they do not, because developed countries design their tariffs in a way that discourages this kind of industrializing, by placing higher tariffs on manufactured goods than on raw materials; the more manufacturing involved, the higher the tariff. This is known as tariff escalation.

Here is how it works. Consider as a hypothetical example an agricultural product, like oranges, that a developed country does not produce itself. Europe may let fresh oranges enter with low duties—assume it is zero—because it has a relatively small domestic orange-growing industry to protect. But it imposes a 25 percent tariff on various forms of processed oranges, from orange marmalade to frozen orange juice. Assume that half of the value of orange marmalade is in the processing, half in the orange ingredient. The tariff is clearly just a tax on processing in the developing country. There is, in effect, a 50 percent tariff on the processing activity, so that the developing country's costs would have to be much, much lower for it even to hope to compete with the canners in the developed country. Through escalating tariffs, Europe continues to receive a supply of cheap oranges while reducing the competitive threat posed to processing industries by developing countries.[45]

The market access proposal—free access for developing countries to the markets of the advanced industrial countries—would obviously solve the problem of escalating tariffs. In recent trade discussions, the developed countries have focused on getting developing countries to lower their high tariffs.[46] The focus should shift: the first priority should be the elimination of escalating tariffs. What matters is not just nominal tariff rates but effective tariff rates—tariffs on value added; and the high effective tariffs on value added by industry in developing countries should be reduced drastically.

Unskilled-Labor-Intensive Services and Migration

Developed countries are rich in capital and technology, while developing ones have an abundance of unskilled labor. What each country produces reflects its resource endowment. A country with skilled labor produces skill-intensive goods and services. The Uruguay Round expanded trade negotiations into the area of services. But, not surprisingly, it covered the liberalization of services such as banking, insurance, and information technology—all sectors in which the United States has an advantage—while leaving unskilled services, such as shipping and construction, entirely off the agenda.

Some forty countries, including the United States, have laws requiring the use of local ships for transporting goods domestically. In the United States, the Jones Act of 1920 requires not only that the ships be owned by Americans but that they be built in American shipyards and manned by Americans. (The history of protectionism goes back much further, to the first session of Congress in 1789.) America does not have a comparative or absolute advantage in shipping—indeed, as long ago as 1986, it was estimated that the Jones Act cost America more than $250,000 for every job it saved.[47] Shipping provides a wonderful opportunity for a pro–poor trade agenda that would focus on unskilled-labor-intensive services.

A similar argument arises for movements of labor and capital themselves. The developed countries are rich in capital, which moves around the world looking for the highest returns. Developing countries have an abundance of unskilled workers, who want to move around the world in search of better jobs. For the past couple of decades, the United States and the EU have pressed, with considerable success, for liberalization of capital markets, which enables investment to flow more freely around the world, arguing that this is good for global efficiency. But even modest liberalization of labor flows would increase global GDP by amounts that are an order of magnitude greater than the most optimistic estimates of the benefits of capital market liberalization. Furthermore, liberalizing migration would benefit developing countries.[48] For one thing, workers employed in the developed world send remittances back home; already billions of dollars are being sent back every year. In 2005, Mexico received an estimated $19 billion in remittances, second as a source of foreign exchange only to oil; for Latin America as a whole, remittances in 2005 were $42 billion.[49] But the cost of sending remittances can be very high, eating up a significant fraction of the amount sent. Developed countries need to facilitate the transfer of remittances to developing countries (as the United States is already doing), so that these countries can reap the full benefits of migration.[50]

Developed countries do, of course, allow the migration to their countries of high-skilled labor, because they see clearly the benefit to themselves of doing this. But as we noticed in the last chapter, this amounts to taking the developing countries' most valuable intellectual capital without compensation: after the developing countries have invested their scarce dollars in education, the developed countries, often inadvertently, try to skim off their best and brightest.

The asymmetry in liberalization of capital and labor flows leads to a further inequity. With capital markets liberalized, countries have to fight to keep capital by lowering taxes on corporations. Because labor—especially unskilled labor—is not as mobile, they don't have to fight as hard to keep it. Hence asymmetric liberalization leads to shifting the burden of taxes on to workers—leading to reduced progressivity in the tax system. The same thing happens in wage bargaining: workers are told that if they do not accept lower wages and reduced protection, the capital (with its jobs) will move overseas.

Nontariff Barriers

The reduction or elimination of tariffs does not eliminate protectionist sentiments or politics; it just forces them to find new outlets. Not surprisingly, as tariffs have come down, the advanced industrial countries have been particularly clever in erecting nontariff barriers. These take a number of forms.

Safeguards

Safeguards are temporary tariffs that can, in principle, play an important role in helping a country adjust as it faces an unanticipated large increase in the level of imports, a "surge." The tariffs keep out, temporarily, the foreign imports, providing the industry needed time to make an adjustment—for instance, to improve efficiency, or for workers to find an alternative job. Developing countries have probably not made as much use of safeguards as they should. At the other end of the spectrum, the United States has repeatedly abused safeguard measures, often employing them to protect an industry in decline—like steel— even when a surge of imports has relatively little to do with the underlying problem.[51]

The justification for invoking safeguards should not be solely the loss of jobs or sales from an increase in imports from a particular country; it ought to be shown that there is a causal link between the import surge and the industry's problems. For instance, an increase in textile imports from China, when matched by a decrease in imports from Bangladesh, should not constitute a situation requiring surge protections. And it should not be left to each country's administrative courts, with all their sensitivities to political pressures, to decide whether a safeguard tariff is justified. There should be international standards, enforced by internationally appointed tribunals. Such a tribunal, for instance, would probably not give a very sympathetic ear to American and European claims for safeguard protection from the surge of textile imports after the elimination of textile quotas in January 2005—given that there had been a ten-year transition period during which the developed countries were supposed to gradually phase out protection in order to ease transition, and during which, in fact, they did nothing.[52]

Dumping duties

The nontariff barrier most preferred by the United States has been dumping duties, which were designed to stop the peculiar unfair trade practice of selling goods below cost. While safeguard measures are temporary, dumping duties can be permanent. America has accused Mexico of dumping tomatoes, Colombia of dumping flowers, Chile and Norway of dumping salmon, China of dumping apple juice and honey. Today, Chilean wine growers worry that should they continue to be successful, California wine producers will demand that the United States impose dumping duties. Dumping duties deter entry and cast a pall over the entire market: any firm worries that, should it succeed in entering the American market with a new product, it will face dumping duties that will render it uncompetitive.

In the 1990s, Vietnam started exporting catfish into the United States, and it quickly became Vietnam's biggest export market. Soon, Vietnam had taken 20 percent of the U.S. catfish market, and furious U.S. catfish producers got Congress to pass a law stating that only U.S. catfish could be sold under the name catfish.[53] But Vietnam outsmarted the United States, reentering the American market with a new name, basa, rebranding their catfish as an upscale and exotic foreign product. Now, not only were they displacing Mississippi catfish; they were also getting a higher price. This time, the United States responded even more aggressively. Since one nontariff barrier had failed, it would use another, dumping duties, charging that Vietnam was selling below costs.

Rational firms do not sell below cost unless they believe they can thereby attain a monopoly position, which they can maintain long enough not only to recover what they have lost but to make a return on their investment (their losses from selling below cost). American anti-trust law recognizes this. Under American law, for charges of predatory pricing (as it is called when a company sells below cost to drive out a domestic rival) to be sustained, it has to be shown that there must be the prospect not only of monopolization but of maintaining that monopoly long enough to recoup the losses. Predation (true dumping) does occur, though it is rare because it is hard to establish a durable monopoly. But American law on competition from international firms does not recognize this basic economic logic. In few of the dumping cases is monopolization—let alone durable monopolization—even a remote possibility. Mexico cannot get a monopoly on tomatoes, Colombia cannot get one on flowers. Yet dumping charges are not only brought; dumping duties are levied. The reason is that costs are measured in ways that often have little to with economic realities or principles. Dumping laws are not designed to discern whether a firm is selling below its (marginal) cost; they are designed to get a high cost number so that dumping duties can be levied. No wonder, then, that rational firms so often are found to be selling below cost.[54]

Matters are even worse when a nonmarket economy is accused of dumping. (China, in spite of its progress toward a market economy, is still treated as a nonmarket economy.)[55] In the case of nonmarket economies, the costs used to calculate whether goods are being dumped are not the actual costs, but what the costs would be in some surrogate country. Those seeking to make a dumping charge stick look for a country where costs will be high, so that high dumping duties can be levied. In one classic case, in the days before the fall of the Berlin Wall, the United States levied dumping duties against Polish golf carts, using Canada as the surrogate country. Costs in Canada were so high that Canada did not produce golf carts, so dumping duties were levied on the basis of a calculation of what it would have cost Canada to produce golf carts, if Canada were to have produced them. In many places, including the EU, the surrogate country can even be the country bringing the charge—in which case, almost by definition, costs are greater, otherwise there would be no trouble competing.

One recent export from the advanced industrial countries is the use of nontariff barriers as protectionist devices. Developing countries are increasingly using them against each other. India, for instance, used Indian costs in bringing a dumping charge against China in the case of an important chemical, isobutyl benzene. In the case of low-carbon ferrochromium from Russia, India chose Zimbabwe as the surrogate country, presumably because of its high electricity prices—the key determinant of costs—and levied dumping duties on that basis.

There is a double standard. If America's own domestic standard for ascertaining predatory pricing were used internationally (when America charges a foreign firm with selling below cost), few, if any, dumping cases would succeed. If the standard the United States uses against foreigners were used domestically, a majority of American firms would be found guilty of dumping. This is an important exception to the principle that the United States heralds as so important: nondiscrimination. Foreign producers are clearly being treated differently from domestic producers.[56]

There should be a single standard for unfair trade practices, which would apply both domestically and internationally. There should be a single law dealing with dumping and with predatory pricing (as there is in the trade agreement between Australia and New Zealand). The presumption should be that firms—whether at home or abroad—do not willingly sell at a loss, and the accuser should be required to show that there was a reasonable prospect of attaining sufficient market power for long enough to recoup the losses.

Part of the problem with dumping duties, as with safeguards, is the procedures by which these duties are levied. I saw this repeatedly while I was in the Clinton administration. We would bring dumping charges (even though selling goods at a low price benefits American consumers). We would be prosecutor, judge, and jury, and the rules of evidence would have made a judge in a kangaroo court blush. The evidence relied on was often that presented by the domestic competitor, who wanted his rivals snuffed out of the market. (In 2000, the Byrd amendment provided an additional incentive: any dumping duties levied would be turned over to the affected industry—i.e., to those who brought the charges.)[57] On the basis of this information, high duties would be imposed preliminarily, causing the exporter to lose sales and go out of business. A year or two later, after a full investigation, revised and often much lower duties would be announced—but by then the damage had already been done.[58]

Again, what is needed is an international tribunal to judge whether a country is guilty of dumping (or engaging in other unfair trade practices). The current system, where each country can set its own standards and do its own cost calculations in such a way as to make a finding of dumping more likely, should be viewed as unacceptable in a world in which there is a rule of law governing trade.

Technical barriers

International trade is complex, with complicated rules that govern it, and these rules often constitute an important barrier to trade—sometimes deliberately so.

Phyto-sanitary conditions are restrictions imposed to protect human or animal life from risks arising from, say, diseases or additives in imported agricultural goods. The difficulty is in determining whether these represent legitimate concerns or are a trade barrier in disguise. The United States claims that other countries' use of such restrictions against its produce—such as genetically modified food—are trade barriers, but its own restrictions—such as the invisible fruit flies that were at one time the justification for excluding Mexican avocadoes from the United States—are reasonable. Brazil claims that restrictions on exports of fresh beef to the United States on grounds of foot-and-mouth disease are unjustified; some areas of that vast country have been certified free from the disease, yet the United States refuses to allow in any Brazilian beef shipments. The Chinese government has estimated that some 90 percent of its agricultural products are affected by technical barriers, costing it some $9 billion in lost trade.

Of all the nontariff barriers, this is the most difficult to deal with. Governments have a right—and an obligation—to protect their citizens, and distinguishing between protectionist uses and legitimate standards is not easy. Some have called for the use of "scientific" standards, but it is not even clear what should be acceptable levels of tolerance of risk. The "scientific" risk from genetically modified foods *may* be low, but a large number of people in the world still think the risk is unnecessary and unacceptable. At the very least, countries should have the right to demand labeling. The United States has argued against this, worried that labeling would discourage purchase—this is strange given its commitment in other contexts to the principles of consumer sovereignty, which is meaningful only if consumers know what they are buying.

While there is no easy answer, a system of international tribunals (as in dumping and safeguards) would at least move the deliberations out of the protectionist environments in which they are now conducted. Judges would be able to ascertain the weight of evidence. Brazilian beef might be required to be labeled as Brazilian beef, but if the scientific evidence suggests that there is no significant risk from foot-and-mouth disease from beef from the certified disease-free areas, then importation should be allowed.

Rules of origin

When developed countries give preferences to developing countries or sign free trade agreements, they want to be sure that the goods admitted are goods actually produced in the country concerned. They don't want the only thing made in Mexico on a shirt with the label "made in Mexico" to be the label itself. The rules that define what makes something Mexican or Moroccan (or any other nationality) are called "rules of origin." But in our complicated global economy, everybody is interdependent. No country makes all the components of what it sells. An apparel maker may import textiles, dyes, or buttons. The machines it uses may be imported too—along with the oil on which the machine is run. If three small countries next door work together—one doing the packaging, another the cutting, another the sewing—none may satisfy the rules-of-origin tests. An apparel manufacturer might only be able to export apparel if he uses textiles produced in his own country; a textile manufacturer might only be able to export textiles if he uses cotton grown in his own country.

Rules of origin can undo the benefits of preferences or free trade. The threshold is sometimes set at a level just high enough to deny benefits. If the exporting country itself imports the cloth, and 50 percent of the value of the shirt is the imported cloth, the importing country sets the rules-of-origin threshold at 55 percent. (Even if it is set at 50 percent, expensive shirts made with high-grade cotton will be excluded.) The United States has even used rules of origin to promote its own exports: countries that make shirts using American cotton are given preferences which those who use the least expensive cotton are not.

Sometimes the problems that arise with rules of origin are ascribed to technical glitches, but the frequency with which they arise suggest that they are used deliberately as a protectionist measure. Complicated calculations and arbitrary rules are used. Exporters are forced by these agreements to choose inputs that satisfy rules-of-origin tests rather than inputs of a given quality at the lowest price. Some producers forgo preferential treatment simply because the cost of documentation is greater than the benefit.[59]

Restricting Bilateral Trade Agreements

After the failure of Cancún, the United States announced that it would push for bilateral trade agreements. These agreements undermine the movement toward a multilateral free trade regime. As was noted, among the most basic precepts that have guided the expansion of trade has been the principle that all nations would be treated the same. The United States' bilateral trade agreements say clearly that the United States will treat some countries better than others. Often these agreements do not even expand trade—they simply divert trade from less favored to more favored countries. Sometimes they are justified by the United States as a precursor to broader multilateral agreements, but in fact these preferential arrangements make it more difficult to reach broader agreements, since inevitably such agreements will take away the privileges—and those favored with the privileges will resist.

In bilateral bargaining, the balance of power between the United States and the developing countries is even more lopsided, and the agreements signed so far reflect that. The United States has succeeded in getting some provisions into bilateral agreements that it failed to get into the Doha Round of talks, such as strengthened intellectual property rights and capital market liberalization. Sometimes developing countries sign these agreements under the illusion that, with such an agreement in place, investors will flock to their country. With Washington's seal of approval and duty-free access to the United States, there will be a boom. But sometimes, developing countries sign these agreements largely out of fear: fear, for instance, that if they don't, they will lose the preferences that they have long had, and that without preferences they will not be able to compete with the flood of imports from countries like China.[60] While a number of agreements have been signed, they are with small countries—such as Chile (population 16 million), Singapore (population 4.3 million), Morocco (population 30.8 million), Oman (population 2.5 million), and Bahrain (population 750,000)—and so involve only a tiny fraction of global trade. The bilateral strategy has thus, so far, largely failed. Meanwhile, developing countries are responding in kind, with agreements already made or in the works within Latin America and Asia. The multilateral system is in the process of fraying.

Bilateral trade agreements should be strongly discouraged; at the minimum, an independent international panel should judge whether a bilateral agreement leads to more trade diversion than trade creation. If it does, the agreement should not be allowed.

Institutional Reforms

Governance—problems in the ways decisions get made in the international arena—are at the heart of the failures of globalization. How decisions get made, what gets put on the agenda, how disagreements are resolved, and how the rules are enforced are, in the long run, as important as the rules themselves in determining the outcome of the international trade regime—and whether it is fair to those in the developing world. This is as true in the arena of trade as it is elsewhere.

The problems of unfairness start in the beginning: with setting the agenda. We have seen how the past focus on manufacturing has moved to high-skill services, capital flows, and intellectual property rights. A development-oriented trade agenda would be markedly different. First, it would remain narrowly focused on those areas where a global agreement is needed to make the international trade system work. The developing countries simply don't have the resources to negotiate effectively on a broad range of topics. And second, it would focus on areas of benefit to developing countries: unskilled-labor-intensive services and migration. There are some new topics that would be added: circumscribing bribery, arms sales, bank secrecy, and tax competition to attract businesses, all of which hurt developing countries, and all of which can only be controlled by international cooperation.[61]

The problems in governance are highlighted by the manner in which negotiations occur. The issue of openness in international discussions has long been a major concern. President Woodrow Wilson put "open covenants . . . *openly arrived at*" (my italics) at the head of his agenda for reforming the international political architecture in the aftermath of World War I, going on to argue that "diplomacy shall proceed always frankly and *in the public view*" (my italics).[62] But this has never been the case—or even a declared objective—in trade negotiations. Typically the United States and the EU would together select a few developing countries to negotiate with—often putting intense pressure on them to break ranks with other developing countries—in the Green Room at the WTO headquarters. (Today, even when the negotiatons occur in Cancún, Seattle, or Hong Kong, the room in which the representatives huddle is still called the Green Room, with all the negative connotations.) Having trade ministers closeted in a room, separated from the experts on whom they rely, negotiating all night, may be a good test of endurance, but it is not a way to create a better global trade regime. Worse still, special interests are far more likely to influence international negotiations when they are conducted under the cloak of secrecy.

The justification for these secret, high-pressure negotiations is that it is impossible to negotiate with dozens of countries at a time. That is certainly true, but there are ways to make the negotiation process fairer and to have the voices of developing countries heard more clearly.[63]

Compounding the problems of an unfair agenda and unfair and nontransparent negotiations is unfair enforcement. As we have noted, the enforcement mechanism is asymmetric. Antigua won a major case against the United States on online gambling, but there was no way that Antigua could effectively enforce the decision. Putting tariffs on American goods would simply raise prices for the people of Antigua, making them worse off. But there is a simple solution, which would go some way toward creating a more effective and fair enforcement mechanism: allowing developing countries, at least, to sell their enforcement rights.[64] Europe, for instance, might have some grievance against the United States in a pending case; rather than waiting for the outcome of that case, it could use the threat of enforcement action in the already-decided case to induce a quicker resolution.

I have laid out an ambitious set of reforms of the international trade regime, one which could make an enormous difference for developing countries. At the Millennium Summit in New York in September, the international community committed itself to reducing poverty; at Monterrey, Mexico, in March of 2002, the advanced industrial countries committed themselves to providing 0.7 percent of their GDP to help achieve this goal. If the world is genuinely committed to doing something about global poverty, and willing to give so much money to help the poor, it should also be willing to enhance opportunities—and especially opportunities for trade. The world needs a true development round, not the repackaging of old promises that the West tried to sell as a development agenda and then didn't even live up to.

Any trade agreement involves costs and benefits. Countries impose constraints on themselves in the belief that reciprocal constraints accepted by others will open up new opportunities, the benefits of which exceed the costs. Unfortunately, for too many developing countries this has not been the case. Unless the direction in which negotiations have been going in recent years is changed dramatically, more and more developing countries are likely to decide that no agreement is better than a bad one.

But what are the prospects of a fairer trade regime? Trade liberalization has not lived up to its promise. But the basic logic of trade—its potential to make most, if not all, better off—remains. Trade is not a zero-sum game, in which those who win do so at the cost of others; it is, or least it can be, a positive-sum game, in which everyone can be a winner. If that potential is to be realized, first we must reject two of the long-standing premises of trade liberalization: that trade liberalization automatically leads to more trade and growth, and that growth will automatically "trickle down" to benefit all. Neither is consistent with economic theory or historical experience.

If there is to be support for trade globalization in the developed world, we must make sure that the benefits and costs are more evenly shared, which will entail more progressive income taxation. We have to be particularly attentive to those whose livelihood is being threatened, and this will require better adjustment assistance, stronger safety nets, and better macro-economic management—so that when individuals lose their jobs, they can find better ones. We have to put in place policies that will lead wages, especially at the bottom—which in the United States have stagnated for years—to rise. Globalization will not be sold by telling workers that they can still get a job if only they lower their wages enough. Wages can rise only if productivity increases, and this will require more investment in technology and education. Unfortunately, in some of the advanced industrial countries, most notably the United States, just the opposite has been happening: taxes have become more regressive, safety nets have been weakened, and investments in science and technology (outside the military) have been declining as a percentage of GDP, as has the number of graduates in science and technology. These policies mean that even the United States and other advanced industrial countries that follow America's lead—the potential big winners from globalization—will gain less than they otherwise would; and these policies mean that more people within these countries will see themselves as losing from globalization.

With these reforms, the prospects of a globalization that will benefit most will be enhanced, and, with that, so too will support for a fairer globalization. With globalization, we have learned that we cannot completely shut ourselves off from what is going on elsewhere. The advanced industrial countries have long benefited from the raw materials they get from the developing world. More recently, their consumers have benefited enormously from low-cost manufactured goods of increasingly high quality. But they have also been affected by illegal immigration, terrorism, and even diseases that move easily across borders. For many, helping those in the developing world, those who are poorer, is a moral issue. But increasingly, those in the advanced industrial countries are recognizing that such help is also a matter of self-interest. With stagnation, the threats of disorder from the disillusioned facing despair will increase; without growth, the flood of immigration will be difficult to stem; with prosperity, the developing countries will provide a robust market for the goods and services of the advanced industrial countries.

I remain hopeful that the world will sooner or later—and hopefully sooner—turn to the task of creating a fairer, pro-development trade regime. Demands for this by those in the developing world will only grow louder with time. The conscience and self-interest of the developed world will eventually respond. When that time comes, the program laid out in this chapter will provide a rich agenda for what can and should be done.

CHAPTER 4

Patents, Profits, and People

I n the Moroccan capital of Rabat and in Paris in late January 2004, demonstrators organized by the AIDS activist group ACT UP took to the streets to protest a proposed new trade agreement between the United States and Morocco that they feared would ban Moroccan companies from manufacturing AIDS drugs. Demonstrations are still an unusual occurrence in the young democracy of Morocco, and the fact that there were protests at all said a lot about the strong feelings of the Moroccans on the matter. When I arrived in Rabat a few weeks later, people were still talking about the arrests that had resulted. A few months later, in July, protests again erupted, this time at the Fifteenth International AIDS Conference in Thailand. Activists stormed the exhibition center, forcing the major drug companies—Bristol-Myers Squibb, Pfizer, Abbott Laboratories, and the Roche Group—to close their booths.

From an economic perspective, Morocco was not the most obvious candidate for a free trade agreement with the United States. Its major export commodity, phosphate (a critical ingredient in fertilizer), which accounts for almost a fifth of its exports, is not even subject to tariffs. But Morocco hoped the agreement would boost its exports of shoes to the United States, and the United States hoped closer economic ties would build friendship.[1] Of the agreement with Morocco, Robert

Zoellick, America's chief trade negotiator, proudly boasted, "This free trade agreement . . . signals our commitment to deepening America's relationship with the Middle East and North Africa."[2] This was especially important in the Middle East, where, in other respects, America's foreign policy was controversial, to say the least. By cooperating with moderate Arab governments, the United States hoped to build good-will in the region.

It turned out, however, that getting the Office of the U.S. Trade Representative to forge an international friendship had its problems, reflected starkly in the protests that ensued. Moroccans involved in the talks told me there wasn't much negotiation involved. The U.S. negotiators were mostly interested in having it their way—and they wanted the new agreement to protect U.S. drug companies. It came down to a matter of life versus profits. The U.S. government, reflecting the interests of its drug companies, insisted that the agreement include provisions that would delay the introduction of generic drugs, and it got what it wanted.

As in the United States and elsewhere in the world, generic drugs in Morocco cost a fraction of brand-name drugs. American drug companies know that as soon as the generics come in, their profits will plummet. So they have devised a number of clever strategies to delay the introduction of generics into the market, including restricting the use of data that proves the safety and efficacy of the drug—and preventing the generic firms from even beginning to produce the drugs until the patent expires.[3] The protestors were especially fearful of delays in the introduction of generic AIDS drugs, delays that would leave most patients unable to afford medicines that could save their lives. Some NGOs argued that the restrictions on generics in the agreement could increase the effective duration of patent protection to nearly thirty years, from its current twenty years, and would make generic drugs even less accessible in Morocco than they are in the United States.[4] It's not clear whether this will happen, or precisely how many people could die as a result.[5] But given how hard the U.S. government pushed, one has to believe that these measures will extend the effective patent life significantly—increasing profits and decreasing access to lifesaving medicines.

This was not the first controversial trade agreement signed in Morocco. It was in Marrakech that the Uruguay Round agreements were finally signed by the trade ministers on April 15, 1994. Among them was an agreement on Trade-Related Aspects of Intellectual Property Rights (TRIPs), which had been long sought by the United States and other advanced industrial countries in order to force other countries to recognize their patents and copyrights.[6] Patents give inventors monopoly rights over their innovations. The higher prices are supposed to spur innovation—whether they do so is a question to which we will turn later in the chapter. But TRIPs was *designed* to ensure higher-priced medicines. Unfortunately, those prices made medicines unaffordable to all but the wealthiest individuals. As they signed TRIPs, the trade ministers were so pleased they had finally reached an agreement that they didn't notice they were signing a death warrant for thousands of people in the poorest countries of the world.

To critics of globalization, the fight over intellectual property is a fight over values. TRIPs reflected the triumph of corporate interests in the United States and Europe over the broader interests of billions of people in the developing world. It was another instance in which more weight was given to profits than to other basic values—like the environment, or life itself. It has also become symbolic of the double standard, the difference in attitudes toward these values domestically and abroad. At home, citizens often demand that their elected representatives go beyond a focus on profits, to look at the effects on other aspects of their society and the environment. Even as the Clinton administration was engaged in a grand battle to enhance access to health care for Americans, by supporting TRIPs it was reducing access to affordable drugs for poor people around the world.

I believe that the critics of TRIPs are right.[7] But the criticism of the intellectual property regime goes even further: it may not even be in the broader interests of the advanced industrial countries. I pointed out in chapter 1 that one of the objections to globalization, *as it was being managed*, was that it foisted on the world, including the developing countries, a particular version of the market economy—a version that might not be well suited to their needs, values, and circumstances. TRIPs presents an example par excellence: it is based on the view that stronger intellectual property rights lead to better economic performance. Particular American and EU corporate interests, using this as a rationale, have attempted to use trade agreements to force developing countries to adopt intellectual property laws that are to their liking.

Innovation is important; it has transformed the lives of everyone in the world. And intellectual property laws can and should play a role in stimulating innovation. However, the contention that stronger intellectual property rights always boost economic performance is not in general correct. It is an example of how special interests—those who do benefit from stronger intellectual property rights—use simplistic ideology to advance their causes. This chapter explains how poorly designed intellectual property regimes not only reduce access to medicine but also lead to a less efficient economy, and may even slow the pace of innovation. The enervating effects are particularly acute in developing countries.

There will always be a need to balance the desire of inventors to protect their discoveries, and the incentives to which such protection gives rise, and the needs of the public, which benefits from wider access to knowledge, with a resulting increase in the pace of discovery and the lower prices that come from competition. In this chapter, I explain what a balanced intellectual property regime—one that pays attention not only to corporate interests but to academia and consumers—might look like. Drug companies claim that without strong intellectual property protection, they would have no incentive to do research. And without research, the drugs that companies in the developing world would like to imitate would not exist. But the drug companies, in arguing this way, are putting up a straw man. Critics of the intellectual property regime are, by and large, not suggesting the abolition of intellectual property. They are simply saying that there is a need for a better balanced intellectual property regime.

It is important to spur innovation, which includes lifesaving drugs designed to combat the diseases that afflict developing countries; I will describe alternatives that would achieve this more effectively than the current system does, and at lower cost. The reforms I suggest will make globalization work better—I believe not only for the developing countries but for the developed world as well.

INTELLECTUAL PROPERTY: ITS STRENGTHS AND LIMITS

Intellectual property rights give the owner of that property the exclusive right to use it. It creates a monopoly. The owner of the property can, of course, allow others to use it, usually for payment of a fee. The protection of intellectual property is designed to ensure that inventors, writers, and others who invest their money and time in creative activity receive a return on their investment. The details of laws that covers different kinds of intellectual property differ. Patents, for instance, give an inventor the exclusive right to market his innovation for a limited period of time, currently twenty years. No one else can sell the product without the permission of the patent holder, even if a second person discovers it on his own. In return for the patent, the patent applicant must provide extensive disclosure of the details of his invention. Copyrights give the writer of a book or the composer of a song the exclusive right to sell that book or song for a much longer period—in the United States, the length of the author's life plus seventy years.

But intellectual property rights are fundamentally different from other kinds of property rights. If you own a piece of land, you can do with it as you please, so long as you remain within the law: obeying zoning requirements, not establishing a brothel, or—most important for our purposes—not conspiring with others who own similar properties to create a monopoly that, left unchecked, may lower economic efficiency and threaten the public welfare. Property rights provide incentives to take care of your property and to put it to its best use, but rights are not unfettered; uses that impede economic efficiency (like monopolization) or infringe on the well-being of others (like using property for a toxic waste dump in the middle of a city) are restricted.[8]

By contrast, intellectual property rights actually create a monopoly.[9] The monopoly power generates monopoly rents (excess profits), and it is these profits that are supposed to provide the incentive for engaging in research. The inefficiencies associated with monopoly power in the use of knowledge are particularly serious, because knowledge is what economists call a "public good": everybody potentially can benefit

from it; there is no cost of usage.[10] Thomas Jefferson, the third president of the United States, put this far more poetically when he
described knowledge as being like a candle—as it lights another candle, the light of the original candle is not diminished. Economic efficiency means that knowledge should be made freely available, but the
intellectual property regime is *intended* to restrict usage. The hope is
that the inefficiencies of monopoly power are counterbalanced by
increased innovation, so that the economy grows faster.

There is another difference between intellectual property and ordinary property. In the case of ordinary property, say a tract of land, there
is normally no difficulty in defining what it is the individual owns. A
property deed specifies it precisely. It may also specify certain covenants
(restrictions on use) or rights of way, detailing the rights of others to
use the land. Defining the boundaries of intellectual property is far
more difficult. Indeed, even determining what is patentable is difficult.
One criterion is novelty: the invention has to be "new." One can't
patent some idea that everyone knows but no one had bothered to
patent. That might provide rewards for patent lawyers, but it does not
spur innovation.[11] What is original? Almost every idea is based on previous ideas. Does a small wrinkle on a well-known idea deserve a
patent, or even a large wrinkle if that wrinkle was obvious? At the turn
of the previous century, George Baldwin Selden applied, and got, a
patent for a four-wheeled self-propelled vehicle.[12] It was, perhaps, an
obvious idea—certainly, if we look around the world, many people
seem to have come up with the same idea at the same time. In Germany, Gottlieb Daimler is widely given credit for the invention.
Should Selden have been given a patent? And if so, should his patent
embrace *any* self-propelled vehicle, or only his particular design?

There is no obvious answer to these questions—but any country
must, in its intellectual property laws, provide answers, and the answers
have enormous consequences. The greater the scope for intellectual
property (the more things that can be patented, and the broader the
patents), the greater the returns to those who get the patent—and the
greater the scope for monopoly, with all its attendant costs. If patents
are made as broad as possible, which is what patent seekers want, there
is a real risk of privatizing what is within the public domain, since some

(possibly much) of the knowledge covered by the patent is not really "new." At least part of what is being patented, and therefore privatized, is knowledge that previously existed—part of common knowledge, or at least the common knowledge of experts in the area. And yet, once the patent has been granted, the owner can charge others for using that knowledge.[13]

Some critics have compared the recent strengthening of intellectual property rights to the earlier enclosure movement in the late Middle Ages in England and Scotland, when common (public) land was privatized and taken over by the local lords. There is one important difference with what is happening today: though the people thrown off the land suffered tremendously, there was some improvement in efficiency as the nobility used the land more carefully and did not overgraze as the peasants had. Economists would describe this as a classic equity/efficiency trade-off. But with the enclosure of the intellectual commons, there is a loss in efficiency.[14]

Indeed, monopolization may not only result in static inefficiency but reduced innovation. A patent that covered all four-wheeled self-propelled cars—that would have granted Selden a monopoly on the automobile—would have left little room for Henry Ford's innovation of an affordable car. Monopolies insulated from competition are not subject to the intensive pressures that drive innovation. Worse still, they can use their power to squash rivals, reducing the incentives of others to do research. The U.S. software giant Microsoft has used the monopoly power that its intellectual property has created to trample innovators like Netscape and RealNetworks.[15] While some innovators are brave, or foolish, enough to think that if they are fortunate enough to come up with a great innovation, they can challenge Microsoft, and others are satisfied simply with the prospect of being bought out, many others, seeing the obvious dangers, are discouraged from producing innovations valuable enough to attract the attention of Microsoft. Even when courts stop the anti-competitive practices, it is hard to re-create a competitive marketplace, especially when powerful patents remain. In these cases, intellectual property results in a lose-lose situation: the economy loses in the short run, as the higher prices of monopoly lower welfare, and in the long run, as innovation too is lowered.

Academics who study intellectual property rights understand the risks and costs of monopolization, because they are familiar with how it has played out in history. For instance, I noted earlier that at the beginning of the last century George Baldwin Selden obtained a patent on all four-wheeled self-propelled vehicles, and in 1903 a group of car manufacturers formed a cartel around this patent, calling themselves the Association of Licensed Automobile Manufacturers (ALAM). As the owner of the patent, ALAM could control who was allowed to manufacture automobiles and who was not—and only those who were willing to collude to maintain high prices were allowed to produce. Were it not for Henry Ford, they likely would have succeeded in controlling automobile production, and the development of the modern automobile industry would have been quite different. Ford's conception of a "people's car"—a vehicle affordable to the masses, selling at far less than the then prevailing prices—ran contrary to ALAM's intention of using the cartel to maintain high prices. Fortunately, Ford had the economic wherewithal to successfully challenge the Selden patent.[16]

More generally, because patents impede the dissemination and use of knowledge, they slow follow-on research, innovations based on other innovations. Since almost all innovations build on earlier innovations, overall technological progress is then slowed.

When there are multiple patents covering various ideas that go into an innovation, the patent system can become an even bigger impediment to innovation. This is sometimes described as a "patent thicket." Progress in the development of the airplane was impeded in the early years of the twentieth century because of the difficulties in sorting out the patents of the Wright brothers and Glenn H. Curtiss. Without the agreement of both, any development risked some patent infringement. With the onset of World War I, the cost of delay became intolerable: airplanes would make a decisive difference in the war's outcome. The government forced a resolution, forming a "patent pool." Anyone using the ideas would pay the pool, and the administrators of the pool would divide the revenues among the holders of the relevant patents, in accordance with their judgment of the relative importance of the various ideas in the final product.[17]

Finally, the patent system may reduce productive innovation by diverting much of a company's expenditure toward either increasing monopoly power or getting around the patents of others. Microsoft has incentives to develop ways to reduce interconnectivity—the ability of others to use its operating system to write competing applications to, for instance, its Office suite, its browser, or its media player. Drug companies expend huge amounts of money coming up with drugs that are similar to existing drugs but are not covered by existing patents; even though these drugs may be no better than the existing ones, the profits can be enormous.[18] This may explain the seeming inefficiency of the big drug companies, which, despite huge total expenditures, have come up with relatively few drugs that are more than a minor improvement on previous drugs.[19]

While we have argued that excessively strong intellectual property rights may slow innovation, advocates of strong intellectual property rights suggest, to the contrary, that they promote research. When they do recognize the dangers of less research (as in the case of Microsoft), they respond by allowing that in abusive cases it should be restricted, as the U.S. government did with AT&T, America's onetime telephone monopoly, when it forced it to license all of its patents to others. But they often go further, arguing that without intellectual property protection there will be no research at all. In this claim, they are clearly wrong: countries without intellectual property rights—Switzerland had none until 1907, the Netherlands until 1912—were highly innovative.[20] Intellectual property is part—but only part—of a country's "innovation system."

Today, the world of innovation is far different from what it was a century ago. The days of the solitary inventor working on his own are, by and large, gone, although there are still apocryphal stories such as that of Hewlett and Packard working in their garage. To oversimplify, basic ideas bubble out of research universities and government-funded research laboratories: both major breakthroughs, like understanding the genetic structure of life or lasers, and smaller ones, such as advances in mathematics, surface physics, or basic chemistry. Sometimes these get translated into specific products and innovations by university researchers; commonly, however, corporations do this work. Traditionally, intellectual property has played little role in promoting basic science. Academia believes in "open architecture," meaning that the knowledge that research produces should be made public to encourage innovation. The great scientists are driven by an inner quest to understand the nature of the universe; the extrinsic reward that matters most to them is the recognition of their peers.

One of the reasons that basic research is advanced most by *not* resorting to intellectual property is that while doing so would have questionable benefits, the costs are apparent.[21] Universities thrive on a free flow of information, each researcher quickly building on the work of others, typically even before it is published. If every time a researcher had an idea, he ran down to the patent office, he would spend more time there—or with his lawyers—than in his lab. Interestingly, even in software, this system of open collaboration has worked. Today we have the Linux computer operating system, which is also based on the principle of open architecture. Everyone who participates is required to accept that it is an open source, a dynamic program that is being constantly improved by thousands of users. A free, viable alternative to Microsoft's operating system, it is expanding rapidly, especially in developing countries. An offshoot of Linux, the browser Mozilla Firefox, has been growing even faster. Not only is it free, but it seems to be less subject to the security problems that have plagued Microsoft's Internet browser.[22] The worry is that inevitably Linux will encroach on one of the hundreds of thousands of patents that have been granted, and the holder of the patent will attempt to hold the entire Linux system up for ransom. Even if the patent is eventually shown not to be valid, the economic costs can be enormous, as Research In Motion (the company that makes the BlackBerry) found out as it was forced to pay over $600 million, not to the inventor, but to a company that had obtained the patent on the cheap—a patent that had already been disqualified in Germany and the U.K.

Designing a balanced intellectual property regime

Designing an intellectual property regime entails answering difficult questions about what can be patented, how long the patent should last, and how broad the patent should be.[23] The answers affect both the extent of competition in the economy and the level of innovation. The longer life of copyright makes sense for two reasons. The monopoly is only over, say, a particular novel, and readers have a multitude of novels among which to choose. Copyright covers just the particular form of expression: another writer can express the same idea in a slightly different way without infringing copyright. The table of contents of a book is not protected by copyright, even though the organization of materials in a textbook may represent its most important intellectual contribution. Normally, copyrights—which apply predominantly to books, artworks, music, and movies—do not give rise to significant monopoly power. Hence, strong intellectual property rights in this arena is appropriate: it provides incentives without significant adverse costs of monopolization.

We noted that many of the most important ideas of basic science and mathematics—mathematical theorems, for instance—cannot be patented, and, I believe, rightly so: the cost in terms of discouraging follow-on innovation would be enormous, and the benefit would be small.

In recent years, there have been attempts to expand the scope of intellectual property, allowing more things to be patented and patents to be broader. It is here that controversies rage. In India there is a lot of anger over the recent patenting of some yoga positions. Is the use of the stroke Q to denote quitting a program enough of an intellectual breakthrough to justify a copyright or patent? Should Amazon.com be able to patent the idea that you can make an order with a single click? These are, to my mind, not the kind of major intellectual breakthroughs that deserve patenting, and they entail a high cost: inhibiting the development of standards that enhance efficiency and competition.

In another example, consider the controversy over patenting a gene: the instructions inside each living being that tell it what proteins to produce, which, for example, determine growth and affect susceptibility to disease. Knowledge of the genetic code can be of enormous benefit in finding cures and vaccines. This was one of the reasons that such importance was placed on decoding the entire genetic structure, which was eventually completed in 2003 by the publicly funded international Human Genome Project (HGP). While the systematic decoding was in progress, there was a race by several private sector firms, including Human Genome Sciences (HGS) and Celera Genomics (headed by Craig Venter, who had earlier worked on the project at the National

Institutes of Health). In the rush to patent, claims were filed on some 127,000 human genes or partial human gene sequences—confronting patent offices around the world with an impossible task and resulting in huge backlogs. HGS filed some 7,500 applications; Celera, 6,500; and a single French firm, Genset, 36,000.[24] Eventually, the U.S. Patent and Trademark Office ruled that while it would grant patents for genes, it would do so only for entire sequences and only if the usefulness of the gene was demonstrated.

Many found the whole idea of patenting genes abhorrent. After all, the researchers did not invent the gene; they only identified what was already there. Moreover, since the publicly financed HGP has succeeded in decoding the entire human genome, there was really little value added by a race to decode a part of or even the whole genome a little bit faster. The lock on knowledge resulting from the granting of a patent might well impede follow-on research, or even applications. Some of these fears seem to have come true: for instance, Myriad Genetics, which has patented two human gene mutations affecting susceptibility to breast cancer, has demanded that even not-for-profit labs screening for mutations pay a license fee, thus discouraging screening.[25] Myriad Genetics' patent, and its willingness to enforce its patent claims, may have discouraged the search for better screening technologies, since anyone discovering a new method faced the uncertainty of how much Myriad would demand in payment.[26]

The answers to questions of what should be patented and how broad and how long the patent should be are not obvious, and there is no reason that answers that are right for one country, for one sector, for one period, should be right for another. More recently, the software industry has begun to rethink its earlier advocacy of intellectual property. The industry has seen how developments by one party risk infringing on another party's patent. The creator of any software program may inadvertently trespass on someone else's ideas—not because he has stolen the ideas, but because he has rediscovered them. With more than 120,000 patent applications every year, it is virtually impossible for any researcher to know every idea that has been patented or for which there is a patent pending.[27] Inherent ambiguities—for instance, in the breadth of the patent (that is, whether, to use our earlier example, Selden's patent did indeed include all cars)—make a difficult task impossible. The result is that even the person usually given most credit for inventing the World Wide Web, Tim Berners-Lee, has concluded that, at least in his field, patents stifle innovation. They present, he says,

> a great stumbling block for Web development. Developers are stalling their efforts in a given direction when they hear rumors that some company may have a patent that may involve the technology.[28]

Over the past hundred years, the laws have changed enormously and differ across countries. The changes and differences reflect changes and differences in the economy, including changes and differences in the trade-offs between monopolization and innovation. A well-designed intellectual property regime balances the costs of monopolization and the benefits of innovation, by, for instance, limiting the period of the patent, requiring disclosure of the details so that others can build on them, and limiting the ability to use patents for "abusive" monopoly power.[29] Earlier, we saw how the U.S. government did this in the case of AT&T. Just as the way those trade-offs are balanced changes over time, they differ between developing and developed countries. When patent systems answer the questions of what can be patented and how broad patents should be in the wrong way, competition is reduced and innovation is inhibited. If the patent is overly broad, there will be less incentive to do research building on the existing innovation.

The changes in intellectual property regimes in recent years reflect not only changes in the economy but also changes in the political influence of corporate interests. Large corporations like monopoly—it is far easier to sustain profits by having a strong monopoly than by continually increasing efficiency; and so to them, monopolization is a pure benefit, not a societal cost. Though one might have hoped that legislatures and courts would have carefully balanced the costs and benefits of each provision, in practice intellectual property law has evolved in a much more haphazard way. But there is one major trend: the corporate interests that care intensely about intellectual property have succeeded in getting more and more of what they want. Many within the United States—myself included—believe it has gone too far.[30]

TRIPs

This was exemplified by the influence of these corporate interests in the adoption of the TRIPs agreement within the WTO. As the TRIPs agreement was being negotiated in Geneva in 1993, the Council of Economic Advisers and the Office of Science and Technology Policy in the White House tried to make the American negotiators understand our deep reservations. What the United States was asking was, we thought, not in its own interests, nor in the interests of the advancement of science, and was certainly not in the interest of developing countries. But American and European negotiators adopted the positions of the drug and entertainment industries, and others who simply wanted the strongest intellectual property rights. (A study by the Center for Public Integrity, a government watchdog group, showed that the drug industry was the single most important influence group at the Office of the U.S. Trade Representative.)[31] They insisted, for instance, on longer patents, without weighing the costs of an extended period of monopolization against the benefits.[32]

Not surprisingly, given the respective bargaining power of those at the table, the agreement that emerged was close to that demanded by special interests in the United States. Time was all the developing countries won—a few years until the intellectual property provisions would come into force—and, seemingly, *some* flexibility in, for instance, compulsory licensing of drugs in the event of a health crisis like AIDS. (With a compulsory license, the generic manufacturer is allowed to manufacture the needed drug without the consent of the patent holder, though typically there is a standard royalty rate. This obviously erodes monopoly power, which is why the patent holder refuses to grant the license voluntarily.)

Intellectual property does not really belong in a trade agreement. Trade agreements are supposed to liberalize the movements of goods and services across borders. TRIPs was concerned with a totally different issue—in some sense, it was concerned with *restricting* the movement of knowledge across borders. So to shoehorn it into the trade agreement, trade negotiators added two words, "trade related." TRIPs may stand for Trade-Related Intellectual Property, but the name is misleading: there is essentially no aspect of intellectual property that, in their view, is not related to trade.

In fact, there already existed an international organization to deal with intellectual property: the World Intellectual Property Organization (WIPO), one of the specialized agencies of the United Nations. It was established in its current form in 1970, although, in fact, international cooperation in this area dates back more than a hundred years, to 1893.[33] But WIPO has a critical limitation: it has no enforcement mechanism. There was little the United States or the EU could do to a country that did not respect intellectual property rights. Under TRIPs, the advanced industrial countries could at last use trade sanctions to legally enforce intellectual property rights, and the drug and media industries were ecstatic.

There are, of course, other international organizations that have achieved international agreements which are hard to enforce without trade sanctions. The International Labour Organization, for example, has forged a global agreement on core labor standards, forbidding, for instance, the use of child and prison labor. Whether a country complies with these labor standards can of course affect trade. For example, we could certainly have had a trade-related labor standards agreement. But the economic interests of developed countries' major multinationals were not as dependent on labor issues as they were on intellectual property. Quite the contrary: it was in the economic interests of American multinationals that an international trade agreement *not* regulate these other areas.[34]

TRIPs imposed on the entire world the dominant intellectual property regime in the United States and Europe, as it is today. I believe that the way that intellectual property regime has evolved is not good for the United States and the EU; but even more, I believe it is not in the interests of the developing countries.

MAKING GLOBALIZATION WORK

Promoting innovation and social justice

Intellectual property is not an end in itself, but a means to an end: it is supposed to enhance societal well-being by promoting innovation. But can we have more innovation with more social justice? Can we have it at lower cost to developing countries? I believe we can. First, however, we have to see more clearly what it is that we seek to achieve. In Geneva in October 2004, the WIPO General Assembly adopted a resolution put forward by Argentina and Brazil for a development-oriented intellectual property regime—just as the international community had, three years earlier, adopted the principle of a development-oriented trade regime.[35] While they agreed that providing incentives for innovation is critical, they had other concerns as well.

One of the most important issues facing the entire world today is poverty in the Third World. Developing countries need more resources—i.e., more assistance—and more opportunity (the focus of the previous chapter, creating a fairer trade regime). But, as I observed in chapter 2, what separates developed from developing countries is not just a gap in resources but a gap in knowledge; and the intellectual property regime can make closing that knowledge gap either easier or more difficult. The developing world's plea was for an intellectual property regime that provided them more access to knowledge. Furthermore, with their limited budgets for health—a dollar spent on drugs was a dollar not spent on education or on development—the cost of medicines matters enormously, which is why access to lifesaving medicines at affordable prices is so important.

New drugs and vaccines can, of course, make a big difference to the well-being of those in the developing countries. But the current system has not been working—it has not been investing in research to produce the drugs to attack the diseases that are prevalent in developing countries, and, not surprisingly, few drugs have been been produced. We need to reform the global innovation system to encourage the development of medicines that treat and prevent such diseases.

Finally, TRIPs did not provide adequate protection for traditional knowledge.

The following program details how these concerns of the developing countries can be addressed.

Tailoring Intellectual Property to the Needs of the Developing Countries

The world has finally learned that one-size-fits-all development strategies do not work. The same is true of intellectual property regimes. There are benefits and costs to standardization. In the United States, many areas of law are left to the states; the benefits of having a national criminal code are thought to be less than the costs. TRIPs attempts to impose a single standard for intellectual property law on the world. I believe that the costs of that standardization far outweigh the benefits. Intellectual property laws always reflect the balancing of the benefits of innovation and the costs of monopolization; and because the circumstances of developed and developing countries differ, how the trade-offs are balanced differs. With, for instance, the dangers of monopolization in small developing countries greater than in large developed countries—because markets are smaller and more frequently dominated by at most a limited number of firms—the costs of an intellectual property regime are greater while the benefits are smaller. We should push for separate intellectual property regimes for the least developed, the middle-income, and the advanced industrial countries. Just as I argued in the last chapter that developing countries should be given more scope in deciding what kind of industrial policies are appropriate—giving them more opportunity to help create new industries—so too should these powers be granted in the arena of intellectual property.

One of the costs of standardization is the risk that a wrong standard will be chosen; when each chooses its own, each jurisdiction can be thought of as a laboratory testing different ideas; those that work best will be imitated. Still, if there is to be a single standard—or at least a minimum standard—imposed on the entire world, it must be adjusted to reflect more of the interests and concerns of the developing countries. The developing countries have been demanding a revision of TRIPs, a "TRIPs minus" agreement, and they are right.[36]

Access to Lifesaving Medicines

Few in the developing world can afford drugs at the monopoly prices that Western pharmaceutical companies charge—prices that are often many times higher than the costs of production. To an economist, this disparity between price and production cost is simply an economic inefficiency; to an individual with AIDS or some other life-threatening disease, it is a matter of life and death. Three reforms would enhance greater access to existing lifesaving medicines. One, discussed at greater length below, is for the advanced industrial countries simply to provide the drugs, or at least subsidize them—in effect paying the "tax," the difference between price and marginal cost.

Medicines at cost to developing countries

One of the simplest ways for the developed countries to help developing countries is to "waive" the tax, allowing them to use the intellectual property for their own citizens, so that their citizens can obtain the drug at cost. Critics might say: But then the developing countries are simply free-riding on the advanced industrial countries. To which the answer is: Yes, and they should. There is no additional cost imposed on the developed countries.[37] And the benefits to the developing countries would be enormous: increased health is not only of value in its own right, but it would contribute to increased productivity.

A start in this direction has already been made. Students at some research universities are arguing that the universities should insist that, as a part of their licensing agreements with drug manufacturers, drugs be provided to developing countries at deeply discounted prices.

Compulsory licenses

In special situations, governments can issue compulsory licenses when they decide there is an urgent need to broaden access to technology or medicines. This right is recognized by almost every government in the world. During the 2001 anthrax scare, the U.S. government threatened to force the drug company Bayer to allow others to produce Cipro, the antibiotic most effective against anthrax at that time.[38] Once they get a compulsory license, firms can produce a drug and sell it competitively at just above cost. Since many generic drug manufacturers in the developing world are highly efficient, licensing makes drugs available at often a fraction of the price at which they would otherwise be sold. For instance, Brazil's state-run drug company, Farmanguinhos, estimates that it can produce the AIDS medicine Kaletra for a fraction of what Abbott charges in the United States. With more than 600,000 HIV-positive patients in the country, at one time, it was estimated that a generic Kaletra would save Brazil some $55 million off even the highly discounted price at which Abbott was then selling the drug to Brazil.

The hope of the big drug companies was that TRIPs would make it more difficult for generic versions of their drugs to be produced.[39] When, in the late 1990s, Brazil and South Africa floated the idea of issuing compulsory licenses for AIDS medicines, the American drug companies were outraged, claiming that TRIPs didn't allow this even for AIDS drugs, and filed a complaint with the World Trade Organization.[40] When a public outcry forced them to compromise, they offered the drugs at a discount that was still far above the price at which generics can be produced, as the example of Kaletra illustrates. But while Brazil has been able to bargain for a better deal for itself by threatening to issue a compulsory license, other developing countries, less astute in their bargaining and without the capacity to produce generics on their own, are left paying very high prices.

The drug companies also argued that TRIPs did not allow trade in generic drugs produced under compulsory licenses. This rendered the licensing provision useless to developing countries—like Botswana, a small country with more than a third of its population afflicted by HIV-AIDs—that have little or no manufacturing capability of their own. They wanted to be able to buy the generic AIDS drugs from neighboring South Africa. Again, public support rallied around these countries and their plight, particularly those in Africa dealing with the AIDS pandemic.[41] Yet even after the rest of the world realized that these policies were unconscionable, the Bush administration continued to hold out for the drug companies' interests. Only shortly before the Cancún meeting, in August 2003, did it concede. Even then, though, the United States insisted on what critics viewed as a cumbersome administrative process.

The United States had, in fact, wanted more: it had wanted to restrict compulsory licenses only to cases of epidemics or similar catastrophes. Of course, the individual about to die due to a lack of access to medicine that could be made available to him at a cost he can afford does not care whether or not he is one of 10,000 or 600,000 who are dying. He only knows that his death is unnecessary. The critical distinction should be between lifesaving drugs and cosmetic and lifestyle drugs, for which there is no compelling reason to issue compulsory licenses. But for the U.S. drug industry, the focus was on profits, which meant doing everything to keep as many generic drugs as possible off the market for as long as possible.[42]

The American drug companies argue in justification for their stance that any attempt to allow trade in generic drugs—for example, allowing South Africa to export to Botswana—will mean the drugs will eventually come into the United States and spoil the market there. But there are already huge disparities in prices (for instance between prices in Europe and the United States), and the problem, while present, is limited. The pharmaceuticals industry is one of the most regulated in the world, with most of the cost of drugs being paid by insurance companies and governments—so incentives to buy drugs at European prices are weak, and it is not easy to do so. It is even less likely that Americans (or Europeans) will get their drugs from South Africa or Botswana.[43]

If developed countries do not sell lifesaving drugs to developing countries at the cost of production of the drugs, then developing countries must be given a green light to use compulsory licenses, producing and trading lifesaving medicines.

Research

Higher prices are supposed to spur research for lifesaving medicines. But in spite of the rhetoric about intellectual property providing incentives, the incentives have not been translated into action. The argument that the monopoly pricing of drugs leads to more innovation is undermined by the fact that most drug companies spend far more on advertising than on research, more on research for lifestyle drugs (e.g., drugs for hair growth or male impotence) than for disease-related drugs, and almost none on research for the diseases prevalent in the poorest countries, such as malaria or schistosomiasis.[44]

The current system of funding research is inequitable and inefficient. Under the current system, basic research is funded by the government and the private sector brings the drugs to market. Once the drugs come to market, the companies make a huge profit. The difference between the price charged and the (marginal) cost of production can be viewed as a tax on their customers. But it's a very regressive tax. Generally, governments levy taxes in relation to the ability to pay, but with medicines the same tax is levied on the poorest in the developing countries and the richest in the developed world. We noted earlier that knowledge is a public good and that restricting knowledge leads to inefficiency—a lower pace of innovation. Here the cost is more serious: life itself. With such a high cost and so little benefit from the current arrangement, we have to ask, can we reform the way we produce and finance research for lifesaving medicines?

The drug companies go so far as to claim that providing developing countries more low-cost access to lifesaving drugs will actually hurt them in the long run. They argue that if they can't get a return on their investments they will do less research, which would ultimately damage everyone. But providing these countries with access to lifesaving drugs will have, at most, a negligible effect on the drug companies' investment in the diseases that affect poor countries. The drug companies garner little revenue from developing regions anyway—African sales represent under 2 percent of the total—because the people are simply too poor to buy expensive drugs; and, as we have seen, they spend little on the diseases that most affect developing countries.

There should, of course, be more research on the diseases afflicting developing countries; but the best and most cost-effective way to promote this is not by implementing more stringent intellectual property rights. It is clear that market incentives haven't been working, and, by themselves, are not likely to do so. Most of the money for financing research will have to come from the governments and foundations of developed countries, particularly in the North. The question is how best to provide the money and organize the research. There are at least two ways in which support can be given.[45]

A market-based incentive: a guarantee fund

One proposal is to have developed-world governments make a purchase guarantee. If a vaccine against AIDS is invented, for instance, the governments and foundations providing the guarantee might pledge to spend at least $2 billion buying the drug. Or if a more effective drug against malaria than what exists at present is discovered, they might pledge to spend at least $3 billion.

The one major problem with this idea is that it would leave the problem of monopoly in place: drug companies would still have an incentive to raise prices and curtail production in order to maximize their revenues, rather than maximizing the social benefits. Also, because no one wants a medicine that is a bit less effective even if it is cheaper, this would be a winner-take-all system: a company that makes a just slightly better product will get all the sales and rewards.

An innovation fund

More effective would be a fund that directly encourages innovations of benefit to developing countries. A prize system, in which researchers are rewarded for the value of their innovations, would move incentives in the right direction. Those who make the really important discoveries—who, for example, tackle diseases with no known cure— would get big rewards. Big rewards too would go to those who research widespread and socially costly diseases, like tuberculosis and malaria, while little would go to a company making a "me-too" drug, which just slightly improves on an already existing medicine.[46] Under this system, drugs could be delivered (through generic producers) *at cost* to those suffering from disease. Not only would the developing countries benefit, but so would the developed ones, as their citizens would benefit from improved knowledge.[47] A bonus is that developed-country governments would be able to aid the developing world without worrying whether the money will be well spent.[48]

Stopping Bio-Piracy and Protecting Traditional Knowledge

The unfairness of the whole intellectual property regime against the developing countries is seen most starkly in the treatment of traditional medicines and drugs based on plant-derived chemicals. I first learned about the problem of bio-piracy in a remote village in the Ecuadorian high Andes, where the village head described how TRIPs was affecting their lives. To Americans and Europeans, TRIPs may be an arcane subject, of relevance mainly to corporate litigators and international trade specialists, but in the developing countries it is much more real. Developing countries see foreign corporations taking their traditional knowledge and their native plants without compensation as a form of piracy—hence the term "bio-piracy." While the United States complains that China is violating TRIPs by not honoring its intellectual property provisions, those in the developing world point out that TRIPs did nothing to protect their intellectual property. Rather it gave U.S. and European corporate interests a license to steal their intellectual property—and then charge them for it.

Traditional medicines have long been used all over the world to treat a wide variety of ailments. While modern science at first looked askance at folk remedies, more recently it has become clear that many of them survived because they really do work—even if those who use them, or the folk doctors who administer them, do not know why. One strand of modern medical research has focused on isolating and then marketing the active ingredients in these remedies, recognizing that there is a wealth of potential cures in the world's flora, particularly in tropical countries. The drug companies, recognizing the profit potential, have followed, "rediscovering" what was long ago discovered by traditional cultures—and in some cases doing no more than rebranding it. The developing countries, seeing the drug companies profit from their rich biodiversity, feel that they should be compensated— for maintaining their forests, for instance. Drug companies, however, while emphasizing the importance of incentives for themselves, dismiss the need for incentives for others. In the international biodiversity agreement signed in June 1992 at the UN Conference on

Environment and Development in Rio, the right to compensation was recognized, but, partially under the influence of the drug companies, the United States has not ratified it.[49] And no wonder: almost half of the 4,000 plant patents granted in recent years by the United States pertain to traditional knowledge obtained from developing countries.[50]

One of the most notorious cases of bio-piracy was the attempt to patent turmeric for healing purposes. Turmeric is a spice used in South Asia, and its healing properties have long been known in the countries where it is found. Nonetheless, the United States issued a patent for the medical use of turmeric in December 1993.[51] The patent was eventually overthrown, but not without expensive litigation.

It is not just drugs that are affected. Basmati rice has been eaten in India for hundreds, perhaps thousands, of years. Yet in 1997 an American company, RiceTec, Inc., was granted patents on basmati rice. India, of course, was outraged, and it had the resources to fight—and win.[52] Smaller and poorer countries, however, don't have those resources, and can't fight back.

Those who defend the granting of these patents say the problem is that the developing countries never published their findings; had they done so, the courts would have given deference to this prior knowledge. The standard for novelty that has sometimes been used in granting patents, though, is not whether the medicinal properties of a certain plant were known, for example, among the indigenous people of the Andes, but whether they were widely known in the United States. So even if the indigenous people had published in their own language (assuming one would even bother to publish what is already well known) the patents might still have been granted. In any case, why should the developing world be forced to conform to the practices of the advanced industrial countries? The United States has taken a more extreme position than the EU on these issues. Consider, for example, the patents on the oil from India's neem tree, which has long been recognized for its cosmetic, medicinal, and pest control properties. Yet, in the 1990s, patents were granted for the tree oil both in Europe and America. By 2000, some ninety patents had been granted in Europe alone. Finally, in May 2000, some of the European patents were withdrawn, not because the properties of neem tree oil were recognized as part of traditional knowledge, but because an Indian entrepreneur was able to show that he had been producing an extract of neem oil for pest control for a quarter century. Still, in 2003 some twenty patents remained in force. And though Europe had withdrawn some of the patents, the United States refused to, on the grounds that the ideas had never been previously patented or published.[53]

We need to do more to protect developing countries' "comparative advantage" in this area. They have a reservoir of knowledge that can be drawn on, such as the medicinal use of plants. Their rainforests provide a wealth of flora from which Western drug companies have been extracting vital medicines. But TRIPs has provided few incentives for developing countries to preserve their rainforests.

Two reforms would go a long way in addressing the concerns of the developing countries:

- There ought to be an international agreement recognizing traditional knowledge, and prohibiting bio-piracy.
- All of the countries of the world—including the United States— must sign the biodiversity convention. Short of that, the guarantees concerning biodiversity property rights incorporated in the convention should be incorporated within international agreements concerning intellectual property rights, especially TRIPs.

Fortunately, there are firms that have acted in a more responsible way—more respectful to the rights of developing countries. One of the most effective recent drugs against malaria, for instance, is derived from the Chinese tree qinghao, which has been used to treat the disease for more than two thousand years. Qinghao has become particularly important as malarial strains resistant to the usual drug treatments have developed. In this instance, a socially responsible Swiss drug company, Novartis, not only looked to traditional knowledge for inspiration but, recognizing the importance of access to the medicine, has provided it free or at cost to developing countries.[54]

Governance

Throughout this book I stress that the way decisions get made—governance—in the international arena suffers from two flaws: the voices of developing countries are heard too little, and the voices of special interests are heard too loudly.

Just as trade is too important to be left just to trade ministers, so too for intellectual property. By now it should be clear: TRIPs was a mistake. A key reform is to change where and how the decisions concerning intellectual property are made. Discussions over global standards for intellectual property should be taken out of the WTO and put back into a reformed WIPO, a World Intellectual Property Organization in which the voices of academia as well as corporations, consumers as well as producers, the developing as well as the developed countries, are all heard. But this is not the only institutional reform that is needed. Among the values held strongly by people all over the world are those related to the rule of law and fairness. The legal system defines the rules of the game, and lawyers are there to ensure that the game is played fairly. We have to be sensitive to the disadvantageous position of developing countries in enforcing their rights in a court of law. Western democracies have government finance legal assistance for the poor. If a poor person cannot afford legal representation, there is a high likelihood he will be treated unfairly unless he has a court-appointed lawyer. This is even more true in the international arena.[55]

Whether we like it or not, intellectual property is likely to remain part of the global trading regime. Poor countries are at a distinct disadvantage when fighting for their rights. Most developing countries simply cannot match the large teams of highly trained and expensive attorneys employed by American and European corporations and governments. Fairness requires that the advanced industrial countries finance strong legal assistance for the developing countries to help them fight claims such as those related to bio-piracy, and to ensure that they can get compulsory licenses for lifesaving medicines when circumstances warrant it.[56]

Trades and Values

Intellectual property laws provide the most dramatic illustration of the conflict between international trade agreements and basic values. But there are many other instances, some of which we noted in the previous chapter in our discussion of nontariff barriers. For example, Europeans have very strong feelings against genetically modified foods: if there is even a tiny health risk from these foods, they don't want them sold in their countries. Under WTO rules, however, it may not be possible to bar them. Foods can be excluded only on the basis of science, and science "says" there is no significant risk. America, accordingly, claims that excluding such foods is unjustified protectionism. Europeans rightly ask, why should they be forced by an international trade agreement to accept that risk, if the majority believes that it is a risk not worth taking?

If genetically modified foods can't be excluded from Europe, those who object to them want full disclosure of the genetically modified content of these foods—labeling, so that consumers can choose what they want to buy. But the United States—normally a believer in free trade and consumer choice—has in this case taken the position that full disclosure would be a trade barrier. A large proportion of America's agricultural exports contains a genetically modified ingredient; America correctly worries that, given the level of concern about genetically modified foods, European consumers would stop buying many American-produced foods. The United States is putting its right to export above European consumers' right to know what they are eating.

Economic interests also often take precedence over cultural identity. Most people put enormous value on their heritage, their language, and their sense of cultural identity. For many, the cinema is important both in contributing to and conveying that identity. But there are large "returns to scale" in reproducing movies; the cost of running off an extra print is negligible relative to the cost of making the movie in the first place. This gives a huge advantage to movies from America and India, both countries with a large cinema-going base. Around the world, many governments find it necessary, and worthwhile, to subsidize artistic enterprises such as opera and theater, and some, including

France and Morocco, provide subsidies for cinema production as well. But the U.S. entertainment industry views these subsidies as unfair competition, and in the Uruguay Round tried (unsuccessfully) to force their elimination.[57] To me, this is a clear example of putting economics above other values. Hollywood's sex- and violence-heavy movies may have a certain universal appeal, but it seems reasonable for governments to want to promote their own artistic traditions, and supporting cinema is one defensible way of doing that. What I find so striking about this example is the social cost-benefit analysis. There is little chance that French-language films, subsidized or not, will make a major dent in Hollywood profits. Whether subsidizing them is a good way for the French government to spend its money should be a matter for the French people to decide. If they spend it well, not only those in France but filmgoers around the world will benefit.

Then there is the question of the environment that I mention here because it speaks to the question of values. In chapter 2, I stressed the importance of a vision of development which goes beyond GDP. For some, treating the environment with respect is a matter of basic values. For others, it is a matter of fairness to future generations: if we despoil the environment and squander our natural resources, we jeopardize the future. Sound environmental policies are essential if development is to be sustainable. For still others, it is a matter of the here and now: living standards today are compromised if the water we drink and the air we breathe are polluted. But whatever the perspective, there is a well-founded worry that badly designed international trade agreements may compromise countries' ability to protect the environment. When, for example, a village in the north-central Mexican state of San Luis Potosi tried to force Metaclad, a U.S. waste disposal company, to close a toxic waste site that was contaminating the local water supply, the Mexican government was forced to pay $16.7 million in compensation, under Chapter 11 of NAFTA. Anti-environmentalists had succeeded in burying in that chapter a provision designed to halt regulation by making it too expensive, by forcing compensation for loss of market value as a result of regulation, including regulations protecting the environment and public health. The irony was that the Clinton administration had devoted enormous energy to stopping the enactment of congressional

legislation that would have done this—and had succeeded; Clinton
and U.S. Trade Representative Mickey Kantor may have known this
was part of the fine print in the NAFTA agreement they were simultaneously pushing, but if so, they neither talked about it publicly nor discussed it privately in the White House NAFTA meetings.[58]

Corporate interests
This chapter has shown how corporate interests have tried to shape
globalization in ways which compromise more basic values. The fact
that one area—intellectual property—has been linked with trade, but
not others, like labor standards, says a lot about globalization as it is
managed today. The job of Western trade negotiators is to get a better
trade deal for their country's industries—for example, gaining more
market access and stronger intellectual property rights—without giving up agriculture subsidies or nontariff trade barriers. Fairness is not
in the lexicon of these trade negotiators. They are not thinking of
American or European taxpayers, who would benefit enormously from
the elimination of agricultural subsidies. They are not thinking of
American or European consumers, who would benefit from lower
prices. They are not thinking of the global environment, which would
benefit enormously from reduced greenhouse gas emissions. They are
not thinking of how to help the poor get access to lifesaving medicines.

Instead they are trying to help the producers, and their job is to get
as much as they can while giving up as little as they can. Trade negotiators have little incentive to think about the environment, health
matters, or even the overall progress of science. The environment is the
problem of the environmental minister, access to lifesaving medicines
is the problem of the health minister, and the overall pace of innovation is a problem of the education, research, and technology ministers.
So while trade agreements affect all of these areas, those who worry
about them are not at the table.

Trade ministers tend to negotiate in secret. Trade agreements are
long and complex, and lobbyists work hard to bury in them self-
serving provisions that they hope will escape attention. But the basic
issues that I have been discussing here—such as the trade-off between
drug company profits and the right to life—are ones that are easy to

understand. If the issue of access to AIDS drugs were put to a vote, in either developed or developing countries, the overwhelming majority would never support the position of the pharmaceutical companies or of the Bush administration.

Conflicts over fundamental values are at the center of democratic debate. Critics of globalization charge that globalization has been managed in such a way as to take some of the most important issues out of the realm of public discourse within individual countries and into closed international forums, which are far from democratic in the usual sense of that term. With the voices of corporate interests heard so clearly and strongly, and without the checks and balances of democratic processes, it's not surprising that the outcomes seem so objectionable, so distant from what would have emerged had there been a more democratic process. The most daunting challenge in reforming globalization is to make it more democratic; a test of success will be in how well it succeeds in ensuring that these broader values triumph more often over simple corporate interests.

CHAPTER 5

Lifting the Resource Curse

A t the turn of the twentieth century, czarist-ruled Azerbaijan was the world's biggest exporter of oil, and its largest city, Baku, on the shores of the Caspian Sea, was like the Wild West. People flooded in from all parts of Russia, intent on making money in the oil rush. Jews, Turkomans, Kazakhs, and assorted Europeans joined the fray. Real estate prices soared as the new arrivals competed for space. Oil rigs and refineries dotted the city. Alfred Nobel worked here for a while, and the park he built still remains. In the course of the century, Azerbaijan's oil made many people rich, yet much of the nation remained very poor. Today, Baku is littered with rusting old factories and equipment in what was known as the "black town," the grimy industrial area on the outskirts of the "white city," where oil millionaires once built vast houses and a boardwalk by the Caspian.[1]

After several decades of Soviet rule and a decline in oil production, new sources of oil and gas were discovered in the 1990s lying underneath the waters of the Caspian. Now, with the construction of new pipelines that began at the turn of the twenty-first century, Azerbaijan is enjoying another oil boom, and billions of dollars are expected to flood into the country. The challenge is whether Azerbaijan can make the most of its windfall before the oil runs out, which is expected to happen in 2030. If it is handled well, per capita income (about $940 in 2004) would double approximately every ten years. That would put Azerbaijan, a quarter century hence, in a league with the richer countries of eastern Europe that have just joined the EU. The danger, however, is that Azerbaijan will succumb to the so-called paradox of plenty, joining the many countries richly endowed with natural resources that have lower growth and higher poverty rates than other countries not so well endowed.[2]

Consider Nigeria. This West African nation, which was ruled by a military government through much of its oil boom, has earned almost a quarter of a trillion dollars in oil revenues over the last three decades. At the same time, its economy decayed and its main commercial city, Lagos, became a dirty, dangerous place. Traffic clogs the streets, unemployment is high, and people stay home at night because crime makes it too risky to go out. In spite of all the oil, per capita income declined by over 15 percent from 1975 to 2000, while the number of people living on less than $1 a day quadrupled from 19 million to 84 million.[3] Saudi Arabia and Venezuela are other examples of countries where oil wealth has not been widely shared. Venezuela has more oil wealth than any other country in Latin America, but two-thirds of the population there live in poverty.[4] It is not surprising that the charismatic Hugo Chavez won the 1998 elections handily after running on a platform of poverty eradication.

Understanding why developing countries that are resource-rich perform so badly—what is sometimes called the "natural resource curse"—is of immense importance:[5] First, because so many developing countries are economically dependent on natural resources: more than a third of the export income of Africa is derived from natural resources; much of the Middle East and parts of Russia, Kazakhstan and Turkmenistan, Indonesia, and substantial chunks of Latin America including Venezuela, Mexico, Bolivia, Peru, and Ecuador all depend heavily on their natural resources for income; Papua New Guinea is dependent on its rich gold mines and on its immense hardwood forests. Second, because resource-rich countries tend to be wealthy countries with poor people, and that paradox provides insights into the broader failures of globalization—and the possible remedies. Most important, reforms in the resource-rich countries—and in the way they are treated by the advanced industrial countries—can perhaps more quickly and easily reduce poverty than changes elsewhere in the global economic system. What these countries need is not more aid from abroad but more help in getting full value for their resources and in ensuring that they spend well the money they get.

The problem is simple: when there is a pile of diamonds sitting in the middle of the room, everyone will make a grab for it. The biggest and strongest are most likely to succeed, and will be reluctant to share it unless they absolutely have to—such as when someone else, even bigger and stronger, tries to grab it away from them, and they need to spend money to buy political support or arms in order to maintain their power. The resources are both the object of the conflict and the source of the financial wherewithal that enables the conflict to go on. Sadly, in the struggle to get as big a share of the pile as possible, the size of the pile itself shrinks as wealth is destroyed in the fighting. Nowhere is this more evident than in parts of Africa, exemplified by the heinous fighting over diamonds between government and rebels in Sierra Leone during the 1990s that killed 75,000 people and left 20,000 amputees, 2 million displaced people, and large numbers of children psychologically damaged by having been forced into combat, or worse.

Once violence has begun, it is hard to stop. Countries fall into a downward spiral, as Congo and Angola both illustrate. Congo has been involved in conflict almost continuously since independence. Typically, all sides claim to represent the will and interests of the people. The conflict in Sierra Leone was an exception: there was hardly any pretense of higher motives, just greed.

Just as there is often conflict between haves and have-nots, there can be conflict between regions that have resources and those that don't. This is especially true of developing countries whose borders were drawn by the former colonial powers and whose national identity is weak. Resource-rich regions—such as the oil centers of Ogoniland in Nigeria, and the Shiite south and Kurdish north of Iraq—have obvious incentives to break away. Why, they reason, should they share their wealth? The rest of the country will be equally determined to hold on to it. The Congolese province of Katanga, which is rich in cobalt, copper, tin, radium, uranium, and diamonds, broke away in June 1960 and was reclaimed by Congo in January 1963 after bitter fighting. Oil-rich Biafra seceded from Nigeria on May 30, 1967, and was reincorporated on January 15, 1970. Bougainville, a small island off the coast of Papua New Guinea that sits atop deposits of gold, silver, and copper, has been fighting for independence since 1989. Of course, the independence movements cloak themselves in more righteous mantles, and, while resource money fuels the conflict, the extent to which the fighting is *just* over resources is not always clear.

The violence that has afflicted these resource-rich countries represents the extreme of the resource curse. More frequently, one sees merely political instability, corruption, and ruthless dictators stealing the countries' wealth. Though resource-rich countries do not have a monopoly on ruthless dictators, they have had more than their share—from Congo's Mobutu Sese Seko to Iraq's Saddam Hussein to Chad's Idriss Déby. Even when the resource-rich countries do not have ruthless dictators, they have a marked aversion to power sharing: not one of the oil-rich countries of the Middle East has anything approaching a democracy.

It is no accident that so many resource-rich countries are far from democratic. The riches breed bad governance. Governments that come to power by grabbing resources and using force have a markedly different sense of responsibility toward their citizens and their country's resources from governments that emerge through the will of the people. In democracies, a leader stays in power by enhancing the well-being of the citizenry; democracies are accountable to their citizens. In undemocratic resource-rich countries, dictators use strength and weapons to remain in power. Arms purchases are funded by control of the revenues from oil and other commodities. There is a vicious circle: with a lack of democracy in so many resource-rich countries—and therefore a lack of accountability—citizens have no effective check against the theft of public funds and the abuse of public trust. Typically, they do not even know how much the government is, or should be, receiving in revenues for their natural resources. They may not even regard it as their money, as they would if they were supporting the government with taxes on their hard-earned incomes.

The political dynamics of resource-rich countries often lead to high levels of inequality: in both developed and less developed countries, those controlling the natural resource wealth use that wealth to maintain their economic and political power—which includes appropriating for themselves a large fraction of the country's resource endowment. Beginning in the 1970s, the elites of the Middle East made their presence felt in London and elsewhere; they bought expensive apartments, stayed in luxurious hotels, and went on shopping sprees. In the 1990s, it was the turn of the rich Russians. Today they are snapping up real estate and luxury goods around the world.

This is a strikingly different outcome from what standard economic theory might suggest. One of the main arguments against creating a more egalitarian society is that progressive taxation weakens incentives. If you tax the rich more heavily, people may not be motivated to work as hard or save as much. But if a country's riches come not from hard work or savings but purely from the good fortune of having oil or other mineral deposits, the country can afford to have much greater equality; government can distribute the wealth fairly without worrying that this will discourage people from working hard and saving what they earn. Such countries can have both greater equality and economic efficiency.

But while resource-rich countries could (and I would argue should) have more equality than others seemingly less fortunate, that is not how it turns out. The distribution of wealth is not determined by a careful balancing of equity-efficient trade-offs. It is not determined by reference to principles of social justice; rather, it is the result of naked power. Wealth generates power, the power that enables the ruling class to maintain that wealth.

And there is a striking difference between riches that arise from hard work and creativity and those which come from grabbing hold, in one way or another, of a nation's natural resources. The latter is particularly enervating for national cohesion. It also undermines faith in the market economy—especially when it is suspected that the wealth is acquired "illegitimately," through underhanded deals with current or previous governments. It is not surprising that discontent seethes beneath the surfaces of these countries.

The Appropriation of Public Wealth

The first challenge facing any resource-rich country is to ensure that the public gets as much of the value of the resources that lie beneath its land as possible. This is far more difficult than it might seem. Even in countries with stable and mature democracies, there is an ongoing struggle by oil, gas, and mining companies to seize as much of the wealth for themselves as possible. Here, though, it is done within the rule of law, often through campaign contributions; grateful candidates, once in office, enact regulations that allow their donors to acquire resources at the lowest possible price, to keep an increased proportion of the revenue they garner through special tax benefits, and to bear the least possible part of the cost of the environmental damage they inflict.

In the United States, mineral resources are essentially given away to the mining companies; when President Clinton tried to have the resources auctioned off, sold to the highest bidder, he was beaten back by lobbyists from mining companies. Even given the already existing preferential tax treatment for oil, gas, and mining companies, and even after high oil prices left them flush with cash, President Bush pushed an energy bill so loaded with subsidies for these companies that Senator John McCain, a member of the president's own party, referred to the bill as one that would "leave no lobbyist behind." By any reckoning, the energy and natural resource sector, which contributed almost $5 million to Bush's 2004 campaign and almost $3 million to his 2000 campaign, obtained a large return on its investment.

When these corporations head overseas to the developing countries, outright bribery comes into play. In the highly competitive world of international oil, it is easier for an oil company to show high profits by bribing government officials to lower the price they have to pay than it is to be more efficient than everyone else while paying full market price. What to an oil company is a small bribe is enormously tempting for the officials involved, who are often civil servants with salaries of only a few thousand dollars a year. The bribes undermine the democratic process as well as the market. Yet the real problem is not the bribes, distasteful as they may be, but their result: when the oil company gains, the local country loses.

The Foreign Corrupt Practices Act of 1977 made it illegal for Americans to bribe foreign governments. While some companies still try to circumvent it, many have tried to comply—although, they complain, this puts them at a competitive disadvantage relative to corporations based elsewhere in the world. The complaints spurred the U.S. government to try to persuade other countries to impose similar statutes. At the Organization of Economic Cooperation and Development (OECD) ministerial meeting in Paris in 1996 (at which I was the U.S. representative), we made great strides in pushing an agreement through—after facing enormous resistance from several countries where bribery was accepted as a way of doing business. At the time, bribes were not only legal but tax-deductible in a large number of countries (France, Switzerland, Luxembourg, Austria, Belgium, Japan, and the Netherlands); in effect, the government was paying a substantial part of the bribe. I was shocked to see governments stand up to defend (with great circumlocutions) the existing system of bribery. There is now an OECD convention on bribery, but enforcement is difficult and incomplete. As of December 2005, there had yet to be a single prosecution outside the United States under national legislation enacted to implement the convention.[6]

Firms, of course, do not necessarily offer the bribes themselves. They hire a "facilitator," who is given enough money to "facilitate" the deal. What he does, how he facilitates, they don't know and don't want to know. Presumably, they know that if they pay the facilitator millions of dollars they are not just buying consulting time. What they are really buying, of course, is deniability, so that they can claim they didn't know the money went for bribes. One of the most famous recent cases is that of James Giffen, who, while working in Kazakhstan on behalf of Mobil in the 1990s, allegedly funneled $78 million to the government, getting the company a 25 percent share in the Tengiz oil field.[7]

Meanwhile, multinationals based outside the OECD operate beyond the OECD strictures. The Malaysian, Russian, Indian, and Chinese oil companies, among others, have become global players. They do not have to follow OECD agreements banning bribes, and as long as there are some companies paying bribes, other companies will have to find ways to compete. The whole market is contaminated.

Whatever the contract that has been signed, corporations are tempted to cheat—to pay less than they are required to—because the amount of money that can be made by cheating is so large. In the 1980s I worked on a case involving cheating by the major oil companies in Alaska. This oil-rich state had leases that generally guaranteed it at least 12.5 percent of gross receipts, less the cost of transporting the oil from the far-flung site at Prudhoe Bay on the Arctic Circle. By overestimating their costs by just a few pennies per gallon—and multiplying those pennies by hundreds of millions of gallons—the oil companies could increase their profits enormously. They could not resist the temptation.[8]

Often the exploitation of developing countries by the mining and oil companies is perfectly legal. Most developing countries are ill prepared to engage in the sophisticated negotiations that are the multinationals' stock in trade. They may not understand the full implications of each contract clause. They will be told that some clause is standard, and it may well be: the oil companies may stand together in demanding contracts that benefit them at the expense of the countries from which they get the oil. For instance, governments have redesigned how they sell, say, the broadcast spectrum (for cell phones, TV, and radio), through auctions that increase government revenues enormously; but in the area of natural resources, the industry has staved off similar reforms, especially in developing countries. Attorney Jenik Radon, an adjunct professor at Columbia University, recalls that when he represented the nation of Georgia in its negotiations with a consortium of oil companies led by BP, he was aghast at the demands they made. Among other things, they wanted to hold Georgia liable for billions of dollars if there were any regulatory delays. At the same time, they want all the risks of environmental damage to be borne by the country rather than by themselves. In many cases where natural gas is concerned, they even demand take-or-pay contracts, designed to shift the ordinary commercial risk—the size of the demand for gas—from business to government. The developing country's government is obligated to pay for a fixed amount of gas, whether or not there are customers for it.

Bribery, cheating, and imbalanced negotiating all cut into what rightfully ought to go to the developing country. The countries get less than they should, the companies get more. A competitive market should mean that oil and mining companies simply get a normal return on their capital; excess returns should belong to the country owning the resources. Economists refer to the value of the resource in excess of the cost of extraction as natural resource "rents." In a competitive market, the oil companies should be paid for their extraction and marketing services, and nothing more; all of the natural resource rents belong to the country. This means that if the price of oil rises, then—since the price of extracting it remains unaffected—the excess returns should belong to the country. This is especially important when the price of oil triples or quadruples—as it did in the 1970s and again in 2004 and 2005. After oil prices skyrocketed in the 1970s, the United States imposed a windfall profits tax on the oil companies. The fact that the typical contract allows the oil companies to walk away with windfall profits suggests that something is wrong with the way these contracts are designed.[9]

It is the strategy of the oil, gas, and mining companies to make sure that the government gets as little as possible—while, at the same time, helping the government find arguments for why it is good or even necessary for the government to receive so little. They may say that there are large social benefits from developing the region, and thus development should be encouraged. Giving away the resources, they claim, does this. In fact, giving away resources simply means the government has less money to pay for infrastructure, schools, and other facilities that are absolutely necessary if the region is to be developed. It may be costly to develop a mine, but that only means that in competitive bidding the government will get less money than it would if the mine were less costly. Too often, the only benefit to the country from a mine is the few jobs it creates, and the environmental damage of the mine may simultaneously destroy jobs elsewhere (for instance, in fishing, as catches in polluted water diminish) and, sometime in the future, impose enormous budgetary costs as the government is forced to pay for the cleanup.

The struggle to get for the country the full value of its resources is particularly pronounced in the sale of government-owned natural assets to the private sector. Whenever the government gets less than the full value of the asset, the country is being cheated; there is a simple transfer of wealth from the citizens as a whole to whoever gets the assets at a "discount."[10] Sometimes the wealth of the state goes to individuals within the country rather than to a multinational corporation; still, wealth slips away that ought to belong to the nation as a whole.

Before privatization, when the oil field (or other resource) is still in the government's hands, the amount government officials can steal is limited by current sales of oil. But with privatization, the future value of the resource is up for grabs, and the stakes increase enormously. By selling a whole company at below fair market value and getting a kickback on the gift they have given the buyer, government officials can, in effect, get a share of all future sales, instead of leaving it to be stolen by their successors. Indeed, in some parts of the world, privatization has been relabeled "briberization." Governments have become expert in maintaining the facade of a fair privatization, by conducting the sale through an auction. But they may pre-qualify bidders—and anybody likely to upset the planned sale at a discount price to the government crony is disqualified. They may say that the unwelcome bid was submitted late, that the bidder has not provided adequate evidence of financial wherewithal, and so forth.

Even without outright corruption, the pressure from the IMF to privatize quickly led to substantially lower revenues for governments. (Developing countries are desperate to please the IMF—not only because the IMF may terminate its own lending if it is crossed but also because others, as a result, may terminate theirs.) As each bidder believes bidding will be less keen, bidders bid less aggressively, and the government ends up accepting a bid that is woefully inadequate. The problems are even worse, of course, in those situations—not infrequent—where the number of bidders is very limited (one, two, or three), in which tacit or explicit collusion may arise.[11]

The argument for privatization is that the private sector is more efficient than the public. This opinion is driven as much by ideology as by hard analysis—there are many examples of highly efficient government oil and mining companies (and examples of inefficient private companies). The inefficiency of some state enterprises is due to the lack of investment caused by the IMF's insistence on treating the debt of state enterprises like any other form of government debt; a private company typically borrows to finance investment, but developing-country state-owned enterprises are effectively forbidden to do so.[12]

Efficiency, however, is not everything. Even if the private sector were more efficient, equally important is how much the public receives for its resources. Typically, when a privatization takes place, countries receive a down payment and then a royalty as the resource is extracted and sold. Poorly executed privatizations result in governments receiving both too little upfront and too little down the line. Malaysia's publicly owned oil company, Petronas, has become a global player, and Malaysia's former prime minister Dr. Mahathir bin Mohamad says his country receives a larger fraction of the value of its resources than countries elsewhere who have privatized, and a larger fraction than it would have received had it privatized.[13] Chile has privatized about half its copper mines, yet the government mines are just as efficient—and because most of the profits from the private mines are sent abroad, the government copper mines provide more revenue to the public.[14]

Russia provides a dramatic case of privatization gone amok. With the end of communism and the decay of an effective state, Russia, once the world's second superpower, became increasingly dependent on its natural resources—by some estimates, some 70 percent of its GDP in recent years related to natural resources. Boris Yeltsin needed help getting reelected in 1996, and a small group of oligarchs had the organizational (and financial) capacity to help him—in exchange for control of the nation's vast natural resources. The critical events occurred in 1995–96, in a sale that *Financial Times* editor Chrystia Freeland called "the sale of the century."[15] There were auctions, but the auctions were rigged. As a result, the oligarchs got the country's vast natural resources legally for a pittance. Some senior government officials believe the amount "stolen" exceeds a trillion dollars.

Later, when Vladimir Putin succeeded Yeltsin, he understood that such concentration of wealth was a threat both to him and to Russian democracy, such as it was. Given that in the early years of transition few of those oligarchs had paid the taxes they owed, it was not hard for Putin to figure out how to use the power of the state—within the rules of the game—to recapture significant amounts of the assets. In the case of Yukos, Russia's largest oil company, he succeeded—even though Mikhail Khodorkovsky, who had got control of Yukos, used his huge wealth to generate a global public relations campaign (joined in by many Western governments, including the Bush administration) objecting to his selective prosecution. Though it was hard to determine what role Khodorkovsky's political opposition played in the prosecution, his supporters seemed to say that it was permissible to use the rule of law to steal assets from the public but not to enforce the law to get back what was legally owed.

Russia's privatizations highlighted a problem that is endemic around the world. In the case of Russia, it was Russians stealing money from their own country; in most other cases, those extracting the resources are foreigners, which only heightens the tension. Governments have been toppled because of this problem, as in Bolivia; and the sense of outrage has given support to those, like Chavez in Venezuela, promising a better deal. Ordinary citizens see rich Venezuelans and foreign companies benefiting from their wealth, but none of it seems even to trickle down to them. Chavez's ability to renegotiate old contracts, to get better terms for his country, simply reinforces the belief that, in the past, Venezuelans have been cheated. Botswana provides another telling example. The country was able to renegotiate the contract with the diamond cartel De Beers to ensure that it got full value for its resources—or, at least, more of the value; it increased its share of the business from 15 percent to 50 percent. Without that renegotiation, Botswana probably would not have been able to enjoy the remarkable economic success it has had since independence.

In the end, too often the country loses twice—first from the unfair contract or privatization, and then from the political turmoil and adverse attention from the international investment community when an attempt is made to set things right.

Using Money Well

Getting a fair share of the value of their natural resources is the first task facing developing countries. The next challenge is to use the money well. The Saudis in London in the 1970s who snapped up expensive property and went on grand shopping sprees provide one of the more ostentatious examples of what not to do with one's newfound wealth. Certainly, the people of Saudi Arabia would have been better off if more of the oil money had been spent on their development and less on London real estate or on arms—since 1988, Saudi military expenditure has been below 10 percent of GDP only three times (the United States, whose defense expenditures equal that of the rest of the world combined, has been spending only 3–4 percent of its GDP). When the oil countries do invest, they often do not invest well. Returns have often been abysmal. Venezuela and Saudi Arabia would have had better returns had they invested their money on the New York or London Stock Exchanges.

Resource-rich governments have a tendency to be profligate. Easy money leads to easy spending. Of course, all governments have problems ensuring that money is well spent. Pork barrel expenditures—money spent on projects that have little value beyond pleasing constituents—are a fixture of many, if not most, democracies. The political forces are as present in developing countries as they are in developed ones—but developing countries simply cannot afford to waste the money.

Added to the problem of spending the money well is the unpredictability of revenues. Resource prices are very volatile. Oil prices, for example, rose from $18 a barrel at the end of 2001 to more than $70 a barrel in 2006. From 2003 to 2005, copper prices have risen by 98 percent, tin by 55 percent. This creates a boom-and-bust pattern in the economy: When prices are high, the country spends freely, failing to anticipate the drop in prices down the line. When prices do drop—as they have repeatedly—there are bankruptcies and an economic slump. The boom is often accompanied by a real estate boom, with banks lending easily, confident in the high value of the real estate collateral that they require. When the collapse of resource prices is accompanied by a collapse in real estate prices, the banking system is weakened and banks are forced to curtail lending, pushing the economy deeper into recession. Even ordinary developed countries find it difficult to manage a market economy in a stable way—periodic recessions and depressions have marked capitalism since its beginning. Managing resource-rich economies is difficult, because of the huge variability of the export income. But the task of managing fragile resource-rich developing countries is truly daunting.

Developing countries do not have the ability to weather the swings in export earnings as well as developed countries do. They do not have the built-in stabilizers—progressive income tax systems, unemployment insurance and welfare programs that pump more money into the economy when the economy is weak. Individuals do not have savings to fall back on. Banks are often not as well capitalized or regulated, so they are more prone to collapse.

Making matters worse, international bankers are always willing to lend to resource-rich countries when the price of their resource is high, and the ruling elite finds it difficult to turn down the offers. That explains the curious phenomenon of several highly indebted countries, which are having a difficult time meeting their debt obligations, being oil-rich, like Indonesia and Nigeria. Even if the projects the banks are backing are no good, a construction boom makes citizens—and especially construction contractors—feel better; the problem of repayment is left to a later date.[16] When resource prices drop, the bankers, of course, want their money back—just when the country needs it most. The boom-and-bust lending exacerbates the economic volatility brought on by boom-and-bust prices.

In several cases when countries understand what needs to be done to stabilize the economy—and even have the resources with which to do so—the IMF has pressured them to adopt policies that actually worsen economic downturns. I saw this vividly in Ecuador and Bolivia in the recessions and depressions that marked the late 1990s. For seventy-five years, the standard prescription for an economy facing recession has been expansionary fiscal policy—spending money on education and especially infrastructure badly needed in any case for growth. Typically, developing countries have difficulty financing the required stimulus, but Ecuador and Bolivia were lucky—they had massive amounts of oil and gas resources that would soon become available, which they could have used as collateral for borrowing. The Bolivians and Ecuadorians argued—rightly, I thought—that the return on investing in the recession was far higher than it would be when global conditions returned to normal levels and their economy was nearer to full employment. In addition to the direct return, there would be a multiplier effect, as the spending would stimulate the entire economy, which was marked by huge underutilization of productive capacity, and help it move toward full employment. Spending money, with natural resources to back the loans, made good economic sense. But the IMF, always worried about government overspending, pressured Ecuador and Bolivia to follow a quite different course. Not only did the IMF not want these countries to stimulate their economies through increased expenditures; they actually demanded cuts in spending in order to offset the decline in tax revenues from the recession. These Andean countries felt they had no choice; they gave into IMF pressure and the policies did, indeed, worsen the downturns.

The IMF even posed problems for one of the world's best-managed economies, that of Chile, when it went into a downturn, along with the rest of Latin America, in the late 1990s. The government had taken to heart the principles of managing its resources well and had established a stabilization fund in 1985. When times were good and the price of copper was high, they put money into the fund, to be drawn on in time of need. When they came to spend money out of their stabilization fund, however, they were told not to by the IMF.[17] Chile wanted only to spend money it had set aside for a rainy day. Now that rainy day had come, but the IMF insisted that it would treat stabilization-fund spending like any other form of deficit spending. Chile rightly raised the question: why have a stabilization fund if you can't spend the money when the economy needs stabilizing? The question fell on deaf ears. But Chile was afraid simply to ignore the IMF. Even though it was not borrowing from the IMF, it worried that financial markets would respond to criticisms from the IMF by raising the interest rates at which it borrowed. Because Chile followed a less expansionary policy than it would have followed if the IMF had encouraged spending out of its stabilization fund—spending that it could well have afforded—it experienced a more marked slowdown in growth than it otherwise would have.

The Dutch disease

Spending money well, and spending money at the right time, are two of the greatest challenges facing resource-rich countries. There is a

third problem, first noticed in the 1970s and early 1980s in the aftermath of the discovery of oil in the North Sea; while they enjoyed this
obvious bounty, the Dutch began to notice that the rest of their economy had slowed. Here was a developed, well-functioning economy
that suddenly faced massive job problems because its firms couldn't
compete. The reason was that the inflow of dollars in payment for the
North Sea oil and gas led to a high exchange rate; at that high exchange
rate, Dutch exporters couldn't sell their products abroad and domestic
firms found it difficult to compete with imports.

The problem, known as the Dutch disease in honor of the country
where it was first analyzed, has plagued resource-rich countries around
the world as they sell their resources and convert the dollars they earn
into local currency. As their currency appreciates, they find it difficult
to export other products. Growth in the nonresource sector slows.
Unemployment increases, since the resource sector typically employs
relatively few people. Before the oil boom three or four decades ago,
Nigeria was a major exporter of agricultural produce. Today it is a
major importer. Before Venezuela became a major exporter of oil, it
was a major exporter of high-quality chocolate (it still produces some
chocolate). In both cases, as in Holland, major natural resources had
the perverse effect of harming the rest of the economy.

It may not be possible to avoid the Dutch disease entirely, but the
magnitude can be reduced. The problem, as we have noted, comes
from converting foreign exchange into domestic currency, which bids
up the value of the domestic currency. Reducing the amount converted
reduces the degree of exchange rate appreciation; that means a country
must spend some of the resource money on imports and keep some of
the rest abroad.

The Dutch disease thus provides another argument in favor of stabilization funds, in which a country can save money when prices are
high and the economy is experiencing a boom, money that they can
then spend when the economy is in a recession. Azerbaijan began putting money into such a fund in 2001; by the end of 2003, more than
$800 million from its oil revenues had been invested.[18] The invest-
ments abroad yield a double return to the economy: there is a direct

return, and by reducing the degree of appreciation of the currency, they help to create jobs and growth.

But, while these policies may represent good economics, they are difficult to achieve in poor democratic countries. Poor people in developing countries cannot understand why their government might want to invest their scant resources abroad when there is such a need for money at home. They fail to understand that while the oil money could, for instance, be used to build a local school, which would create jobs, even more jobs would be lost elsewhere in the economy as a result of the appreciation of the currency—the Dutch disease. There is a simple lesson: countries need to finance local expenditures—say, for teachers or workers employed in road construction—with locally raised revenues, for example through taxes, saving the dollars earned from the sale of natural resources for buying the necessary imported goods, or for some future time. This, of course, requires the government to raise taxes to finance the domestic content of its expenditures. The problem is that no government likes to raise taxes, and in countries with high unemployment there is enormous political pressure to spend the oil money at home and at once.

MAKING GLOBALIZATION WORK: THE RESOURCE CURSE IS NOT INEVITABLE

The natural resource curse is not fate; it is choice. The exploitation of natural resources is an important part of globalization today, and in some ways the failures of the resource-rich developing countries are emblematic of globalization's failures. The West is heavily dependent on the natural resources it receives from developing countries, and its short-run, self-interested incentives—and more especially, the interests of the extractive resource industries—do not always coincide with the well-being of the developing countries. But if globalization is to work in the long run, the developing countries—and their citizens—must be given a better deal. Fortunately, there are also successes that give us reason to be optimistic that we can make globalization work.

Among the developed countries, Norway stands out as a model of good practices. Oil generates almost 20 percent of GDP and 45 percent of exports. The state oil company (recently partially privatized) is efficient; more important, the country has recognized the limited amount of its resources—its oil and gas are expected to run out in seventy years—and has set much of the revenue aside in a stabilization fund of $150 billion, which today amounts to some 50 percent of the country's GDP.[19]

Botswana, though recently ravaged by AIDS, stands as one of the few success stories in the developing world, especially in the way it handled its wealth of diamonds. The country's economy has grown at an average rate of 9 percent over the past thirty years, rivaling those of the East Asian tigers. It did this with a democratic government committed to building consensus among the population on the policies required for successful growth, which included using a stabilization fund to handle the consequences of the volatility of diamond prices. Malaysia is another resource-rich country that used its natural resource wealth as a basis to join the club of newly industrialized countries.

The major responsibility for getting as much value as possible from their natural resources and using it well resides with the countries themselves. Their first priority should be to set up institutions that will reduce the scope for corruption and ensure that the money derived from oil and other natural resources is invested, and invested well. It may be desirable to have some hard and fast rules for that investment—a certain fraction devoted to expenditures on health, a certain fraction to education, a certain fraction to infrastructure. Procedures need to be put into place for independent evaluations of the returns on investments. Stabilization funds are essential, but governments must be allowed to use them in appropriate circumstances—and especially to help stabilize the economy. Most important, developing countries need to view their natural resources as their endowment, of which the current government and generation are trustees for future generations.

I believe, however, that the international community can do more than just provide pious lectures on what developing countries can and should do to get more for their resources and to use their resources better. More effective would be for developed country governments to provide role models, give advice and provide assistance in ways that

change incentives and opportunities, and do what they can to circumscribe the enormous forces for corruption that come from the developed world.

Corruption and conflict

The political forces in developing countries that lead to persistent corruption and entrenched elites using natural resource wealth to increase
their own wealth will not go away simply through pointing out the
consequences of their actions or their lack of moral underpinnings.
They hear the lectures from the West, but they see Western oil companies sending monthly checks to bolster repressive regimes—in, for
example, Sudan and Chad—and Western governments providing the
arms that maintain the repression. Naturally this calls into question
Western priorities: money is seen to reign supreme. The seeming lack
of commitment to democracy is, of course, reinforced by events such
as the violation of basic human rights at Guantánamo Bay and Abu
Ghraib. When one of the less than fully democratic premiers in one of
the developing countries was asked about the lessons of 9/11, his
immediate response was the importance of the right to detain people
without trial.

It is not only in these particular, and publicized, scandals that Western governments set a bad example. The United States' natural resource
policy—which gives away mineral rights and is controlled by special
interests—is a model for how things should not be done. The secrecy
with which the Bush administration formulated its energy policy—
even refusing to disclose the names of the industry representatives who
participated—also makes for a dismal role model. Bush's arguments for
executive privilege are loved by those trying to keep secret what they
are doing—whether to benefit themselves, their cronies, or a wider circle of friends who have helped them stay in power.

Transparency has long been recognized as one of the strongest anti-
dotes to corruption; as the expression has it, "Sunshine is the strongest
antiseptic." If citizens are to provide a check against corruption, they
have to know what is going on. Citizens' right-to-know laws (like
America's and Britain's Freedom of Information Acts) are necessary to
promoting both meaningful democracy and accountability. The Initia-

tive for Policy Dialogue, which I founded at Columbia University, for enhancing the understanding of policies affecting development, has as one of its major goals the passage of such legislation in developing countries. It has been marked by considerable success, culminating in a global conference co-sponsored with the Mexican government, which has recently passed legislation, matched by corresponding legislation in the majority of Mexican states. Thailand enshrined its citizens' right to know in its constitution. In every country, with full disclosure of how much the country is selling and what it is receiving for its natural resources, citizens can do a better job of assessing whether the country is getting full value for its resources, or whether, somehow, it is being cheated.

Sometimes governments claim that they cannot disclose information because it violates business confidentiality. Typically, such claims are nothing more than an excuse, a veil behind which government officials and the company can continue in their corrupt practices. The government *can* set the rules, and there are enough honest companies willing to play with rules of transparency. The citizens' right to know should trump any claims to business confidentiality.

The advanced industrial countries, while lecturing the developing countries on their levels of corruption, do not understand the role that their advice—and even the policies that they foist on developing countries—sometimes plays, unintentionally, in weakening the forces for the creation of a rule of law. Economic policies can shape, or at least affect, political processes. For instance, the likelihood that a country adopts a rule of law depends in part on the demand for it—on political support, especially among the wealthy elite. But in Russia, those who obtained wealth through illegitimate privatizations had little interest in establishing a rule of law conducive to investing (as opposed to one conducive to stripping assets).[20]

Those who advised Russia to privatize quickly, focusing on speed above all else, thus contributed to its current problems. Other economic policies too undermined the demand for a rule of law. As we noted earlier, with capital market liberalization, the oligarchs could easily take their money out of the country; they could enjoy the benefits of the rule of law abroad while stealing and stripping assets at home.

By the same token, when the IMF encouraged, or even demanded, that Russia have very high interest rates as a condition for assistance, that too may have had political consequences: at the high interest rates, new investments were not profitable; and those who had gained control of Russia's wealth were provided further incentives to strip assets. Politically, the oligarchs' interests were in a legal framework that allowed this, rather than in one that supported wealth creation for all Russians, and the toxic combination undermined support for the rule of law in the country as a whole.

What the IMF did mattered more than what they said: they weakened the politics of reform by ignoring the effects that their policies had on economic and political behavior.

While sometimes the advice from the West has proven counterproductive, there are some areas where advice can be helpful: in achieving greater transparency (as discussed earlier) and in helping developing countries think through carefully how best to manage their resources, both for short-run stability and long-term growth. For instance, the commonly used accounting frameworks need to be reformed, removing the false sense of growth on the part of countries that are living off their inheritance of resources. As I stressed in chapter 2, output measures should focus on sustainability.

Think of the oil in the ground as an asset: a country's natural resources are its endowment, located below the ground; like any other asset, they need to be managed. When they are removed, the asset is gone. Unless the proceeds are invested, the country is poorer. Just as a company's books show the depreciation of its assets, so too should a nation's accounting framework reflect the depletion of its scarce resources. But the most commonly used measure of output, gross domestic product, does not do this. It shows only that the more oil it extracts, the higher its income—regardless of how it is spent, regardless of the fact that such spending *without investment* is unsustainable. As a result, a country with a high GDP may actually be getting poorer and poorer—its seeming prosperity is not sustainable. Matters can be even worse: the extraction of some natural resources leads to environmental degradation, a liability. It may cost billions to remedy the damage—as Papua New Guinea found when it closed the Ok Tedi gold mine.

Green net national product (Green NNP) is a measure that subtracts out not just the depreciation of capital but also the depletion of natural resources and the degradation of the environment. It focuses on the income of those within the country—excluding the profits from a mine that go to the overseas owners. In extreme cases, the costs of repair may equal or exceed the return on the resources extracted; GDP may be increasing, but Green NNP may be decreasing.

Accounting is important because it affects decisions. A focus on Green NNP would induce countries to spend more on conservation. It would ensure that natural resource contracts are good for the citizens of the country; no matter how much GDP is increased, any contract lowering Green NNP should be rejected. When I was chair of the Council of Economic Advisers, I pushed for the creation of these accounts for the United States as a supplement to the usual GDP accounts, but the coal industry, recognizing how thinking—and action—might be affected, pushed legislation that cut off funding for their development.

There must also be changes in accounting for deficits. All countries worry about deficits. But accounting frameworks that look just at deficits, at liabilities—without looking at the other side of the balance sheet—are particularly dangerous. Countries need to create capital accounts that look at both assets and liabilities, and make especial note of situations where asset sales (including sales of natural resources and privatizations) are misleadingly being used to make deficits look lower than they otherwise would be. Countries can reduce their deficits by cutting down forests, selling national assets, giving away their natural resources at a fraction of the full value. In IMF accounting the country is then given good marks; and IMF accounting is important not just because bad marks from it means that it and other donors may curtail financial assistance but also because capital markets may refuse to extend credit. But the reality is that the country is poorer, not richer, as a result. By the same token, investments that might enable more natural resources to be extracted efficiently—say, from an oil field—are effectively discouraged, because if the country has to borrow to finance the investment, even if the return is high, it will be chastised for the increased deficit spending. To get around the shackles of these accounting frameworks, many countries privatize at disadvantageous terms, impoverishing themselves and imperiling their future unnecessarily.

These accounting reforms would help in another way. Countries should be encouraged to create stabilization funds—buffers to fall back on when times are hard, to help insulate natural resource countries against the volatility of natural resource prices. But as we noted earlier in the case of Chile, IMF accounting frameworks, which treat spending out of stabilization funds just like any other form of deficit spending, discourage countries from setting up these funds. Stabilization funds are an important tool in helping developing countries to achieve macro-stability. Without that, the economic turmoil so prevalent in resource-rich countries will continue, and market economics will not be given a fair chance to work its wonders.

An Action Agenda for the International Community

In addition to giving better advice and being better role models, developed countries can undertake concrete actions to help resource-rich countries. Three of these are discussed elsewhere in this book: chapters 2 and 7 show how enacting anti-bribery laws and curtailing bank secrecy will reduce opportunities for corruption; chapter 6 explains how developing countries provide enormous environmental services— tropical forests help preserve biodiversity and reduce greenhouse gas concentrations—from which the entire world benefits, but for which they are not compensated; chapter 7 details a set of legal reforms that would prevent multinational corporations from despoiling the environment of developing countries as they extract its resources—or if they do, make them pay the consequences. Here I describe seven additional measures.

1. The Extractive Industries Transparency Initiative[21]

I described earlier how greater transparency would discourage corruption, making it more likely that developing countries would receive full value for their natural resources. The advanced industrial countries can help ensure transparency by simply saying: no one gets a tax deduction for money spent on royalties or other payments to foreign governments unless they fully disclose what was paid and how much of the resource in question was extracted. Without such a broad agreement, there will continue to be a race to the bottom, and the companies and countries most willing to engage in corrupt practices, and least willing to be transparent, will have an advantage over the others.

2. Reducing arms sales

Even worse than corruption is the armed conflict that mineral and oil resources finance. Again, the international community could do more to make it more difficult and more expensive to acquire arms. We have a responsibility to choke off supply at the source—the manufacturers of arms who profit from this nasty business—or at least impose a heavy tax on the sale of arms and to check the source of the money which pays for them.[22]

3. Certification

On July 5, 2000, the United Nations Security Council imposed a ban on the import (direct or indirect) of rough diamonds from Sierra Leone not accompanied by a certificate of origin from the Sierra Leone government. Uncertified Sierra Leone diamonds are now known as "conflict diamonds"; this public recognition of the role of resources in financing a conflict, and the acknowledgment that it must be curtailed, is a move in the right direction. Amnesty International, Partnership Africa Canada, and Global Witness, along with other NGOs, are spearheading the effort to enforce the ban.[23]

A similar certification system should be established for tropical hardwood. Here, the problem is not so much the financing of conflict (though sometimes illegal logging does that too) but rather that illegal logging leads to rapid deforestation—with little benefit to the country.[24] What the Papua New Guineans receive for their lumber is typically under 5 percent of its value once it reaches the developed world. Certified lumber would be harvested in an environmentally sustainable way, so that not only the current generation but future generations too could benefit from the forests—and the world as a whole could be saved from rapid deforestation (I will discuss this at length in the next chapter). With a system of certification in place, lumber that is taken out in a sweetheart deal between, say, a Papua New Guinean chieftain and a Malay timber baron would not find a ready market. The higher prices received for certified lumber—and the lower prices for uncertified lumber, as outlets are cut off—would provide a natural incentive for countries to sign on to certified lumber programs. Indeed, the beginnings of such programs already exist in Indonesia and Brazil; the warm welcome they have received from consumers and some retailers suggests that there would be a positive response in the developed countries, especially from socially responsible firms such as Home Depot.

4. Targeting financial assistance

Developed country governments can provide further incentives by limiting aid, both through the World Bank and through their own assistance programs, to countries that are not getting full value for their resources. While there is considerable debate about the effectiveness of conditionality (imposing conditions as a prerequisite for the receipt of aid), in the case of resource-rich countries there is a fundamental question: why should taxpayers in the developed world subsidize a government that is itself in effect giving away its resources? The debate was particularly intense in the early days of Russia's transition to a market economy. Some argued that the West should be giving the country more money.[25] But at the same time, there was a massive flow of money out of the country. If the government could have stemmed that outflow—facilitated by its corrupt privatization—there would be little need for outside money. And if the government was so corrupt and incompetent that it could not get enough money from the sale of its natural resources to manage its transition to a market economy, why should one think that a few billion dollars more from the West would be well spent, or even make much of a difference?[26] While one cannot buy good policies (aid given on the condition that countries fulfill a long list of conditions does not accomplish what was intended), selectivity—or giving aid to countries that have demonstrated their ability to pursue appropriate policies—does provide incentives, and there is at least a hope that this will help countries move in the right direction.

5. Setting norms

One of the key problems we have identified is that developing countries do not get even close to full value for their resources. There is a role for some international body—perhaps the World Bank—to help ensure that developing countries are treated well by the oil and other extractive industries, to develop auction procedures that make it more likely that the developing countries get a larger fraction of the value of their resources, to design model contracts that ensure developing countries are treated fairly (for instance, sharing more fully in the benefits when oil prices boom), and to assess what fraction of the value of the resource is being received by the developing countries. It could try to create a race to the top, by comparing the fraction of the net value that actually goes to each developing country.[27]

6. Limiting environmental damage

Multinationals need better incentives not to despoil the environment. And unless they are forced to pay for the environmental damage that results from their actions, their incentives will be in the opposite direction. Today, international investment agreements are one-sided: they are designed to ensure that developing countries do not expropriate investors' assets, but they pay little attention to the converse problem that has plagued so many developing countries, companies that spoil the environment and then leave. There is a need for an international agency to monitor environmental damage. Just as developing countries guarantee that they will not expropriate investments, the developed countries where the oil companies are registered would guarantee that any environmental damage will be fully repaired, with clear and high standards set out for what that means.

7. Enforcement

We have described a variety of good practices, ways in which developed countries can help the developing world ensure that citizens reap the benefits of the resources that lie within their countries—by enhancing transparency, discouraging bribery and corruption, and protecting the environment. But these measures cannot and should not be left to goodwill. The amounts of money at stake are too large, the incentives for a race to the bottom too great. There must be effective enforcement. Trade agreements can be used to force "good behavior." Trade sanctions can be used against companies and countries that engage in unfair trade practices—and failing to subscribe to the extractive industries transparency initiative and other anti-bribery measures should be treated as an unfair trade practice.

We can make globalization work, or at least work better, for those in resource-rich countries. If it cannot work for them, what hope do we have of it working for those who live in the many far poorer countries of the developing world? The international community must not only work to ensure that the resource-rich countries get the full value of their resources, but help them manage their economies in ways which ensure stability and growth—and ensure that the fruits of that growth are shared widely.

We can lift the natural resource curse and turn bountiful natural resources into what they should be—a blessing.

There is one overriding problem: the well-being of the resource-rich developing countries depends on how much they get for their resources; the well-being of the rich corporations of the advanced industrial nations depends on how little they pay for them. This is the natural and inevitable conflict that we have identified at the center of the paradox of plenty. Where will the people of the developed countries and their governments stand? In support of the few in those countries who own and run the rich corporations, or in support of the billions in the developing nations whose well-being, in some cases, whose very survival, is at stake?

CHAPTER 6

Saving the Planet

The world is currently engaged in a grand experiment, studying what happens when you release carbon dioxide and certain other gases into the atmosphere in larger and larger amounts. The scientific community is fairly sure of the outcome, and it is not pretty. The gases act like a greenhouse, capturing solar energy in the atmosphere—which is why they're called greenhouse gases—and gradually the earth warms up. Glaciers and the polar ice caps melt, ocean currents change and ocean levels rise. It is not yet clear how long this will take to happen, but it appears that the northern polar ice cap will be gone within seventy years, and that America's famed Glacier National Park—a million-acre reserve in the state of Montana—will be without glaciers much sooner than that.

If we had access to a thousand planets, it might make sense to use one to conduct such an experiment, and if things turn out badly—as I believe this experiment will—move on to the next. But we don't have that choice; there isn't another planet we can move to. We're stuck here on Earth.

Unlike the other problems of globalization, global environmental problems affect developed and developing countries alike. And globalization, as it has so far been managed, has—with a few exceptions—not dealt adequately with the global environmental problem. In this chapter, I explain both why it has proven so difficult and what can be done: how we can take the economic forces of globalization—which have so far been injurious to the environment—and make them work to preserve it.

The Underlying Problem: The Tragedy of the Commons

In chapter 4 we talked about enclosure of the commons, and what happens when something that should be owned by all in common becomes sequestered as private property. But there is another aspect to common property—what is sometimes called the "tragedy of the commons."[1] When there is a common resource that can be used freely by all, each user fails to think about how his actions might harm others; each loses sight of the common good.

The expression first arose in a description of the common land upon which peasants in England and Scotland grazed their sheep in the late Middle Ages. As each farmer put more sheep on the commons, the amount of grass available diminished. But each farmer looked only at his own benefit, not at the costs that were inflicted on others, and so the problem grew.

Today, the problem is most simply apparent in the global fishing industry. Each country has an incentive to send out a larger fishing fleet in order to catch more fish—which, after all, are free to anyone who can catch them. But as more and more fishing boats are sent out, the stock of fish gets depleted, and the costs of fishing go up for everyone. Indeed, there is now evidence that, thanks to modern industrial fishing, boats are taking fish out far faster than the fish can reproduce.

The underlying economic principles are both simple and clear. When an individual or a country does something that hurts someone else, and for which they do not pay, there is a negative externality.[2] Generally, markets produce too much of things that generate negative externalities. Markets by themselves lead to too much pollution of the atmosphere and water; without government intervention, there will always be overgrazing of sheep on the commons.

The problem of the commons is easy to understand, and so, in some sense, is the solution: in one way or another, individuals have to be restricted in their use of it. There are two approaches. The first, which was used in Scotland in the sixteenth and seventeenth centuries, was privatizing the commons: the Scottish lords simply took the commons for themselves. As the owner, each one had an incentive to make sure that the land was not overgrazed. Of course, privatization had enormous impacts on the distribution of income. There may have been some gains in efficiency, but those farmers who were thrown off the commons were made far worse off, the Scottish lords reaping for themselves all the gains in efficiency—and more.

The privatization approach cannot, however, be realistically extended to the problems of global fisheries and global warming. It was relatively easy to enforce the privatization of grazing land through enclosures; but even if the fisheries could somehow be privatized, even if the enormous distributional issues that privatization raises could be solved, it would be close to impossible for any private owner to enforce his property rights. When enforcement problems arise, the state will inevitably become strongly involved in the management of resources; the question then is only the form of involvement. The second approach—and the only practicable one for global natural resources—involves government itself managing the common resource, restricting the amount of grazing or fishing. Throughout history, this is the way that common resources have often been managed. Communities impose social and legal controls that prevent the kinds of negative externalities represented by overfishing and overgrazing.

In principle, either approach—privatization or social control—can lead to an efficient and equitable outcome. The community could have calculated the "efficient" number of sheep that could be allowed to graze without damaging the common land just as well as a private owner could. Alternatively, the commons could have been privatized by being sold to the highest bidder with the proceeds divided equally. In practice, however, privatizations have always been marked by grave inequalities. In the enclosure movement, this was part of the rationale, as the rich and powerful saw an opportunity to redistribute wealth in their favor.

Nor has privatization always resulted in efficiency. Often private ownership itself is associated with environmental externalities, such as when the excessive use of fertilizer pollutes the watershed. When privatizations lack full political legitimacy, the owners have an extra incentive for excessive utilization, since they may not hold their property for long. As we have seen, this was the case in most of the Russian privatizations. In Brazil, forest privatizations have led to rapid deforestation, as the owners realize, perhaps rightly, that the government may recognize the importance of the forests as a national treasure and will in the future impose restrictions on cutting. With public management, on the other hand, officials may allow their relatives and friends to graze more sheep than others, while politicians may allow overgrazing in order to increase their vote, reckoning that the consequences will not become apparent for years. This is the fundamental dilemma of the management of the commons: historically, neither the private nor the public solution has consistently promoted both efficiency and equity.

Most environmental resources are not global in nature. The quality of ground water, lakes, or air usually affects only those nearby. If there is excessive air pollution in Los Angeles or Mexico City, it is local people who suffer. Sometimes, of course, effects go from one area to the next: my neighbor is hurt by the smoke when I burn leaves; Canada is hurt by the acid rain from midwestern American power plants. While there are some bilateral and regional agreements that attempt to deal with these cross-border environmental externalities (such as the 1991 U.S.-Canada Agreement on Air Quality), they cannot control the truly global environmental problems.

As imperfect as our ability to manage scarce natural resources and reduce negative externalities within a country may be, our ability to manage global natural resources and to reduce global negative externalities is even more circumscribed. The most important tools that are used domestically are not available. Within a country, if one person harms another, the injured party can sue. Forcing individuals to pay for the consequences of their actions is necessary for economic efficiency. Internationally, no such recourse is available. Even when the actions of one country damage the well-being of another, there is little that the injured party can do. China's pollution affects Japan. The Maldives and Bangladesh are almost certainly going to be seriously harmed by the rising sea level caused by global warming, to which the United States' pollution is contributing significantly. Japan can't sue

China, and the Maldives and Bangladesh cannot sue the United States and the other countries whose greenhouse gas emissions are leading to rising sea levels.

Within a country, problems of the commons can sometimes be dealt with, even if imperfectly, by privatization. To remedy the problem of the global commons, however, no one is seriously proposing the privatization option. The only sensible and workable remedy is some form of global public management of global natural resources, some set of global regulations on usage and on actions giving rise to global externalities. This is, of course, the way we deal domestically with many problems of negative externalities—when the actions of one person hurt another. You can't burn leaves in U.S. suburbs, because homes downwind will suffer from the smoke. You can't put a garbage dump on your land, because the smell makes your neighbor's life miserable. There are strong regulations restricting air and water pollution and toxic waste.

Democratic political processes have recognized the need for collective action. There are losers and winners—the polluters see their profits decrease, while those who might have got cancer, for instance, as a result of the pollution are better off. In spite of the opposition from those who see their profits diminished, most democracies have succeeded in passing some kind of regulation to limit pollution, recognizing that social benefits far exceed the costs.

Those who pollute the most always tend to minimize the problem. It is not surprising that the world's worst polluter, the United States, which adds almost 6 billion tons of carbon dioxide to the atmosphere every year, pretends that it does not believe the evidence that there is a need to curtail its greenhouse gas emissions. If greenhouse gases stayed only over the United States, America could conduct its own experiment; unfortunately, however, carbon dioxide molecules do not respect borders.[3] And though U.S. emissions affect the global atmosphere, the United States (or China, or any other country) does not have to pay for the consequences outside its borders. As a result, it has insufficient incentives to reduce its emissions—to curtail, for instance, its oil usage—and not surprisingly, has not reduced them.

While the extent to which the other advanced industrial countries have embarked on policies reducing pollution is both commendable and remarkable, it is hard to do anything really significant unless all the major countries, including the United States and China, participate. The central question, to which we turn in the next section, is: how can we marshal the cooperation of all to solve our most pressing global issue? I will show how we may be able to use the economic forces of globalization to achieve a better global environment.

GLOBAL WARMING

No issue is more global than global warming: everyone on the planet shares the same atmosphere. There are seven almost incontrovertible facts concerning global warming: (1) the world is warming—by about 1 degree Fahrenheit (0.6 degrees Celsius) in the last century; (2) even small changes in temperature can have large effects; (3) this rate of warming is unprecedented, even going back millions of years; (4) sea levels are rising—by some four to eight inches (ten to twenty centimeters) in the last century; (5) even small changes in sea level can have large effects—for example, a one-meter rise would inundate low-lying areas around the world, from Florida to Bangladesh; (6) there have been huge increases in greenhouse gases in our atmosphere, to a level that is estimated to be the highest in at least 20 million years, and which has been increasing at the most rapid rate seen for at least the past 20,000 years; and (7) it is possible that the pace of change in temperature could accelerate, with small increases in the concentration of greenhouse gases leading to even larger changes in climate than in the recent past.[4]

Virtually all scientists agree that greenhouse gases have contributed to global warming and rising sea levels, and they believe that most of this is a result of human activity (80 percent from burning fossil fuels, 20 percent from deforestation). Most agree, too, that there will be significantly more warming—between 2.5 and 10.4 degrees Fahrenheit (1.4 and 5.8 degrees Celsius) by the end of this century, and a further rise in sea level of eighty centimeters to one meter. The experts say we can expect more droughts and floods, cyclones and hurricanes, and that Europe's basic climate may change drastically, as the Gulf

Stream—the current off the east coast of North America that now warms it—shifts course.

In chapter 2, I described the great successes that Bangladesh is having with some of its development programs. But much of Bangladesh is a low-lying delta, which is great for rice growing but vulnerable to even small changes in sea level, and is frequently buffeted by deadly and destructive storms. If, as a result of global warming, those storms get more intense, the death toll will soar. Rising sea levels will leave one-third of the country—and half of the rice-growing land—submerged, and the 145 million Bangladeshis will be even more crowded than they already are. Their incomes, already barely above subsistence, will fall still further.

Nor is Bangladesh the country most likely to be worst hit by global warming. The Maldives, a small nation of 1,200 islands in the Indian Ocean with a population of 330,000—a tropical paradise—will be totally submerged in as little as fifty years, according to reliable predictions. Along with many other low-lying islands in the Pacific and elsewhere, it will simply be lost—our own twenty-first-century Atlantis.

Bangladesh and the Maldives are facing a fate far worse than that caused by even the worst of wars. Forces beyond their control, set in motion by the polluting actions of others—actions not intended to be harmful, but whose effects are global and destructive—threaten them with annihilation.

While a broad scientific consensus has emerged on global warming, there is still some uncertainty. It is true that things might not be as bad as today's doomsayers claim; on the other hand, they may prove to be far worse. This is no different from most of life: one always has to make decisions based on imperfect information. If, fifty or seventy years from now, the polar ice caps melt and parts of New York and London lie under water, along with some island nations in their entirety, it will be too late to reverse course. Even if we quickly reduced our emissions, the atmospheric concentration of greenhouse gases would be reduced only very, very slowly. This is why we need to start planning and acting now: it is far better to plan for the worst-case scenario than to wait and find that we didn't do enough.

As we think about whether the world can summon the strength and resources to tackle the threat posed by global warming, we should note that this kind of mobilization has been accomplished before. In 1946, in response to concerns that whales would become extinct, the International Convention for the Regulation of Whaling was signed. The agreement held, despite protests, and whale populations have largely recovered. Another agreement involved chlorofluorocarbon gases (CFCs), commonly used as refrigerator and air-conditioner coolants, which, it was found, were destroying the ozone layer and allowing cancer-inducing ultraviolet radiation to penetrate the atmosphere. The international community's reaction was swift. It took little more than a decade between the discovery of the problem and the signing, in 1987, of the Montreal Protocol. The convention was successful, and the phase-out of CFCs occurred faster than anticipated.

These examples show that the international community has been able, in the past, to respond to the challenge posed by a threat to the global environment. Can it respond to the enormous challenge posed by global warming?

The Rio Earth Summit

Some twenty years ago, as scientists first became aware of the changes taking place in the global climate, the world recognized that there was a potential problem and decided to study it. In 1988, the UN created the Intergovernmental Panel on Climate Change (IPCC), asking the world's leading experts to assess the scale of climate change and its likely impact.[5] The IPCC published three major studies between 1990 and 2001, concluding in each of them that there is indeed mounting evidence of the dangers of global warming. The evidence has also been reviewed in innumerable studies by the academies of science in individual countries, including one in the United States after President George W. Bush had seemingly cast doubt on the seriousness of global warming. The discussion here reflects the broad consensus on the basic findings.

As more and more scientific evidence came in, pressure mounted on politicians. In 1992, more than 100 heads of state gathered in Rio de Janeiro and resolved to do something about the problem. With the United Nations Framework Convention on Climate Change, they set up a procedure to develop a treaty that would restrict emissions. They did not agree on a specific target but committed themselves to "stabilization of greenhouse gas concentrations in the atmosphere at a level that would prevent dangerous anthropogenic interference with the climate system . . . within a time-frame sufficient to allow ecosystems to adapt naturally." The United States and 152 other countries signed the agreement, which became the cornerstone of the international community's attempt to come to grips with one of the most serious threats to our planet. A series of technical meetings followed, culminating in the next major worldwide conference on global warming, held in Kyoto.

The Kyoto Protocol

In 1997, more than 1,500 delegates, lobbyists, and heads of state from over 150 countries gathered in the historic Japanese city of Kyoto for the purpose of coming up with a treaty to cut greenhouse gas emissions worldwide. Their task was to devise a way of cutting emissions that was fair and efficient, that minimized the economic costs of reducing emissions and shared the burden equitably among the countries of the world. The resulting Kyoto Protocol made no immediate demands on the developing countries but called on each of the developed countries to cut back their emissions by specified amounts from 1990 levels—Europe as a whole by 8 percent, the United States by 7 percent, Japan by 6 percent—by 2012.[6]

The countries that came together at Kyoto recognized that the agreement constituted only rough justice, but that rough justice was better than the whole world suffering from the failure to do anything at all. Although there was some sensitivity to differences of circumstances—Norway, for example, which produces most of its electricity through hydropower, has little leeway to reduce pollution and was actually allowed to increase its emissions by 1 percent—other countries that had already made efforts to move out of polluting fossil fuels by using nuclear energy, like France, were required by the protocol to reduce their emissions just the same as those countries which had made no efforts.

The developing countries, including India, China, and Brazil, took the view that the high levels of greenhouse gas accumulations in the world's atmosphere are largely the result of the past sins of the developed countries, whose factories, cars, and power plants have been burning fossil fuels for decades; it is the profligate consumption of the advanced industrial countries that is largely responsible for the one-third increase in greenhouse gases in the atmosphere over the last 250 years. Not only would it be unfair to make developing countries pay for the past sins of the developed world, they argued, but—given their struggle to grow and pull their citizens out of poverty—they should not be forced to bear any of the economic burden of reducing pollution.

To enhance the efficiency of the overall system of reducing emissions, a trading mechanism was introduced; such tools had already been successfully applied in the United States in reducing sulfur dioxide emissions. If it was more expensive for one country to reduce its pollution than for another, the high-cost country could buy pollution reduction credits from the low-cost country; through these "carbon trades," the surplus reduction by one would offset the shortfall in the other. While some environmentalists disliked the notion that pollution could be bought and sold like any other goods, economists argued that this was necessary if pollution reduction was to be achieved efficiently, and the economists eventually prevailed. Potential cost savings as a result of carbon trading are enormous—for the United States, for instance, the cost of meeting its commitments could be reduced by 60 percent.[7] Today, such a carbon trading system is actually working.

The United States and Kyoto

Since the United States is the world's largest economy, it is no surprise that it is the world's largest polluter. When economies produce more, they pollute more. But some economies pollute more per dollar of GDP; that is, the way they produce is worse for the environment than the way other countries produce. Developing countries often pollute a great deal per dollar of GDP, because they have old and inefficient cars and machines. Among the developed countries, the United States is one of the worst. As of 2003, the United States was about as energy efficient as Uruguay and Madagascar. Britain, Ireland, Denmark, and Switzerland use two-thirds as much energy per dollar of GDP; Japan uses half as much.[8] Given the United States' relatively high level of energy utilization per dollar of GDP, and its high level of technological capability, it should have been relatively easy for the United States to meet its Kyoto commitment. Simply matching Japan's energy efficiency would have reduced U.S. emissions by more than half.[9]

Instead, the United States refused to play ball. Even before the United States signed the protocol in Kyoto, the Senate passed (with no dissensions) the Byrd-Hagel Resolution, stating that it was the sense of the Senate that the United States should not be a signatory to any protocol that did not include binding targets and timetables for developing as well as for industrialized nations, or which "would result in serious harm to the economy of the United States." Given the strong opposition of the Senate, the Clinton administration did not submit the Kyoto Protocol for ratification, and on March 13, 2001—only two months after taking office—President Bush released a letter addressed to four Republican senators assuring them of his opposition to the protocol and reneging on a campaign promise to regulate carbon dioxide emissions. Nonetheless, the rest of the world went ahead, and with Russia's ratification on October 22, 2004, the treaty came into force. By February 16, 2005, the date it went into effect, 141 countries, accounting for 55 percent of greenhouse gas emissions, had ratified the protocol.

With the United States out of the picture, however, progress in reducing greenhouse gases will be severely limited. The United States emits close to 25 percent of all greenhouse gases. Wyoming, the least populous state, with only 495,700 people, emits more carbon dioxide than seventy-four developing countries with a combined population of nearly 396 million. The carbon dioxide emissions of Texas, with a population of 22 million, exceed the combined emissions of 120 developing countries with an aggregate population of over 1.1 billion people.[10] Part of the reason for the refusal of the United States to go along with Kyoto is clear: doing anything about global warming imposes costs on some influential industries—automobiles, oil, and coal. The United States also bears less of the brunt of global warming; some economists and businessmen have noted that parts of the United States may be better off, as growing seasons in the northern states lengthen. At the 2006 annual meeting in Davos, those from the oil industry talked about the new opportunities that global warming is providing: the melting of the polar ice cap will make the oil beneath the Arctic Ocean more accessible. Though water levels on the eastern seaboard may rise and some land may be submerged, there is no comparison with the devastating effects global warming is having on countries like Bangladesh and the Maldives. However, Hurricane Katrina revealed a major flaw in this selfish calculus: because of the United States' vast wealth, the value of the potential damage, even if it is less extensive, will be enormous.

The Bush administration argued that the cost of restricting emissions is just too high relative to the benefits. To most of the world, this argument was outrageous: here was the richest country in the world complaining that it could not afford to implement sound environmental policies, at the same time as other developed countries are managing to reduce their own levels of pollution to a fraction of that of the United States, even on a per dollar of GDP basis. Japan, Germany, France, and Sweden are all emitting greenhouse gases at a rate no more than half that of the United States, yet these countries' citizens live comfortable, satisfying lives—by some measures, their living standards are higher than those in the United States.[11]

It is understandable that corporations do not want to spend money to reduce emissions, but it is unacceptable to let them sabotage global efforts to curb global warming.[12] Rather, U.S. firms would do well to learn from their Japanese competitors. During the oil price shock of the 1970s, when the price of oil more than quadrupled, Americans began to buy Japanese cars because they were more fuel efficient; Detroit, continuing to produce gas guzzlers, could not compete. Rather than turning to its engineers to produce a more fuel-efficient car, Detroit turned to its lawyers and lobbyists to ensure that the government did not force it to do so. With the Iraq war leading to soaring gasoline prices (they increased by 114 percent from 2002 to 2006), it appears now that Detroit bet the house and lost; its losses were so great that the bonds of those great bastions of American capitalism, Ford and GM, were downgraded to junk status. Their strategy—ignoring global warming in order to increase their profits by selling more gas guzzlers—was immoral; it also turned out, ironically, to be unprofitable.

Of course, even if there are significant costs to reducing emissions, as the world's richest country America is in the best position to afford it. Instead (unlike Europe and Japan), it used the exemption of the developing countries from the Kyoto strictures as another excuse for doing nothing. The developing countries point out that the United States emitted, over the course of the twentieth century, 50 percent more greenhouse gases than all of the world's developing countries combined.[13] The argument is not that developing countries should not work to reduce their greenhouse gas emissions. It is in their economic interests: many are profligate in their use of energy, and both their economy and the environment would benefit if they were more energy efficient. Many, including China, have high energy subsidies, which make no sense in today's world; there are better ways of encouraging industrialization.[14] But while I believe that it is in their interests to do so—and I believe that there is a moral obligation on the part of every one to protect our precious and irreplaceable atmosphere—I do not believe that it is unfair to put the brunt of the cost of adjustment on the world's richest country and the world's greatest polluter.

As America refuses to live up to its global moral responsibility, there are many, including those in the Bush administration, who believe—or perhaps simply hope—that somehow technology will come to the rescue. Somehow, innovation will so increase efficiency that emissions will go down on their own; or even better, someone will discover a better alternative to energy from coal, oil, or gas. This may happen, but we simply cannot let the survival of the world depend on our good luck. Moreover, the likelihood of better technologies being developed depends in part on incentives. Kyoto, with its strict limits on emissions, provides the appropriate incentives.

We saw in chapter 4 how the Bush administration, in advocating strong intellectual property rights, stresses the importance of incentives. In the context of global warming, it seems to ignore them. It called for voluntary reductions in energy usage: people should just behave better. Normally, we do not rely on voluntarism as a basis for using resources well. We do not say, when the supply of oranges has decreased because of a frost, "Please, voluntarily reduce your consumption of oranges." We rely on the price system. People conserve on their use of resources because they have an incentive to do so, because they have to pay for that usage. A clean atmosphere is a resource just like any other; there is a social cost to polluting, and people should have to pay that cost.

Finally, in 2006, the Bush administration seemed to recognize that the production of knowledge is a public good, justifying government support. It provided some money for research into alternatives to fossil fuels. Its support, however, was very limited; and there is a need for public research to be complemented with private, which is why "getting the prices right"—that is, making households and firms pay the social costs of emissions—is so important.

MAKING GLOBALIZATION WORK: CONFRONTING GLOBAL WARMING

Global warming is a global problem, yet no one wants to pay to fix it. Everybody wants a free ride on the efforts of others. But it is in everybody's interest that the world act collectively to do something.

If we stay within the Kyoto framework, and if it is to work, three problems have to be addressed. First, if the United States is to be brought along, it is clear that the developing countries must be included also, but we need to find a fair system of setting targets for them. Second, if there is an agreed set of targets, there must be some way of enforcing them; otherwise, so long as there is a cost to reducing emissions, there will be incentives not to meet obligations. Third, compliance will be much easier if the cost of reducing emissions is lowered, so we need to find ways of lowering these costs.

Emissions Targets for Developing Countries

While under the Kyoto Protocol developing countries have no obligations, it is clear that if the world is to do something meaningful about global warming they too will have to reduce their emissions. The "business as usual" approach simply will not work anymore: a world in which everyone pollutes at the rate that the United States currently does—let alone the rate at which it will pollute in twenty years' time unless something is done—is a world writing the script for its own doomsday scenario. Already, in 2005, developing countries are expected to account for nearly 40 percent of global greenhouse gas emissions, and by sometime around 2025, on current projections, developing countries will be emitting more greenhouse gases than the developed world.[15] Though their emissions on a per capita basis are much less, their incomes and populations are rising, and so their aggregate emissions are rising too.[16]

Under the Kyoto Protocol, each developed country is obligated to reduce emissions by a certain amount, and so there has to be agreement about the target for each country. The current system focuses on reduction of emissions relative to 1990: the more a country polluted in 1990, the more it is entitled to pollute in the future. The United States polluted more, so according to the system it should have the right to continue to pollute more.

The basic principle of the Kyoto Protocol—targets based on reductions from 1990 levels—makes no sense to the developing countries. By this logic, the poor countries, which polluted less in 1990, have less of an entitlement to pollute in the future. They naturally ask, "By what right are the developed countries entitled to pollute more than we are, simply because they polluted more in the past?" Their argument goes the other way: because the United States polluted more in the past, it should be made to pollute less in the future. At the very least, they argue, they should have the right to emit the same amount per capita as the United States. But with U.S. emissions presently some seven times higher per capita than those of China—twelve times that of the developing world as a whole—such an agreement would mean that it will be decades before emission restrictions on developing countries are binding.[17] Even if the United States kept the level of emissions per capita to its 1990 level (which it has so far failed to do), China, at its current rate of growth, will have more than 200 years before its emissions per capita catch up.[18]

The United States has not provided any coherent defense of why it should be entitled to pollute more than others; no one has really provided a reasoned defense of the premise underlying Kyoto. America might take a stance that the level of pollution allowed should be related to production, and since America produces more it should be allowed to pollute more. If the Kyoto approach is to work, a compromise will have to be found between targets based on emissions per dollar of GDP and targets based on emissions per capita. If the standard is based even partly on emissions per capita, the United States will have to increase its energy efficiency at a far, far higher rate than it has done so far. Right now, there seems to be little prospect of the United States voluntarily doing this, and thus the targets approach is destined, I believe, quickly to reach an impasse. The United States remains intransigent, and the developing countries can see no good reason why they should sacrifice their incomes and growth to help Americans. We are in a stalemate—and, meanwhile, the world is getting rapidly more polluted.

Carrots and Sticks: Improving Compliance

Whatever targets are set, there will have to be incentives—carrots and sticks—to ensure that countries first join the protocol and then comply with it. The renunciation of the Kyoto Protocol by the United States shows that we need some way of pressuring countries to participate. If moral suasion does not work (which it hasn't) and we cannot find enough carrots, there are some effective sticks—and their very existence means that they may not even have to be used. There is already a framework for doing this: international trade sanctions. The Montreal Protocol on ozone-depleting gases employed the threat of trade sanctions—though they never had to be used. Unfortunately, trade sanctions were not built into the Kyoto Protocol.

Still, within the World Trade Organization, we have the precedents we need. When the United States tried to force Thailand to use turtle-friendly nets for catching shrimp—the nets then being used were killing endangered species of turtles—by threatening to prevent shrimp caught in the old-fashioned nets from entering the United States, the WTO sustained the U.S. position. It established the principle that maintaining the global environment is important enough that normal access to markets, which the WTO guarantees for its members, can be suspended when a country's export industries endanger it. When the United States brought its case, it apparently did not consider the long-term implications, but some on the WTO appellate body were aware of the far-reaching consequences of their decision. The precedent set by this case should apply to U.S. companies that pollute through high levels of greenhouse gas emissions during the manufacturing process; Europe, Japan, and others adhering to the Kyoto Protocol should restrict or tax the import of American goods that are produced in ways that unnecessarily pollute the atmosphere. Preserving endangered turtles is valuable, but preserving our planet's atmosphere is infinitely more important. If, as the United States argues, trade actions are justified in the former case, they are even more justified in the latter.

One can look at America's energy profligacy another way: by not paying for the damage they do to the environment, U.S. businesses are in effect getting a subsidy. One of the main purposes of the WTO is to create a level playing field; subsidies distort the playing field, which is why countries are allowed to offset subsidies through countervailing duties; and this should be the case for hidden subsidies—not forcing firms to pay for the environmental damage they inflict—as well as for open subsidies.

There are several ways this could be done. Under the current WTO regime, the countries of Europe and elsewhere could impose countervailing duties to make up for the subsidies that American producers, using energy-intensive technologies, implicitly receive when they degrade the global environment without paying the costs. Assume, for instance, that American-produced steel sells for $500 per ton, and that in the process of producing that ton of steel two tons of carbon are emitted. The price of a ton of carbon is, say, $30 (its price in the European carbon trading system in early 2006). Because America did not join the Kyoto Protocol and its firms are under no obligation to reduce carbon emissions, they are in effect being subsidized to the tune of $60 per ton of steel. Thus, European and other countries could levy a tax on American steel of $60 per ton (just over 10 percent). Energy-intensive products like aluminum would face higher duties. This would provide strong incentives for America to sign on to the Kyoto Protocol and to reduce emissions. Even widespread discussion of the possibility of imposing these tariffs might induce the United States to act.[19]

I have discussed this idea with senior officials in many of the advanced industrial countries that are committed to doing something about global warming. And while, almost to a person, they agree with the analysis, almost to a person they also show a certain timidity: the proposal is viewed by some as the equivalent, in the trade arena, of declaring nuclear war. It is not. It would, of course, have large effects on the United States, but global warming will have even larger effects on the entire globe. It is just asking each country to pay for the full social costs of its production activities. Following standard practice, the pressure of trade sanctions could gradually be increased; and almost surely, as America recognizes the consequences, its policies would be altered—as they have been in other instances where the United States has been found in violation of WTO rules.

Much is at stake. The United States and the other Western countries have shown that they are willing to risk a great deal to prevent nuclear proliferation—in the case of North Korea, they even faced the possibility of war. Surely the dangers to the world from global warming are important enough to warrant risking the displeasure of a rogue nation that seems willing to put the well-being of the planet in jeopardy simply in order to maintain its emissions-profligate lifestyle.

The Rainforest Initiative: Improving Efficiency

Efficiency requires atmospheric greenhouse gas concentrations to be reduced in the most cost-effective way. Most attention has been focused on the reduction of emissions, but there is another way: to remove carbon dioxide from the atmosphere and store it. That is what trees do. In photosynthesis, plants take carbon dioxide out of the atmosphere, emitting oxygen and storing the carbon. Thus, planting forests reduces the concentration of greenhouse gases, while deforestation makes matters worse. Deforestation is bad for the atmosphere for two reasons: first, there are fewer trees converting carbon dioxide into oxygen; second, carbon that is stored in the wood is released into the atmosphere as the wood is burned. In recent years, about 20 percent of the increase in atmospheric concentrations of greenhouse gases came from deforestation. In other words, the damage done by deforestation is comparable to the damage done by the world's largest polluter, the United States.

But the 2.7 billion people in the over sixty developing countries that are home to these tropical forests are not being compensated for their valuable environmental services. Tropical rainforests not only reduce the level of carbon in the atmosphere; they help to preserve biodiversity. As we noted in chapter 4, many medicines have been derived from this precious resource. Compensation would not only be fair and help the economies of the rainforest countries; it would provide incentives for them to maintain their forests, which would be of enormous environmental benefit to all.

We can obtain rough calculations of the carbon benefits of reducing the annual rate of deforestation by, say, a modest 20 percent. At the price of $30 per ton of carbon, the annual value of the avoided deforestation—the value of the increase in atmospheric carbon that would have occurred as a direct result of those trees being cut down—is between $30 billion and $40 billion a year. (By comparison, all foreign assistance to developing countries is around $60 billion.) In addition, as we have noted, the forests "clean" carbon dioxide out of the atmosphere. The annual "negative emissions" of the rainforest countries are estimated (using the $30 a ton figure) at some $100 billion a year.[20]

While Kyoto recognized the role that planting forests could play—countries are given "credit" for planting trees—it did nothing about deforestation. This was a big mistake, for it makes countries like Papua New Guinea doubly better off if they cut down their ancient hardwood trees and replant: they get money from both the cut trees and the replanting. But this makes no sense—countries should be given incentives to maintain their forests.[21]

In principle, this would be relatively easy to do under the carbon trading system. Energy companies in Europe are allowed to buy "carbon offsets" (allowing them to emit more carbon than they otherwise would be allowed to do) by paying for the planting of a forest ("carbon sequestration") in some developing country. Led by Papua New Guinea and Costa Rica, a group of developing countries calling themselves the Rainforest Coalition put forward an innovative proposal in January 2005, offering to commit to greenhouse gas limits but asking in return that they be able to sell carbon offsets not just for new forests but for avoided deforestation.[22] Countries would, under this proposal, be paid for not cutting down their forests. Adopting it would ensure the most efficient use of these resources from the global perspective—which is to maintain them as forests rather than to harvest them for timber.

Without some form of compensation for maintaining their forests, developing countries have, unfortunately, neither the means nor the incentive to continue underwriting conservation. Cutting down the hardwood forests—even when they presently receive only a small fraction of the final price the wood fetches in, say, New York—is the only way impoverished people in these countries can make ends meet. Much of the logging in Indonesia, Papua New Guinea, and other tropical countries is, in fact, illegal or the result of corrupt contracts. The countries do not currently have the resources to stop illegal logging; carbon offset payments would provide them with the resources and the incentive to stop it, and the countries of the Rainforest Coalition have made a commitment that they would.

Some have suggested waiting to address the issue until 2012, when a revised protocol is supposed to come into effect. But can we wait? At current rates of deforestation, the combined contribution to greenhouse gas concentrations from deforestation in Brazil and Indonesia alone will offset some 80 percent of the emissions reductions gained through the Kyoto Protocol. (Moreover, some of the ancillary damage— the loss of old hardwood forests and biodiversity—may be irreversible if we do not act soon.) It is urgent that we fix the problem now and not accede to yet another impulse to delay.

What is so impressive about the new rainforest initiative is that it comes from the developing countries themselves, demonstrating their creativity and social commitment. For the first time, developing countries seem willing to undertake the kinds of commitments that Europe, Japan, and other advanced industrial countries—though not the United States—are making to avoid a global disaster.

An Alternative Framework

Kyoto was the natural approach to global warming. The problem: excessive emissions. The solution: reduce the emissions. But life is never so simple or easy. The principal difficulty with Kyoto, as we have noted, is agreeing by how much each country should reduce its emissions. Underlying Kyoto were two broad principles: all developed countries would be asked to make approximately the same reductions; and the developing countries would be treated differently—though discussion of exactly what that meant was postponed for the future.

It was an achievement that the rest of the world put aside their quibbling and reached an agreement; it was a disappointment that the United States walked away. The momentum behind the Kyoto Protocol gives us good reason to stay within that system, but I doubt that we will find an agreement acceptable to both the United States and the developing countries within the Kyoto approach. There is no set of generally accepted principles for allocating rights to usage. Should those who have polluted more in the past be entitled to pollute more in the future? Or should they face larger reductions in their emission allowances, to compensate the world for past damages? Should allowances be set on a per capita or a per dollar of GDP basis? This problem of distribution is at the core of the international community's failure to deal with global warming.

There is an alternative framework for approaching the reduction of emissions that employs the market mechanism more directly, and thus, perhaps, has a better prospect of appealing to the United States. There is a social cost associated with any activity that emits greenhouse gases, which those engaged in the activity do not pay. That is why, of course, they emit too much. The simple solution: make people pay for the full costs of what they do; that is, make them pay for their pollution.

The way to do this is to have all the countries of the world impose a common tax on carbon emissions (that is, taxing the externality of emissions) or, equivalently, a tax on oil, coal, and gas at rates reflecting the emissions they generate when burned. Firms and households would respond to this tax by reducing usage, and thereby emissions. The tax would be set high enough to achieve a global reduction in emissions equivalent to that envisaged in the common targets approach of Kyoto. But the level of emissions could well differ from country to country, depending on their circumstances. A very hot country might, for example, use more energy for air-conditioning than a country with more moderate temperatures.

What, then, is the advantage? It avoids setting national target levels. The reason that setting target levels is so difficult is that each country's circumstances differ. The United States might claim that, because distances within the country are greater and GDP higher, it should be allowed to pollute more. France might claim that, because its pollution rate per capita is already one-third of the United States', it is unreasonable to demand that it should cut its emissions any further. The developing countries claim that since they are poor and racing to catch up with the living standards of the developed world, it is difficult for them to reduce emissions.

Setting target levels is so contentious because allowing a country high emission levels is tantamount to giving it money—a fact that has become more obvious with the advent of carbon trading. As I have noted, countries that exceed their reduction targets can sell the excess (the amount of pollution they are allowed to generate but don't) to countries that have fallen short. A higher emissions target (that is, a target involving a smaller reduction) means that a country either has more emission rights to sell or has to pay less to other countries to compensate for its shortfall.

Under the common tax proposal, all of these issues are avoided. Each country would keep the revenue it receives from the tax, rather than having to give the money to another country. As a result, the costs of pollution reduction are relatively small. In fact, the country as a whole might be better off; it can use the revenue from the carbon tax to reduce other taxes, such as those on savings, investment, or work. These lower taxes would stimulate the economy, with benefits far greater than the cost of the carbon tax. This is consistent with a general economic principle: it is better to tax things that are bad (like pollution) than things that are good (like savings or work).[23]

Of course, the energy industries in almost every country will not like this. All companies prefer getting a subsidy, which is what allowing countries the unfettered right to pollute amounts to. I do not want to suggest that it will be easy to overcome the weight of the energy-producing and energy-using lobby. It may only be possible under the threat of the kinds of trade sanctions described earlier.

A Way Forward

From an economic point of view, both the common tax and the targets approach can achieve the necessary reductions in emissions, and both can do so efficiently as long as there is carbon trading. With the world having invested so much in the development of the targets approach, it is understandable that there will be reluctance to abandon it. Yet there is not even a glimmer of an idea at the moment of how targets can be set that will be acceptable both to the United States and to the developing countries. Global warming is too great a threat to the well-being of our planet for us just to ignore this crisis and pray that a resolution will eventually emerge.

There is a third alternative that synthesizes the distributive advantages of the common tax measure with the forcefulness of the targets approach. The big advantage of the common tax approach is that it avoids the most difficult issues of figuring out how much each country should reduce its emissions; each country agrees to provide appropriate tax incentives not to emit, but garners for itself the revenues from the taxes. We can easily estimate the resulting reductions in carbon emissions for each country that would result, and use those estimates as a basis of determining appropriate targets assigned to each country. The country could, if it chose, use taxes to achieve those targets. But it could use alternative measures, like direct controls on technology, such as requiring higher mileage standards for cars.

Any system, whether of targets or taxes or a combination of the two, will require periodic revision. Technology may one day enable us to reduce emissions faster, at a lower cost, than we anticipate today; in that case, we should tighten the targets. A commonly imposed tax on emissions may yield more or less reduction than anticipated, in which case we may want to lower or raise the tax rate.[24] While the burden of adjustment on most countries—other than the producers of oil and gas—is likely to be limited, some countries may be more seriously affected than others; a periodic review could identify circumstances in which some countries would be given longer times for adjustment (just as, as I argued in chapter 3, some developing countries need a longer time to adjust to the opening of trade).

And any system, whether of targets or taxes, will require enforcement—including action against countries that refuse to cooperate. Global warming is too important to rely on any country's goodwill. If the United States continues to refuse to reduce its emissions, trade sanctions should be imposed. If this is done, I feel confident that America will respond to the economic incentives provided. (I hope this is not just my bias as an economist.)

Making economic globalization work will be of little use if we cannot solve our global environmental problems. Our atmosphere and oceans are global resources; globalization and so-called economic progress have enhanced our ability to exploit these resources more ruthlessly and at a pace faster than our ability to manage them has grown.

Jared Diamond, in his best-selling book *Collapse,* puts it most clearly. After describing how numerous other civilizations faced their demise as a result of ignoring the environment, he goes on to explain that:

> Our world society is presently on a non-sustainable course. . . . [B]ecause we are rapidly advancing along this non-sustainable course, the world's environmental problems *will* get resolved, in one way or another, within the lifetimes of the children and young adults alive today. The only question is whether they will become resolved in pleasant ways of our own choice, or in unpleasant ways not of our choice, such as warfare, genocide, starvation, disease epidemics, and collapses of societies. While all of those grim phenomena have been endemic to humanity throughout our history, their frequency increases with environmental degradation, population pressure, and the resulting poverty and political instability.[25]

In this chapter, I have made three appeals. For the developing countries, doing something about global warming is in their own interests: indeed, among them are the countries that will be most hurt by global warming. Curtailing their energy usage can be good for both the environment and the economy.

For the United States, there is a moral imperative to join the rest of the world in addressing the problem of greenhouse gases. The devastation the United States risks bringing on other countries is as bad as any war it might wage against them. It may mean them no harm—just as it means no harm with its cotton subsidies—but there are costs to its actions, and it must take responsibility for those costs. As the world's leader, if it evades its responsibility, it cannot expect others to live up to theirs; and if we all fail, we all suffer—including the United States. Some interests within America will be hurt if the United States deals forcefully with global warming, but I believe the country as a whole will actually be better off. Even if it costs the United States something, the United States can afford it. Far better to make small expenditures now in order to reduce the risks of much larger expenditures down the line.

Finally, even as I commend Europe and Japan for making commitments, on their own, to reduce their emissions and working hard to fulfill those commitments (though they will have to work still harder), I argue that these commitments by themselves will remain largely gestures unless the rest of the world can be brought along. This may entail significant assistance to developing countries; it also entails getting tough with the United States.[26] I have argued that simply as a matter of fairness in trade, it is intolerable for one country to provide, in effect, emission subsidies to its firms. Globalization has meant the increasing interdependence of the countries of the world. Withholding the benefits of globalization through trade sanctions can be an effective instrument for bringing accountability to those that despoil the global environment. We have created an international trade law that was designed to ensure that trade is fair; while critics worried that the WTO would place commercial interests over the environment, the WTO has in fact shown that it can be used to force better environmental behavior. But the WTO does not act on its own. Europe must use the foundations of the international trade law we have created to force any recalcitrant country, any rogue state—including the United States—to behave responsibly. Europe has to be willing to use the enormous power of economic globalization to address the world's most important global environmental problems.

In the aftermath of the Christmas tsunami of 2004, there was much discussion of the importance of having an early warning system, so that people could take action to avoid the next disaster. We are getting early warnings about global warming loud and clear. But we have yet to respond.

CHAPTER 7

The Multinational Corporation

The Left (and the not-so-Left) often vilifies corporations, portraying them in documentaries such as *The Corporation* and *Wal-Mart: The High Cost of Low Prices* as greedy, heartless entities that place profit above all else. Many instances of corporate evildoing have rightly become infamous, the stuff of legend: Nestlé's campaign to persuade Third World mothers to use infant formula instead of breast milk to feed their children; Bechtel's attempt to privatize Bolivia's water (documented in the film *Thirst*); the U.S. cigarette companies' half-century conspiracy to persuade people that there was no scientific evidence that smoking is bad for health even as their own research confirmed that it was (wonderfully dramatized in the film *The Insider*); Monsanto's development of seeds that produced plants which in turn produced seeds that couldn't be replanted, thereby forcing farmers to buy new seeds annually; Exxon's massive *Valdez* oil spill and the company's subsequent attempts to avoid paying compensation.

For many people, multinational corporations have come to symbolize what is wrong with globalization; many would say they are a primary cause of its problems. These companies are richer than most countries in the developing world. In 2004, the revenues of U.S. car company General Motors were $191.4 billion, greater than the GDP of more than 148 countries. In its fiscal year ending 2005, U.S. retailer

Wal-Mart's revenues were $285.2 billion, larger than the combined GDP of sub-Saharan Africa. These corporations are not only rich but politically powerful. If governments decide to tax or regulate them in ways they don't like, they threaten to move elsewhere. There is always another country that will welcome their tax revenues, jobs, and foreign investment.

Businesses pursue profits, and that means making money is their first priority. Companies survive by getting costs down in any way they can within the law. They avoid paying taxes when possible; some skimp on health insurance for their workers; many try to limit spending on cleaning up the pollution they create. Often the bill is picked up by the governments in the countries where they operate.

Yet corporations have been at the center of bringing the benefits of globalization to the developing countries, helping to raise standards of living throughout much of the world. They have enabled the goods of developing countries to reach the markets of the advanced industrial countries; modern corporations' ability to let producers know almost instantly what international consumers want has been of enormous benefit to both. Corporations have been the agents for the transfer of technology from advanced industrial countries to developing countries, helping to bridge the knowledge gap between the two. The almost $200 billion they channel each year in foreign direct investment to developing countries has narrowed the resource gap.[1] Corporations have brought jobs and economic growth to the developing nations, and inexpensive goods of increasingly high quality to the developed ones, lowering the cost of living and so contributing to an era of low inflation and low interest rates.

With corporations at the center of globalization, they can be blamed for much of its ills as well as given credit for many of its achievements. Just as the issue is not whether globalization itself is good or bad but how we can reshape it to make it work better, the question about corporations should be: what can be done to minimize their damage and maximize their net contribution to society?

Before answering that question, I want to dispose of one charge that is largely, though not totally, unfair. Corporations are often blamed for the materialism that is endemic in developed societies. For the most part, corporations simply respond to what people want—for instance, the need to get from one place to another, which cars and motorbikes make easier; if cars and motorbikes are fancier or larger than they need to be, it is mainly because consumers like ones that are fancier or larger, and buy them. Still, it must be admitted that corporations have sometimes worked to shape those desires in ways that enhance their profits, and at least some materialistic excesses can be attributed to their efforts. If advertising did not enhance desire, they would not spend billions of dollars on advertising every year.[2] Food companies teach children to want sugary cereals that are bad for their teeth; auto companies campaign against public transportation—and in some cases actively removed it—regardless of the effect on the environment. Los Angeles once had the world's largest urban rail system (1,100 miles of track), until a group led by General Motors bought it out, dismantled it and replaced it with GM buses.[3]

One or two instances of corporate misbehavior might be overlooked, but the problems are clearly systemic. Whenever there are systemic problems, economists look for systemic causes. The primary one is obvious: corporations are in the business of making money, not providing charity. Therein lies both their strength and their weakness. Money is a powerful incentive, and the desire to make it can bring enormous benefits to everyone. When things go well, international corporations can marshal enormous resources, spread the most advanced technology, and increase available markets exponentially. But too often they are encouraged to do the wrong thing. Corporate incentives can be reshaped. If we are to make globalization work, they will have to be.

Here again, the eminent eighteenth-century economist Adam Smith has often been misunderstood. He argued that individuals, in pursuing their self-interests, would advance the broader interests of society: that incentives to outcompete rivals would lead to lower costs and to the production of goods consumers wanted, and that consumers, and society more generally, would benefit from both. In Smithian economics, morality played no role (though Smith himself was intensely concerned about moral issues, as evidenced in *The Theory of Moral Sentiments*, a work that preceded *Wealth of Nations*). Individuals did not have to think about what was right or wrong, only about what was in their own self-interest; the miracle of the market economy was that, in doing so, they promoted the general welfare. Building on this logic, many economists believe that the first—some go so far as to say the only—responsibility of corporations is to their shareholders. They should do whatever it takes to maximize stock market value or profits. In this extension of Smithian economics, if morality enters the picture at all, it does so only to enjoin firms to think about the interests of shareholders above all else—in fact, to think *only* of shareholders.

Sometimes, markets do work in the way that Smith argued—the large increases in living standards over the past two centuries are, in part, testimony to his insights. However, even Smith realized that in an unfettered market economy private incentives are often not aligned with social costs and benefits—and when that happens, the pursuit of self-interest will not result in the well-being of society. Modern economists call these misalignments "market failures." Market failures arise whenever there are externalities, consequences of an individual's or a firm's actions for which they do not pay the cost or receive the benefit. Markets, by themselves, lead to too little of some things, like research, and too much of others, like pollution.[4]

Much of public policy and economic theory in the last hundred years has been directed at identifying major market failures and analyzing the most efficacious and least costly ways of correcting them, for instance through regulations, taxes, and government expenditures. Modern economics has shown, similarly, that social welfare is not maximized if corporations single-mindedly maximize profits. For the economy to achieve efficiency, corporations must take into account the impact of their actions on their employees, on the environment, and on the communities in which they operate.

The environment provides one obvious instance in which private and social costs may differ, with enormous consequences. It costs more money to refine oil or generate electricity in ways that do not pollute the air. It costs more money to dispose of waste or to mine in ways that do not pollute the water supply. These are real environmental costs to society, but—at least before strong government regulations were established—they were not costs to the corporations involved. Without government regulation and pressure from civil society, corporations lack incentives to protect the environment sufficiently; they actually have an incentive to despoil it if doing so saves them money.

Bribery and corruption represent another area where social and private interests clash. Mining and oil companies can often reduce the cost of acquiring natural resources by bribing government officials for concessions. It is far cheaper to pay a government official a large bribe than to pay market price for oil or some other natural resource. In practice, companies in many industries pay bribes to get all manner of favors, such as protection from outside competition, which allows them to raise prices, or the overlooking of violations of environmental or safety regulations. In the amoral view of the modern corporation, if they can get away with it—if the expected return exceeds the risk and costs of being found out—then, were it not illegal, they would practically have an obligation to bribe, for that would increase the profits of the company and the return to shareholders.

In sophisticated economies such as that of the United States, outright bribery has been largely replaced by political campaign contributions, and the return may not be simply a road construction contract at above-market prices but a change in policy whose ramifications cost society far, far more.[5] Forty-one companies (including General Electric, Microsoft, and Disney), which invested—"contributed"—$150 million to political parties and campaigns for U.S. federal candidates between 1991 and 2001, enjoyed $55 billion in tax breaks in three tax years alone.[6] Pharmaceutical companies spent $759 million to influence 1,400 congressional bills between 1998 and 2004; the pharmaceutical industry ranks top in terms of lobbying money and the number of lobbyists employed (3,000). Their success reflects their investment: as we saw in chapter 4, the U.S. government has made their interests paramount in international trade negotiations, and under the new Medicare drug benefit the government is proscribed from bargaining for lower prices—a provision worth billions of dollars just by itself.[7] The "big five" U.S. accounting firms contributed $29 million to federal candidates and parties between 1989 and 2001, partly to shield themselves from threatened regulations. It worked—at least until the Arthur Anderson–Enron scandal made clear why such regulations are so necessary.

As a final example of the social impact of global corporations on developing countries, consider the impact on local communities. Corporate giants like Wal-Mart do not intend to weaken the communities in which they open their stores. They intend only to bring goods at lower prices—and it is these lower prices that have earned them such success. But as they drive out small businesses, they may, at the same time, hollow out the town. Small businessmen are often the backbone of a community, and as Wal-Mart squelches its competitors, it breaks that backbone. A few donations to local charities do little to compensate. Chapter 2 emphasized the important role that communities play in successful development; by weakening communities, corporations may, in the long run, even weaken development.[8]

Some of Wal-Mart's success is based on greater efficiency (better inventory management and logistics), but much is based simply on its market power, its ability to squeeze its suppliers and its workers. Its strict policy against union organizing means that its workers are often low-paid, and their low wages force down wages at Wal-Mart's competitors, so not only Wal-Mart workers are affected. Only about half of its 1.4 million employees are covered by health-care benefits. The U.S. state of Georgia's public program providing coverage for children who would otherwise be uninsured found that more than 10,000 of the 166,000 children it covers had a parent working for Wal-Mart—more than any other employer. Wal-Mart's health-care plan does not cover preventive care such as children's vaccinations, flu shots, or eye exams. As a result, taxpayers pick up costs that elsewhere are borne by employers.[9]

The problems of corporations pursuing policies that impose costs on society which the firm itself does not bear arise in all businesses—multinational and domestic, large and small. But there are several distinct reasons that large multinational corporations pose greater problems—problems which Smith, writing more than two hundred years ago, could not have fully anticipated. In Smith's time, businesses were relatively small and usually run by individuals who could be held accountable for any damage they did. The corporations of today are vast enterprises, some with tens of thousands of employees; though it is individuals within the corporation who make the decisions that determine what the firm does, these individuals are often not easily held responsible for the consequences of those decisions. While they seldom reap the full value of the increase in profits that follow from their good decisions, even more seldom do they pay the full social costs of their bad decisions.[10]

It is too easy for corporate managers to hide behind the corporate veil. Even after he admitted that he had been drinking prior to boarding the ship, Joseph Hazelwood, the captain of the ship responsible for the 1989 *Exxon Valdez* oil spill—a spill which did environmental damage valued in the billions—was given only a slap on the wrist, with a fine of $51,000 and 1,000 hours picking up garbage along Anchorage-area highways. The Indian government did try to prosecute Union Carbide executives for the thousands of deaths at Bhopal, where a chemicals plant exploded in 1984, but Union Carbide was an American company and the United States refused to cooperate. Charges against the executives, including CEO Warren Anderson, were brought before an Indian court in 1991; when they did not appear to face charges, India pressed for their extradition. Finally, in September 2004, the U.S. State Department refused the extradition request without explanation.

There are exceptions, but they are rare. Former WorldCom CEO Bernard Ebbers was convicted of responsibility for the $11 billion fraud that triggered the largest bankruptcy in U.S. history—because too many Americans had lost too much to let him just go free. He was sentenced to twenty-five years in prison, which is the longest sentence ever for a CEO found guilty of corporate crime while running a Fortune 500 company.

Making matters worse is limited liability, which essentially defines corporations. Limited liability is an important legal innovation, and without it modern capitalism almost surely could not have developed. Investors in corporations with limited liability are at risk for only the amount of money they invested in the company, and no more. This is quite different from partnerships, in which all partners in a firm are jointly responsible for the actions of the others. If a partnership makes a major mistake—say, in the case of an accounting firm, certifying the books of a company when it should not have done so (as Arthur Anderson did in the case of Enron)—then in principle all of the partners can be sued and may lose not only what they have invested in the partnership but their homes, cars, and savings as well, possibly forcing them to take refuge in personal bankruptcy. The theoretical advantage of an unlimited liability partnership is that each partner has a strong incentive to monitor the others, and that customers, knowing this, will have more trust. But when hundreds of partners are involved, the ability to monitor one another closely disappears, and the advantages of partnership are outweighed by the disadvantages. In fact, many accounting firms, which were traditionally organized as partnerships, have restructured themselves as "limited liability partnerships," combining the tax advantages of partnership with the protection of limited liability.

Limited liability has a major advantage: it allows huge amounts of capital to be raised, since each investor knows that the most he can lose is his investment. But limited liability can have large costs for society. A mining company can mine gold, making huge profits for shareholders, but leave behind poisonous tailings of arsenic-ridden waste. From both the social and financial point of view, the cost of cleaning up the mess may exceed the value of what is mined. But when the problem is discovered and the government demands a cleanup, the mining company declares bankruptcy and the public is left holding the bag. Thus, the people suffer doubly—from the environmental degradation and from the cost of the cleanup.

The list of companies that have inflicted costly damage—especially in developing countries—for which they have not had to pay, or for which they paid a fraction of what they should have paid, is long. The explosion at the Union Carbide plant in Bhopal is probably the most dramatic example: more than 20,000 people were killed and some 100,000 more bear lifelong health damage, including respiratory illness, eye disease, neurological and neuromuscular damage, and immune system impairment.[11] The total number affected was even larger; those eventually receiving compensation, including dependents, will probably number close to 600,000. The disparity between the terrible damage and what the company was forced to pay—an estimated

$500 per person—is also huge, by any reckoning. Dow Chemical has since bought the Bhopal plant, taking all of Union Carbide's assets but assuming none of the liability.

In Papua New Guinea, a large gold and copper mine, Ok Tedi, dumped 80,000 tons of contaminated material daily into the Ok Tedi and Fly Rivers over the course of a dozen or so years, as it extracted some $6 billion worth of ore. Once the mine was exhausted, the Australian-majority ownership, after admitting that it had vastly underestimated the environmental impact, just walked away, turning over its shares to the government—leaving the government, already strapped for funds, with the cleanup costs. What those will eventually amount to is still hard to assess, but it is clear that they will be vast and will be borne by the Papua New Guinean people.

Incentives are misaligned when a corporation does not bear these downside costs; this is the result of limited liability. When we add in the size of multinational corporations relative to the developing countries in which they operate, and the poverty of developing countries, we see a set of opportunities in which this misalignment can lead—and has led—to a host of serious problems. Developing countries need the jobs the corporations bring in, even if the environment, or the health of workers, is harmed. The mining and oil companies exploit this imbalance of power.

In Thailand and Peru, corporations threatened to move elsewhere if environmental regulations were enforced; in Peru, one mining company went so far as to pressure the government not to test children living near their mining operations to see if they had been exposed to health hazards. At one point, Papua New Guinea passed a law making it illegal to sue international mining companies outside the country even for the enforcement of health, environmental, or legal rights, fearing that such suits would discourage investment in the country. In a perfectly competitive market a threat to leave would not be a problem; if one mining firm pulled out, others would step in. But there are large barriers to entry—the development of a mine can cost more than a billion dollars, and entails a great deal of risk. If one company leaves, another may not fill the gap—or if it does, it may demand even more unfavorable terms.

Globalization has compounded the problems arising from the misalignment of incentives in modern corporations. Competition among developing countries to attract investment can result in a race to the bottom, as companies seek a home with the weakest labor and environmental laws.

As the case of Bhopal illustrates, the ability to hide behind borders makes it even more difficult to hold corporations and their officers accountable. Furthermore, the speed with which assets can be moved from one country to another means that even if there is a monetary judgment against a firm in one country, it may be impossible to collect.

At home, where companies are part of the fabric of the community, individuals often take some moral responsibility for their actions; they do the right thing even if they are not compelled to do so by laws or regulations or the threat of suit and even if there might be some short-term loss in profits. But when multinationals operate overseas, moral responsibility is weakened. Many executives would not even contemplate treating their workers or the environment at home the way they routinely do abroad. They may reason that overseas regulations are lax, that workers are lucky to have jobs, or that overall the country benefits from their investment. Despoiling the environment or ignoring basic working conditions is easier thousands of miles from the head office, and because the local people are poor, it is easy to consider their lives and land as being worth less than lives and land at home. Dow Chemical and Union Carbide executives may actually feel that $500 is ample compensation for a death or a life maimed in Bhopal. After all, with so much poverty and death in developing countries, to outsiders life may seem cheap; and accountants can note that life expectancy in India is shorter than in the United States, and incomes just a fraction.

Corporations often claim it is not their responsibility, but that of governments, to align private and public interests—by, for instance, passing regulations restricting pollution. But this lets corporations off the hook, by ignoring the fact that they routinely use their money to get laws and regulations passed that free them to pollute at will—thus ensuring that social and private interests are not aligned.[12] Politics is part of business strategy; corporations lobby strongly against environmental standards that cost them profits, and the payoff for these political investments is often higher than on any other investment.

While money speaks loudly in all countries, it speaks especially loudly in developing countries. With many corporations having more resources at their disposal than developing country governments, it is not surprising that corporate efforts to construct favorable regulatory environments are often successful. Unfortunately, it is all too easy for desperately poor countries—especially countries where governments are not democratically accountable—to succumb to corporate enticements.

Worse still, multinationals have learned that they can exert greater influence in designing international agreements than they can in designing domestic policies. Within Western democracies, there has been an attempt to temper the worst abuses of the market economy, and increasingly firms have become subject to environmental regulations. But the secrecy that surrounds trade negotiations provides a fertile medium for corporations wishing to circumvent the democratic process to get rules and regulations to their liking. For example, hidden in Chapter 11 of the North American Free Trade Agreement—a chapter designed to protect U.S. investors from expropriation of their investments—was a provision stipulating that American investors in Mexico could be compensated for any loss in value of their assets as a result of regulation; they are even given the right to sue in special tribunals, with damage payments coming directly from the Mexican treasury, even when the losses are a result of legitimate local regulations. To date, suits with claims in excess of $13 billion have been filed. The provision also applies to foreign investors in the United States— giving them protection that the courts and Congress have repeatedly and explicitly refused to provide for American investors.[13] Thus, through trade agreements, social and private incentives have become even more misaligned.

MAKING GLOBALIZATION WORK

It is easy to understand why multinational corporations have played such a central role in globalization: it takes organizations of enormous scope to span the globe, to bring together the markets, technology, and capital of the developed countries with the production capacities of the developing ones. The question is how to ensure that developing countries get more benefits—and face fewer of the costs. In the following pages, I set out a five-pronged agenda that, though it will not eliminate all instances of corporate abuse, will I believe lessen them. Underlying most of these reforms is a simple objective: to align private incentives with social costs and benefits.

Corporate social responsibility

Though many corporations, especially in the United States, continue to argue that their sole responsibility is to shareholders, many do recognize that their responsibility goes further. There is an element of self-interest here: doing good can be good for business, and doing bad can subject companies to expensive lawsuits. Bad behavior also can harm a company's image: the negative publicity surrounding the U.S. shoe company Nike after its suppliers in Vietnam mistreated local workers and the furor after Ken Saro-Wiwa was killed in Nigeria amid accusations that the Anglo-Dutch oil company Shell supported the military junta that murdered him were wake-up calls. Executives realized that they could be blamed for problems thousands of miles away from headquarters. Events like these have led to a number of voluntary initiatives by companies to improve the lot of their workers and the communities where they do business.

While increasingly more corporations see business social responsibility (BSR) as a matter of good business (and some studies suggest that socially responsible firms have performed better in the stock market than others), for many firms, their executives and employees, social responsibility is as much a moral issue as an economic one. Companies can be thought of as communities, people working together in a common purpose—say, to produce a product or provide a service. And as they work together, they care about each other, the communities in which they work, and the broader community, the world, in which we all live. This means that a company may not fire a worker the moment he is no longer needed, or that it may spend more money to reduce pollution than it is absolutely required to do by law. These companies may gain, of course, not just by avoiding the negative publicity described earlier; they may benefit from the higher quality labor force that they attract and improved morale: their workers feel better about working for a company that is socially responsible.[14]

The BSR movement has helped bring about a change in the mindset of many corporations and of the individuals who work for them. It has also worked hard to develop tools to ensure that companies live up to their ideals: accounting frameworks are being developed that track contributions to the community and environmental impact, and these are helping firms think more about the full consequences of their actions.

Regrettably, in a world of ruthless competition, incentives often work against even those with the best of intentions. A mining company that is willing to skimp on safety and environmental safeguards will be able to underbid one of comparable efficiency that pursues sound environmental policies. The oil company that is willing to engage in bribery to obtain oil at a lower price will show higher profits than a comparable company that does not. The bank that is willing to help its clients avoid or evade taxes may do better—at least if it's not caught—than the one that discourages them from doing so.

There is a further problem. Today, all companies, even the worst polluters and those with the worst labor records, have hired public relations firms to laud their sense of corporate responsibility and their concern for the environment and workers' rights. Corporations are becoming adept at image manipulation, and have learned to speak in favor of social responsibility even while they continue to evade it.

As a result, important as it is, the BSR movement is not enough. It must be supplemented by stronger regulations. Those who are really serious about higher standards should welcome regulations that support the codes of conduct they publicly endorse, for such regulations would protect them from unfair competition from those who do not adhere to the same standards. Regulations will help prevent a race to the bottom.

Limiting the power of corporations

Corporations strive for profits, and one of the surest ways of garnering sustainable profits is to restrict competition—buying up competitors, squashing competitors by driving them out of business, or colluding with competitors to raise prices. The problem of anti-competitive behavior has been evident since the birth of economics: as Adam Smith put it, "People of the same trade seldom meet together, even for merriment and diversion, but the conversation ends in a conspiracy against the public, or in some contrivance to raise prices."[15] When there is a lack of competition, the potential for abuses of multinationals grows much worse.

For more than a century, the advanced industrial countries have recognized the dangers of monopolies and anti-competitive behavior, enacting laws to break up the former and to punish the latter. Collaborating with supposed competitors to fix prices is a criminal act in most advanced industrial countries, with stiff penalties in both criminal and civil actions: in the United States, those who are convicted in a criminal action may go to jail and those who can show that they have paid higher prices as a result of monopolization receive triple damages (three times the amount overcharged by the monopolists).

With the advent of globalization and globally traded commodities, monopolies, and cartels—and the problems they create—often have become global in scope.[16] Globalization has unleashed a new potential for anti-competitive behavior that may be harder both to detect and to curtail.

The nature of global monopolies was revealed by a rash of global pricing cases uncovered in the early 1990s, including two involving U.S. giant Archer Daniels Midland (ADM). One case involved vitamins; another, lysine (an essential amino acid fed to pigs); a third, corn fructose. In the lysine case, the cartel fixed prices, allocated market share, and fixed quotas, managing to increase prices by 70 percent within three months. ADM was fined $100 million; Michael Andreas, the son of the CEO, and one other executive were sent to jail. In the corn fructose case, ADM faced damage claims of up to $2 billion and agreed to pay $400 million. In the vitamin case, criminal penalties imposed by the United States and the EU on the conspirators amounted to more than $1.7 billion; though the civil suits have not all been settled yet, almost $600 million has been paid out so far and there are further claims in excess of a billion dollars. Those outside the

United States and the EU, however, have little prospect of receiving significant compensation.

This reflects a general problem: while the benefits to the monopolists are global, enforcement remains fragmented, with each jurisdiction looking after its own citizens—meaning in practice that no one looks after consumers in small and developing countries. Worse still, home nations frequently fight in favor of their own global monopolies. This is natural; harm done to consumers and firms abroad is not their concern. When, in July 2001, the EU found that a proposed merger between the two U.S. giants GE and Honeywell would significantly reduce competition, the U.S. government vociferously complained. But the EU was right, and it took courage for the EU competition commissioner, Mario Monti, to stand up to the United States, fulfilling his obligation to enforce EU competition laws. His decision effectively blocked the merger.

Perhaps worse are instances where governments actually help to create global cartels to advance the interests of their own national companies. This happened while I was serving in the White House. In the face of weakening aluminum prices, Paul O'Neill, later to be secretary of the Treasury under President George W. Bush but at the time head of Alcoa, the world's largest producer of aluminum, pleaded for a global aluminum cartel to stabilize the market and protect America against "destructive" competition from Russia, then making its transition to a market economy. In a dramatic meeting, with the Council of Economic Advisers and the Department of Justice both strenuously opposing the proposal, the Clinton administration decided to take the lead in creating a global cartel—such a clear violation of competitive market principles that Assistant Attorney General Anne Bingaman announced as the meeting ended that she might have to subpoena those at the meeting for violating anti-trust laws. The cartel resulted, as O'Neill had hoped it would, in higher prices and profits for Alcoa—but also in higher prices for consumers.[17] Indeed, the cartel worked so well from O'Neill's perspective that after he became Treasury secretary he proposed another, for steel, to raise prices and restore profits in the U.S. steel industry. But with so many more countries and firms involved in steel production than in aluminum, the complexity of establishing and maintaining a global steel cartel was far greater, and the attempt failed.

Perhaps the most successful global monopoly is Microsoft, which has succeeded in gaining global market power not only in PC operating systems but in key applications such as browsers. A firm is said to monopolize a market if it has an overwhelming share; as of August 2005, Microsoft operating systems accounted for 87 percent of the total PC market and 89.6 percent of the Intel-based PC market. The personal computer, the Internet, word processing, and spreadsheets almost define the modern economy—and a single company has obtained dominance in these key areas. When Microsoft bundles a program such as Media Player with its operating system, it is effectively selling the program at a zero price. No company can compete with that. Courts in the United States as well as in Europe found not only that Microsoft had monopoly power but that it had abused this power. The only controversy was over the appropriate remedy. Microsoft has had to pay billions to settle anti-trust claims; as a result of a 2004 ruling in Europe, Microsoft must offer a version of its operating system there without Media Player included. Still, with Microsoft's monopoly so entrenched, it is unlikely that, without much stronger action, a competitive marketplace will be restored.

Microsoft's monopoly power leads not only to higher prices but to less innovation. Innovators saw what happened to Netscape, the first major Internet browser, as it was squashed by Microsoft—a powerful warning to anyone discovering a major innovation that might compete with or be integrated into Microsoft's operating system. One possible solution might involve limiting Microsoft's intellectual property protection for its operating system to, say, three years. That would provide strong incentives for it to provide innovations of the kind that users value and for which they would be willing to pay. If it failed to innovate, others could innovate off its *old* operating system—it would become a free platform, on top of which innovations in applications could be built.

The failure to develop a global approach to global cartels and monopolies is yet another instance of economic globalization outpacing political globalization. The current piecemeal approach, with each country looking after its own citizens, is costly and inefficient, and especially ineffective in protecting those in developing countries, whose resources, we have noted, are no match for those of large multinationals. Even if they dared to take on Microsoft, there is an imbalance of legal resources; and in the end, Microsoft might threaten to leave (as it did to South Korea)—and without Microsoft's operating system, they would lose interconnectivity with the rest of the world.

Globalization of monopolies requires a global competition law and a global competition authority to enforce it, allowing both criminal prosecution and civil action in any case in which anti-competitive behavior affects more than one jurisdiction. This does not require the dismantling of national competition authorities. The risks and costs of monopolization are sufficiently great, and the dangers of large firms using political influence wherever they can to suppress prosecution are sufficiently large, that there is a need for multiple oversight. Both the United States and the EU have kept in place multiple oversight—in the United States, at the level of both state and federal government; in the EU, at the level of the EU itself and national governments.

Improving corporate governance
A third set of reforms focuses on the laws governing corporations themselves. How do we make corporations, and their officers, act in ways that are consistent with the broader public interest? What reforms in the legal system can help align private incentives with social costs and benefits?

One step in the right direction would be to have companies take into account all stakeholders—employees and the communities in which they operate, not just their shareholders. It should not, for instance, be a violation of their fiduciary responsibility to their shareholders for them to pursue good environmental policies, even if profits are thereby hurt.[18]

Limited liability law was intended to limit the liability of investors, not to absolve employees, however senior, of responsibility. But, as we have seen, sometimes that is the result. Executives should be held personally responsible for more of their actions, making it more difficult for them to hide behind the veil of their corporations. Recently, there have been some moves in this direction, among them the agreement by the board of directors of WorldCom to provide some compensation to investors who lost as a result of WorldCom's misrepresentations. In publicly owned corporations, financial penalties typically have little effect on the incentives of managers. Even a large payout by the corporation as compensation for damages will have little direct effect on them, and with managers and boards of directors protected by insurance, even when fines are levied on them directly the costs are borne by others.

Just as the effective enforcement of competition policy has been found to require criminal sanctions—prison—so too is it necessary in other arenas. In 2002, following the corporate accounting scandals in the United States, the U.S. Congress passed the Sarbanes-Oxley Act, which makes the CEO responsible for the company accounts. Sarbanes-Oxley has been criticized for being excessively stringent and costly to comply with; there is often a danger of overreaction, and with experience the legislation may get fine-tuned. But the costs of the abuses—the misallocation of resources, the loss of confidence in the market economy—were also large, almost surely of an order of magnitude greater than the costs of the regulation. Moreover, many of the costs are start-up costs; once firms have adjusted to the new system, annual costs will be lower.

If there is a case for making corporate officers individually responsible in the area of accountability to shareholders and other stakeholders, then there is an even stronger case in other areas. It is no less a crime to ruin the environment (stealing the heritage of the entire community) than to cheat investors by manipulating the books. Environmental damage done by corporations is longer lasting, and those injured are innocent bystanders who were neither party to any agreement nor stood to gain from investment. When a company has egregiously violated a nation's environmental laws, the CEO and others who made the decisions and took the actions should be held criminally liable.

Another important step in achieving congruence between private and social interests is to make it easier for compensation to be obtained when damage has been done. Making firms pay for the damage they inflict—injury to workers or to the environment—provides firms with greater incentives to act more responsibly and to ensure that their employees do so. Of course, legal systems are imperfect. Large corporations can hire the best lawyers, against whom the lawyers that (often poor) injured parties can afford are no match. Sophisticated legal tactics often enable clearly culpable American firms to go free; until recently, few of the cigarette companies responsible for millions of deaths had been made to pay compensation. But, as we have already seen, the problems of making an American company pay for the consequences of its actions in a developing country are even greater. Even when the corporation is found guilty, it may be difficult to enforce the judgment. The company may well have protected itself by limiting its assets within the country, and attaching assets outside the country may be nearly impossible.

Several changes would go a long way toward repairing the system. The first is to allow those in other countries to sue in the home country of the offending corporation. The United States has allowed such suits since 1789 under the Alien Tort Claims Act, which allows those injured abroad to bring suit in the United States for any injury "committed in violation of the law of nations or a treaty of the United States." There have been attempts in recent years to bring actions in U.S. courts against multinational corporations, with some small measure of success. Of course, corporations would like to restrict such suits, but, if we are to make globalization work, there is a need to establish such legal provisions worldwide. This is the only way that there can be effective enforcement, especially when the offending corporation has few assets in the country where the damage occurred. A further advantage of these suits is that an American or European firm can no longer complain that it lost because the plaintiff had a home-court advantage.

A complementary reform would be to allow judgments made in foreign courts to be enforced by courts in the advanced industrial countries. If a court in, say, Brazil finds that an American mining company has done a billion dollars' worth of damage but does not have a billion dollars' worth of assets in Brazil, Brazil could use U.S. courts to help it collect damages. This is the case today in most international commercial arbitrations—but these are directed at protecting investors. Once again, there is an asymmetry: there is less concern about protecting countries against damage done by footloose international firms, who limit their assets within a country as a way of controlling their liability exposure.

Some firms are wary about being subject to foreign courts, claiming that the courts are stacked against them. This is simply one of the prices that one has to, and should, pay if one wants to do business in a country—including, in particular, extracting that country's natural resources. Alternatively, any firm claiming, as a defense against the enforcement of an adverse judgment, that a proceeding abroad was unfair could be automatically subject to suit in its own country's courts, to be judged according to the higher environmental and other regulatory standards of the two countries. This is not double jeopardy in the usual sense: the firm could have accepted the first judgment; it subjects itself to a second court only because it refuses to accept the findings of the first. The stipulation that the company should be judged by the environmental standards of the home country reflects a presumption increasingly recognized by the business social responsibility movement—that there should not be a double standard, with, say, lower environmental standards in developing countries than in the United States and the EU.

In the lore of America's West, bandits would cross the state line to seek a safe haven. For international environmental bandits, there should be no safe haven. Any country in which the corporation (or the substantial owners of the corporation) has assets should provide a venue in which suits can be brought or in which enforcement actions to ensure payment of liabilities can be undertaken. The corporation may incorporate where it wants, but this should not make it any less accountable for its actions in other jurisdictions.

To make this effective, it may be necessary to pierce the corporate veil. Mining companies, for example, often incorporate subsidiaries to run a particular mine, so that when the mine is exhausted—and all that remains are the costs of cleanup—the subsidiary goes bankrupt, leaving the parent unscathed. A simple rule would be that in certain classes of liabilities, such as those associated with environmental abuses, any entity owning more than, say, 20 percent of the shares of a company could be held liable even if the corporation itself went bankrupt. Limited liability should not be sacrosanct. Like property rights—including intellectual property—it is a creation of man, to provide appropriate incentives; when that artifice fails to fulfill its social function, it needs to be modified.

Global laws for a global economy

Eventually, we should be working toward the creation of international legal frameworks and international courts—as necessary for the smooth functioning of the global economy as federal courts and national laws are for national economies.

When consumers within the United States and certain other countries are hurt by price-fixing, they can band together, file what is called a "class action" suit, and if they succeed, they receive an amount that is triple the damages they incurred. This provides a strong incentive for firms not to engage in price-fixing. With global price-fixing, the harm done has become global, so consumers around the world need to band together and perhaps sue in, say, American courts. A recent Supreme Court decision gives the perpetrators, however, an easy way out. Once they have paid off the Americans who are injured, which may be just a fraction of the global liability, the plaintiffs have to find another venue.[19] By the same token, a single injured individual—say, in Bhopal—cannot afford to bring a suit; the maximum he or she can collect would be too small to pay any but the poorest of lawyers. But by acting collectively, the injured have some hope of redress. Those injured in Bhopal may have received far too little, but that they got as much as they did was a result of class action.

Not surprisingly, defense lawyers try to stop class actions by saying that the injured parties are sufficiently different that their cases cannot be consolidated. Insisting on a large number of separate cases against the same corporation for the same injury obviously imposes an enormous—in many cases, an impossible—burden on the legal system.

When a large number of individuals have been injured in a similar way, they should be able to band together to bring a single suit. We need to make it easier to pursue global class action suits, either in newly established global courts, or in national courts. Justice is far better served by recognizing the common element, to establish culpability and a base level of compensation, which can be supplemented if necessary by separate trials focusing on adjustments for unusual situations. For instance, price-fixing raises costs for all those who buy the product. A class action suit would establish that there has been price-fixing and calculate the amount prices have been raised from what they otherwise would have been. Of course, the magnitude of the injury suffered by a large producer in a developed country and a small consumer in a developing country will be very different. Having determined, however, the cartel's liability for price-fixing and ascertained the magnitude by which prices were increased, it would be a relatively easy matter to determine how much each should receive (which might have to be done in a series of mini-trials).[20]

And just as we recognize that access to justice for the poor requires the government to finance legal aid, this should be the case internationally as well: advanced industrial countries should provide legal assistance to those in developing countries.

Reducing the scope for corruption

There are several other actions that advanced industrial countries can undertake in order to make it more difficult for corporations to get away with the worst kinds of misdeeds. As we noted earlier, there is now widespread recognition of the corrosive effects of corruption and the need to attack it at both the supply and demand sides. The United States' passage of the Foreign Corrupt Practices Act in 1997 was a major step in the right direction. Every government needs to adopt a foreign corrupt practices act, and penalties should be imposed on governments that do not enact or enforce such laws. This is the kind of new issue that should have been introduced as part of the development round of trade negotiations (see chapter 3); it was not even broached. Bribery should be viewed as an unfair competitive practice and, just like any other unfair competitive practice outlawed under WTO rules, be subject to sanctions.

Bank secrecy aggravates the problems of corruption, providing a safe haven for ill-gotten gains. In the aftermath of the East Asian crisis, there were calls from the IMF and the U.S. Treasury for greater transparency in the Asian financial markets. When the developing countries pointed out that one of the problems in tracing the flow of funds was bank secrecy in offshore Western banks, there was a decided change in tone. The money is in these so-called offshore accounts not because the climate in the Cayman Islands is more conducive to banking; money goes there precisely because of the opportunities it affords for avoiding taxes, laws, and regulations. The existence of these opportunities is not an accidental loophole. The secrecy of the offshore banking centers exists because it is in the interests of certain groups in the advanced industrial countries.

There was an accord among the advanced industrial countries to do something about bank secrecy, but in August 2001 the Bush administration vetoed it. Then, when it was discovered that bank secrecy had been used to finance the terrorists involved in the September 11 attacks, the United States changed its views—but only where fighting terrorism was involved. Other forms of bank secrecy, as corrosive as they are to societies around the world, as bad as they are for development, are evidently still permissible; after all, bank secrecy is another way by which corporations increase the after-tax profits that are enjoyed by corporation owners. The international community should quickly broaden the rules against bank secrecy to areas beyond terrorism. The G-8 could itself bring this about, simply by forbidding any of their banks to have dealings with the banks of any jurisdiction that did not comply. The United States has shown that collective action can work: it has been effective in stopping the use of banks for financing terrorism. The same resolve should be used against corruption, arms sales, drugs, and tax evasion.

I have argued throughout this book that politics and economics are intricately interwoven: corporations have used their financial muscle to protect themselves from bearing the full social consequences of their actions. Why should we expect them to respond any more enthusiastically to these reforms than to any of the more modest attempts to temper their abuses?

One thing that makes me hopeful is the corporate social responsibility movement. There is an increasing number of firms who do not want to see a race to the bottom. It is firms like these, in the United States and other countries, that supported the Foreign Corrupt Practices Act. Civil society too is playing a more active role, by monitoring the actions of the large mining companies and of manufacturing firms that abuse their workers. The new technologies that have helped bring about globalization have been used to bring these abuses to the attention of the world, so that even those who have little moral compunction have been forced to account for their actions.

These are the realities, and they will not be easily changed: we should neither take corporations for the villains that they have often been portrayed as, or for munificent benefactors of developing countries. Limited liability has underpinned the growth of modern capitalism; but with globalization the abuses of limited liability have become global in scale; without the reforms suggested here, they could become far worse. The lesson here, as in much of the rest of this book, is simple: incentives matter, and governments and the international community must work harder to ensure that the incentives facing corporations are better aligned with those they touch, especially the less powerful in the developing world.

CHAPTER 8

The Burden of Debt

In August 2002, I visited Moldova, a small, largely agricultural, landlocked country with 4.5 million inhabitants squeezed between Romania and Ukraine. It had been one of the richest of the Soviet Union's republics, but since the beginning of its transition from communism in 1991 its GDP had plummeted some 70 percent. While the situation there had been dire since the collapse of the Soviet Union, when the ruble devalued in 1998[1] to one-fourth of its pre-crisis level, things became even worse. Moldova's currency devalued too, and the cost of servicing its foreign debt soared—rising to 75 percent of the government's budget. This left little money for social services and infrastructure. I saw roads in disrepair and broken-down villages. Even in the capital, Chişinău, the streets were filled with potholes, and, with no money to pay for street lights, the city was completely dark at night. I was deeply saddened by what I saw, but I was horrified when, during our trip, the daughter of a colleague was hospitalized. She died when the hospital ran out of oxygen. While those in the West take a ready supply of oxygen for granted, in Moldova it was an unattainable luxury.

At the same time, Argentina was dealing with the consequences of its January 2002 default on its debt, one of the largest defaults in history, rivaling the Russian default three and a half years earlier.[2] Before the default, foreign debt (including that owed to the IMF and the

World Bank) of almost $150 billion had been crushing the economy, with debt service on public and publicly guaranteed loans alone amounting to $16 billion in 2001, or 44 percent of exports and 10 percent of GDP.

Around the world, from Argentina to Moldova, from Africa to Indonesia, debt poses a burdensome problem for developing countries. Occasionally, the consequences of debt are dramatic, as with debt crises, but more commonly the debt burden shows its face as countries struggle to avoid default. Paying their debts often requires countries to sacrifice education and health programs, economic growth, and the well-being of their citizens. Money should flow from rich countries to poor, but partly because debt repayments have become so large in some years the flow of funds has been moving in the opposite direction. Obviously, with money bleeding out of developing countries, it is all the more difficult for them to grow and reduce poverty.

The problem is easy to state: developing countries borrow too much—or are lent too much—and in ways that force them to bear most or all of the risk of subsequent increases in interest rates, fluctuations in the exchange rate, or decreases in income. Given this, it is not surprising that they often cannot repay what is owed. Sometimes, even a country that has borrowed moderately and pursued good economic policies finds itself facing hard times—a tsunami or other natural disaster, the collapse of the market for its exports, a sudden rise in interest rates.

Often the debtor country is blamed for borrowing too much when, in fact, the lenders share the blame; they lent excessively, not looking carefully to see whether the borrowing country would be able to repay. Developing countries are poor; they make easy marks for anyone selling loans. The imbalance between the sophisticated lender and the less knowledgeable recipient could not be starker. Because they so often result in a struggle for repayment, international loans become the portal through which a developing country encounters the power of the IMF and other global institutions. The country is often torn between two unpleasant choices: defaulting, which brings with it fear of economic collapse, or accepting assistance, which brings with it the loss of economic sovereignty.

The bias against developing countries is reflected not only in that the onus is typically put on their "overborrowing" (rather than the creditor countries overlending), but in the lack of a solid framework of laws determining what happens when countries cannot reasonably meet their debt obligations. While every advanced industrial country has recognized the importance of bankruptcy laws that help individuals and firms to restructure overbearing debt, we have no parallel set of laws governing the restructuring of sovereign debt, ensuring that it is done fairly, efficiently, and expeditiously.

This chapter proposes a set of reforms: an expedited process of restructuring for private debts—money owed by private firms to foreign creditors—and a new, more balanced approach for public debts. The worry, though, is that even if debts are forgiven, new debts will occur: all the problems will reappear in a few years' time. So we must also ask the more basic question: what can be done to ensure that debt burdens do not again grow to levels that are beyond the ability of poor countries to pay? I argue that developing countries should borrow less—much less—than they have in the past, but also that, when they do borrow, they ought to be able to do so in ways that shift more of the risk—including the risk of exchange and interest rate fluctuations—to developed countries.

We've come a long way from the nineteenth century, when Western governments had an easy way of dealing with countries that didn't meet their financial obligations. They used brute force: invasion, occupation, and regime change.

A little more than a hundred years ago, Britain, Germany, and Italy sent a joint naval expedition to the Venezuelan coast and blockaded and shelled its seaports. They had the express consent of the United States to force Venezuela to pay its international debts. Dr. Luis María Drago, Argentina's foreign minister, came to the support of his neighbor, stating what has come to be known as the Drago Doctrine in response to the attack. In this kind of "unfortunate financial situation," he argued in a letter to Martín García Merou, Argentine minister to the United States, "the public debt cannot bring about a military intervention or give merit to the material occupation of the soil of the American nations by a European power."

He went on to say what is as true today as it was in 1902:

> In the first place the lender knows that he is entering into a contract
> with a sovereign entity, and it is an inherent qualification of all sov-
> ereignty that no proceedings for the execution of a judgment may be
> instituted or carried out against it, since this manner of collection
> will compromise its very existence and cause the independence and
> freedom of action of the respective government to disappear. . . .
> [T]he summary and immediate collection at a given moment, by
> means of force, would occasion nothing less than the ruin of the
> weakest nations, and the absorption of their governments, together
> with all the functions inherent in them, by the mighty of the earth.[3]

This was not the first time, nor the first place, that the strong
nations of the world had used military means to enforce repayment of
debt. France invaded Mexico in 1862, installing Napoleon III's relative
Archduke Maximilian of Austria as emperor, using the unpaid debt the
country had accumulated in the years since independence in 1821 as
an excuse.[4] In 1876 France and Britain jointly took charge of Egypt's
finances; six years later, Britain occupied the country.[5] The United
States used debt defaults as part of the justification for its interventions
in the Caribbean, for example in 1904 when the Dominican Republic
defaulted and President Theodore Roosevelt forced the Dominican
Republic to give the United States supervision of customs revenues so
that they could be used to pay foreign creditors. As recently as 1934,
Newfoundland, then not part of Canada, had to give up its parliament
as it went into "receivership."[6] During the heady days of the 1920s it
had borrowed heavily, and with the Great Depression—when a quarter of the population went on relief and government revenues
decreased by a third—it could no longer service the debt. It did not
really become self-governing again until it became part of Canada on
March 31, 1949.

Views of default have changed considerably in the course of a century. At the level of personal debt we've made progress: bankruptcy
laws have replaced the debtor's prisons that Charles Dickens portrayed
so graphically. Debtor's prisons gave the debtor few opportunities to

earn money to repay what was owed (although inhumane prison conditions did often elicit help from family members in repaying debts), but this disadvantage, it was thought at the time, was more than outweighed by the strong disincentive to default. So, too, views have changed about how to respond to the inability or unwillingness of a sovereign country to repay its debt. The Drago Doctrine is now universally accepted. But while there is a consensus about what should not be done—forcible debt collection by military means—there is less consensus about what should be done instead.

When countries can't pay what they owe, there are three alternatives: debt forgiveness, debt restructuring—where the debt is not written down, but payments are postponed in the hope that things will be better sometime in the future—and default (the borrower simply does not pay). This was the course taken by Argentina: after announcing that it would pay only a fraction of what was owed, it negotiated with creditors in an attempt to persuade them that something is better than nothing. In the end Argentina prevailed; in March 2005, 76 percent of its creditors agreed to a settlement of approximately 34 cents on the dollar. Some have concluded that the case of Argentina proves that the current system works, but I would argue otherwise. Years went by before an agreement was reached, and delay can be costly, with investors reluctant to make decisions while the economy is in limbo. Argentina demonstrated immense negotiating skills and immense resolve; most countries are lacking in both, and are more likely to cave under pressure from global financial markets and the IMF, agreeing to an inadequate debt reduction that leaves the country still overly burdened. And the fear of default leads countries to postpone default, putting their people through enormous sacrifices; default is undertaken only when it is the last remaining option. To my mind, the case of Argentina simply reinforces the conclusion that an orderly way of restructuring and reducing debt is needed.

THE ROAD TO CRISIS

There is a simple cause of the debt crisis of Argentina and that of the other emerging markets: too much debt. But why would well-functioning markets seem so often to lead to such a situation?

Overborrowing or Overlending?

Every loan has a lender and a borrower; both voluntarily engage in the transaction.[7] If the loan goes bad, there is at least a prima facie case that the lender is as guilty as the borrower. In fact, since lenders are supposed to be sophisticated in risk analysis and in making judgments about a reasonable debt burden, they should perhaps bear even more culpability.

Does it make a difference if we say there is overlending rather than overborrowing? The difference in where we see the problem affects where we seek the solution. Is the problem *more* on the side of the lenders, that they are not exercising due diligence in judging who is creditworthy? Or on the borrowers, being profligate and irresponsible? If we consider the problem to be overborrowing, then we naturally think of making it more difficult for borrowers to discharge their debts; on the contrary, if the problem is overlending, we focus on strengthening incentives for lenders to exercise due diligence.

The political economy of overborrowing is easy to understand. The current borrowing government benefits, and later governments have to deal with the consequences. But why have sophisticated, profit-maximizing lenders so often overlent? Lenders encourage indebtedness because it is profitable.[8] Developing country governments are sometimes even pressured to overborrow. There may be kickbacks in loans, or even more frequently in the projects that they finance. Even without corruption, it is easy to be influenced by Western businessmen and financiers. They wine and dine those responsible for borrowing as they sell their loan packages, and tell them why this is a good time to borrow, why their particular package is particularly attractive, why this is the right time to restructure debt.[9] Countries that aren't sure that borrowing is worth the risk are told how important it is to establish a credit rating: borrow even if you really don't need the money. I saw this firsthand in Vietnam, which had borrowed extensively from the World Bank, the Asian Development Bank, and other official sources but was reluctant to borrow from private sources. For years, foreign bankers told the country to issue a Eurobond as a benchmark, and for years the Vietnamese resisted doing so; eventually, they gave in.[10]

Excessive borrowing increases the chances of a crisis, and the costs of a crisis are borne not just by lenders but by all of society (a negative externality). In recent years, IMF programs may have resulted in significant further distortions in lenders' incentives. When crises occurred, the IMF lent money in what was called a "bail-out"—but the money was not really a bail-out for the country; it was a bail-out for Western banks. In both East Asia and Latin America, bail-outs provided money to repay foreign creditors, thus absolving creditors from having to bear the costs of their mistaken lending. In some instances, governments even assumed private liabilities, effectively socializing private risk. The creditors were let off the hook, but the IMF's money wasn't a gift, just another loan—and the developing country was left to pay the bill. In effect, the poor country's taxpayers paid for the rich country's lending mistakes.

The bail-outs give rise to the famous "moral hazard" problem. Moral hazard arises when a party does not bear all the risks associated with his action and as a result does not do everything he can to avoid the risk. The term originates in the insurance literature; it was deemed immoral for an individual to take less care in preventing a fire simply because he had insurance coverage. It is, of course, simply a matter of incentives: those with insurance may not set their houses on fire deliberately, but their incentive to avoid a fire is still weakened. With loans, the risk is of default, with all of its consequences; lenders can reduce that risk simply by lending less. If they perceive a high likelihood of a bail-out, they lend more than they otherwise would.

Lending markets are also characterized by, in the famous words of former chairman of the U.S. Federal Reserve Bank Alan Greenspan, "irrational exuberance," as well as irrational pessimism. Lenders rush into a market in a mood of optimism, and rush out when the mood changes. Markets move in fads and fashions, and it is hard to resist joining the latest fad, especially when international financial organizations and the U.S. Treasury give their imprimatur, as they did in Argentina. If only one firm were affected by a mood of irrational optimism, it would have to bear the cost of its mistake; but when large numbers share the same mood, in a fad, there are macro-economic consequences, potentially affecting everyone in the country—as happened during the East Asian crisis.

Failures in Risk Markets

Overborrowing, or overlending (depending on one's perspective), has something to do with many of the crises that have marked the last three decades. But the problems go deeper. Debt contracts providing for the borrowing country to pay back a certain amount in dollars or euros, and in which interest rates adjust to market circumstances (typically the case with short-term loans) place the burden of the risk of interest rate and exchange rate volatility squarely on developing countries. Worse, the IMF and the World Bank encouraged many countries to sign contracts for the construction of power plants that transferred all the risk of demand volatility to themselves; in these take-or-pay contracts, the government would guarantee to buy whatever electricity was produced, whether or not there was a demand for it.

If a country owes, say, almost $2 billion denominated in dollars, and its exchange rate collapses, say to one half of its value, then the amount of the debt in its own currency has doubled. A debt-to-GDP ratio of, say, 75 percent—high but still manageable, by international standards—suddenly becomes 150 percent, beyond the country's ability to pay. How did Moldova get into the desperate situation described in the beginning of the chapter, when only a few years earlier it had no debt at all? Part of the responsibility lies with the lenders who provided loans to facilitate Moldova's transition to a market economy. But the burden increased vastly when the value of Moldova's currency, the leu, depreciated enormously following the devaluation of the Russian ruble in 1998, more than doubling Moldova's debt-to-GDP ratio. The country was, in part, an innocent victim of the Russian crisis, precipitated by Russia's inability to meet its debt obligations.

Similarly, if interest rates increase from 7 percent to 14 percent, a country's repayments will double. Perhaps, before the increase, it was paying 25 percent of export revenues to service its debt; after the increase, it is paying 50 percent—which means it will have insufficient amounts left to pay for vital imports. That is what happened to

Argentina. Largely as a result of interest rate increases for emerging markets, Argentina's debt service more than doubled from 1996 to 2000.

In these cases, the major factor contributing to an unrepayable level of debt came from outside the country's borders. The consequence of developing countries having to bear so much risk—and global markets being so volatile—is, as we have seen, that even moderate levels of borrowing can, and often do, turn into an insurmountable debt burden. Making matters more difficult is the fact that because the loans have been primarily short-term (sometimes payable simply on demand), foreign banks can—and do—pull money out of developing countries at any sign of a downturn. A well-functioning global financial system would, on the contrary, provide money *to* countries in their times of need, thereby contributing to global economic stability, rather than demanding money *from* them at such times.

Technical aspects of Western banking regulations actually encourage short-term lending. Banks use short-term loans in part because this makes it easier for them to meet what are called "capital adequacy requirements." Regulators, concerned about the soundness of the banks for which they are responsible, require them to have a certain amount of capital relative to their outstanding loans, and less capital is required to back a short-term loan than a long-term one. The rationale is that, when lending is short-term, the bank can quickly pull its money out if circumstances change. But to a large extent this position of greater safety is a mirage. What may be true for one individual bank is not true for the banking system as a whole. When all lenders lend short, and then all decide to pull their money out simultaneously, they can't. The rules actually encourage panic: each bank knows that if it can beat the others, it may be able to get its money out before the problem becomes widely recognized and the money gets locked in. Once a problem is suspected, therefore, there is a race to be out first—a race in which almost everyone, and especially the developing country, winds up being a big loser.

Credit rating agencies panic as well; they do not want to be caught short as a country goes into default. In East Asia, they shared the optimism of the rest of the market in the days before Thailand's crisis on

July 2, 1997, but at that point they downgraded East Asian debt below investment grade. Because many mutual and pension funds are not allowed to hold funds below investment grade, they stampeded to the exit as well, exacerbating the crisis.

Rather than working to reduce these problems in the way markets function—that is, to help markets develop debt contracts in which the rich bear more of the risks associated with exchange rate and interest rate fluctuations—or to offset the consequences, the IMF and the governments of creditor countries have done what they could to make sure that those who have entered into these unfair contracts fulfill them, whatever the costs to their people. Among the policies which they pushed were high interest rates designed to stabilize exchange rates. At higher exchange rates, it was thought, debtors could more easily repay foreign-denominated debt. And while it was not always clear whether high interest rates stabilized the exchange rate, it was clear that they pushed the countries into recessions and depressions.[11]

The Case of Argentina

Capital is at the center of capitalism; if we are to have a global market economy, we must have well-functioning global capital markets. It is clear, however, that a key element of these capital markets—the market for debt—has not been working well, at least from the perspective of emerging market economies.[12] Repeatedly, they wind up with crushing levels of debt, leading to crises that result in economic recessions and depressions and increased poverty. Argentina's crisis illustrates the cost of mismanaging debt—and the need to reform the system.

Argentina suffered its debt crisis a century after Dr. Luis Drago came to the defense of Venezuela. It was not Argentina's first crisis. Like other Latin American countries, Argentina had been persuaded in the 1970s to borrow enormous amounts of money at a time when real interest rates were low, or sometimes even negative (real interest rates take account of inflation; the real interest rate is the nominal rate minus the rate of inflation). When in the late 1970s and early 1980s the United States raised interest rates to nearly 20 percent in a battle to throttle back its persistent inflation, Argentina found itself unable to meet its debt repayments. Debts were restructured, but there was inadequate debt forgiveness, and for much of the 1980s money flowed from Latin America to the United States and other advanced industrial countries. Latin America stagnated. It was not until the end of the decade that there was serious debt forgiveness—and only then did growth resume.[13]

Argentina had an episode of very high inflation at the end of the 1980s, hitting a peak annual rate of 3,080 percent in 1989; to fight inflation, the country pegged its exchange rate to the U.S. dollar. The strategy worked: inflation came down. But it was a risky strategy; a volatile international economy requires frequent adjustments of exchange rates, which Argentina's new economic regime did not allow. The consequences would unfold over the next decade.

With the burden of debt lifted, for a while, in the early 1990s, Argentina had a boom. New confidence in the economy meant that banks and other lenders were willing to lend, even to finance consumption. The consumption boom was sustained too as the country privatized state enterprises, selling them to foreigners. Had anybody bothered to look at the country's balance sheet, they would have realized that it was worsening, as it sold assets and accumulated liabilities; but the IMF focused only on the deficit, and was so pleased by the adoption of its Washington Consensus policies that it ignored the problems. Foreigners were encouraged to lend to Argentina, as the IMF continually singled out the country, praising it for its low inflation and other policies that were in accord with its advice, even going so far as to parade its president, Carlos Menem—shortly thereafter to be widely accused of corruption—before its annual meeting in Washington in 1999 as a paragon of economic virtue.

But suddenly, Argentina's fortunes changed. The precipitating event was the East Asian crisis in 1997, which by 1998 had become a global financial crisis. Global interest rates to emerging markets soared. Largely as a result, Argentina's debt service increased from $13 billion in 1996 to $27 billion in 2000. These problems were compounded by the strong dollar; since the Argentine peso was tied to the dollar, it was increasingly overvalued. The misalignment of its exchange rate increased further when Brazil, its largest trading partner, devalued its currency because of its own crisis. Argentina was flooded with imports and, at the high exchange rate, found it difficult to sell its own goods abroad. With fewer exports and more imports, its balance of payments deteriorated, and it had to borrow more abroad.

There began a vicious circle in which the IMF played a critical role. As global interest rates increased, Argentina's loan payments increased, so its fiscal deficit increased. The IMF, focusing on the deficit, demanded tighter fiscal and monetary policies: increasing taxes, cutting expenditures, and raising domestic interest rates. These had the predictable effect of lowering Argentina's output—and tax revenues.

There were other ways in which the IMF was responsible for the emerging crisis. The IMF had encouraged Argentina to privatize social security—which resulted in a reduction in revenues coming into the government (through social security taxes) faster than it resulted in a reduction in expenditures (for the retired); had Argentina not privatized social security, even at the time of crisis its deficit would have been close to zero.[14] The IMF had not only insisted on the privatization of public utilities like water and electricity but insisted that when they privatized, prices be linked to those in the United States; this meant that when prices rose in the United States, Argentineans had to pay more and more for basic necessities—making the country less and less competitive and increasing the level of social unrest.

That which is not sustainable will not be sustained; and Argentina's high exchange rate and mounting debt was not sustainable. Finally, in late 2001 and early 2002, the country's economic crisis came to a head; it defaulted on its debt—it simply did not pay what was owed—and let its exchange rate float. The value of the peso quickly fell by a third. In the economic chaos that ensued, the official unemployment rate soared to over 20 percent, and GDP fell by 12 percent.

By then, Argentina owed an enormous amount to the IMF. The IMF is supposed to help countries in their time of need—and this was a true time of need for Argentina. Private creditors typically call in their loans when the economy goes into a downturn—just when the government is in especial need of funds. The IMF was created in part in recognition of this market failure, but rather than offering to lend more to Argentina, it too demanded that the country repay what was owed, and that if it wanted the IMF to roll over its loans (in effect, extend the due dates) it would have to accede to its conditions—more of the same conditions which had contributed to the crisis in the first place. In closed-door, heated negotiations between Argentina and the IMF, Argentina did not cave in. Argentina bargained hard, recognizing that any further loans from the IMF would never reach Buenos Aires; the money would simply stay in Washington to repay what Argentina owed the IMF. (The IMF had even boasted of this achievement in the case of a loan to Russia after that country's default.) Argentina knew too that if it gave in to IMF conditions, its economic downturn would deepen. Finally, Argentina also recognized that the IMF and other international lenders had as much to lose as it did if they did not roll over their loans. While the country had defaulted on its private loans, whether it defaulted on its loans to the IMF and other official lenders depended on whether the IMF rolled over the loan. If it did not, it would have to declare Argentina in default, making its books look terrible. Argentina was right on this score as well; though it paid only a fraction of what was owed and refused to go along with the conditions the IMF demanded, the IMF did not declare the country in default.

The IMF too bargained hard. One former IMF staffer explained that his institution was simply reflecting the collective interests of the creditors, of which it was a principal one, who wanted to instill the fear of bankruptcy. They wanted any sovereign country considering default to think long and hard before doing so. They knew that there was no court which could force a sovereign country to repay what was owed; there were typically no or few assets that could be seized (by contrast with private bankruptcies, where creditors can take over the company or collateralized assets). It was only fear that drove repayment; without fear loans would not be repaid, and the sovereign debt market would simply dry up. The IMF "refused to take yes for an answer." If Argentina agreed to a particular demand, the IMF would impose new demands, wanting to prolong Argentina's agony and make default as costly as possible.

With no IMF program in place, Argentina then did something that no one had expected. It began to grow. Without IMF-style contractionary policies, without the flow of money out of the country to repay creditors, and helped by the large devaluation of its currency, Argentina racked up three years of growth of 8 percent or more. As growth was restored, it even managed to turn around its fiscal deficit—something the IMF program had never achieved. Had Argentina continued to send money to Washington and continued to accept the dictates of the IMF, it almost surely would have fared far worse.

While Argentina managed to recover in spite of—or, more accurately, because of—not having an IMF program, the failure to restructure its debt quickly made recovery more difficult than necessary. The creditors, including many ordinary savers in Italy who had been induced to buy Argentine bonds without fully realizing the risk associated with them, also suffered as a result of the long delay. Many could not hold out and had to accept large losses, selling to speculators who were gambling that in the end Argentina would improve its settlement offer.

Argentina made it clear from the beginning that it wanted a new IMF program, that it was not just walking away from its obligations; but it also recognized its obligations to its citizens, and that it was better to have no IMF program than to have one that would stifle its economy or use its scarce resources to bail out Western banks.

Argentina's story has many lessons for what should, and should not, be done both by countries and by the international community (especially the IMF). It shows, once again, that even countries that seem to be behaving well and borrowing moderately can wind up with crushing debt as a result of forces beyond their control and beyond their borders; it shows how easy it is for one debt crisis to be followed by another; it shows that outside assistance can come at a very high price—and that following the IMF's advice, being its A+ student, neither protects a country from crisis nor immunizes it against later criticism from the IMF. Most important, Argentina's successful recovery without the IMF's help has raised questions elsewhere. Should this country or that follow its lead? Would Brazil have been better off had it defaulted rather than following the tight-budget austerity policies which led to so little growth during President Luiz Inácio Lula da Silva's first term, in spite of enormously strong exports?

Argentina has also shown that there is life after default: a country can even grow faster afterward. But few countries are as brave as Argentina.

It is a fear of the consequences if they do not repay that drives countries to repay, imposing enormous hardships on their citizens.

The strength of these fears was brought home to me during my visit to Moldova. Though debt payments were taking three-quarters of their already meager budget, officials there kept saying that if they defaulted, they would not have access to money. I pointed out that they weren't getting any money. The flow of funds was from them to Europe and the United States, not the other way around. Moreover, it would be, at best, many, many years before they would ever get any funds anyway from the private sector. With all the debt service, they, like other highly indebted countries, could not make the investments needed for growth, and without growth, they were a poor prospect for lending. At least default would stop the hemorrhaging of money out of the country.

For most of the countries overburdened by debt, so long as their economies remain stagnant—as they will, so long as they are shackled by debt—they will not be able to gain access to capital markets, no matter how faithful they are in servicing their debt. But once they start to grow, they will gain access to international capital markets again, even if they have defaulted. Russia regained access within two years of its 1998 default. Financial markets are forward-looking. They ask about a country's prospects of repaying. An economy at full employment and stronger because it has rid itself of a huge overhang of debt is a better bet.[15] In other words, default can, in a relatively short time, actually lead to an enhanced net inflow of capital.

MAKING GLOBALIZATION WORK:
WHAT TO DO ABOUT DEVELOPING COUNTRY DEBT

Discussions on debt relief are confused and confusing partly because there are four largely (but not totally) distinct categories. There are the "normal" very poor countries that have mostly borrowed from other governments and multilateral institutions like the IMF. Then there are countries that have suffered under corrupt and oppressive governments, who among their many adverse legacies have left a legacy of debt. Third, there are emerging markets where largely private lenders have lent too much to private borrowers, so much that the problem has national consequences; the case of Korea, Thailand, and Indonesia, where private debts precipitated a regional crisis, provides the dramatic illustration. And finally, there are middle-income countries, like Argentina, that have been lent too much (or, depending on one's perspective, have borrowed too much), mostly by private lenders, but also by the IMF, World Bank, and regional development banks, and cannot repay what is owed without wrenching adjustments.

Debt Relief for the Poorest

The very poor countries are so desperately poor that they take money in any form that they can get. Typically, private lenders will not lend to them; but in the past the World Bank, the IMF, and advanced industrial countries have often provided loans at low interest rates. The hope was that the loans would finance projects and programs which would lead to growth—enough growth that the country would find it easy to repay the loans. But this is often not how matters turned out. Even when there has been growth, it has been so feeble that it has not offset the increase in population; twenty years after the loan was granted, the country is even poorer, and in no position to repay.

In 1996, the international community finally recognized the need for debt relief for highly indebted poor countries. But the program (referred to by its acronym, HIPC, for "highly indebted poor countries") has had a rocky history. Over the ensuing four years, only three countries got relief. The IMF was in charge of setting conditions for debt relief, and it set the bar so high that few qualified. Countries had to follow closely what the IMF recommended. They were given little leeway. Critics claim that this was no accident: the offer of debt relief was a powerful tool for the IMF to compel these countries to go along with almost anything it demanded, but once debt relief was granted the IMF's stranglehold was greatly diminished. The IMF is, however, not the only source of the problem; some loans were made bilaterally, and debt relief has to be agreed to by all the major creditors.

In response, in the year 2000, a movement called Jubilee 2000 (commemorating the biblical Jubilee that granted debt relief every fifty years) mobilized public opinion behind the issue of debt relief, and there was an agreement to expand the HIPC program. As of July 2005, twenty-eight countries had been granted more than $56 billion of debt relief, reducing the debt this small set of countries owed to foreigners by approximately two-thirds. Of the twenty-eight countries, nineteen have been granted debt service relief amounting to $37 billion; for the others, full debt relief requires the countries to fulfill certain conditions, less onerous than those of the past and focused on reducing poverty. The pace is better than before, but not fast enough. There are still many countries waiting for debt relief: many, such as Indonesia, are not part of the HIPC program because, though they are very poor, they are considered too rich for debt write-off; Moldova is not eligible simply because debt relief was not extended to the countries of the former Soviet Union. And while relief gets delayed, the magic of compound interest works so that debts continue to grow.[16]

Something more was needed, and, as we noted in chapter 1, there was a response. The leaders of the advanced industrial countries, the G-8, at their summit meeting in June 2005 at Gleneagles, Scotland, agreed to provide up to 100 percent debt relief for the eighteen poorest countries of the world, fourteen of which are in Africa.[17]

As the situation in Moldova demonstrates, without debt relief the highly indebted poor countries will not be able to meet the basic needs of their citizens, let alone make the investments necessary if they are to grow out of their poverty. For the poorest countries of the world, there needs to be an expedited form of debt relief, an extension and expansion of the current HIPC initiative to more countries. And as the G-8 countries recognized at Gleneagles, the debt relief has to be deep: any dollar sent to Washington or London or Bonn is a dollar not available for attacking poverty at home. Shallow debt relief simply leaves the country struggling on, with another debt crisis looming in the not too distant future.

Debt relief has to be done in ways that do not detract from the availability of other forms of assistance. Help for the very poor should not come at the expense of the poor. Already, debt relief has been criticized for rewarding not just the unlucky but the irresponsible. Countries that have gone to great efforts to keep their debt under control should not be effectively punished by getting less aid than those that have been profligate. Today, the developing countries that have repaid what was owed, at least to the point where they no longer qualify for debt relief, worry that debt relief is commandeering money that might otherwise have been available to them—especially at the World Bank, where repayment of loans provides a major source of money for lending. Only time will tell whether the advanced industrial countries will make up for the shortfall, so that the World Bank can maintain its lending programs. This is especially important because there is often less to debt relief than meets the eye: much of it is simply a matter of accounting, a recognition of the reality that the country would never have been able to repay the amounts owed anyway.[18] If, therefore, money that would have gone to other forms of assistance is accounted as debt relief, it will mean in practice that the total amount of assistance is reduced.

Many worry that these poor developing countries will soon again become highly indebted. In one sense, the onus should be on the lenders. Most of these countries are so desperately poor that it is not reasonable to expect them to turn down loans.[19] Lenders should make sure that any loan is limited to the amount the country can repay. In practice, this means that there should be relatively little lending. Most of these countries are not only desperately poor now; they will be desperately poor when it comes time to repay the loan. Even if the money lent has a high return, it will be difficult for governments to raise the revenues required to finance repayments; and money spent repaying loans inevitably comes partly at the expense of education, health, and other vital social and growth expenditures.

Combining more assistance in the form of grants with more diligence on the part of lenders will make it less likely that so many of the poorest countries in the world will, in the future, be burdened with excessive debt.[20]

Odious Debt

In one category of lending, the moral case for debt forgiveness is especially compelling. These are referred to as "odious debts"; they were incurred by a government that was not democratically chosen, and the borrowed money may even have helped a brutal regime stay in power. Whatever the motivation of the lender—whether political (to buy favor in the Cold War) or economic (to get access to rich mineral resources)—it is immoral to force the people of these debtor countries to repay the debts.

Iraq's debt incurred under Saddam Hussein is in this category, as is that of Ethiopia which, until 2006, was still paying back the debts incurred by the hated Mengistu regime and its Red Terror, which brutalized the country from the fall of Haile Selassie in 1974 to its overthrow in 1991. Mengistu Haile Mariam used the money to buy arms to suppress those who opposed his tyranny. The current government has actually been paying for the arms that were used to kill its fellow fighters as they struggled to establish a new regime.

By 2005, Nigeria had a debt of some $27 billion—much of it cumulative interest on borrowings made by corrupt military dictators during the periods 1964–79 and 1983–99, when the country's wealth was pillaged, even as some quarter of a trillion dollars in oil was being pumped. During the Cold War, Congo was lent money by Western powers and the international financial institutions. The money was shipped by its military dictator, Mobutu Sese Seko, to secret bank accounts in Switzerland and elsewhere; the lenders knew, or should have known, that the money was not being spent on development. That was not its purpose: its purpose was to buy friendship in the Cold War, or at least to stave off Congo selling its friendship to Russia, and to ensure access for Western companies to that country's rich natural resources. By the end of his regime, the country had amassed $8 billion of external debt, and Mobutu had amassed a personal fortune estimated between $5 billion and $10 billion. And now, unless the debt is forgiven, it is not the citizens of the Western powers who pay for the support given to Mobutu but the citizens of Congo who are left to pay his debts.

Chileans today are repaying the debts incurred during the Pinochet regime, South Africans the debts incurred during apartheid. Had Argentina not defaulted on its debt, Argentineans would still be paying down the loans that financed the "dirty war" from 1976 to 1983, in which an estimated 10,000 to 30,000 Argentineans disappeared.

There is a simple solution to the problem of odious debt: there should be a presumption that these countries should not repay the loans. This simple solution not only solves the problem of the current debt overhang but also of its recurrence: if creditors are on notice that if they lend to such regimes they risk not being repaid, then they will be unlikely to lend. "Credit sanctions" are likely to be much more effective than trade sanctions (where the international community tries to get countries to behave "well" by threatening to cut off trade). For, as Foreign Minister Drago pointed out a hundred years ago, there is no court of law that can force countries to repay; and if there is a broad consensus in the international community that a particular debt is odious and that the country has no obligation to repay it, then there are unlikely to be adverse consequences to not repaying; and with no adverse consequences, there will be no incentive to repay. Following this reasoning, lenders will not want to lend; in contrast, trade sanctions are often ineffective, because trade with the sanctioned countries is profitable, so firms always try to circumvent the sanctions.[21]

Going forward, the United Nations could keep a list of countries for which contractors and creditors would be put on notice that their contracts and debts will be reexamined once the regime is gone. Governments and banks that lend money to oppressive regimes would know that they risk not getting repaid. Guidelines for what are acceptable contracts and debts could be established: loans to build schools would be permitted, while loans to purchase arms would not be. (Some argue that since funds are fungible—money lent to finance a school frees up money for the government to spend on arms—any loans to repressive regimes should be treated as odious, but there is evidence that lending to, say, education does result in more educational expenditure than would otherwise be the case.) An International Credit Court could be established to make the required judgments. For existing loans, it would ask, should the lender have recognized when the loan was being made that it was in fact odious debt? Clearly, the many private lenders to apartheid South Africa, especially after sanctions were imposed by the UN, should have known that these debts were odious, just as today anyone lending to Sudan's regime, which both the United States and the UN have judged to be engaging in genocide, should realize that the loans are odious.

Analogous issues are raised with respect to contracts. Should governments be forced to compensate private contractors for breaking a contract, when that contract was made with a corrupt and dictatorial regime? Should those contracts be treated like odious debt—especially when the contracts may have helped maintain the regime in power? And does the fact that there is often corruption in the contracting process itself make a difference? In the case of Iraq, the United States argued that honoring contracts with Saddam Hussein was rewarding corruption. In the case of Indonesia, after the overthrow of Suharto, the U.S. ambassador argued that the sanctity of contracts was inviolable. (The ambassador was duly rewarded, upon his retirement from the State Department, by being put on the board of a U.S. mining company active in Indonesia that has been accused of both corruption and despoiling the environment.)

To many, the issue is not just whether the debts should be repaid or the contracts honored but whether Western institutions should be liable for some of the damages that resulted from the continuation of the regimes they helped perpetuate.

Private Cross-Border Debt

Until the East Asian crisis of 1997, many believed that only public borrowing could be a problem. After all, it was reasoned, private parties would only borrow if they could repay, and creditors would only lend if they were confident that the private parties could repay. Moreover, it was argued that if there was a problem with repayment, only the lender would bear the consequences. The East Asian crisis showed that this reasoning was wrong. Underlying the crisis was excessive indebtedness of private companies. As creditors refused to roll over their dollar-denominated loans, the entire region was plunged into crisis.

What happened then was what happened in so many other places: private liabilities were in effect nationalized. The IMF provided the governments with the dollars to repay the Western creditors. The creditors were protected, the borrowers were let off the hook—and taxpayers in developing countries were left with the burden of repaying the IMF.

There was an alternative: the private borrowers could have simply defaulted on their loans—declaring bankruptcy. The problem was that few of the countries had good legal frameworks to deal with what happens then. There was almost universal agreement that developing countries needed better bankruptcy laws, and the IMF tried to foist a particular set of bankruptcy laws—a creditor-friendly set of laws—on those countries that turned to it for money and advice. Not surprisingly, the IMF's macro-economists did not really understand the micro-economics of bankruptcy. They did not, in particular, recognize that there is no single, "right" approach to bankruptcy. Indeed, the design of bankruptcy law has been among the most contentious topics within the American political scene. To think that one can rely on some international technocrats for the solution to what is a quintessentially political issue is not just nonsense but dangerous, for those seeming technocrats may well reflect particular interest groups. But bankruptcy law reflects more than just the balance between creditor and debtor interests; it says something about a society's views of social justice.

There are a host of considerations that go into the design of bankruptcy law. There is of course a need to get the right balance between the interests of creditors and debtors. An excessively debtor-friendly bankruptcy law will provide insufficient incentives for borrowers to repay; without this, credit markets will not be able to function. But an excessively creditor-friendly bankruptcy law will provide insufficient incentives for creditors to engage in due diligence, to ascertain whether the borrower can repay. One American bank advertises its credit cards with the slogan "qualified at birth"—suggesting a certain lack of effort in distinguishing between good and bad borrowers.

If bankruptcy procedures are prolonged, companies may remain in limbo for an extended period during which ownership is not clear; it will be difficult to borrow, and management may have an incentive to strip assets—selling them quickly to get hold of the cash. But tough bankruptcy laws can force liquidation, destroying jobs and organizational capital. (The value of a firm's goodwill—the value of a firm beyond that of its physical assets and which includes the value of its reputation—is often far greater than the value of its physical assets.) All of these concerns play an important role in modern bankruptcy law. In the United States, Chapter 11 of its bankruptcy code provides for fairly rapid corporate reorganization—a discharge of debts, a conversion of debts into equity, with existing equity owners being largely or totally squeezed out as the creditors become the new owners. Companies continue to operate throughout the bankruptcy period. While some criticize it for being too debtor-friendly, it has not impeded firms from getting access to credit—even when, like several still-functioning airlines (Continental and US Airways), they have gone into bankruptcy more than once.

During the East Asian crisis, as chief economist for the World Bank, I argued for the creation of a "super Chapter 11," a special bankruptcy provision for countries where bankruptcy is brought on by a major macro-economic calamity—the collapse of the exchange rate, a major recession or depression, or an unanticipated spike in emerging market interest rates. In these circumstances it is even more imperative to have a quick resolution. Additionally, the presumption that the problem facing the company was not the result of bad management, but of forces beyond its control, would be greater than in a normal personal or business bankruptcy. Therefore, the super Chapter 11 would be more debtor-friendly and allow for more expeditious restructuring than the ordinary Chapter 11.

But a quick resolution to the problem of companies not being able to repay what they owe is no substitute for avoiding the problem in the first place. Again, this means avoiding borrowing—and reducing exposure to risk and volatility so that "reasonable" debt does not quickly turn into unmanageable debt. With both borrowers and lenders alike ignoring the macro-economic consequences of excessive indebtedness, it is not surprising that foreign indebtedness is often too high, which is why government intervention is required.[22] Since foreign short-term borrowing in particular exposes countries to a risk of a crisis, governments should discourage it, for instance by putting taxes and restrictions on short-term capital flows.

Sovereign Bankruptcy

Very poor countries and countries recovering from corrupt regimes are not the only ones to face overwhelming debt problems. Mexico, Brazil, Argentina, Russia, and Turkey are on the long list of countries that have recently not just had a problem but faced an economic crisis because of difficulties in meeting debt obligations. No one talks about debt forgiveness for these countries, partially because, at one level, these countries do have the capacity to repay: they could presumably raise taxes and cut expenditures enough to generate the required revenue. The value of the country's assets exceeds by a wide margin the value of what is owed. But the cost to the country can be enormous, beyond what its citizens are willing to pay. Even if creditors are not willing to forgive debt on their own initiative—which, typically, they are not—there is an alternative: default and renegotiation. This, as we have seen, was the route taken by Argentina. But as we have also seen, Argentina's debt restructuring was unnecessarily difficult.

Five key reforms are required.

Do no harm

The first is for the developed countries to do no harm. Debt relief should not be an occasion for holding countries to ransom, or for undermining their democratic institutions. Debt relief is supposed to provide a fresh start. The Paris Club is an informal group of nineteen creditor countries, including the United States, Japan, Russia, and many European countries; they collectively decide on who gets how much debt relief and under what conditions. When the Paris Club insists as a condition for debt relief that Iraq subscribe to shock therapy and adopt Washington Consensus economic policies, it is taking away Baghdad's economic sovereignty.[23] In November 2004, they agreed to forgive 30 percent of Iraq's $40 billion debt, and another 30 percent in three years' time, if Iraq complied with an IMF program that would entail adopting the privatization and liberalization program that the Bush administration had wanted Iraq to adopt all along. At the time, prospects for shock therapy working in Iraq appeared to be even bleaker than in Russia, where the IMF had imposed the same recipe and produced a 40 percent decline in GDP. Iraq's economy similarly has not fared well, though part of the blame lies with the insurgency, part with the inadequacy of U.S. efforts to reconstruct the infrastructure.

By the same token, during its negotiations for debt relief, Nigeria was asked to have an IMF program as a condition for debt forgiveness. Critics asked why, when it had already shown that it could, on its own, manage its economy well, having brought down inflation, managed its budget, and increased transparency.

Whatever the conditions imposed by the IMF, they will be objected to simply because they are imposed—they come from outside the country. But IMF conditions are especially objectionable because they are often so ill-suited for the country. The IMF has become so obsessed with inflation that it often seems to forget about growth and real stability—paying little attention to volatility in output and employment. As a result, rather than remedying the deficiencies in private capital markets or offsetting the effects of these deficiencies, it has often worsened them. Rather than providing funds to finance counter-cyclical policies, it has typically demanded that countries undergoing a downturn impose contractionary policies. One of the most important advances in economics over the past century was the insight of John Maynard Keynes that government, by spending more and lowering taxes and interest rates, could help countries recover from a recession. The IMF rejected these Keynesian policies, adopting instead pre-Keynesian policies focusing on government deficits; these entail raising taxes and cutting expenditures in recessions, just the opposite of what Keynes recommended. In virtually every case where they were tried, IMF policies worsened the downturn. Economists do not, after all, have to rewrite their textbooks, but what was good news for academic economists was devastating for millions of people living in these countries.

Especially problematic are the high interest rate policies that the IMF pushed to stabilize exchange rates; while the high interest rates failed to do that, they quickly led to an explosion of the debt burden. Governments had to borrow more and more just to make the interest payments on what was owed.

The policies the IMF pushed as a condition of loans hurt the borrowing countries in other ways. I have repeatedly noted that even countries that borrow moderately may face a problem as a result of the high level of economic volatility, including volatility of exchange rates and interest rates. Capital market liberalization (which the IMF urged, or forced, upon developing countries) exposed countries to more risk and volatility, and limited their ability to respond. (If they lowered interest rates, for instance, in an economic downturn, capital could bleed out of the country.)

Return to counter-cyclical lending

The pattern of pro-cyclical private lending—demanding that money be repaid just when the country needs money the most—will surely continue. Banks are in the business of making money, and the old adage that banks lend only to those that do not need money is based on hard experience. But it was market failures like this that provided a key rationale for the establishment of the IMF and the World Bank in the first place; as we have already noted, they were created in part to help avoid another global disaster like the Great Depression. Counter-cyclical lending (lending more when the economy is weak) was within their original mission. By offsetting the pro-cyclical pattern of private lending, such counter-cyclical lending can contribute enormously to stability. It can help developing countries finance expenditures in recessions, providing needed fiscal stimulus. The IMF, the World Bank, and the regional development banks in Africa, Asia, eastern Europe and the former Soviet Union, and Latin America must return to counter-cyclical lending.

Risk reduction

Third, the risk of borrowing must be reduced. I emphasized earlier that many debt problems are caused by the fact that developing countries are forced to bear the risk of exchange rate and interest rate volatility. Wall Street prides itself on its ability to slice and dice risk, enabling risk to shift from those less able to bear it to those more able. Yet, in the case of developing country debt, it has largely failed to do this.

Until private financial markets step in, showing that they are able and willing to absorb more of the interest rate and exchange rate risks facing developing countries, the international financial institutions need to take a more active role in risk absorption. This is especially so in the case of their own loans; debt contracts can be designed to protect developing countries from the ravages of fluctuations in interest rates and exchange rates.[24] They can also help in loans from others. The World Bank already provides insurance against the risk of nationalization; it could extend this insurance to include risks of interest rate and exchange rate changes and even of default. The premium would make borrowing more expensive, and thus might discourage borrowing (which I have suggested may in itself be good), but the cost of the premium would be far less than the cost of the volatility facing borrowers today.

The risks of borrowing can be lowered if countries borrow in their own currency, which is why it is important to develop local currency debt markets. The World Bank and other multilateral development banks can help to strengthen these markets by borrowing in them as they raise funds.[25] Several Asian countries, led by Thailand, are actually trying to create an Asian bond market, in which borrowing occurs in a basket of local currencies. The sound macro-economic policies of these countries, as reflected in low inflation and (with the exception of the 1997 crisis) relatively stable exchange rates, provide an environment conducive to such a market; and the fact that so much savings originates within Asia should also help in creating one.

The advanced industrial countries must be sensitive to how policies designed to provide greater stability to their own economies, such as treating short-term lending abroad as safer than long-term lending, may have exported the instability to the developing world. The regulations, and the institutional arrangements by which they are formulated, need to be changed. For instance, banking regulations and standards are set by the Bank of International Settlement (BIS), an institution that is even less democratic and transparent than the IMF; in setting these standards, at least in the past, it has paid little attention to the impact on developing countries.

Conservative borrowing
The fourth reform mirrors what should be done in the case of the highly indebted poor countries: countries should borrow very conservatively, and when they do borrow they should do so in their own currencies. If markets or governments can't—or won't—do anything to shift the burden of risk, then developing countries should be especially conservative in borrowing.

Borrowing brings more problems than it's worth. Historically, it is apparent that for many developing countries the costs of debt have exceeded the benefits. Latin America grew rapidly in the early 1990s, supported by debt, but what it lost later in the decade almost surely exceeded the benefits of the earlier growth; much of the debt went to finance a consumption binge, with much of the benefit going to those who were already doing very well, and with much of the cost of the ensuing crisis being borne by workers and small businessmen. The costs and benefits of debt are inequitably distributed. Debt and its aftermath contribute to poverty and inequality.

The hard lesson of the last fifty years is that, even when there are high social returns on investments—say, in education, health, and roads—it is hard for a government to raise money to repay loans. This means, of course, that countries will need to rely more on their own savings to finance their capital accumulation—reemphasizing the importance of high national savings rates. East Asia did many things right; one of those things was to save a great deal and to borrow little. It was only when they began to borrow abroad, in the late 1980s and early 1990s, that South Korea and Thailand ran into problems. For them, the debt problem was truly unnecessary, given their high savings rate. Any reasonable calculus would put the costs far greater than the benefits.

International bankruptcy laws
No matter how responsible the borrower, bad things happen—the price of exports may plummet, a crop may fail several years running, international interest rates may soar, there may be a global recession resulting in the disappearance of export markets. In any of these contingencies, a country may not be able to repay its debts, or only with great sacrifice from its citizens. In these circumstances, there needs to be a systematic way of restructuring—and forgiving—debt, a form of international bankruptcy. This is the final major reform.

Today, we have an informal system in which countries negotiate and beg for debt forgiveness. Success is based on bargaining skills and politics. The United States argued hard on behalf of Iraq (though only a small amount, some $4.5 billion, was owed to the United States). Under American sponsorship, Iraq eventually got debt relief. There were many other countries, equally or more deserving on almost any account, which did not. By the same token, Argentina knew how to bargain hard and had confident and informed political and economic leadership; as a result, they got a much better deal from their creditors.

The idea that firms and individuals faced with overwhelming debt need a fresh start is now universally accepted. But it is even more important for countries with overwhelming debt to have a fresh start. Keynes recognized this in his book *The Economic Consequences of the Peace*, written immediately after the Treaty of Versailles imposed enormous reparations—effectively a debt burden—on Germany at the end of World War I; he predicted correctly that it would lead to recession and depression in Germany, and social and political turmoil.[26] When a single firm or individual has a problem, the social and political consequences are limited; when a country faces an unbearable debt burden, everyone in society is touched. Argentina's default in 2002, and the long drawn-out process of negotiation that followed, demonstrated the need for a better mechanism for dealing with sovereign defaults.

The United States has, unfortunately, not joined the consensus about the need for a better mechanism, and has, so far successfully, blocked any action, contending that an international bankruptcy procedure is unnecessary; all that is required is a slight modification in debt contracts.[27] This includes a collective action clause, which means that if, for example, 80 percent of a country's creditors agree to a debt-restructuring proposal, it can be adopted. (Under prevailing practice, all creditors have to agree, leading to the problem of holdouts who can veto a restructuring unless they are paid in full.) The fact that every advanced industrial country has found it necessary to have a bankruptcy law reinforces the conclusions of economic theory, that collective action clauses will not suffice; some judicial process is required.

A systematic way of engaging in debt forgiveness/restructuring would ensure fairer and faster restructuring. Several principles should guide this. First, enough debt should be forgiven so that the country will not face a high probability of being back in default in, say, five years' time. In the past, the IMF used rosy growth scenarios in order to minimize the extent of debt forgiveness, and countries following their guidelines in debt restructuring often found themselves back in trouble within a few years. Restructuring without adequate debt write-down means debt still casts a shadow over growth.

There is obviously considerable uncertainty about future growth, and here Argentina has come forward with an ingenious solution: a GDP bond, which pays more if growth is stronger. This has the further advantage of aligning the interests of creditors and debtors; creditors now have an incentive to help the economy grow faster.

Second, any resolution must recognize that foreign creditors are not the only claimants. There are many public claimants in addition to formal creditors—including, for instance, those owed retirement payments by the government, as well as health services and education. This is one major difference between sovereign debt restructuring and private bankruptcy. In private bankruptcy, a list of creditors and assets is drawn up, and bankruptcy law and debt contracts determine who has the most senior claim; in sovereign bankruptcy, however, there is no well-defined set either of creditors or of assets. A determination needs to be made in advance: that the primacy of a government's obligations to its citizens is inviolable.[28]

Third, restructuring needs to be fast and debtor-friendly. There is tremendous cost in delay—the delay in providing adequate debt forgiveness in the early 1980s for Latin America led to a decade of stagnation. Earlier, we explained how Chapter 11 of America's bankruptcy regime provides speedy restructuring on terms proposed by the debtor; today, many countries are considering following America's Chapter 11 example. An international system of debt restructuring must similarly incorporate some expedited procedures.

Fourth, whatever the process of determining the extent of debt restructuring and/or forgiveness, it must not rest in the hands of the creditors, including the IMF. They simply cannot act as an impartial judge.

I believe that an international bankruptcy organization will have to be created eventually, just as every advanced industrial country has had to create bankruptcy law, and some have created special bankruptcy courts. But in the short run, it may be useful to create an international mediation service to establish norms. After all, since the abandonment of military intervention, moral suasion plays an important role in inducing repayment, and in determining what fair repayment is. The creation of a set of norms and expectations might go a long way toward smoothing the restructuring process.

Two factors besides the ability to repay should be taken into account. Some weight should be given to the extent to which the lender knowingly lent money in a situation of high risk of not being repaid. When, as in the case of Russia, lenders were getting 150 percent interest, it was because there was a strong likelihood of default. At that rate, if a lender lends money in January, by October he has fully recouped what has been lent. Everything after that is pure profit. If the loan is for five years, the creditor would obviously like the restructuring to continue to pay him the promised 150 percent interest, but to most people that would be unreasonable. He might complain that getting only a 7 percent return is cheating him—that the value of the bond has been written down enormously. But the high interest rate meant that he knowingly undertook a risk of getting back substantially less than the bond's face value.

A second factor is the extent of culpability of lenders for the problems facing the country. I have already discussed one case: odious debts, and suggested that there should be a presumption for complete debt write-off. The discussion of Argentina highlighted the extent to which the IMF was responsible for that country's problems, including its inability to repay. That is why many inside Argentina thought that the IMF should take at least as large a debt write-down as the private creditors took (66 percent); they were sorely disappointed when the government repaid the IMF in full early, in 2006. But the government took a pragmatic approach: it simply wanted to get the IMF off its back. Paying back the IMF in full was a small price to pay to regain its economic sovereignty.

There are a host of situations where there is shared blame. It makes sense, in these situations, to adjust the extent of repayment in accordance with the degree of culpability. Economists emphasize the importance of incentives: making lenders (including the IMF) bear the consequences of their actions (including their advice) would provide incentives for improving the quality of advice and engaging in more care in lending.

Consider, for instance, the IMF loan to Russia in July 1998. It was intended to support the ruble. However, at that time the ruble exchange rate was overvalued, making it difficult for Russia to export anything beyond oil and other natural resources. I, along with most of my colleagues at the World Bank, believed that the loan would not sustain the exchange rate for very long, and that it would almost certainly do little more than leave the country more deeply in debt. Moreover, there was a strong likelihood of corruption—that the money would quickly flow out of the country, quite likely into the pockets of the oligarchs. The lending was largely politically motivated, as at the time the United States was eager to keep President Yeltsin in power. Nor did it want to face the fact that policies that it, together with the IMF, had pushed had, by 1998, left Russia's GDP over 40 percent lower—and poverty over ten times higher—than it had been at the beginning of the transition from communism to a market economy. (Ironically, even as the United States was lecturing Russia on the dangers of corruption, there was a major corruption scandal involving Harvard University, which had been given the contract for administering U.S. assistance for privatization.)[29] Even if the loan eventually failed, it was a small price to pay—a price, in any case, to be paid by the Russian people—to postpone the discussion of "who lost Russia."[30]

The loan did fail. The money left the country for Swiss and Cypriot bank accounts faster than critics had thought possible. This case is more complicated than those of the odious debts of Congo and other countries discussed earlier, because the Russian government was democratically elected. Still, the question is, ethically, who should bear the consequences—the people of Russia, who had no say in the loan, or the lender, the IMF, who designed it?

Earlier, I argued that in Argentina the IMF had particular culpability, because Argentina viewed itself as dependent on IMF loans, which it could only get if it followed IMF advice, and that advice exacerbated its economic problems. The same thing was true in Russia. The IMF advised Russia, prior to its default, to convert more of its debt from ruble- to dollar-denominated loans. The IMF knew—or should have known—that this would expose the country to enormous risk. With an overvalued exchange rate, a devaluation was in prospect. With dollar-denominated debt, the benefit Russia got in increased export revenues and reduced import costs would be offset by the losses on balance sheets. What it owed would (in terms of rubles) increase enormously. The IMF saw lower interest rates on dollar loans, but it should have known that this simply reflected the markets' expectation of a ruble devaluation.

Indonesia provides another telling example, where the IMF, together with others including the World Bank and the Asian Development Bank, provided some $22 billion in loans during the East Asian crisis. The money was characterized as a bail-out for Indonesia, but a closer look shows that, as is typically the case, it was really the Western banks that were being bailed out. The extent to which Western creditors, not Indonesia, were the real beneficiaries became clear when the IMF insisted that food and fuel subsidies to the poor be cut back, arguing that although there were billions of dollars available to repay Western banks, there simply wasn't enough money to help Indonesia's poor (though the costs were a mere fraction of what was provided to the country). This came after unemployment had soared tenfold and real wages had plummeted—partly because of the policies that the IMF had insisted upon. Inside Indonesia, there is widespread sentiment that, since the IMF is so much to blame for that country's economic problems, there should be substantial debt forgiveness. But until the tsunami hit on December 26, 2004, those pleas fell on deaf ears. The tsunami gave debt forgiveness a humanitarian rationale, and payments on some $3 billion of debt due in 2005 were postponed for a year.

In other cases, one might argue for an even greater degree of culpability on the part of the lender—for example, when a World Bank project fails because insufficient attention has been paid to environmental impact or because there has been an inadequate economic analysis. The World Bank is supposed to have the experts, and—particularly in the past—developing countries relied on its expertise. But when the project fails or does not perform up to expectations, it is not the World Bank that bears the consequences but the developing country, which is still responsible for repaying the loan.

Clearer guidelines on the circumstances in which debt would be forgiven would have two effects. The process of debt restructuring would be smoothed and be less expensive, reducing the chances of a costly crisis such as that which afflicted Argentina, and with costs contained, countries would more willingly go to court to have their debts restructured. The long periods in which debt overhang slows growth and impedes development might be shortened. At the same time, incentives for lenders would be strengthened: they would be put on notice to lend more carefully. Greater caution in lending might lead to lower growth in the short run, but the long-run benefits would be enormous. The crises that have plagued developing countries would not be eliminated, but their frequency and magnitude would be reduced. As a result, long-run growth would actually be enhanced.

Resistance to these ideas will be great. As we have seen, the United States has opposed the establishment of an orderly process of debt restructuring. Some in the financial markets do not want to have an orderly process; they want the costs of going through a default to be high, so that few will do it. They object that debt relief will lead to more defaults, higher interest rates, and therefore less borrowing. Given that one of the underlying problems is overborrowing, reducing borrowing would actually be desirable.[31] And even many emerging markets will vocally oppose it—especially those that are looked at in the financial markets as suspects for default. They are putting on a brave show for the benefit of the creditors, showing by their willingness to undergo enormous pain, were a default to occur, that default is, for them, simply not an option. (Whether they are really against these reforms is another matter.)

Many of the problems in meeting debt payments arise not from mistakes on the part of developing countries but from the instabilities of the global economic and financial system. The need for better mechanisms for sharing risk and for resolving debt problems will continue to be great so long as international financial markets continue to be marked by such instability. Making globalization work will require doing something about this instability—the subject of the next chapter.

CHAPTER 9

Reforming the
Global Reserve System

The global financial system is not working well, and it is especially not working well for developing countries. Money is flowing uphill, from the poor to the rich. The richest country in the world, the United States, seemingly cannot live within its means, borrowing $2 billion a day from poorer countries.

Some of these dollars from the developing to the developed world go to pay off their enormous debts—the subject of the last chapter. Others go to buy bonds from the United States and other "strong" currency countries; these bonds will be added to the developing country reserves. They have an enormous advantage: they are highly liquid, so they can be sold quickly whenever the country needs cash; but they have an enormous cost: they earn a very low interest rate. Most of the bonds are short-term U.S. Treasury bills (usually referred to as "T-bills"), which in recent years have yielded as low as 1 percent interest. There is something peculiar about poor countries desperately in need of capital lending hundreds of billions of dollars to the world's richest country. In 2004, the flow from China, Malaysia, the Philippines, and Thailand alone, mostly to build up reserves, amounted to a whopping $318 billion.[1]

We saw in the last chapter the harm that excessive debt brings to developing countries. We saw too that the huge volatility in the global economy—including interest rates and exchange rates—may quickly convert moderate debt into an unbearable burden. While money *should* be flowing from the rich to the poor and risk from the poor to the rich, the global financial system is accomplishing neither.

With poor countries left to bear the brunt of risk, crises have become a way of life—with more than a hundred crises in the last three decades.[2] It is the failings in the global reserve system that lie behind many of the failings in the global financial system, and a simple reform of this system would lead to a stronger and more stable global economy. Reform would also solve one of the world's biggest problems: the lack of funds to promote development, fight poverty, and provide better education and health for all.

All countries in the world hold reserves. They serve a multiplicity of purposes. Historically, reserves were used to back up a country's currency. Those who held South African rand or Argentinean pesos might feel more confident in the currency knowing that behind the currency the country held dollars or gold, that they might in fact be able to convert the currency into gold or dollars—which in turn can be used to purchase goods and services. Historically, gold was used as "money"— the medium of exchange in which people traded. People would buy and sell food or clothing in exchange for pieces of gold. Then it was discovered that "fiat money"—pieces of paper that could be converted into gold—was far more convenient, and governments and central banks issued this money. At first, it was thought that there had to be full backing—for every dollar of fiat money issued, the government or the central bank had to hold a dollar's worth of gold. Then it was discovered that this was not necessary; all that was required was confidence in the currency. Confidence meant that other individuals would be willing to accept the money in payment, and confidence could be achieved with only partial backing. At first, it was thought that confidence could only be achieved by using gold as backing; then it was realized that the currency (or debt) of strong economies—initially Britain's sterling, and for much of the period after World War II the U.S. dollar— could be used.

Reserves also help countries manage the risks they face, and this bolsters confidence in both the country and its currency. They can be drawn upon in times of need. Reserves form a buffer against unexpected changes in the cost of debt caused by an increase in interest rates. There may be a sudden hardship, such as a crop failure, and the country can use reserves to import food. The amount of reserves a country needs varies, but a rule of thumb is that countries should have enough reserves to cover at least a few months of imports. Historically, developing countries held reserves to the value of three to four months' imports; more recently, they have held as much as eight months' imports.

In the last chapter I discussed another risk: many countries have borrowed in dollars from abroad short-term. Short-term lenders are often fickle. If a sudden fear that the country cannot meet its debt obligations sweeps the market, lenders demand their money back simultaneously, and so their fears turn out to be self-fulfilling as countries are usually unable to repay all their debts on such short notice. If a country has large reserves, investors are less likely to panic; and if they do panic, it is more likely that the country will be able to meet its debt obligations. Today, prudence requires countries to maintain reserves at least equal to their short-term dollar debts or debts denominated in other hard currencies, such as the yen or euro.[3]

Reserves can also be used to manage the exchange rate; without reserves, the exchange rate can fall, often quite dramatically, as fickle investors or profit-seeking speculators or currency manipulators sell a country's currency. Instability in exchange rates can lead to enormous economic instability. By countervailing these moves—buying the country's currency when others are selling or selling the country's currency when others are buying—governments can stabilize the exchange rate, and thereby stabilize the economy. But they can only sell dollars to buy the local currency if they have a reserve of dollars to sell.[4]

While countries have always held reserves, the amount they hold has been soaring. In just the four years between 2001 and 2005, eight East Asian countries (Japan, China, South Korea, Singapore, Malaysia, Thailand, Indonesia, and the Philippines) more than doubled their total reserves (from roughly $1 trillion to $2.3 trillion). But the real superstar was China, which by mid-2006 had accumulated reserves of approximately $900 billion—amounting to well over $700 in reserves for every man, woman, and child in the country. That accomplishment is all the more astounding given that China's per capita income at that time was less than $1,500 per year. For the developing countries as a whole, reserves have risen from 6–8 percent of GDP during the 1970s and 1980s to almost 30 percent of GDP by 2004.[5] By the end of 2006, developing country reserves are estimated to reach $3.35 trillion.

While there is no agreement on the explanation for this huge increase, two factors are clearly important: the high level of global economic and financial instability, and the manner in which the East Asian crisis of 1997 was handled by the IMF. Countries felt a loss of economic sovereignty; worse, the policies the IMF imposed made the downturns far worse than they would have been otherwise. The East Asian countries that constitute the class of '97—the countries that learned the lessons of instability the hard way in the crises that began in that year—have boosted their reserves in part because they want to make sure that they won't need to borrow from the IMF again. Others, who saw their neighbors suffer, came to the same conclusion: it is imperative to have enough reserves to withstand the worst of the world's economic vicissitudes. Exchange rate management also plays a role in the buildup of reserves; a low exchange rate promotes exports, and a country can keep it low by selling the local currency and buying dollars.

The High Cost of Reserves to Developing Countries

As I have noted, historically, reserves were held in the form of gold, and some countries still do this. However, virtually all reserves today are held in dollar-denominated assets, sometimes dollars themselves but, as we have noted, more likely U.S. Treasury bills, which can easily be converted into dollars. The popularity of the dollar in international reserves stems mainly from the dominance of the United States in the world economy and the fairly stable history of the currency. Whether the dollar can and should remain the basis for the international reserves system is one of the questions I will address. First, however, we need to come to grips with the staggering cost of reserves to developing nations.

For all the advantages of holding these accounts, countries pay for the insurance they provide. Today, developing countries earn on average a real return of 1–2 percent or less on the $3 trillion plus of reserves.[6] Most developing countries are starved for funds. They have a myriad of high-return projects. If the money weren't being put into reserves, if it weren't being lent to the United States at such low returns, it could have been invested in these other projects, earning some 10–15 percent.[7] The difference between the interest rates can be viewed as the cost of the reserves. Economists call these costs—the difference between what could have been earned and what was actually earned—"opportunity costs."

Using a conservative estimate of 10 percent as the average percent difference between the two, the actual cost to developing countries of holding the reserves is in excess of $300 billion per year. That's huge.[8] To put it into perspective: it represents four times the level of foreign assistance from the whole world. It represents more than 2 percent of the combined GDP of all developing countries; it corresponds roughly to estimates of what the developing countries need in order to achieve the Millennium Development Goals, including reducing poverty by half.[9] It is much larger than the gains to developing countries from a *successful* pro-development Doha Round trade agreement. (As we noted in chapter 3 what is likely to emerge, at best, will be of limited value to the developing countries.)

The costs to developing countries of the global reserve system can be seen another way. Assume an enterprise within a poor country borrows $100 million short-term from an American bank, paying, say, 20 percent interest. Following the prudential guideline that countries should maintain reserves equal to short-term dollar-denominated debt, the government then—if it doesn't want to face the threat of an imminent crisis—must add $100 million to its reserves: by buying $100 million worth of T-bills, paying 5 percent interest. There is, in fact, no net flow of funds from the United States to the developing country as a result of the loan; it is simply a wash. But the U.S. bank charges much more for the $100 million it sends than the U.S. government gives for the $100 million it receives. There is a net transfer of $15 million to the United States. This is a great deal for the U.S. bank and the United States generally, but a bad deal for the developing country. It is hard to see how the net transfer of $15 million to the United States by the developing country will enhance its growth or its stability.

In addition, there is in effect a transfer from the public sector in the developing country to the private. The private sector is better off (otherwise it would not have borrowed the money, even if the rate is high), but the government has had to spend money on building reserves that it could have used to build schools, health clinics, or roads.

In spite of these large costs, the developing countries benefit from reserves—if they work as intended, an economy is less volatile than it otherwise would be. (That they are willing to pay such a high price indicates the huge costs of instability to developing countries.) But the real beneficiaries of the global reserve system are those in whose currency the reserves are held. They get low-cost loans; were it not for the demand for reserves, their costs of borrowing would likely be markedly higher. With nearly two-thirds of reserves being held in dollars, the United States is, in this sense, the major recipient of these benefits.[10] If the interest rate America has to pay is just one percentage point lower than it otherwise would be on these $3 trillion of loans from poor countries, what America receives from the developing countries via the global reserve system is more than it gives to the developing countries in aid.

A Weaker Global Economy

The cost of the current global reserve system to the developing countries is the most conspicuous, but it is not actually the most important cost to the global economy. The global reserve system depresses the global economy and makes it more unstable. The current reserve system makes it difficult to maintain the world economy at full employment. The money put into reserves is money that could be contributing to global aggregate demand; it could be used to stimulate the global economy. Instead of spending the money on consumption or investing the money, governments simply lock it up.

To see the magnitude of the problem, note that the world's economies hold more than $4.5 trillion of reserves, increasing at a rate of about 17 percent a year. In other words, every year some $750 billion of purchasing power is removed from the global economy, money that is effectively buried in the ground.[11] A strong global economy requires that there be a strong demand for goods and services—strong enough that it can meet the world's capacity to produce. The total demand for goods and services (the sum of the demand by households for consumption, by firms for investment, and by government) around the world is called global aggregate demand. If the world is not to face an insufficiency of aggregate demand—leading to a weak global economy—this has to be made up somehow. In the old days, many developing countries counteracted this through lax monetary and fiscal policy, leading to spending that was beyond the country's means. While this spending made a "contribution" to global aggregate demand, loose fiscal policies gave rise to increasing government debts, which often precipitated costly crises, as we saw in the last chapter. With more than a hundred crises in the last three decades, most developing countries have learned their lesson.

There is one country that can make up for the inadequacy of aggregate demand that comes from burying purchasing power in the ground: the United States has become the consumer of last resort. It is able and, especially since 2000, willing to run huge deficits. There is a seeming unending appetite for reserve country bonds, and it is all too easy for governments of reserve currency countries to get more and more into debt to feed this appetite. The fact that others are willing to lend at a low interest rate creates a situation politicians find hard to resist. It is easy to run fiscal deficits, to spend more than one has. Since the dollar became the major reserve currency, the United States has twice—in 1981 and 2001—financed huge tax cuts through deficits. This helps to explain our peculiar observation earlier—that the United States is the world's richest country, yet is living beyond its means. In this respect, it is doing the world a service. Without America's profligacy, the fears of a weak global economy, possibly so weak that prices might actually start to fall—the fears of deflation that surfaced in the early years of this century, and which have plagued Japan for a decade—might have been realized.[12] The question is, for how long can

America continue to provide this service; that is, can it continue its spending spree? And are there alternative, more equitable ways of avoiding the global downward bias?

Insufficiency of aggregate demand in the reserve currency country
We have seen how the global reserve system leads to a problem of inadequacy of global aggregate demand. It also presents a special problem of inadequate aggregate demand in the reserve currency country.

A country whose currency is being used as a reserve must—if it is to continue to be used as a reserve—"sell" its currency (or more accurately, its T-bills or bonds) to other countries, who hold on to them.[13] When a country sells a T-bill to another country, it is, of course, simply borrowing from that country. A government borrows when it spends more than it takes in; and it borrows abroad when its own citizens are not saving enough, at least relative to what they are investing. In this case, because there are not enough funds at home to finance government spending, it must turn to foreigners to finance its fiscal deficit.

Put it another way: a country, as a whole, borrows from abroad when the country as a whole is spending more than its income. This, in turn, means that the country is importing more than it is exporting—it is borrowing to finance the difference.

Trade deficits and foreign borrowing are two sides of the same coin. If borrowing from abroad goes up, so too will the trade deficit. This means that if government borrowing goes up, unless private savings goes up commensurately (or private investment decreases commensurately), the country will have to borrow more abroad, and the trade deficit will increase.

That is why economists often talk of the twin deficit problem: when government borrowing increases—that is, when the fiscal deficit increases—so too is it likely that the trade deficit will increase.[14]

The reserve country can be thought of as exporting T-bills; but the export of T-bills is different from the export of cars or computers or almost anything else: it does not generate jobs. That is why countries whose currency is being used as a reserve, and are exporting T-bills rather than goods, often face a problem of insufficiency of aggregate demand. Or, to put it another way, we saw that the counterpart of borrowing from abroad (issuing T-bills) is a trade deficit, with imports exceeding exports. And just as exports create jobs, imports destroy them, and when imports exceed exports there is a real risk of insufficiency of aggregate demand.[15] Aggregate demand that would have been translated into jobs at home is translated into demand for goods produced abroad.

Most democratic governments cannot sit idly by as unemployment grows. They intervene, typically by lowering interest rates or increasing government expenditure. Unfortunately, as America's slowdown of 2001–03 showed, even interest rates close to zero may not be sufficient to restore robust growth and full employment. Large deficit spending may be necessary.[16] In this view, it is the trade deficit that leads to the fiscal deficit, not the other way around. Support for seeing the world of deficits through this lens is provided by looking at the pattern of trade and fiscal deficits during the past quarter century. What is remarkable about America is that it has had trade deficits through thick and thin—when the government has had a fiscal deficit and when it has not. The 1990s can be thought of as an exceptional period: an investment boom meant that the economy could remain at full employment even without a fiscal deficit, but the gap between investment and savings remained—the elimination of the fiscal deficit may have increased national savings, but national investment increased almost in tandem. So even as the fiscal deficit disappeared, the trade deficit remained strong as America continued to supply the world with the T-bills other countries wanted for their reserves.

From this perspective, underlying America's persistent trade deficit is its role as a reserve currency: others persistently stockpile America's T-bills. The problem is that the system is not sustainable. The mounting debt eventually undermines the confidence that is required to maintain the dollar as a reserve currency. Of course, America is able to pay back what is owed. But with increasing indebtedness, there is an increasing risk of a reduction in the real value of the debt through inflation. Even a slight increase in the rate of inflation can have large effects in "writing down" the real value of the debt. As I travel around the world, talking to investors and central bankers, I hear this worry increasingly openly expressed. And with confidence in the dollar fragile, the value of the dollar becomes more volatile.

Instability

This brings me to the final set of major costs of the global reserve system, the instability to which it gives rise. Reserves are intended to reduce the costs of instability. But the irony is that, while the costs of instability for each country are reduced, directly and indirectly, the current global reserve system is a major factor behind the high level of global instability. And the level of global instability has been truly enormous. For instance, in less than two years, between February 2002 and December 2004, the value of the dollar relative to the euro plummeted by some 37 percent. This immense decline shook the financial world and debunked the then widely held notion that the almighty dollar was unassailable.

That unassailability had been questioned before. Too long ago for the memories of the young traders who determined the fortunes of exchange rates in the early 2000s, a previous crisis, in the early 1970s, provides a backdrop to today's anxieties. The United States had, in the years after World War II, felt that a speculative attack might be a problem for the weak countries of Europe, but not a problem that it would ever have to face. That was just wishful thinking. At the time, the United States had a fixed exchange rate—the dollar could be converted into gold at the rate of $35 to the ounce. A speculative attack on the dollar forced the United States to give up on its commitment to the convertibility of dollars to gold; it let the dollar float, let the market by itself determine the exchange rate.

The system has been working, if not working well. But there is a fundamental problem underlying the whole reserve system: it is self-defeating. The reserve currency country winds up getting increasingly into debt, which eventually makes its currency ill suited for reserves.

Already, the current system is fraying at the edges. In early 2005, China announced that it is no longer committed to holding reserves in dollars. It had, in fact, already moved substantial amounts out of dollars (about a quarter of its reserves), but the announcement had immense symbolic value. Other central bankers, more in keeping with their tradition of staying out of the public eye, quietly confided to me that they too were moving out of dollars.

These changes in central bank policies—the move out of the dollar—make sense. When it was believed that reserves had to be held in gold or in gold-backed dollars, no one thought about managing them. Since 2000, a major change in mindset has occurred. Central bankers have realized that they don't need dollars to back their currencies. With currencies freely convertible into one another, what is important is not the number of dollars but the amount of wealth in reserves. Then the question becomes how best to manage that wealth—and the principles of wealth management, including diversification, are well known. With so much of reserves having been held in dollars, diversification means movement out of the dollar.

This change in mindset came, in part, because central banks had discovered that dollars were a bad store of value. Traditionally, central bankers have focused on inflation—no one wants to hold a currency whose value, in terms of the goods it can purchase, is being greatly eroded. With its low inflation, the dollar would seem an excellent store of value. But for those outside the United States, its value depends on the exchange rate. Central bankers and the IMF have failed—and failed miserably—to create a system of stable exchange rates. When the value of the dollar relative to the yen was relatively stable, the dollar was a good store of value for those in Japan. But as the volatility of the dollar has increased, as the exchange rate between the yen and the dollar fluctuates enormously, the dollar has lost its ability to be a good store of value for Japan. Similar arguments apply to Europe and elsewhere: the increasing volatility of the dollar has meant that it is no longer a good global store of value.

For instance, in the span of a few months in 1995, the dollar lost 20 percent of its value relative to the yen. There was little inflation in the United States, but those in Japan who had put their money into dollars discovered that they could buy far fewer goods in Japan in April 2005 than they had been able to buy in January. There have been even larger losses over a longer period of time relative to the euro. The opportunity cost also was huge—had they held their money in euros rather than dollars, reserve holders would have been much better off. The opportunity cost perspective becomes particularly important for the countries of East Asia, who held some $1.6 trillion of hard currency reserves (mostly in dollars) at the end of 2003. Had they held their reserves in euros during the following year, rather than in dollars, their balance sheets would have been some 11 percent larger—some $180 billion. That's a lot of money to throw away.

Of course, no one can predict exchange rate movements, but that's why the modern theory of portfolio allocation emphasizes diversification: don't put all your eggs in one basket. A dynamic has been set in motion that is not good for the dollar: as central banks move out of the dollar, the dollar weakens, reinforcing the view that the dollar is not a good store of value.

The emergence of the euro has accelerated the fraying of the dollar reserve system. Although Europe has been plagued with problems such as low growth, high unemployment, and a constitutional crisis, the euro has been a strong currency. The logic of diversification says that, however one assesses the prospects of Europe versus America, one should carry significant amounts of one's reserves in each.

Early on, Europe was pleased by this development. It had looked with relish at the prospects of the euro becoming a reserve currency because it wanted the new currency to be treated with respect, and its adoption as a reserve currency signaled this. But as the reality of what this status entails has become increasingly clear, not everyone in Europe has been so enthusiastic. As central banks hold more euros as reserves, the value of the euro will increase, making it harder for Europe to export and opening it up to a flood of imports.[17] It will have an increasingly difficult time maintaining full employment. And with unemployment already so high, and with its central bank focusing exclusively on inflation and not at all on unemployment or growth, there is good reason to be worried about Europe's macro-economic prospects.[18]

Scenarios—from evolving instability to crisis
That there is a problem with the global reserve system seems clear. There is less certainty about how all of this will unfold. There are several different scenarios—from crises to gradual evolution.

Here is a picture of what might happen over the next few years: As American debt mounts, doubts about the soundness of the dollar increase. At first, a few investors think they would be better off putting their money elsewhere. As they do this, the dollar falls. (The partial recovery of the dollar in 2005 is at least partly due to the repatriation of corporate profits; profits repatriated during the year were given specially low tax rates, which induced abnormally high levels of repatriation. By mid-2006, the dollar has started to weaken again.) When the losses in the value of the dollar are taken into account, keeping money in dollars appears foolish; returns are just too small to justify the risk. There is, of course, no such thing as a safe bet; but, perceiving the riskiness of the dollar, more and more investors will decide to shift more and more of their money out of dollars into euros, yen, or, where possible, the yuan, China's currency. (In spite of capital controls, there was an inflow of some $100 billion into China, in addition to foreign direct investment, in 2004.) As this happens, more and more downward pressure is put on the dollar. Simultaneously, as investors pull their money out of American securities, stock prices will fall or stagnate. Keeping money in the United States will look increasingly like a bad bet.

The consequences of increases in medium- and long-term interest rates may be particularly serious, given the high level of indebtedness of individual households, many of whom took out large mortgages in response to the unusually low interest rates. What matters is not the average level of indebtedness but the number of households that will face difficulties in meeting their debt obligations. The increasing fraction of mortgages having interest rates that are variable makes this particularly worrisome.

The march out of the dollar may be orderly and smooth, occurring over a period of months, perhaps even years. Or it may be disorderly, a crash. In the former case, the U.S. stock market may simply go through a malaise; it may even continue to climb, but simply at a lower rate than otherwise would have been the case. In the latter case, the U.S. economy would go into a downturn. If there is to be a crash, it is, as always, difficult to predict what kind of event might precipitate it. Even in retrospect, it is hard to identify any single event that caused the crash of October 1987, which wiped out close to 25 percent of the value of U.S. equities in a day. But there are plenty of events, including baseless rumor, that could do the trick. Events in the Middle East might turn out even nastier than they have been. A new terrorist attack in the United States might show that, for all the money that has been spent, America is still vulnerable.

While America's increased indebtedness—the predicted historical course for the reserve currency country—is a major source of the global financial instability facing the world today, the counterpart to this indebtedness—the large holdings of dollars by China and Japan—has been a force for stability. Together, they have increased their holdings of reserves enormously, by over $1 trillion from 2000 to 2006 alone. As I have already noted, sound portfolio management suggests moving out of dollars, putting more into euros—and China has already been moving in that direction. But here's where China and Japan have a problem: their dollar holdings are so large that were they to sell significant amounts quickly, it would put downward pressure on the dollar—causing losses on their remaining holdings. China and Japan's central banks have an interest in maintaining stability, and they are not subject to the panics, the attacks of irrational pessimism and optimism, that characterize markets.

Moreover, there is a political dimension to all exchange rate policy, especially that of China. There is an element of mutual hostage in U.S.-Chinese economic relations. China has a huge bilateral trade surplus with the United States, selling far more than it buys. But China makes it possible for the United States to sustain its deficit spending, by buying billions and billions of dollars' worth of America's bonds. America and China know the nature of their mutual dependence; that's why matters seldom get beyond the rhetoric.

America has been highly vocal, blaming China's unfair exchange rate policy for its trade deficits. Though China has let its exchange rate appreciate slightly, it knows that even a more significant increase will only decrease the bilateral trade surplus a little. A change in the exchange rate would not, moreover, affect the United States' overall trade deficit, which is related to its macro-economic imbalances—the fact that it is saving less than it is investing, a problem exacerbated by the huge fiscal deficit. Americans would simply buy more textiles from, say, Bangladesh. At the same time, a significant appreciation of the currency would lower prices for agricultural goods, which are depressed due to the distortion of global prices by U.S. and EU subsidies, as we saw in chapter 3; this would make life more difficult for those in the rural sector—a part of China that is already falling behind. China could offset the effects through subsidizing its farmers, but this would divert money badly needed to promote its development. In short, China knows that there would be high costs to it—and little benefit to the United States—were it to allow its exchange rate to appreciate. And presumably, America understands this too.

Although China and the United States need each other, there is, of course, always the fear that political forces will get out of control: some American politician, in a district where there is an especially large loss of jobs as a result of Chinese imports, might try to make hay of China's allegedly unfair trade policies; or America might come to the side of Taiwan, as some Taiwanese politician stirs the murky waters of Taiwan-China relations. Will it be acceptable, under these circumstances, and given China's political system, for China to be seen as helping the United States by lending it several hundred billion dollars? Will there be pressure on the Chinese government to divest itself of at least significant amounts of U.S. dollars, even if there is a cost to doing so? Though central banks strive for stability, politics can trump economics, forcing actions that might not be in the best economic interests of anyone. The possibility of political forces inducing a sell-off of dollars cannot be dismissed, and if that happens, we could see the dollar plunge. Economists might like to believe that economic forces underlie all prices, but the prices of national currencies, at least, are determined as much by politics as by economics.

Though reasonable people in both countries understand the facts, there is an important asymmetry: China doesn't really need to send its goods to the United States in return for pieces of paper of diminishing value used to finance America's deficits. There is a certain irony in China having, in effect, funded a tax cut for the richest people in the richest country on earth. Rather than lending money to the United States to increase consumption by these people, it could lend its money to its own people or it could finance investment in its own country. It would be far easier for China to redirect production toward its own consumers or investment than it would be for the United States to find an alternative source of cheap funding for its deficits.

Fortunately, however, the long-term economic consequences of tensions in U.S.-China relations are today but a shadowy cloud on the distant horizon. They merely add one further layer of uncertainty in a global financial system that is already straining.

MAKING GLOBALIZATION WORK: A NEW GLOBAL RESERVE SYSTEM

The dollar reserve system may not be the only source of global financial instability, but it contributes to it. The question is, will the global economy lurch from the current system to another—such as the two-currency reserve system toward which the world now seems to be moving—equally beset with problems? Or will something be done about the underlying problem?

There is a remarkably simple solution, one which was recognized long ago by Keynes: the international community can provide a new form of fiat money to act as reserves. (Keynes called his new money "bancor.")[19] The countries of the world would agree to exchange the fiat money—let's call it "global greenbacks"—for their own currency, for instance in a time of crisis.

Not only is this a theoretical possibility, but at the regional level, in Asia, there is already an initiative underway that employs some of the same concepts. The origins of the initiative go back to the East Asian crisis. At the peak of the crisis, Japan proposed establishing an Asian Monetary Fund, a cooperative movement among the countries of Asia, and generously offered to put in $100 billion to help finance it—funds badly needed to help restore the economies in the region. The United States and the IMF did everything they could to stop this; both were worried that an Asian Monetary Fund would undermine their influence in the region, and both were willing to put their own selfish concerns above the well-being of those countries. They succeeded in scuppering the proposal, but only a few years later, in May 2000, the members of the Association of Southeast Asian Nations (ASEAN), plus

China, Japan, and South Korea, meeting in Thailand, signed the Chiang Mai Initiative, agreeing in effect to exchange reserves, to set up the beginnings of a new regional cooperative arrangement that would enhance their ability to meet financial crises.

The IMF's management of the 1997 crisis laid bare the divergence of interests between it—and, by extension, the United States—and the countries of the region. These countries naturally asked, why should we put the money in our sizable reserves in Western countries that treated us so badly, when we could just keep reserves in the region, with each holding the currencies of the others? We need more investment, and if we are going to lend to enhance someone's consumption, why not lend to support the low level of consumption of our people, rather than the profligate consumption of the United States?

There were both economic and political dimensions to the initiative. The fact that, in the dollar reserve system, they received lower interest rates on what they lent than on what they borrowed was particularly galling, given that they were saving more and pursuing far more prudent fiscal policies than the United States and other advanced nations. They have, moreover, repeatedly been on the losing side of exchange rate instability. As debtors, their falling exchange rate in the 1990s meant that, in terms of their own currency, they had to repay far more than they borrowed; in 2000, with the falling dollar, as creditors they would be repaid in real terms far less than what they had lent.

As of November 2005, around $60 billion in currencies had been made available for exchange between various Asian nations, with agreements in place to expand that amount even further. As this initiative illustrates, reserves can be viewed like a cooperative mutual insurance system. The holdings of one another's currencies in reserves has the same effect as a line of credit, a commitment on the part of other countries to allow the country access to resources in times of need.

The international community has already recognized that it can provide the kind of liquidity that Keynes envisioned, in the form of special drawing rights (SDRs). SDRs are simply a kind of international money that the IMF is allowed to create.[20]

The global greenbacks proposal simply extends the concept. I refer to the new money as global greenbacks to emphasize that what is being created is a new global reserve currency, and to avoid confusion with the existing SDR system, which has two problems: SDRs are only created episodically, while global greenbacks would be created every year; and SDRs are given largely to the wealthiest countries of the world, while global greenbacks would be used not only to solve the world's financial problems but also to combat some of the deeper problems facing the world today, such as global poverty and environmental degradation.[21]

Here is a simplified description of how the system might work. Every year, each member of the club—the countries that signed up to the new global reserve system—would contribute a specified amount to a global reserve fund and, at the same time, the global reserve fund would issue global greenbacks of equivalent value to the country, which they would hold in their reserves.[22] There is no change in the net worth of any country; it has acquired an asset (a claim on others) and issued a claim on itself. Something real, however, has happened: the country has obtained an asset that it can use in times of an emergency. In a time of crisis, the country can take these global greenbacks and exchange them for euros or dollars or yen; if the crisis is precipitated by a harvest failure, it can use the money to buy food; if the crisis is precipitated by a banking failure, the money can be used to recapitalize the banks; if the crisis is precipitated by an economic recession, the money can be used to stimulate the economy.

The size of the emissions each year would be related to the additions in reserves. This will undo the downward bias of the global reserve system. Assuming that, going forward, the ratio of reserves to GDP remains roughly constant, and that global income grows at 5 percent a year, with a global GDP of approximately $40 trillion, annual emissions would be approximately $200 billion. On the other hand, if the ratio of reserves to imports stays constant, with imports growing at roughly twice the rate of GDP, annual emissions would be as much as $400 billion.

Normally, of course, these exchanges of pieces of paper make no difference. Each country goes about its business in the same way as it did before. It conducts monetary and fiscal policy much as it did before.

Even in times of emergency, life looks much as it did before. Consider, for instance, an attack on the currency. Before, the country would have sold dollars as it bought up its own currency to support its value. It can continue to do that so long as it has dollars in its reserves (or it can obtain dollars from the IMF). Under this new regime, it would exchange the global greenbacks for conventional hard currencies like dollars or euros and sell those to support its currency.

(There is an important detail: the exchange rate between global greenbacks and various currencies. In a world of fixed exchange rates [the kind of world for which the SDR proposal was first devised], this would not, of course, be a problem; in a world of variable exchange rates, matters are more problematic. One could use current market rates; alternatively, the official exchange rate could be set as the average of the exchange rates over, say, the preceding three years. In such a case, to avoid central banks taking advantage of discrepancies between current market rates and the official exchange rate, restrictions could be imposed on conversions [for instance, conversions could only occur in the event of a crisis, defined as a major change in the country's exchange rate, output, or unemployment rate]. I envision global greenbacks being held only by central banks, but a more ambitious version of this proposal would allow global greenbacks to be held by individuals, in which case there would be a market price for them and they could be treated like any other hard currency.)

Because each country is holding global greenbacks in its reserves, each no longer has to hold (as many) dollars or euros as reserves. For the global economy, this has enormous consequences, both for the former (current) reserve currency countries and for the global economy.

We noted earlier the self-destructive logic of the current system, where the reserve currency country becomes increasingly in debt, to the point at which its money no longer serves as a good reserve currency. This is the process that is currently in play with the dollar. Because the global reserve system would no longer rely on the growing debt of a single country—the basic contradiction of the current system, which makes instability almost inevitable—global stability would be enhanced.

There is a second reason that the system of global greenbacks would bring greater global stability. A major factor in the repeated crises of recent decades has been trade deficits; when countries import more than they export, they have to borrow the difference. So long as trade deficits continue, foreign borrowing continues; but at some point lenders worry that the country is too much in debt, that it may not be safe to continue to lend. When these questions start to get asked, there is a good chance that there will be a crisis around the corner.

Obviously, if one country exports more than it imports, then other countries must import more than they export. In fact, apart from statistical discrepancies, the sum of the world's trade deficits and surpluses must equal zero. Put another way, trade deficits must collectively match trade surpluses. This is the iron law of global trade deficits. Accordingly, in order for a country like Japan, which insists on running a surplus, to achieve that surplus, some other country or countries must have a corresponding deficit. Similarly, if some countries get rid of their deficit, either the deficits of other countries must increase or the surpluses of other countries must decrease, or a combination of the two.

In this sense, deficits are like hot potatoes. As South Korea, Thailand, and Indonesia eliminated their trade deficits after the East Asian crisis and turned them into surpluses, it was almost inevitable that some other country or countries would wind up with a very sizable deficit to offset their gains. In this case, the country was Brazil. But just as South Korea and Thailand could not sustain a trade deficit, neither could Brazil. As investors saw Brazil's deficit rise, they acted as they had so often before: debts were recalled, precipitating a crisis. As Brazil's economy plunged into recession, imports contracted, and Brazil's deficit was converted into a surplus; again, that surplus means a same-sized deficit was created somewhere else in the global system.

While the IMF—and the financial community generally—has focused on countries with trade deficits as the problem giving rise to global instability, this analysis suggests that trade surpluses are just as much the problem. In fact, Keynes, thinking about the problems of the global financial system sixty years ago, went so far as to suggest that there should be a tax levied on countries running a trade surplus, to discourage them from letting the trade surplus grow too large.[23]

As badly as the system has functioned, matters could have been worse. There is one country that can—so far—maintain a trade deficit without precipitating a crisis, and that is the United States. The United States has become not just the consumer of last resort, but also the deficit of last resort. It has been able to get away with this because it is the richest country in the world and because other countries have wanted to hold dollars in their reserves. But even if the United States can mount deficits longer than other countries, it cannot do so indefinitely. There will be a day of reckoning.

The global greenback system breaks the zero-sum logic that has resulted in one crisis following another. Of course, it would still be the case that the sum of the trade deficits equals the sum of the surpluses, but there would be an annual emission of global greenbacks to offset— to pay for—the deficits. So long as deficits remained moderate, there would be no problem. There would be a cushion equal to the emission of global greenbacks. The game of hot potato deficits would effectively be stopped, and a buffer would be created, to stabilize the global economy in the face of the inevitable shocks it faces.

The United States might think that the global greenbacks system would make it worse off because it would no longer effectively get cheap loans from developing countries. There is, of course, something unseemly about the poorest countries providing low-interest loans to the richest. However, the United States would benefit from the greater global stability, along with the rest of the world. The global greenbacks system would make it easier for the United States to maintain its economy at full employment without massive fiscal deficits (undoing the forces described earlier, by which increased dollar reserve holdings abroad lead to weaker aggregate demand within the United States).[24]

If the United States cannot be persuaded to join the new Global Reserve System, there is another, tougher approach. The rest of the world could agree to move to this system, a form of cooperative mutual help, and, in doing so, agree that they would gradually shift more of their reserves to countries that are part of the co-op. As the benefits to the United States from exploiting the developing countries diminish, the United States would face increasing incentives to join.

Reform and the Broader Globalization Agenda

Who would receive the annual emissions of global greenbacks? The answer to this has great consequences for global well-being.

Here is an opportunity for the global community to make globalization work so much better.[25] Globalization entails the closer integration of the countries of the world; this closer integration entails more interdependence, and this greater interdependence requires more collective action. Global public goods, the benefits of which accrue to all within the global community, become more important. These include, for instance, health (finding a vaccine against malaria or AIDS) and the environment (reducing greenhouse gas emissions, maintaining biodiversity in rainforests). These should be first priorities for the funds.

The new global reserve system could not only solve the problem of how to finance global public goods; it could demonstrate the global community's commitment to global social justice. After providing funds for global public goods, the bulk of the remaining funds could go to the poorest countries of the world. This would be a major change in philosophy from that underlying the IMF, which recognized the need for greater liquidity, through the issuance of SDRs, but based it on the principle "to he that hath, more shall be given." The rich got the lion's share.

There are many ways in which the funds could be administered. Inevitably there will be disagreements about the best way, but we should be careful not to let the perfect be the enemy of the good.[26] Probably it makes the most sense to have a combination of approaches. One approach would be to allocate funds to different countries on the basis of their income and population (consistent with principles of social justice, poorer countries would get a larger allocation per capita). Given the failure of conditionality in the past, the only condition that should be imposed relates to global externalities—costs that countries impose on others. The most important is probably nuclear proliferation—only countries that commit themselves fully to a non-nuclear regime would be eligible. Other conditions might involve global environmental externalities, such as greenhouse gas emissions, emissions of gases that destroy the ozone layer, ocean pollution, abiding by international agreements on endangered species, etc.

In a second approach, funds would be distributed through international institutions, either existing ones or newly created "special trust funds" established under the auspices of the United Nations. They could be issued to individual countries, who would agree in turn to make a contribution of an equivalent amount to the UN trust funds. A portion might be used to help achieve the Millennium Development Goals—the goals the international community set for itself in reducing poverty by 2015, including promoting health, increasing literacy, and improving the environment in developing countries.[27] Take, for instance, the area of health. The record of the World Health Organization is impressive. Some diseases, including smallpox, polio, and river blindness, have been virtually eliminated. With more money, much more could be done at relatively low cost.[28] We already know that the incidence of malaria can be greatly reduced by draining stagnant pools of water and using impregnated mosquito nets. I have visited smoke-filled huts around the world, where indoor pollution leads to lung and eye diseases; they need only chimneys. As I mentioned in chapter 2, great advances in public health can be achieved simply by teaching people to build latrines downhill from sources of drinking water. These are small changes that could make big differences in the lives of millions of people, and more money would help enormously.

Some of the money could, similarly, be used to achieve literacy for all. Today, some 770 million people around the world remain unable to read or write; one of the Millennium Development Goals calls for every child in the world to complete primary education by 2015. The cost would be small, between $10 billion and $15 billion a year,[29] but so far the international community has not been forthcoming with the money needed. Issuing some of the new global greenbacks to a special UNICEF education trust fund could make a big difference.

As we saw in chapter 6, global warming is a global problem. The international community has established a global environment lending facility to help pay for the incremental costs associated with reducing greenhouse gases and other good environmental policies, but it is vastly underfunded. Some of the global greenbacks could go there.

A third approach might involve competitive allocations for development-oriented projects, for which governments and NGOs could apply. Competition might spur innovation in schemes to enhance the well-being of those in the developing world.

A fourth alternative, direct distribution to individuals, is perhaps too problematic to be practical. Aside from the difficulties of getting the money to the poorest individuals, it makes little sense to give money to people and then charge them for basic health and education. Both sides of the transaction are wasteful and imperfect. Better simply to use the money to provide education and health services for the poorest.[30]

As I have broached the idea of a reform in the global reserve system in seminars around the world, I have been heartened by the extent of support. George Soros has advocated the onetime use of SDR emissions for financing development.[31] But why restrict emissions to a onetime event?

The problems of the global financial system are systemic and have much to do with the global reserve system. The world is already moving out of the dollar system, but that doesn't mean that it is moving toward a better system—and, sadly, little thought has been given to where it is going or how it should evolve. This single initiative could do more to make globalization work than any other. It would not eliminate the problems faced by developing countries, but it would make things better. It would enhance global stability and global equity. It is not a new idea, but it is an idea perhaps whose time has come.

CHAPTER 10

Democratizing Globalization

G lobalization was supposed to bring unprecedented benefits to all. Yet, curiously, it has come to be vilified both in the developed and the developing world. America and Europe see the threat of outsourcing; the developing countries see the advanced industrial countries tilting the global economic regime against them. Those in both see corporate interests being advanced at the expense of other values. In this book, I have argued that there is much merit in these criticisms—but that they are criticisms of globalization as it has been managed. I have attempted to show how we can remake globalization, to make it more nearly live up to its promise.

This book has been mainly about the economics of globalization, but as I noted in chapter 1, the problems have much to do with economic globalization outpacing political globalization, and with the economic consequences of globalization outpacing our ability to understand and shape globalization and to cope with these consequences through political processes. Reforming globalization is a matter of politics. In this concluding chapter, I want to deal with some of the key political issues. Among them are the prospects for unskilled workers and the impact of globalization on inequality; the democratic deficit in our global economic institutions, which weakens even democracy within our own countries; and the human tendency to think locally even while we live in an increasingly global economy.

Growing inequality and the threat of outsourcing
When in February 2004 President Bush's chief economic adviser, N. Gregory Mankiw, praised the opportunity that outsourcing, with its lower costs and hence higher profits, provided for U.S. companies, he was widely criticized. Americans were worried about jobs, in manufacturing—in which some 2.8 million jobs were lost from 2001 to 2004—and even in the high-tech and service sectors.[1] In some sense, outsourcing is not new: U.S. companies have been sending jobs overseas for decades. The number of manufacturing jobs in the United States has been shrinking since 1979, and the fraction of Americans working in manufacturing has been declining since the 1940s. (In 1945, 37 percent of working Americans were employed in manufacturing, while today the figure is less than 11 percent.)[2]

A dynamic economy is, of course, characterized by job loss and job creation—the loss of less-productive jobs and the shift of workers to areas of higher productivity. The production of horse carriages declined with the arrival of the automobile. During the debate over the North American Free Trade Agreement, 1992 presidential candidate Ross Perot warned that there would be a "giant sucking sound" as jobs were pulled out of the United States. The response from the Clinton administration was that America didn't want those low-wage, low-skill jobs, and that the market would create better-paid, higher-skill jobs. And during the first few years of NAFTA unemployment in the United States actually declined, from 6.8 percent, at the beginning of NAFTA, down to a low of 3.8 percent.

Just as the United States and European countries made the transition from agriculture to manufacturing more than a hundred years ago, more recently they have made the move from manufacturing to services. The share of manufacturing in employment and output has fallen not just in the United States but also in Europe and Japan (to 20 percent).[3] As America and Europe lost jobs in manufacturing, they gained jobs in the service sector, a sector that includes not only low-skill jobs flipping hamburgers but high-paid jobs in the financial services sector. It was thought that America, with its high level of skills and its service-sector dominated economy would be protected from competition from abroad. What made outsourcing so scary was that even highly skilled jobs began to go abroad. The strategy of "upskilling" and education, though clearly valuable and important, does not provide a full answer for how to respond to global competition.

The scale and pace of the competitive threat, of the job loss in a relatively short time, is beyond anything that has happened before. This is the flip side of another unprecedented change: two countries, China and India, that were once desperately poor and economically isolated are now part of the global economy. Never before have the incomes of so many people risen so fast.[4]

Standard economic theory, which underlies the call for trade liberalization, has a scenario for what should happen with full liberalization—a scenario that its advocates seldom mention, but which we noted briefly in chapter 3. With full global economic integration, the world will become like a single country, and the wages of unskilled workers will be the same everywhere in the world, no matter where they live. Whether in America or in India or in China, unskilled workers of comparable skills performing comparable work will be paid the same. In theory, the actual wage will be *somewhere* between that received today by the Indian or Chinese unskilled worker and that received by his American or European counterpart; in practice, given the relative size of the populations, the likelihood is that the single wage to which they will converge will be closer to that of China and India than to that of the United States or Europe.

Of course, taking down all tariff and trade barriers will not lead instantly to full integration or to the equalization of wages. There will still be transportation costs, and in the case of very poor and remote countries, these remain important. In the past, at least two factors played a part in enabling wage differences to persist. The first is the scarcity of capital in developing countries. This matters because with less capital (such as new machines and technology) workers are less productive. Handlooms are less productive than machine looms—and because they are less productive, workers' wages will be lower. The second is the gap in knowledge between the developed and the less developed countries. Skills and technology have lagged in the developing world, and that has lowered productivity and depressed wages.

However, these impediments to wage equalization are disappearing. International capital markets have improved enormously. Today, while China is saving 42 percent of its GDP, it is also receiving more than $50 billion every year in foreign direct investment, an amount close to 4 percent of its GDP.[5] And in recent years, the flow of knowledge from the developed to the undeveloped countries has accelerated.

It will take decades to fully overcome the knowledge gap and the capital shortage in the developing world. The good news is that there will be a strong force pulling up wages in China and India. The downside is that there will be a strong force pushing down wages for unskilled workers in the West. So, while Americans and Europeans can rejoice in the rising living standards of unskilled workers in the developing world, they will be worrying about what is happening at home. The issue is not just the total number of jobs that will be outsourced—lost—to China or India. The real problem is that even a relatively small gap between the demand for and the supply of labor can create large problems, leading to wage stagnation and decline, and creating high levels of anxiety among the many workers who feel their jobs are at risk. That is what appears to be happening.

Of course, as we have seen, globalization and trade liberalization will increase overall incomes (if the country can manage to maintain full employment, a big "if"). But it follows that with incomes on average increasing, and wages, especially at the bottom, stagnating or falling, inequality will increase. Those in the industries who find themselves outcompeted especially will suffer; they may find their "human capital," the investments made in particular skills, no longer of much value. For the past five years, real wages in America have been basically stagnant; for those at the bottom, real wages have stagnated for more than a quarter of a century.[6] Whole communities may find themselves in difficult straits. As businesses shut down and jobs are lost, real estate prices will fall, which will hurt most people in those areas, since their main asset is their home.

Responding to the Challenges of Globalization

There are three ways in which the advanced industrial countries can respond to these challenges. One is to ignore the problem and accept the growing inequality. Those who take this position (many of them proponents of the now-discredited theory of trickle-down economics, which holds that so long as there is growth, *all* will benefit) emphasize the underlying strengths of a market economy and its ability to respond to change: we may not know where the new jobs will be created, they say, but so long as we allow markets to work their magic, new jobs will be created. It is only when, as in Europe, a government interferes with market processes by protecting jobs, that there are problems with unemployment.

But in both Europe and America, this approach is not working. While there are winners from globalization, there are numerous losers. Globalization is, of course, only one of the many forces affecting our societies and our economies. Even without it, there would be increasing inequality. Changes in technology have increased the premium the market places on certain skills, so that the winners in today's economy are those who have or can acquire those skills. These changes in technology may in the end be more important than globalization in determining the increase in inequality, and even the decline in unskilled wages. Voters can do little about the march of technology; but they can—through their elected representatives—do something about globalization. Protectionist sentiment has been increasing almost everywhere. In the United States, even a small trade bill, free trade with Central America, attracted enormous opposition, barely passing the House of Representatives by a 217 to 215 vote in July 2005. I do not believe it is tenable to pretend that everything will be fine if we just leave the markets alone. Nor is it tenable to ask workers to have faith that, with enough patience, globalization will make them all better off, even though now they must accept lower wages and decreased job security. Even if they were to accept on faith the proposition that globalization will lead to faster GDP growth, why should they believe that it would lead to faster growth in *their* incomes or an overall increase in *their* well-being? While politicians may refer obliquely to lessons of economics to reassure their constituents, both standard economic theory and a wealth of data is consistent with workers' own intuitions: without strong government redistributive policies, unskilled workers may well be worse off.

Similar issues arise with migration. I explained in chapter 3 how migration may lead to an increase in global efficiency, and how it may be of particular benefit to those in the developing world. But migration of unskilled labor leads to lower wages for unskilled workers in the developed world. With both trade liberalization and migration, the country as a whole may benefit, but those at the bottom are likely to be made worse off.

The second tack is to resist fair globalization. In this view, now is the time for America and Europe to use their economic power to make sure that the rules of the game favor them permanently—or at least for as long as possible. Power begets power; and by using their current combined economic power, they can at least protect their position, and perhaps even enhance it. This is a view based not on what is right or fair but on realpolitik.

In this logic, the United States, while continuing to pay lip service to fair trade, should protect itself from the onslaught of foreign goods and from outsourcing, while at the same time doing what it can to get access to foreign markets. America's seeming brazenness in doubling its agricultural subsidies while preaching the rhetoric of free trade is an example. As a sop to those who insist on fairness, some effort is put into finding "legal" ways of providing these subsidies, such as devising concepts like "non–trade distorting subsidies," getting other countries to agree that such subsidies are allowed, and then claiming that one's subsidies are of that sort. The presumption seems to be that because something is legal, it is morally right.

I believe that this approach is both morally wrong and economically and politically unviable. America's standing in the world has long been based not just on its economic and military power but on its moral leadership, on doing what is right and fair. But for those who believe in realpolitik, this is of little concern. More to the point, this option is not really possible, given how far we already are down the path of globalization. While the Uruguay Round trade agreement may not be fair to the developing countries, it created the beginnings of a semblance of an international rule of law in trade, which the United States has to obey.

Moreover, one of the successes of the last three decades has been the creation of strong democracies in many parts of the developing world. Their citizens know what is going on, and they know when a proposed trade agreement is fundamentally unfair. American citizens may not care about the hypocrisy of its leaders in talking about free trade and maintaining agricultural subsidies, but Brazil's and Argentina's citizens do.

Too much is at stake—and there are too many who have already benefited from globalization—to allow America and Europe to pull back from globalization, to walk away from it. There are too many losers from globalization in the developing world to allow the developed world to try to shape globalization unfairly in its favor.

That leaves but one course—coping with globalization and reshaping it. For America, coping means recognizing that globalization will mean downward pressure on unskilled wages. The advanced industrial countries have to continue upskilling their labor forces, but they also have to strengthen their safety nets and increase the progressivity of their income tax systems; it is the people at the bottom who have been hurt by globalization (and, probably, by other forces, like changing technology); it seems the right thing to do, to lower taxes on them and to increase taxes on those who have been so well served by globalization. Regrettably, in America and elsewhere, policies have been moving in precisely the opposite direction. Investments in research, which will increase the productivity of the economy, are also important. These investments yield high returns. Increased productivity is likely to lead to increased wages and incomes; and if even a portion of the higher income that results is spent on a social agenda of education and health, the well-being of all citizens will be enhanced.

The critics of globalization are right: as it has been managed, there are too many losers. And I think the optimists among these critics—those who, at meetings like the World Social Forum at Mumbai with which I began this book, claimed that "another world is possible"—are also right. This book has laid out a number of reforms that would enable globalization more nearly to live up to its potential of benefiting those in both the developed and less developed countries: a reformed globalization that could receive the support of those in both.

THE DEMOCRATIC DEFICIT

I have argued in this book that we have to learn how to cope better with globalization (in both the developed and the less developed countries). We also have to learn how to manage it better, with a greater concern both for the poor countries and for the poor in rich countries, and for values that go beyond profits and GDP. The problem is that there is a democratic deficit in the way that globalization has been managed. The international institutions (the International Monetary Fund, the World Bank, the World Trade Organization), who have been entrusted with writing the rules of the game and managing the global economy, reflect the interests of the advanced industrial countries—or, more particularly, special interests (like agriculture and oil) within those countries. This imbalance is in some cases the result of distorted voting rights;[7] at other times, it comes from the sheer economic power of the countries and interests involved. The imbalance is seen both in the agenda and in the outcomes in every arena of globalization, from trade to the environment to finance. We see it in both what is on the agenda and what is not.

Over the past two centuries, democracies have learned how to temper the excesses of capitalism: to channel the power of the market, to ensure that there are more winners and fewer losers. The benefits of this process have been staggering, and have given many in the First World wonderfully high standards of living, much higher than were conceivable in 1800.

At the international level, however, we have failed to develop the democratic political institutions that are required if we are to make globalization work—to ensure that the power of the global market economy leads to the improvement of the lives of most of the people of the world, not just the richest in the richest countries. Because of the democratic deficit in the way globalization is managed, its excesses have not been tempered; indeed, as we noted in earlier chapters, globalization has sometimes circumscribed the ability of national democracies to temper the market economy.

The need for global institutions has never been greater, but confidence in them, and their legitimacy, has been eroding. The IMF's repeated failures in managing the crises of the past decade was the coup de grâce, following years of dissatisfaction with its programs in Africa and elsewhere, including the excessive austerity it forced upon these countries. The failure of the countries that followed the IMF–World Bank ideologically driven Washington Consensus policies and the contrast with the ongoing success of the East Asian countries, which I described in chapter 2, has not helped to restore confidence in these institutions. Neither did the arrogance with which the IMF demanded that it be allowed to force developing countries to open up their markets to speculative capital flows, followed a few years later by a quiet recognition that capital market liberalization might lead to instability but not growth. And while they pushed an agenda that led to financial market instability, they did nothing about one of the root causes of global instability, the global reserve system. At the WTO, in the trade front, matters are no better. After admitting at Doha, in November 2001, that the previous round of trade negotiations was unfair, the advanced industrial countries eventually effectively reneged on their promise of a development round.

The institutions themselves are, in some sense, not to blame: they are run by the United States and the other advanced industrial countries. Their failures represent failure of policy by those countries. The end of the Cold War gave the United States, the one remaining superpower, the opportunity to reshape the global economic and political system based on principles of fairness and concern for the poor; but the absence of competition from communist ideology also gave the United States the opportunity to reshape the global system based on its own self-interest and that of its multinational corporations. Regrettably, in the economic sphere, it chose the latter course.

Just as the international institutions cannot be fully blamed—the responsibility must lie partly with the governments that govern them—the governments themselves cannot be fully blamed. The responsibility lies partly with their voters. We may increasingly be part of a global economy, but almost all of us live in local communities, and continue to think, to an extraordinary degree, locally. It is natural for us to value a job lost at home far more than two jobs gained abroad (or in the context of war, a life lost at home far more than those lost abroad). Part of the mindset of thinking locally is that we don't often think of how policies that we advocate affect others and the global economy. We focus our attention on the direct effect on our *own* well-being. Cotton growers in the United States think of how they gain from their subsidies, not how millions in the rest of the world lose.

To make globalization work there will have to be a change of mindset: we will have to think and act more globally. Today, too few have this sense of global identity. There is an old aphorism about all politics being local, and, with most people living "locally," it is not surprising that globalization is approached within the very narrow framework of local politics. Local thinking persists even as the world grows more economically interdependent. It is this disjunction between local politics and global problems that is the source of so much of the dissatisfaction with globalization.

The contrast between analysis and advocacy for policies at the national and global level is stark. Within each country, we are aware that laws and regulations affect different people differently. Economists carefully calculate, for each tax, rule, or regulation, the extent to which different income groups are affected. We argue for and against different policies on the basis of whether they are just, whether they hurt the poor, whether their burden falls disproportionately on those less well off.

In the international arena, not only do we fail to do the analysis, we almost never argue for a policy on the basis of its fairness. Trade negotiators are told to get the best agreement they can, from the perspective of their country's own interests. They are not sent off to Geneva (where the trade negotiations generally occur) with the mandate to craft an agreement that is fair to all. Special attention is not given, as it should be, to the poorest, but to the strongest—such as the special interests that are the largest contributors to the campaigns of the American president and the party in power. In fact, often the special interests are elevated to be national interests: doing what is best for America's drug companies, for Microsoft and for ExxonMobil, is viewed as equivalent to doing what is best for the country in general. This is encapsulated in the famous quote of Charles Wilson, the head of GM, in 1953 that "what was good for our country was good for General Motors, and vice versa."[8] In the era of globalization, this is no longer true—if it ever was.

Even within the international institutions, seldom is global policy discussed in terms of social justice. There is a pretense that there are no trade-offs, and that, accordingly, decision making can be delegated to technocrats, who are assigned the complex task of finding and managing the best economic system, and who are thought to be better equipped than politicians to make objective decisions. There are, of course, some problems which should be delegated to technocrats—like choosing the best computer system for running the social security system. But delegating the writing of the rules of the economic game to technocrats can be justified only if there is a single best set of rules, one that makes everyone better off than any other set of rules. This is simply not the case; this view is not only wrong, but dangerous. With a few exceptions, there are always trade-offs. The existence of trade-offs means that there are choices to be made. It is only through the political system that those choices can be properly made, which is why it is so important to remedy the global democratic deficit.

Depoliticizing the decision-making process paves the way for decisions that are not representative of broader social interests. By removing decisions about the right trade regime or the right intellectual property regime from the *overt* political process, the door is opened to *covert* shaping of those decisions by particular interests. The drug companies can shape intellectual property agreements; producers, not consumers, can shape trade policy. Monetary policy provides another example. No economic issue affects people more than the macro-economic performance of the economy. Increasing the unemployment rate makes workers worse off, but the resulting lower inflation makes bondholders happy. Balancing these interests is a quintessentially political activity, but there has been an attempt by those in financial markets to depoliticize the decision, to turn it over to technocrats, with a mandate to pursue the policies that are in the interests of financial markets. The IMF has been encouraging, sometimes even forcing (as a condition of assistance), countries to have their central banks focus *only* on inflation.

Europe succumbed to these doctrines. Today, throughout Euroland, there is unhappiness as the European Central Bank pursues a monetary policy that, while it may do wonders for bond markets by keeping inflation low and bond prices high, has left Europe's growth and employment in shambles.

Responding to the Democratic Deficit

There are two responses to the problem of the democratic deficit in the international institutions. The first is to reform the institutional arrangements, along the lines suggested earlier in this book. But this will not happen overnight. The second is to think more carefully about what decisions are made at the international level.

Globalization means that events in one part of the world have ripple effects elsewhere, as ideas and knowledge, goods and services, and capital and people move more easily across borders. Epidemics never respected borders, but with greater global travel diseases spread more quickly. Greenhouse gases produced in the advanced industrial countries lead to global warming everywhere in the world. Terrorism, too, has become global. As the countries of the world become more closely integrated, they become more interdependent. Greater interdependence gives rise to a greater need for collective action to solve common problems.

The agenda for collective action should focus on those items that represent the most essential areas for benefiting the entire global community. Other items should not be on the agenda.[9] In chapter 4, I argued that there is no need for a uniform set of intellectual property rights rules; excessive standardization not only takes away important degrees of political sovereignty but is actually counterproductive. A focused agenda is especially important because the expansiveness of the agenda itself puts developing countries, which cannot afford large staffs, at a disadvantage in negotiations. Global collective action should focus upon the need to halt negative externalities—actions by one party that adversely affect others—and on the opportunity to promote, by acting together, the well-being of all through the provision of global public goods, the benefits of which are enjoyed around the world.

As the world becomes more globalized, more integrated, there will be more and more areas in which there are opportunities for cooperative action, and in which such collective action is not only desirable but necessary. There is an array of global public goods—from global peace to global health, to preserving the global environment, to global knowledge. If these are not provided *collectively* by the international community, there is a risk—indeed, a likelihood—that they will be underprovided.[10]

Providing global public goods requires some system of finance. Chapter 9 described how a reform of the global reserve system can provide a large source of finance, in the order of magnitude of $200 billion to $400 billion a year. A second idea is to use revenues from the management of global resources—auctioning off fishing rights, or the right to extract natural resources beneath the sea, or carbon emissions permits—for providing global public goods. Finally, there are some instances in which taxation can actually contribute to economic efficiency. Such taxes, levied to overcome problems of negative externalities, are called corrective taxes. Taxation on global negative externalities, such as arms sales to developing countries, pollution, and destabilizing cross-border financial flows, can provide a third source of revenues for financing global public goods.

In the long run, the most important changes required to make globalization work are reforms to reduce the democratic deficit. Without such changes, there is a real danger that any reforms will be subverted. In chapter 3, for instance, we saw how as tariffs have come down, nontariff barriers have been erected. This is not the place to provide a detailed description of how each of the international institutions needs to be changed. Instead, I list the major elements of any reform package:

- *Changes in voting structure* at the IMF and the World Bank, giving more weight to the developing countries. At the IMF, the United States remains the single country with an effective veto. At both institutions, votes are largely on the basis of economic power—and too often, not economic power today but, to a too large extent, economic power as it existed at the time these institutions were created more than a half century ago.[11]

- *Changes in representation*—who represents each country. So long as trade ministers determine trade policy and finance ministers determine financial policy, other related concerns, like the environment or employment, will be given short shrift. One possible change is to insist that when there are areas of overlapping concerns, all the relevant ministries be represented. When intellectual property provisions are being discussed, surely the science and technology ministries—who may not only have a more balanced position but will even know something about the matter—should be at the table.

- *Adopting principles of representation.* It is difficult to make decisions, or to engage in negotiations, when 100 or more countries are involved. But the way, for instance, that trade negotiators have responded to this problem in the past should be viewed as totally unacceptable.[12] No matter what is done, there will be an imbalance of economic power, and there is little that can be done to stop the powerful from exercising that power; but at the very least, the formal processes should be more in accord with democratic principles. The major countries should be joined in negotiations by representatives of each of the various major groups: the least developed countries, the small agricultural exporters, and so on. In fact, some progress in this direction is already taking place.

Given that it will be difficult to make these changes, it is all the more important to make the following reforms in the way international institutions operate:

- *Increased transparency.* Because there is no direct democratic accountability for these institutions (we do not vote for our representatives to these institutions or for their leadership), transparency, enforced through strong freedom of information acts, is vital. Ironically, these institutions are *less* transparent than the more democratic of their member governments.

- *Improvements in conflict-of-interest rules* will not only increase confidence in, and the legitimacy of, international governance but (if economists are correct and incentives do matter) might actually lead to policies that are more in the general interest.

- *More openness, including improvements in procedures* to ensure not only more transparency but that more voices are heard. NGOs have taken on increased importance in ensuring that voices other than those of the multinational corporations get heard in the process of global economic decision making. In democracies like the United States, when regulatory agencies propose rules, interested parties are given an opportunity to comment, and the regulatory agency must respond. It should be the same for global institutions and regulatory agencies.

- *Enhancing the ability of developing countries to participate meaningfully in decision making,* by providing them with assistance in assessing the impact on them of proposed changes. The U.S. Treasury and the finance ministries of some of the other advanced industrial countries can make their own assessments, but developing countries typically do not have the resources to do so. The deliberative discussions of the WTO and other international economic organizations would also be helped if there were an independent body to evaluate alternative proposals and their impact on developing countries.

- *Improved accountability.* Even if there is not direct electoral accountability, there can be more independent evaluations of the performance of the international economic institutions. While the World Bank and the IMF presently do this—and, indeed, spend a considerable amount of money on such evaluations—the evaluation units have typically relied heavily on temporary staff supplied by the Fund or the Bank. Though this has an advantage in that they are well informed about what is going on, it is hard for them to provide a fully independent evaluation. The task of evaluation should be moved—to the UN, for instance. Assessments must be made of the disparity between predicted consequences and what actually happens: Why, for instance, did the IMF bail-out packages not work in the way predicted during the crises? Why was there money available to bail-out international banks, but not money to pay for food subsidies to the poor? Why were the benefits received by many of the poorest countries from the last round of trade negotiations so much less than had been promised?

- *Better judicial procedures.* The need for this was highlighted by our discussion in chapter 3 of the process by which dumping duties are imposed by the United States, where it is simultaneously the prosecutor, judge, and jury in assessing dumping duties. Such a judicial procedure is obviously flawed. There needs to be an independent global judicial body to determine, for instance, whether dumping has occurred, and if so, what the dumping duties should be.

- *Better enforcement of the international rule of law.* I have repeatedly commented on the great achievement of the Uruguay Round in creating the beginning of a semblance of international law. It means that principles, not just power, can govern trade relations. The law may be imperfect, but it is better than no law at all. There are, however, still many areas where the law would make for a better globalization *if it were enforced.* One important instance was noted in the last chapter: America's refusal to do anything about global warming can be considered a major and unwarranted trade subsidy. The enforcement of regulations against such subsidies could be an important instrument both in creating a fairer trading system and in addressing one of today's most important global problems.

We have an imperfect system of global governance without global government; and one imperfection is the limitations on our ability to enforce international agreements and stop negative externalities. We must use what instruments we have—including trade sanctions.[13]

In chapter 3, I noted another major problem: the fragmentation of the global trading system into a series of bilateral and regional trade agreements. The great achievement of the multilateral trading system over the past sixty years, the most favored nation principle under which each country gave to every other country the same terms, is now being undermined by the United States, followed by others. Such agreements are legal under WTO rules only when they create more trade than they divert; almost surely, some bilateral agreements would fail this test. There should be an international tribunal to determine whether, as each agreement is proposed, it is legal, with the burden of proof lying with the countries trying to fragment the global trading system. The tribunal would determine, for instance, whether Mexico's gains under NAFTA, to the extent that they exist, arose largely from diversion of the trade in textiles that the United States might have bought from Latin American countries other than Mexico. This might slow down, or even put a stop to, the rash of bilateral agreements that threatens to undermine the multilateral trade system.

Finding a New Balance

What is needed, if we are to make globalization work, is an international economic regime in which the well-being of the developed and developing countries are better balanced: a new *global social contract* between developed and less developed countries. Among the central ingredients are:

- A commitment by developed countries to a fairer trade regime, one that would actually promote development (along the lines outlined in chapter 3).
- A new approach to intellectual property and the promoting of research, which, while continuing to provide incentives and resources for innovation, would recognize the importance of developing countries' access to knowledge, the necessity of the availability of lifesaving medicines at affordable prices, and the rights of developing countries to have their traditional knowledge protected.
- An agreement by the developed countries to compensate developing countries for their environmental services, both in preservation of biodiversity and contribution to global warming through carbon sequestration.
- A recognition that we—developed and less developed countries alike—share one planet, and that global warming represents a real threat to that planet—one whose effects may be particularly disastrous for some of the developing countries; accordingly, we all need to limit carbon emissions—we need to put aside our squabbling about who's to blame and get down to the serious business of doing something; America, the richest country on the earth, and the most energy profligate, has a special obligation—and one of its states,

California—has already shown that there can be enormous emission reductions without eroding standards of living.

- A commitment by the developed countries to pay the developing countries fairly for their natural resources—and to extract them in ways that do not leave behind a legacy of environmental degradation.

- A renewal of the commitments already made by the developed countries to provide financial assistance to the poorer countries of 0.7 percent of GDP—a renewal accompanied this time by actions to fulfill that commitment. If America can afford a trillion dollars to fight a war in Iraq, surely it can afford less than $100 billion a year to fight a global war against poverty.

- An extension of the agreement for debt forgiveness made in July 2005 to more countries: too many countries' aspirations of development are being thwarted by the huge amounts they spend on servicing their debt—so large, in fact, that, as we noted, net flows of money in some recent years have been going from developing countries to the developed.

- Reforms of the global financial architecture that would reduce its instability—which has had such a crushing effect on so many developing countries—and shift more of the burden of the risk to the developed countries, which are in such a better position to bear these risks. Among the key reforms is a reform in the global reserve system, as discussed in chapter 9, which, I believe, would not only lead to enhanced stability, from which all would benefit, but could also help finance the global public goods that are so important if we are to make globalization work.

- A host of institutional (legal) reforms—to ensure, for instance, that new global monopolies do not emerge, to handle fairly the complexities of cross-border bankruptcies both of sovereigns and companies, and to force multinational corporations to confront their liabilities, from, for instance, their damage to the environment.

- If the developed countries have been sending too little money to the developing world, they have also been sending too many arms; they have been part and partner in much of the corruption; and in a variety of other ways, they have undermined the fledgling democracies.

The global social compact would entail not just lip service on the importance of democracy but the developed countries actually curtailing practices that undermine democracy and doing things to support it—and especially doing more to curtail arms shipments, bank secrecy, and bribery.

For globalization to work, of course, developing countries must do their part. The international community can help create an environment in which development is possible; it can help provide resources and opportunity. But in the end, responsibility for successful, sustainable development—with the fruits of that development widely shared—will have to rest on the shoulders of the developing countries themselves. Not all will succeed; but I believe strongly that with the global social contract described above, far more will succeed than in the past.

Elements of this new global social contract are already in place. At the international meeting on finance for development convened by the UN in Monterrey, Mexico, in March 2002, the advanced industrial countries made a commitment to increase their aid to 0.7 percent of GDP, but the meeting was also important because it recognized—at last—that development is too important and too complex to be left to finance ministers. Finance ministers and central bank governors bring a particular perspective to the discussion—an important perspective, but not the only one. Consider, for instance, the issue of sovereign debt restructuring. No government would entrust legislation setting forth the framework for bankruptcy to a committee dominated by creditor and creditor interests; however, putting the IMF in charge of the bankruptcy proceeding, as the IMF argued should happen, would have created an equivalent situation. Such decisions have to be approached with greater balance.

One way of achieving greater balance is to strengthen the Economic and Social Council at the UN. The Council could play an important role in defining the global economic agenda, in ensuring that attention gets focused not just on issues that are of interest to the advanced industrial countries but on those that are essential to the well-being of the entire world. It could encourage discussions of global financial reform which address the problems of the developing countries—the fact, for instance, that they are left to bear the brunt of exchange rate and interest rate risk. It could push for a reform of the global reserve system, or for new ways of handling sovereign debt restructuring—in which the bankruptcy process is not controlled by creditor countries. It could have a particularly important role in the many issues that cross the "silos" in which so much of international decision making is confined. It could push for the rainforest initiative that I described in chapter 6, which would simultaneously provide developing countries with incentives to maintain their rainforests (with enormous world-wide benefits for reducing global warming and maintaining biodiversity) and with money to promote their development. It could push an intellectual property regime that advances science and pays due respect to other values, like life and access to knowledge. It could make sure that any international oversight of a country's economic policies ("surveillance," as it is often called) focuses not just on inflation, which is of such concern to financial markets, but also on unemployment, which exerts such a toll on workers.

Discontent with globalization as it has been managed has partly reflected the discontent with outcomes, and partly the discontent with the lack of democratic process. Reducing the democratic deficit would be a major step forward in making globalization work on both counts. I have faith that policies and programs that have been subject to democratic scrutiny are likely to be more effective and more sensitive to the concerns of the citizenry.

Much is at stake

The globalization debate has become so intense because so much is at stake—not just economic well-being, but the very nature of our society, even perhaps the very survival of society as we have known it. The globalizers of the past twenty years may have thought that the economic doctrines they pushed for through the international institutions would by now have succeeded so well in enhancing the well-being of everyone that all would be forgiven. Perhaps they hoped that even if there were growing inequality, so long as there was enough money trickling down the poor could be placated. Even if a few were denied access to lifesaving medicines, if overall enough people saw their health improve they might be satisfied. As we have seen, for too many the promised benefits did not materialize.

But even had there been more *economic* success, unhappiness with some aspects of globalization would persist—and if more people realized what was going on, that unhappiness might have been even greater. The United States has argued that maintaining open trade borders is more important than the preservation of culture or the protection of food safety, at least against what it views as irrational fears over genetically modified foods. But even the United States has recognized that there are other values more important than economic globalization—or at least one value, security. The United States argues forcefully for trade restrictions that it claims will enhance its national security. It subsidizes oil and does not allow foreign ships to transport goods within its borders—in both cases arguing national security concerns. It even argues for "secondary boycotts": not only does it not allow its firms to sell products that might be of military use to China, but it has put enormous pressure on Europe to follow suit. The United States' Helms-Burton Act of 1996 imposes sanctions against foreign firms that trade with Cuba, even when the laws of those countries allow them to do so. The anthrax scare of 2001 (which in the end was never traced to terrorism from outside the United States) led to the passing of a bioterrorism law that imposes registration and record-keeping requirements on those wishing to export goods to the United States. The United States says the requirements are not onerous and costly; many foreign firms claim they are. At the very least, they are an added cost to selling to the United States. The increased difficulties of getting visas also make it more difficult for foreign companies to do business in the United States, including providing services. If other countries reciprocate, it will be clear that as one set of man-made barriers to trade is coming down, a new set is being erected.

Yes, a country's first responsibility to its citizens is protection, and national security must be given priority. The concerns are real; the worries about security are not just hypothetic exercises. Europe has become dependent on gas imports from Russia, the United States on oil imports from abroad. The challenge in making globalization work is to universalize these concerns and to democratize the procedures. The

United States cannot be allowed to pursue its security concerns without allowing others to do the same; it should not be allowed to be the
sole arbiter of which countries European firms may trade with or what
products they can sell.

The full potential implications of security for globalization are enormous. Worries about the availability of anything essential (like energy
or food) bought from abroad in times of emergency are a rationale for
restricting imports and subsidizing domestic production. When followed to its logical conclusion, the entire framework of trade liberalization is put into jeopardy. Does each country simply accept these
risks as part of the price we face for a more efficient global economy?
Does Europe simply say that if Russia is the cheapest provider of gas,
then we should buy from Russia regardless of the implications for its
security, or is it allowed to intervene in the energy market to reduce
dependence? Do we welcome the increased interdependence and the
risk that it brings, as a further incentive for peaceful resolution of international political disputes? Should we create an international procedure to judge when trade interventions for national security purposes
are to be allowed? Or should we simply allow each country to use the
national security card as a justification for protectionism at will?

The debate about security and globalization highlights—even for
those who have been among globalization's cheerleaders—that values
other than economic well-being are at stake. But these other values
have been given short shrift in the way that globalization has been proceeding. The reason is simple: the democratic deficit means that issues
that are, or should be, of importance to ordinary citizens don't get the
attention they deserve. The richest country, the United States, knows
it can get what it wants—it can do what it wants whenever its concerns, especially its security concerns, are at risk. The rest of the world,
at least so far, has not been willing to stand up. Too many have just
been swept along in a U.S.-orchestrated euphoria for globalization,
regardless of how it has been designed and managed. But the time will
come when the United States cannot do whatever it wants. The forces
of global economic, social, political, and environmental change are
more powerful in the long run than the capacity of even the mightiest
nation to shape the world according to its interests or perspective.

The debate about security and globalization highlights a second theme of this book: economic globalization has been outpacing political globalization.

We have become economically interdependent more quickly than we have learned how to live together peacefully. Though the bonds that economic globalization forges—both the mutual interdependence that it implies and the greater understanding that arises from daily interactions—are a powerful force for peace, by themselves they are not enough; and without peace, there cannot be commerce. Once before, a century ago, the turmoil of war set back the pace of globalization; it would take more than half a century for globalization (as measured, for instance, by global trade relative to global GDP) to resume where it had left off.[14] Once before, at the end of World War I, the United States, already the world's strongest country, turned its back on multilateralism when it walked away from the League of Nations, the international institution created to help ensure global peace. The Bush administration, too, having previously announced its rejection of the Kyoto Protocol, the International Criminal Court, and major agreements designed to contain the arms race, also walked away from the UN when it went to war in Iraq with a preemptive attack in violation of international law.

The UN proved the value of deliberative democracy: after carefully weighing the evidence presented of an imminent threat from weapons of mass destruction, it concluded that the evidence was insufficient to justify a departure from long-standing precepts and embark on preemptive warfare. The conclusion proved correct; no weapons of mass destruction were found. The world's sole superpower has simultaneously been pushing for economic globalization and weakening the political foundations necessary to make economic globalization work. It has justified its actions as strengthening democracies globally, but it has undermined global democracy. It has talked about human rights, but has trod on those rights in its brazen defense of its right to use torture in contravention of the UN Convention Against Torture, and in a myriad of other ways.

If there was ever a country that should have been responsive to the calls of those seeking a fairer globalization based on an international rule of law, it should have been the United States: its Declaration of Independence does not say, "all Americans are created equal," but "all men are created equal." The Founding Fathers were concerned with the universality of the principles that they were articulating so well, and the Declaration of Independence, the Constitution, along with the Bill of Rights, the first ten amendments to the Constitution, provided the model for much of the rest of the world; the creators of those documents would have been pleased with the adoption by the UN of the Universal Declaration of Human Rights on December 10, 1948. From its beginnings as a nation, the United States benefited from globalization: the massive migration of workers to its shores, supported by capital and ideas from abroad. Today it is among the biggest beneficiaries of economic globalization. It is in the interests of the United States to make sure that there is no retrenchment; but if that is the case, it is also in its interest to make sure that the gap between economic and political globalization is reduced.[15]

For much of the world, globalization as it has been managed seems like a pact with the devil. A few people in the country become wealthier; GDP statistics, for what they are worth, look better, but ways of life and basic values are threatened. For some parts of the world the gains are even more tenuous, the costs more palpable. Closer integration into the global economy has brought greater volatility and insecurity, and more inequality. It has even threatened fundamental values.

This is not how it has to be. We can make globalization work, not just for the rich and powerful but for all people, including those in the poorest countries. The task will be long and arduous. We have already waited far too long. The time to begin is now.

NOTES

Preface

1. This is especially important given the attempt by the IMF to discredit me rather than to engage in intellectual debate, both during the period in which I served as the World Bank's chief economist and after. The IMF tried to give the impression that what I was saying in *Globalization and Its Discontents* was a departure from what I had said during my years at the Bank. Nothing could be further from the truth. (I should be thankful for their vehement reaction to my book, for, in most of the world, it led to increased sales—one country's publisher even pasted a quote from the IMF attack on the book's cover.)

2. I should be clear: while the intellectual groundings have been taken away from market fundamentalism, newspaper columnists and pundits—and occasionally, even a few economists—sometimes still invoke economic "science" in defense of their position.

3. This research was cited when I was awarded the Nobel Prize.

4. See Bruce Greenwald and Joseph E. Stiglitz, "Externalities in Economies with Imperfect Information and Incomplete Markets," *Quarterly Journal of Economics*, vol. 101, no. 2 (May 1986), pp. 229–64.

5. Joseph E. Stiglitz, *The Roaring Nineties* (New York: W. W. Norton, 2003).

6. An expression used by the philanthropist George Soros.

7. Matthew Miller, *The Two Percent Solution: Fixing America's Problems in Ways Liberals and Conservatives Can Love* (New York: PublicAffairs, 2003). This is how Miller phrases the issue in the prologue to his book: "We'll first step back and lay a little philosophical groundwork by examining the pervasive role of luck in

life, and how taking life's 'pre-birth lottery' seriously can bring the consensus we need to make progress."

8. By the same token, one cannot delegate these key societal decisions to technocrats. A key criticism of globalization discussed in this book is that it has attempted to "depoliticize" decisions that are quintessentially political.

9. As we shall explain in chapter 3, the underlying economic forces for globalization too may change over time, as the composition of production and trade changes.

10. These are views already incorporated into corporate governance in many European countries. The views expressed here are, I should note, as reasonable as they may seem to the laymen, highly controversial—particularly within American academia. There are some extreme conditions under which one can show value- (or profit-) maximizing behavior of firms lead to economic efficiency, and it is upon these extreme models that much of the economics literature focuses. But so long as there is imperfect information or an incomplete set of markets, then maximizing the well-being of shareholders does *not* lead either to economic efficiency or general well-being. See, for instance, Sanford J. Grossman and Joseph E. Stiglitz, "On Value Maximization and Alternative Objectives of the Firm," *Journal of Finance*, vol. 32, no. 2 (May 1977), pp. 389–402.

Chapter One

1. World Commission on the Social Dimension of Globalization, *A Fair Globalization: Creating Opportunities for All* (Geneva: International Labour Office, 2004), p. x; available at www.ilo.org/public/english/fairglobalization/report/index.htm.

2. World Commission on the Social Dimension of Globalization, *A Fair Globalization: Creating Opportunities for All*, op. cit., p. 44; and Giovanni Andrea Cornia and Tony Addison with Sampsa Kiiski, "Income Distribution Changes and Their Impact in the Post–World War II Period," World Institute for Development Economics Research Discussion Paper 2003/28, March 2003. Inequality is measured by the Gini coefficient, one of the standard measures.

3. Even the $2-a-day standard is less than a fifth of the poverty standard used in the United States and western Europe.

4. Shaohua Chen and Martin Ravallion, "How Have the World's Poorest Fared since the Early 1980s?," World Bank Development Research Group, World Bank Policy Research Working Paper 3341, June 2004. The $1-a-day standard actually is defined as $1.08 in 1993 "real" (or purchasing power parity) dollars; the $2-a-day standard is defined as $2.15. China's poverty reduction has been truly remarkable. At the $1-a-day standard, the number in poverty has fallen from 634 million to 212 million—more people have been brought out of absolute poverty than the total number living in Europe or America.

5. The *Voices of the Poor* project was undertaken while I was chief economist of the World Bank as part of the preparation for the decennial report on poverty (*World Development Report 2000/2001: Attacking Poverty*). It entailed an unprecedented effort to understand poverty from the perspective of the poor themselves. The results were published in three volumes: *Can Anyone Hear Us?* (Vol. 1), *Crying Out for Change* (Vol. 2), and *From Many Lands* (Vol. 3) (Washington, DC: World Bank, 2002).

6. From 1985 to 2000, the share of the world's population living in democratic countries increased from 38 percent to 57 percent, while the share living under authoritarian regimes dropped from 45 percent to 30 percent. See Figure 1.1 of United Nations Development Programme (UNDP), *Human Development Report 2002: Deepening Democracy in a Fragmented World* (New York: Oxford University Press, 2002), p. 29; available at http://hdr.undp.org/reports/global/2002/en/.

7. There were eight overarching Millennium Development Goals: to eradicate extreme poverty and hunger; to achieve universal primary education; to promote gender equality and empower women; to reduce child mortality; to improve maternal health; to combat HIV/AIDS, malaria, and other diseases; to ensure environmental sustainability; and to develop a global partnership for development. See www.un.org/millenniumgoals/.

8. In 2004, the Organisation of Economic Co-operation and Development (OECD) nations contributed only 0.25 percent of GDP to development assistance, with Japan, the United States, and Italy all contributing less than 0.2 percent (0.7 percent is less than the United States has been spending on the war in Iraq). Only Norway, Luxembourg, Denmark, Sweden, and the Netherlands had met the 0.7 percent commitment. See OECD, "Preliminary Official Development Assistance (ODA) by Donor in 2004, as Announced on April 11, 2005," at www.oecd.org/document/7/0,2340,en_2649_34485_35397703_1_1_1_1 ,00.html.

9. See HM Treasury, "G-8 Finance Ministers' Conclusions on Development, London 10–11, June 2005," at www.hm-treasury.gov.uk/otherhmtsites/g7/news/conclusions_on_development_110605.cfm.

10. See Table A.24 of World Bank, *Global Development Finance: The Development Potential of Surging Capital Flows* (Washington, DC: World Bank, 2006); available at http://siteresources.worldbank.org/INTGDF2006/Resources/GDF06_complete.pdf.

11. See UNDP, *Making Global Trade Work for People* (London and Sterling, VA: Earthscan Publications, 2003).

12. See Oxfam, "Running into the Sand: Why Failure at the Cancun Trade Talks Threatens the World's Poorest People," Oxfam Briefing Paper 53, September 2003.

13. Eswar Prasad, Kenneth Rogoff, Shang-Jin Wei, and M. Ayhan Kose, "Effects of Financial Globalization on Developing Countries: Some Empirical Evidence,"

IMF Occasional Paper 220, March 2003. Even the *Economist*, long a committed advocate of deregulated markets in general and capital market liberalization in particular, conceded the issue in their excellent article "A Fair Exchange?," September 30, 2004.

14. The term "Washington Consensus" was originally coined by a distinguished economist, John Williamson, to describe policy reforms in Latin America. His list was longer (including ten points) and more nuanced. See John Williamson, "What Washington Means by Policy Reform," chapter 2 in *Latin American Adjustment: How Much Has Happened?*, ed. John Williamson (Washington, DC: Institute for International Economics, 1990); and Joseph E. Stiglitz, "The Post Washington Consensus Consensus," IPD Working Paper Series, Columbia University, 2004, presented at the From the Washington Consensus Towards A New Global Governance Forum, Barcelona, September 24–25, 2004.

15. The ideas of the late great Harvard philosopher John Rawls have been influential. He has urged thinking about social justice "behind a veil of ignorance," before we know what position into which we would be born. See John Rawls, *A Theory of Justice* (Cambridge, MA: Harvard University Press, 1971); and Patrick Hayden, *John Rawls: Towards a Just World Order* (Cardiff, UK: University of Wales Press, 2002).

16. Some of the changes have to do with changing patterns of production, and this may happen again, as the economies of the world become more based on services.

17. See Karl Polyani, *The Great Transformation: The Political and Economic Origins of Our Time* (Boston: Beacon Press, 2001). In the preface to the 2001 reissue of this 1944 classic work, I describe the parallels between these two historical changes.

Chapter Two

1. See William Easterly, *The Elusive Quest for Growth: Economists' Adventures and Misadventures in the Tropics* (Cambridge, MA: MIT Press, 2001).

2. This alternative view has some semblance to the "Third Way" commonly associated with U.K. prime minister Tony Blair, U.S. president Bill Clinton, and German chancellor Gerhard Schroeder. The annual *Economic Report of the President* in the early years of the Clinton presidency articulated these views, relating what the government should do closely to the limitations of the market.

3. The quest for understanding the circumstances under which Adam Smith's idea that markets do or do not lead "as if by an invisible hand" to economic efficiency has been at the center of economic research for two centuries. Kenneth J. Arrow and Gerard Debreu won Nobel Prizes for their rigorous mathematical analyses. They defined the ideal conditions under which Smith was right, but also identified the numerous instances of market failures, where he was not—when, for

instance, there are externalities (like pollution) where the actions of one individual have effects on others for which they are not compensated. My own work added to the list of situations in which market failures lead to inefficiency—where information was imperfect and/or asymmetric (that is, where some individuals know something that others do not). Arrow and Debreu's analysis also assumed that technology was unchanging, or at least unaffected by actions of market participants; yet changes in technology are at the center of development.

4. Gunnar Myrdal, *Asian Drama: An Inquiry into the Poverty of Nations* (New York: Pantheon, 1968).

5. Its performance in the past fifteen years has been a little bit better—a measly annual increase in per capita income of 0.2 percent.

6. See World Bank, *China 2020: Development Challenges in the New Century* (Washington, DC: World Bank, 1997), p. 3; available at http://www-wds.world bank.org/servlet/WDSContentServer/WDSP/IB/1997/09/01/000009265_398 0625172933/Rendered/PDF/multi0page.pdf.

7. Since 1970, the total (average annual rate of) increase in income per capita has been: China, 923 percent (6.8 percent); Indonesia, 286 percent (4.0 percent); Korea, 566 percent (5.6 percent); Malaysia, 283 percent (3.9 percent); Thailand, 347 percent (4.4 percent). Though data on poverty over such longtime spans are unreliable and spotty, it appears that in less than two decades, using the $2-per-day measure of poverty, China's poverty rate dropped from 67 percent to 47 percent between 1987 and 2001, Indonesia's poverty rate dropped from 76 percent to 52 percent between 1987 and 2002, Malaysia's poverty rate dropped from 15 percent to 9 percent between 1987 and 1997, and Thailand's poverty rate dropped from 37 percent to 32 percent between 1992 and 2000. At the $1-a-day standard, poverty eradication has been even more dramatic. See World Bank, World Development Indicators, GDP per capita (constant 2000 US$) and Poverty headcount ratio at $2 a day (PPP) (percentage of population). World Bank, Development Data and Statistics; available by subscription at www.worldbank.org/data/onlinedatabases/onlinedatabases.html.

8. Using the $1-a-day method. See World Bank, World Development Indicators. World Bank, Development Data and Statistics; available by subscription at www.worldbank.org/data/onlinedatabases/onlinedatabases.html.

9. Source: IFS data, 1963–2003; available by subscription at http://ifs.apdi.net.

10. From 64 percent in 1981 to 16 percent in 2001. See Chen and Ravallion, "How Have the World's Poorest Fared since the Early 1980s?," op. cit.

11. Author's calculations based on Table 1 of Leandro Prados de la Escosura, "Growth, Inequality, and Poverty in Latin America: Historical Evidence, Controlled Conjectures," Universidad Carlos III de Madrid, Departmento de Historia Economica e Instituciones, Working Paper 05-41, 2005; available at http://docubib.uc3m.es/WORKINGPAPERS/WH/wh054104.pdf.

12. Numerous studies suggest a threefold increase in poverty in Russia, from 11.5 percent in 1989 to 34.1 percent in 1999. (See Anthony Shorrocks and Stanislav Kolenikov, "Poverty Trends in Russia During the Transition," World Institute for Development Economics Research, May 2001, Table 1.) Increases in poverty in other economies in transition were even worse, so that for the region as a whole, there was an almost tenfold increase in poverty. See Chen and Ravallion, "How Have the World's Poorest Fared Since the 1980s?," op. cit.

13. Poland is often thought of as a country that followed the shock therapy route but was relatively successful (not in comparison to China, but in comparison to Russia). Poland did follow macro–shock therapy policies, quickly bringing down its inflation; that done, it took a more gradual approach, for instance, to privatization.

14. Explaining the difference in performance between China and Russia has spawned a large literature. (No one really disputes China's relative success in growth and in reducing poverty, and no one really disputes that Russia adhered much more closely to the Washington Consensus policies than did China.) Some claim that, after all, China really did engage in its own version of shock therapy; some claim that other factors accounted for China's relative success; some claim that had China followed shock therapy, it might have grown even faster. See, for instance, chapters 7 and 8 of Jeffrey D. Sachs, *The End of Poverty: Economic Possibilities for Our Time* (New York: Penguin, 2005); Jeffrey D. Sachs and Wing Thye Woo, "Structural Factors in the Economic Reforms of China, Eastern Europe, and the Former Soviet Union," *Economic Policy*, vol. 9, no. 18 (April 1994), pp. 101–45; or the range of views on the transition posted on the IMF's Web site, https://www.imf.org/External/Pubs/FT/staffp/2001/04/.

15. Dani Rodrik and Arvind Subramanian, "From 'Hindu Growth' to Productivity Surge: The Mystery of the Indian Growth Transition," NBER Working Paper 10376, March 2004. They identify as critical the change in attitude from anti-business to pro-business—but that was a far cry from the Washington Consensus free market policies.

16. Discussed more fully in chapter 10 of Stiglitz, *The Roaring Nineties*, op. cit.

17. The UNDP, in its annual *Human Development Report*, provides a summary measure, called the "Human Development Indicator" (HDI), that combines measures of income, health, and other aspects of human well-being. In the 2005 HDI, the United States ranked tenth, behind Norway, Iceland, Australia, Luxembourg, Canada, Sweden, Switzerland, Ireland, and Belgium.

18. World Bank, Papua New Guinea Environment Monitor 2002, at http://www wds.worldbank.org/external/default/main?pagePK=64193027&piPK=641879 37&theSitePK=523679&menuPK=64154159&searchMenuPK=64187514& theSitePK=523679&entityID=000012009_20030729110929&searchMenu PK=64187514&theSitePK=523679.

19. By 2000, CEO pay was more than 500 times the wages of the average employee, up from 85 times at the beginning of the decade, and 42 times two decades earlier. See Stiglitz, *The Roaring Nineties*, p. 124.

20. See Polanyi, *The Great Transformation*, op. cit.

21. Roderick Floud and Bernard Harris, "Health, Height, and Welfare: Britain, 1700–1980," in *Health and Welfare During Industrialization*, ed. Richard Steckel and Roderick Floud (Chicago: University of Chicago Press, 1997), pp. 91–126.

22. Economists say that individuals are "risk averse." The fact that they are willing to pay a considerable amount to reduce key risks they face shows the importance of security.

23. The former World Bank economist William Easterly has written about these changes in thinking. See Easterly, *The Elusive Quest for Growth*, op. cit.

24. Though some free market advocates might have claimed that this was because of intervention by colonial powers to prevent development—for example, the notorious restrictions imposed on India.

25. The intellectual framework for this new approach was laid out in "Towards a New Paradigm for Development: Strategies, Policies, and Processes," my Prebisch Lecture delivered to UNCTAD (United Nations Conference on Trade and Development) on October 19, 1998, available at http://ww2.gsb.columbia.edu/faculty/jstiglitz/papers.cfm.

26. Since 1976, the end of the Cultural Revolution, China's annual growth rate of per capita income has averaged 7.8 percent. Since 1990, the growth rate has been 8.3 percent. Source: World Bank, World Development Indicators, GDP per capita (constant 2000 US$); available by subscription at www.worldbank.org/data/onlinedatabases/onlinedatabases.html.

27. See Amartya Sen's powerful book *Development as Freedom* (New York: Oxford University Press, 2001).

28. I view the highlighting of the importance of knowledge in development, including redressing the imbalance in education, as one of the major changes while I was chief economist at the World Bank. See *World Development Report 1998–1999: Knowledge for Development* (Washington, DC: World Bank, 1998).

29. In chapter 6, we will describe the enormous contribution that developing countries are making to the global environment—services valued at tens of billions of dollars—for which they are not compensated.

30. This is called "peer monitoring." I developed the underlying economic theory explaining the success of these lending institutions almost two decades ago. See Joseph E. Stiglitz, "Peer Monitoring and Credit Markets," *World Bank Economic Review*, vol. 4, no. 3 (September 1990), pp. 351–66.

31. See, for instance, Deepa Narayan, *The Contribution of People's Participation: Evidence from 121 Rural Water Supply Projects* (Washington, DC: World Bank, 1995), which found that local participation in rural water supply projects

increased significantly the number of water systems in good condition, the portion of target populations reached, and overall economic and environmental benefits. For more current information, see the World Bank's participation Web site at www.worldbank.org/participation.

32. Thomas L. Friedman, *The World Is Flat: A Brief History of the Twenty-First Century* (New York: Farrar, Straus and Giroux, 2005).

33. Friedman himself is aware that the world is not flat, devoting a chapter to "The Un-Flat World."

Chapter Three

1. Of course, most of that reflected the size of the U.S. economy. After its enlargement in 2004, the EU has a population in excess of 450 million; the size of its economy is comparable to NAFTA.

2. OECD, *OECD Economic Surveys*, "OECD Economic Surveys Mexico: Migration: The Economic Context and Implications," vol. 2003, suppl. no. 1 (Paris: OECD, 2003), pp. 152–212.

3. Early on in the Clinton administration, the Council of Economic Advisers (of which I was a member at the time) was asked its view on NAFTA. Many in the administration thought that, given the opposition to NAFTA and the controversy concerning other priority items in its agenda (health care, welfare reform), efforts to get NAFTA approved should be at least temporarily postponed. We concluded that the United States would be little affected—it simply wouldn't make much difference for our economy. The main effect would be reduced pressure on immigration—which we did think would be of considerable value. We thought that Mexico would benefit enormously, and we thought that hemispherical "solidarity" would be enhanced if we could narrow the income disparity. In retrospect, we were wrong in our estimates of how much Mexico would gain. I explain below some of the reasons for our failed judgment.

   While we were wrong about the effects on Mexico, we were right about the effects on America. Ross Perot, in his presidential campaign, had alleged that there would be a "giant sucking sound" as jobs left America to Mexico. I was not surprised that NAFTA had so little effect on America's economy. Tariffs were already low, and given the strength of American markets, the economy was fully capable of adjusting. Indeed, in the months and years following NAFTA, unemployment fell, from 6.6 percent to 5.5 percent, and eventually to 3.8 percent.

4. Growth statistics depend greatly on how output is measured, which is particularly problematic in periods of large exchange rate fluctuations. If the exchange rate appreciates, the value of a country's output, in dollar terms, increases, even if the country is producing no more than it was before. As a result, economists focus on what happens to real income, measured in purchasing power. Growth

in per capita real income fluctuated between 3.5 percent of the 1960s, 3.2 percent of the 1970s, and 2.7 percent of the 1950s.

5. From Instituto Nacional Estadística Geografía e Informática, cited in William C. Gruben, "Was Nafta Behind Mexico's High Maquiladora Growth?," *Economic and Financial Review* (Third Quarter, 2001), pp. 11–21.

6. Overall, employment in the domestic manufacturing sector declined in the decade after NAFTA. Export manufacturing employment increased slightly, but these gains were largely overset by losses of jobs in agriculture, and it was not clear how permanent the jobs created would be: by the end of the first decade, 30 percent of the jobs created in the maquiladora in the early 1990s had disappeared. See Sandra Polaski, "Mexican Employment, Productivity, and Income a Decade after NAFTA," Carnegie Endowment for International Piece, brief submitted to the Canadian Standing Senate Committee on Foreign Affairs, February 25, 2004.

7. See Gruben, "Was Nafta Behind Mexico's High Maquiladora Growth?," op. cit. In the case of Mexico, the debate is complicated by its 1994–95 financial crisis. A World Bank study concluded that without NAFTA, Mexican income per capita would have been 4 percent lower. (Daniel Lederman, William F. Maloney, and Luis Servén, *Lessons from NAFTA for Latin America and the Caribbean Countries: A Summary of Research Findings,* World Bank, December 2003.) But there were serious flaws with that study. See, for instance, Mark Weisbrot, David Rosnick, and Dean Baker, "Getting Mexico to Grow with NAFTA: The World Bank Analysis," Center for Economic Policy Research, September 20, 2004, available at www.cepr.net/publications/nafta_2004_10.htm. Even putting this statistical debate aside, it is striking that even NAFTA advocates suggest that it has had at most a small effect on growth, even in a period in which, because of the Mexican crisis, trade was vital.

Mexico's joining the WTO in January 1995 may have made more of a difference in some respects than NAFTA, because it limited what the government could do in the aftermath of the 1994–95 crisis. (In earlier crises, the government had imposed numerous quantitative trade restrictions, which critics say had long-lasting adverse effects.).

NAFTA proponents sometimes argue that NAFTA's real contribution was opening up investment, not trade. But, critics say, while the effect on overall foreign investment is uncertain, some aspects of foreign investment may have contributed to Mexico's slow growth. As international banks took over all but one of Mexico's banks—acquisitions that NAFTA effectively encouraged—the supply of credit to small- and medium-sized domestic enterprises became constrained, and growth (outside firms linked with international exports) diminished. Moreover, as we shall see later, the lopsided investor protection—foreigners were provided better protections than domestic investors—put into jeopardy environmental and other regulations.

8. See Instituto Nacional Estadística Geografía e Informática, "Personal ocupado en la industria maquiladora de exportacion segun tipo de ocupación"; available at www.inegi.gob.mx/est/contenidos/espanol/rutinas/ept.asp?t=emp75&c=1811.

9. In 1993, Mexico's per capita PPP (purchasing power parity) income was 3.6 times that of China; by 2003, the ratio was cut in half, to 1.8. China had a distinct wage advantage over Mexico—wages are one-eighth of those in Mexico. But over the period of NAFTA, China's wages have increased, while Mexico's wages have stagnated. Thus, China's relative success must be based on other factors.

10. Some simple models—where there are no transportation costs and where everyone has access to the same *knowledge* (technology)—predict that there will be complete factor price equalization. That is, wages of skilled workers, of unskilled workers, and the return to capital will be the same everywhere in the world. It is as if the total global economy is fully integrated—so that wages of workers of any given skill level are the same anywhere in the world. See the classic paper by the great twentieth-century economist Paul A. Samuelson, "International Trade and the Equalization of Factor Prices," *Economic Journal*, vol. 58 (June 1948), pp. 163–84, in which he shows that even short of free trade, trade liberalization leads toward the equalization of factor prices. See also Wolfgang F. Stolper and Paul A. Samuelson, "Protection and Real Wages," *Review of Economic Studies*, vol. 9, pp. 58–73.

11. Politicians, as they wax poetic about the virtues of trade liberalization, often talk about how exports *create* jobs. But by that logic, imports destroy jobs. And that leads to the incoherent positions of many governments that, while they speak in favor of trade, argue against imports.

12. Louis Uchitelle's recent book, *The Disposable American: Layoffs and their Consequences* (New York: Knopf, 2006), provides a convincing analysis of the large costs faced by displaced workers—and the costs borne by society as a whole. The loss in wages are not just a consequence of loss of higher-than-normal wages enjoyed by unionized workers in protected sectors; there are also large costs that follow from the loss of effective human capital—skills that were no longer relevant in their new jobs.

13. John Maynard Keynes, *A Tract on Monetary Reform* (London: Macmillan, 1923).

14. See note 10 above. Indeed, once the consequences of imperfect risk markets are taken into account, free trade, rather than making everybody better off, can actually make everybody worse off. The reason is that it increases the risks that households and firms face. See Partha Dasgupta and Joseph E. Stiglitz, "Tariffs versus Quotas as Revenue Raising Devices under Uncertainty," *American Economic Review*, vol. 67, no. 5 (December 1977), pp. 975–81; and David M. Newbery and Joseph E. Stiglitz, "Pareto Inferior Trade," *Review of Economic Studies*, vol. 51, no. 1 (January 1984), pp. 1–12.

15. The evidence also suggests that globalization has been associated with increasing inequality in developing countries, for reasons that are not yet fully understood.

16. For instance, the U.S. International Trade Administration on January 13, 2006, imposed dumping duties against various Brazilian orange juice producers at rates ranging from just under 10 percent to as much as 60 percent. It imposes safeguard duties depending on the price. In some years average duties have exceeded 50 percent. See Hans Peter Lankes, "Market Access for Developing Countries," *Finance & Development* (a quarterly publication of the IMF), vol. 39, no. 3 (September 2002); available at www.imf.org/external/pubs/ft/fandd/2002/09/lankes.htm.

17. For instance, as a percentage of GDP, tariffs are fourteen times larger in Africa than they are in the OECD (advanced industrial) countries. They represent almost 5 percent of GDP in Pakistan, 6.7 percent in Mauritius, and 3 percent in Costa Rica, but only 0.27 percent in the United States, 0.13 percent in France, 0.35 percent in the U.K., and 0.21 percent in Japan and Germany. Data are for 1995; from Liam Ebrill, Janet Stosky, and Reint Gropp, *Revenue Implications of Trade Liberalization,* IMF Occasional Paper 180 (Washington, DC: International Monetary Fund, 1999).

18. For a more extensive discussion of the arguments for aid for trade, see Joseph E. Stiglitz and Andrew Charlton, "Aid for Trade: A Report for the Commonwealth Secretariat," delivered at a meeting at the WTO in Geneva, March 24, 2006; available online at http://www2.gsb.columbia.edu/faculty/jstiglitz/download/2006_Aid_For_Trade.pdf. Summary available at Papers and Proceedings of the Annual Bank Conference on Development Economics, Tokyo, 2006 (forthcoming).

19. The infant industry argument for protection has a pedigree almost as long and distinguished as that of the free trade argument. It was developed in the nineteenth century by Friedrich List in *The National System of Political Economy* (1841; translated by Sampson S. Lloyd [London: Longmans, Green, 1909]). See Ha-Joon Chang, "Kicking Away the Ladder: Infant Industry Promotion in Historical Perspective," *Oxford Development Studies*, vol. 31, no. 1 (2003), pp. 21–32; and Partha Dasgupta and Joseph E. Stiglitz, "Learning by Doing, Market Structure, and Industrial and Trade Policies," *Oxford Economic Papers*, vol. 40, no. 2 (1988), pp. 246–68. The general theory of "learning"—and why government action may be required—was developed by Nobel Prize–winning economist Kenneth Arrow in "The Economic Implications of Learning by Doing," *Review of Economic Studies*, vol. 29, no. 3 (June 1962), pp 155–73.

20. A dramatic illustration was provided by America's illegal imposition of steel tariffs on March 20, 2002, in response to political pressure from steel producers. (They were ended on December 4, 2003, after an adverse WTO ruling.) It was

estimated by the Consuming Industries Trade Action Coalition that the steel tariffs led to the loss of nearly 200,000 American jobs—while total employment in the steel-producing sector is only 190,000. Joseph Francois and Laura M. Baughman, "The Unintended Consequences of U.S. Steel Import Tariffs: A Quantification of the Impact During 2002," CITAC Foundation, 2003; available at www.citac.info/steeltaskforce/studies/attach/2002_Job_Study.pdf.

21. See Bruce Greenwald and Joseph E. Stiglitz, "Helping Infant Economies Grow: Foundations of Trade Policies for Developing Countries," *American Economic Review*, vol. 96, no. 2 (May, 2006), pp. 141–46.

22. See UNDP, *Making Global Trade Work for People* (London and Sterling, VA: Earthscan Publications, 2003). For arguments that globalization and/or trade would lead to more growth, see Martin Wolf, *Why Globalization Works* (New Haven: Yale University Press, 2004); Jagdish N. Bhagwati, *In Defense of Globalization* (New York: Oxford University Press, 2004); World Bank, *Globalization, Growth, and Poverty: Building an Inclusive World Economy* (Washington, DC: World Bank, 2002); Jeffrey D. Sachs and Andrew M. Warner, "Economic Reform and the Process of Global Integration," in *Brookings Papers on Economic Activity 1995*, vol. 1, *Macroeconomics*, ed. William C. Brainard and George L. Perry (Washington, DC: Brookings Institution Press, 1995), pp. 1–95. A compelling critique of the econometric studies is provided by Dani Rodrik and Francisco Rodríguez, "Trade Policy and Economic Growth: A Skeptic's Guide to the Cross-National Evidence" in *NBER Macroeconomics Annual 2000*, ed. Ben S. Bernanke and Kenneth S. Rogoff (Cambridge, MA: MIT Press, 2001), pp. 261–325.

23. There is a large "fair trade" movement, which has been particularly influential in Europe. It focuses on a slightly different set of questions: it worries that farmers in the developing world get such a small share of the ultimate price paid by consumers, with middlemen taking most of the money—a tiny percentage of the cost of the cup of coffee actually goes to the coffee grower—and it seeks ways to ensure that the farmers are treated more fairly. My focus here is on the rules of the game—and how the rules of the game are unfair to those in the developing world.

24. Nowhere are the inequities of the international trade regime more evident than in the process by which new countries are allowed to join the WTO. While most countries were members at the start, there are a number of countries, such as Cambodia, Russia, and Vietnam, that were not. Any country can veto their admission, so any country has the power to enforce whatever rules it wants—never mind what is fair. There is no economic rationale for differential treatment of the new applicants; it is just another manifestation of realpolitik. The United States has the power, and therefore it uses—and abuses—that power, so much so that Oxfam, the international aid agency, has referred to the practice as

"extortion at the gate." (See Oxfam, "Extortion at the Gate," Oxfam Briefing Paper 67, November 2004.) Even very poor countries like Cambodia are told that if they wish to join the WTO they have to comply with strictures tougher than those imposed on existing members. Cambodia, for instance, must comply with the intellectual property requirements of the WTO far faster than better-off members, like India. What should be done is simple: any country willing to adhere to the WTO trade agreements (with adjustment periods corresponding to their stage of development) should be admitted.

25. Under the multifiber agreement (MFA), which expired on January 1, 2005, countries negotiated quotas on a product-by-product, country-by-country basis. This is why so many garment factories opened up all over the world in places you wouldn't expect them. China might be the low-cost producer, but when China's quota was exhausted, importers had to turn to the next cheapest place with quota. Once the agreement ended, many companies began buying from China. Not only producers in the EU and the United States lost out, but so did producers in other developing countries. In an obvious reneging on the spirit of trade liberalization, pressure was brought to bear for China to limit its exports.

26. Another problem is that, like many a judicial process, it is long and drawn out— while Brazil brought the cotton case in September 2002, and a ruling against the United States occurred in April 2004, the cotton subsidies remain in effect as this book goes to press. There is a marked contrast with the dumping duties described later, where the United States routinely imposes high dumping duties on a preliminary basis, which are often revised downward after a careful look at the evidence.

27. Trade rounds are named after the city in which they were begun, or the president under which they were begun. Perhaps Clinton hoped that, like the round that began at Geneva on May 4, 1964, that came to called the Kennedy Round, the round that was to begin in Seattle would be known as the Clinton Round. Today, what is remembered are the Seattle riots.

28. See UNDP, *Human Development Report 1997: Human Development to Eradicate Poverty* (New York: Oxford University Press, 1997).

29. Upper-middle-income countries are defined by the World Bank as countries with per capita incomes of $3,256–$10,065. Lower-middle is defined as having per capita incomes of $826–$3,255. Low-income countries are those with incomes below $826.

30. See United States International Trade Commission, "Interactive Tariff and Trade Dataweb," at http://dataweb.usitc.gov/.

31. See chapter 3 of Joseph E. Stiglitz and Andrew Charlton, *Fair Trade for All: How Trade Can Promote Development* (New York: Oxford University Press, 2005).

32. As chief economist of the World Bank, I had called for such a "development round," in a speech delivered to the WTO in March 1999, where I laid out the

many ways in which the Uruguay Round had disadvantaged the developing countries.

33. As this book goes to press, the Doha Round has not come to an end. But the parameters of any potential agreements are sufficiently clear that one can make these conclusions, with considerable confidence.

34. See Stiglitz and Charlton, *Fair Trade for All*, op. cit.

35. One of the main forms that differential treatment takes is giving developing countries longer to adjust. Well-functioning markets facilitate adjustment by helping to redeploy resources. When markets work well, it doesn't take long for an unemployed worker to find an alternative job; when markets don't work well, it may take an extended period, during which he may remain unemployed. This is one of several reasons why less developed countries need longer to adjust, and will need financial assistance to help them in the adjustments toward a more liberalized trade regime.

36. The proposal for opening up markets to all countries smaller and poorer is contained in Stiglitz and Charlton, *Fair Trade for All*, and elaborated in Andrew Charlton, "A Proposal for Special Treatment in Market Access for Developing Countries in the Doha Round," in *Trade Policy Research 2005*, ed. John M. Curtis and Dan Ciuriak (Ottawa: Department of International Trade, 2005).

37. Europe's new policy, announced in February 2001, was called the "Everything But Arms" (EBA) initiative. Critics referred to it as the "Everything But Farms" initiative, since it did little to address many of the concerns of the developing countries over agriculture. A brief overview of the EBA initiative is contained in World Bank, *Global Economic Prospects 2004: Realizing the Development Promise of the Doha Agenda* (Washington, DC: World Bank, 2003). A fuller explanation is at http://europa.eu.int/comm/trade/issues/global/gsp/eba/ug.htm. There has been remarkably little expansion of trade under the EBA initiative. Complicated technical provisions (rules of origins, which detail how much of the "value added" in the good have to be produced within the country) seem partially responsible, highlighting the importance of the fine details within a trade agreement.

38. I have, accordingly, dubbed the proposal the "EBP" initiative—opening up markets to everything but what you produce.

39. OECD, *Agricultural Policies in OECD Countries: Monitoring and Evaluation* (Paris: OECD, 2005).

40. In 2004, OECD subsidies were $279 billion, including water subsidies and other indirect subsidies. See ibid.

41. In purchasing power parity, the farmer's income is somewhat higher, between $1,100 and $1,200.

42. The International Cotton Advisory Committee (ICAC), an association of forty-one cotton-producing, -consuming, and -trading countries formed in 1939, estimates that the elimination of American cotton subsidies would raise the global price by between 15 percent and 26 percent. Oxfam estimates the losses

to Africa at $301 million a year, with the bulk of these losses ($191 million a year) happening to eight West African countries. In Mali, Burkina Faso, and Benin, American subsidies led to losses in excess of 1 percent GDP yearly. See Kevin Watkins, "Cultivating Poverty: The Impact of US Cotton Subsidies on Africa," Oxfam. Briefing Paper 30, 2002.

43. Since benefits are proportional to sales, small farmers get little benefit. The data are for the period 1995–2004. This describes the distribution of benefits among farmers who receive subsidies. But 60 percent of all farmers and ranchers do not collect government subsidy payments, largely because they do not produce subsidized commodities. Source: Environmental Working Group's Farm Subsidy Database, "Total USDA Subsidies in United States," available at www.ewg.org/farm/progdetail.php?fips=00000&progcode=total&page=conc.

44. The EU and especially the United States sometimes claim that they have "delinked" the subsidies from production, that is, designed them so that they do not lead to increased production, but those claims are suspect—as the WTO panel ruled in the case of cotton. But even purportedly delinked subsidies can have effects on output, as they provide farmers with more income with which to buy fertilizer, higher quality seeds, and other output increasing inputs.

45. According to the Harmonized Tariff Schedule of the United States (2005), the general tariff on imported oranges is 1.9 cents/kg (0805.10.00); on citrus fruit preserved in sugar, it is 6 cents/kg (2006.00.60); on orange marmalade, 3.5 cents/kg (2007.91.40); on orange pulp, 11.2 cents/kg (2008.30.35); on oranges packed in liquid medium in airtight container, 14.9 cents/kg (2008.92.90.40); and on frozen orange juice, it is 7.85 cents/liter (2009.11).

46. This is called nonagricultural market access, or NAMA, in the technical jargon of the WTO.

47. Sometimes the Jones Act is defended on grounds of national security—America needs its own shipping fleet. The irony was that in America's most recent emergency, when Hurricane Katrina struck, the Jones Act had to be suspended. (For a slightly more extended discussion of globalization and security, see chapter 10.)

48. It would also benefit the developed countries as a whole, but low-wage workers would lose. The effects are analogous to those discussed earlier for trade liberalization (not surprising, because as we noted, trade in goods is a substitute for the movement of people). And the necessary responses—in terms of helping the losers from globalization discussed elsewhere in this book—are similar. Other benefits include the transfer of knowledge and access to markets that such migration facilitates. The story of Infosys illustrates this: among its founders were several who had spent extensive time in the United States.

49. Data from the Inter-American Development Bank show that for each of the twenty-three Latin American countries, remittances substantially exceed assistance. They account for at least 10 percent of GDP in Haiti, Nicaragua, El Salvador, Jamaica, the Dominican Republic, and Guyana. Central America and the

Dominican Republic combined received over $10 billion, and the Andean countries over $7 billion. In 2004, remittances to Latin America totaled $41 billion, an amount nearly identical to the $45 billion the continent received in net foreign direct investment. See remittances data in World Bank, *Global Economic Prospects 2006: Economic Implications of Remittances and Migration* (Washington, DC: World Bank, 2006); FDI—World Bank, World Development Indicators, foreign direct investment, net (BoP, current US$).

50. The Mexican government has been working with the United States to reduce these costs, and by 2004 the U.S. Treasury was claiming that it had reduced the costs by 60 percent. See USINFO, "Treasury Official Notes Importance of Remittance in the Americas," October 7, 2004, at http://usinfo.state.gov/wh/Archive/2004/Oct/08-233308.html.

51. As we noted in note 20 above, the WTO declared illegal the safeguard steel tariffs it had imposed in March 2002, and the United States eventually complied, removing them in late 2003.

52. Such courts would also not likely show much sympathy for America's claim after NAFTA was signed to protection from a "surge" of brooms from Mexico—which threatened to destroy between 100 and 300 jobs! Surely America had the ability to cope without protection.

53. The 2002 appropriations bill contained an amendment saying only the American-born Ictaluridae could be called catfish.

54. Cost is a more elusive concept that noneconomists often grasp. What is relevant is marginal cost, the extra cost associated with producing an extra unit of the product, and this (or an attempt at a surrogate) is the standard in domestic cases. In international cases, there is no such conceptual clarity. They often use long-term average costs, and in cyclical industries, in downturns, marginal cost is usually less than average cost. What matters is the extra cost of producing a ton of steel today, not the cost of the plant and equipment, costs which will have to be borne whether more steel is produced or not. For agricultural products, the honest way to calculate the pricing of tomatoes would be to analyze prices for the whole season and compare those prices with costs. However, the Americans looked at only the first two months of 1996 and the end of 1995—the time when prices were at their lowest. By analyzing only these months, the United States was able to justify high anti-dumping protection.

     Dumping law even allows an artificially high profit margin to be added as part of the calculation of the cost.

55. Chinese officials sometimes tease their American counterparts with a simple syllogism: Americans believe that the most successful economies (or the only successful economies) are market economies. China has clearly been successful. Therefore, China *must* be a market economy.

56. For instance, Lester C. Thurow, formerly dean of MIT's Sloan School of Management, concluded that "if the law were applied to domestic firms, eighteen

out of the top twenty firms in the Fortune 500 would have been found guilty of dumping in 1982." Lester C. Thurow, *The Zero-Sum Solution: Building a World-Class American Economy* (New York: Simon & Schuster, 1985), p. 359. A study by Princeton University economist Robert D. Willig found that "in more than 90 percent [of the cases] the indicators seem inconsistent with the hypothesis that interventions were needed to protect competition from international predation, or to protect competition at all." Robert D. Willig, "Economic Effects of Antidumping Policy," in *Brookings Trade Forum: 1998*. ed. Robert Z. Lawrence (Washington, DC: Brookings Institution Press, 1998) pp. 57–79.

57. The WTO ruled in January 2003 that the Continued Dumping and Subsidy Offset Act of 28 October 2000 (the Byrd amendment) was against both the letter and the spirit of the WTO. Retaliation was authorized in the fall of 2004. So far it has not occurred.

58. When dumping charges are filed, countries have a short time span in which to respond to lengthy questionnaires intended to establish their production costs. They must respond in English, and if they fail to respond, or fail to respond in a way that satisfies the Department of Commerce, then the Department of Commerce imposes duties based on the "best information available," which is typically provided by the industry bringing the dumping charges. Not only does the foreign producer have to fill out a questionnaire, so do the importer and purchaser. One of the sample questionnaires on the United States International Trade Commission Web site itself ran to twenty-two pages. After a public hearing, the foreign producer accused of dumping files another brief. For a more extensive discussion, see, for example, J. Michael Finger, *Antidumping: How It Works and Who Gets Hurt* (Ann Arbor: University of Michigan Press, 1993); and Joseph E. Stiglitz, "Dumping on Free Trade: The U.S. Import Trade Laws," *Southern Economic Journal*, vol. 64, no. 2 (October 1997), pp 402–24.

59. Another nontariff measure, countervailing duties, allows countries to impose tariffs to "undo"—countervail—the effects of subsidies (though not, unsurprisingly, agricultural subsidies, or other subsidies such as those provided indirectly to aircraft manufacturers in defense contracts). It is not frequently used.

60. CAFTA, the Central American Free Trade Agreement, barely passed the House of Representatives in July 2005 by a 217 to 215 vote, which occurred minutes after Republican representative Robin Hayes switched his vote from "nay" to "aye." His reason for surrendering his long-standing opposition to the bill? House Speaker J. Dennis Hastert promised Hayes, who represents North Carolina and its textile industries, that he would push for restrictions on imports of Chinese clothing, thus trading one set of tariffs for another. See Edmund L. Andrews, "Pleas and Promises by G.O.P. as Trade Pact Wins by 2 Votes," *New York Times*, July 29, 2005, p. A1. Its reception within the region was similarly mixed. The countries of the region saw themselves facing high drug prices (see chapter 4) and new competition from the giant to the north; but they were

not sure what they were getting out of it in return. It has been beset with problems—including ratification in the Central American countries and implementation issues, often arising from rules-of-origin tests.

61. Countries compete to attract firms by offering tax breaks. As they bid against each other, the real winners are the businesses who manage, as a result, to avoid most of the taxes. While it may be in the interest of each country to compete, together they lose. Later chapters will discuss the enervating effects of bribery and bank secrecy.

62. "Fourteen Points Speech," delivered to Joint Session of Congress, January 8, 1918.

63. After Cancún, representation and voice of the developing countries improved. What is needed is some more systematic approach to representation.

64. The notion is very much like tradable emission rights, which have become part of the system of managing global warming under the Kyoto Protocol. See chapter 6.

Chapter Four

1. See, for instance, Robert B. Zoellick, "When Trade Leads to Tolerance," *New York Times*, June 12, 2004, p. A13.

2. Statement of U.S. Trade Representative Robert B. Zoellick following Senate approval of the Morocco Free Trade Agreement.

3. In the United States, for instance, generic producers can manufacture their product and have it on the shelves, ready to sell the day the patent expires. Under the Moroccan agreement, generic producers in Morocco may not be able to do this. The drug companies have also been arguing for "data exclusivity," in which "clinical information that is essential to the approval of a pharmaceutical product" is considered protected for a period of time. Drug companies have demanded restrictions on the use of data, even when it has been published and made publicly available, and even when the research has been partially supported with public funds. It is, of course, inefficient simply to replicate research that has already been done. But worse, drug testing requires a fraction of the population be subject to a placebo or an alternative drug. But it would arguably be unethical to conduct a test in which some of the patients were provided a product that was known to be less effective than a product available on the market. (The countries could, presumably, change their regulations, so that any drug approved in the United States would be automatically approved in their country; all that one would have to show is that the generic chemical is in fact the same. This is what is done in the United States. One suspects, however, that the U.S. government would bring to bear enormous pressures against the change in regulations in this way. It is the results—delayed introduction of generics— that the U.S. drug industry wants.)

4. See Khabir Ahmad, "USA-Morocco Deal May Extend Drug Patents to 30 Years," *Lancet*, vol. 362 (December 6, 2003), p. 1904.

5. The agreements are complex and difficult to interpret, so there remains uncertainty about the consequences. The ambiguities may be deliberate. The developing country can claim that it won "flexibility" in, say, applying the intellectual property protection—they could allow the production of generics if there was a valid health need—while the Office of the U.S. Trade Representative can report to its client, the pharmaceutical industry, that it won major concessions in extending the effective life of the patent. When the developing countries seek to use the "flexibilities" they thought they had bargained into the agreement, the United States brings its enormous economic power to bear to stop them—as Brazil and South Africa found out when they tried to produce generic versions of AIDS medicines in the years following the signing of the Agreement on Trade-Related Aspects of Intellectual Property Rights, usually referred to as "TRIPs."

6. Under TRIPs, all members of the WTO were compelled to have an intellectual property regime that met certain "high" standards—essentially the standards set by the advanced industrial countries. Each country would still be responsible for running its own patent and copyright offices.

7. There was another criticism of TRIPs: it was unfair to developing countries, in two respects. While it provided the advanced industrial countries the protection they wanted, it did not provide developing countries protection for their traditional knowledge. (See the discussion below.) And while TRIPs would reduce developing countries' access to knowledge and force them to pay billions in royalties, it was meant to be part of the "Grand Bargain" described in chapter 3, in which the developing countries would get greater access in agriculture and reduced agricultural subsidies by the advanced industrial countries. The developed countries did not keep their side of the bargain.

8. These examples illustrate the general proposition: all property rights are circumscribed. By the same token, in June 2005, the U.S. Supreme Court once again reaffirmed the government's right of eminent domain: the individual cannot do with his property what he wants (including selling the property to whom he wants) if the state decides otherwise. Obviously, there are important restrictions to prevent abuses of these enormous powers.

9. Of course, intellectual property often does not lead to a true monopoly—a single firm producing a product facing no effective competition—but it does alter the intensity and nature of competition, and the results may be even worse than with pure monopoly. Consider, for instance, the drug industry, where the basic research is publicly provided. Drug companies play a role in bringing the results of this research to the market; but under current arrangements, they compete more through marketing and product differentiation. If one firm discovers a drug, others try to use a variant of the idea (a "me too" drug) not covered by the

original patent, but of limited benefit to consumers. Profits are at least partially dissipated in this form of inefficient competition.

10. Paul A. Samuelson formalized the concept a little more than a half century ago in his classic paper "The Pure Theory of Public Expenditures," *Review of Economics and Statistics*, vol. 36, no. 4 (November 1954), pp. 387–89. The key distinction with ordinary goods is called "nonrivalrous consumption": if I eat a bowl of rice, you can't eat it; while if I know something, your knowing it doesn't detract from my knowing it (though it obviously has an effect on what rents I can receive from the knowledge).

11. For a patent to be sustained, it has to satisfy a number of conditions described earlier (for example, novelty); obviously, there cannot exist a previous patent for essentially the same idea. Many patent applications get rejected, and sometimes patents, after being granted, are not sustained after (typically, very expensive) litigation. In the European patent system there is an opportunity for others to oppose a patent application before it is granted. In the United States, other voices are heard only afterward.

12. Selden first filed for the patent on May 8, 1879, in an application that included both the engine and its use in a four-wheeled car. He then proceeded to file so many amendments to the patent application that he stretched the process to sixteen years. The patent (#549,160) was finally granted on November 5, 1895.

13. These concerns have been particularly clear in patenting traditional knowledge. One of the issues raised by the World Commission on the Social Dimension of Globalization in its report *A Fair Globalization: Creating Opportunities for All*, op. cit., was "the adverse impact of international rules for intellectual property rights, which open the door to the privatization of indigenous knowledge" (p. 20).

14. See, for instance, "A Tragedy of the Public Knowledge 'Commons'?" by Paul A. David, a distinguished professor of economic history at Stanford and Oxford, at www.cepr.stanford.edu/papers.html; James Boyle, a law professor, has written extensively on legal aspects, for example, "The Second Enclosure Movement and the Construction of the Public Domain," *Law and Contemporary Problems*, vol. 66 (Winter/Spring, 2003), pp. 33–74. See also Richard Poynder, "Enclosing the Digital Commons," *Information Today*, vol. 20, no. 5 (May 2003), pp. 37–38; and Lawrence Lessig, *The Future of Ideas: The Fate of the Commons in a Connected World* (New York: Random House, 2001).

15. In negotiations with AOL, for example, Microsoft demanded that AOL drop RealNetworks' RealPlayer, which was in direct competition with Microsoft's Windows Media Player. RealNetworks's anti-trust lawsuit against Microsoft quoted a Microsoft executive as saying Microsoft would target RealNetworks "for obliteration." See John Markoff, "RealNetworks Accuses Microsoft of Restricting Competition," *New York Times*, December 19, 2003, p. C5. For more details on Microsoft's attacks on Netscape and the subsequent lawsuit, see

Paul Abrahams and Richard Waters, "You've Got Competition," *Financial Times*, January 24, 2002, p. 16. The Council of Economic Advisers became concerned about the economic impact of Microsoft's dominant position while I served as member and chair. Later, I served as an expert witness in several of the instances of litigation against Microsoft's anti-competitive practices in the United States, Europe, and Asia; based on a careful review of the evidence, the repeated findings by courts in the United States and elsewhere that Microsoft had abused its monopoly position was not a surprise.

16. For a more extensive discussion of this episode, see William Greenleaf, *Monopoly on Wheels: Henry Ford and the Selden Automobile Patent* (Detroit: Wayne State University Press, 1961).

17. For a discussion of this story, see, for instance, Tom D. Crouch, *The Bishop's Boys: A Life of Wilbur and Orville Wright* (New York: W. W. Norton, 1989). The patent pool was in place by July 1917. Joel Klein referred to it in a speech, given on May 2, 1997, during his tenure as acting assistant attorney general of the Anti-trust Division of the U.S. Department of Justice; see www.usdoj.gov/atr/public/speeches/1118.htm.

18. In some cases, through better marketing, follow-on drugs have sometimes done as well as or better than the original drug. For instance, Zantac was a "me-too" anti-ulcer drug that followed on from the pathbreaking drug Tagamet (based on research that received the Nobel Prize). While some research suggests that Zantac did not, in general, outperform Tagamet, because of better marketing it outsold it. (Its success may also be related to its having fewer side effects.)

19. Total spending has been enormous: the combined R&D spending of the seven largest pharma companies in America alone was $17 billion in 2001. See "Industry Dominates R&D Spending in US," *Chemical and Engineering News*, October 28, 2002, pp. 50–52. R&D spending has risen enormously with little to show for it. From 2000 to 2004, the average annual number of new drug applications (NDAs) submitted to the FDA was 107. This average has been declining more or less since the start of data collection in 1970. More telling, of the forty-six NDAs in the first part of 2005, only seven have been for a new molecular entity. In 2004, of the 113 NDAs, only thirty-one were for new molecular entities. Most of the others were either for a new formulation or a new manufacturer. See U.S. Food and Drug Administration Center for Drug Evaluation and Research, "CDER Drug and Biologic Approval Reports," at www.fda.gov/cder/rdmt/default.htm.

20. See Ha-Joon Chang, *Kicking Away the Ladder: Development Strategy in Historical Perspective* (London: Anthem Press, 2002); and Eric Schiff, *Industrialization without National Patents: The Netherlands, 1869–1912; Switzerland, 1850–1907* (Princeton: Princeton University Press, 1971). Even today, many countries remain innovative without as strong intellectual property protections as the

United States provides. (Japan's copyright protection, for instance, has been weaker.) See H. Stephen Harris Jr., "Competition Law and Patent Protection in Japan: A Half-Century of Progress, a New Millennium of Challenges," *Columbia Journal of Asian Law*, vol. 16 (Fall 2002), pp. 71–140.

21. Of course, basic research is provided only because it finds some sources of funding, such as the government or foundations. Even a dedicated scientist needs a laboratory, and laboratories are expensive. The monopoly profits arising from intellectual property provide an alternative source of funding, one that has some large costs for society, which have to be weighed against the benefits.

22. The Firefox browser was written as part of the Mozilla project. The project produces open source software and is supported by the Mozilla Foundation, which received start-up support from AOL's Netscape division. (More information available at www.mozilla.org.) As of March 2006, eighteen months after its launch, Mozilla Firefox was estimated to have 10 percent of the market. See Antony Savvas, "Firefox Reaches One in Ten," ComputerWeekly.com, April 5, 2006, at www.computerweekly.com/Articles/2006/04/05/215224/Firefoxreachesonein ten.htm.

23. President Jefferson, who as secretary of state was one of the original drafters of the 1793 Patent Act, envisioned patents being granted only to physical, useful inventions. When the 1952 law was passed, though, a congressional report said that "anything made by man under the sun" was patentable. Since then, there has been a vast expansion of what is patentable. In 1980, the Supreme Court, in *Diamond v. Chakrabarty*, ruled that genetically modified bacteria were patentable. Since then patents have been extended to business processes.

24. See James Meek, "The Race to Buy Life," *Guardian*, November 15, 2000, available at www.guardian.co.uk/genes/article/0,2763,397827,00.html; and the Center for the Study of Technology and Society, "Genome Patents," at www.tec soc.org/biotech/focuspatents.htm.

25. The firm is based in Salt Lake City, Utah, and was co-founded by Walter Gilbert, who won the 1980 Nobel Prize in Chemistry for his contributions to the development of DNA-sequencing technology.

26. See Claude Henry, *Patent Fever in Developed Countries and Its Fallout on the Developing World*, Prisme N° 6 (Paris: Centre Cournot for Economic Studies, May 2005); and Andrew Pollack, "Patent on Test for Cancer Is Revoked by Europe," *New York Times*, May 19, 2004, p. C3. Myriad eventually developed a screening technology, and asks $3,000 for a complete screen; it refuses to let other firms perform the screen. The province of Ontario is ignoring this, allowing its citizens to be screened for free.

27. The global number has been soaring—up by 14 percent in just three years. The Intellectual Property Statistics Database is available at www.wipo.int/ipstatsdb/en/stats.jsp.

28. Tim Berners-Lee, *Weaving the Web: The Original Design and Ultimate Destiny of the World Wide Web* (New York: HarperCollins, 2000).

29. The provision on data exclusivity—designed to limit the use of information—that the United States has been insisting upon in recent bilateral trade agreements clearly goes completely against the spirit of this traditional requirement.

30. There is a curious tension in the position of some of the most ardent free market advocates of intellectual property rights: while the liberalization/privatization agenda that they support in general entails minimizing the role of the government, this new set of reforms calls for a more active government and a new and restrictive set of regulations on the use of knowledge.

31. Their investment in lobbying has yielded high returns. See Stephanie Saul, "Drug Lobby Got a Victory in Trade Pact Vote," *New York Times*, July 2, 2005, p. C1.

32. Congress changed the term of patents in 1994 to conform with GATT standards, so that a patent now lasts twenty years from its earliest filing date, whereas it used to last seventeen years from its grant date. The twenty-year term is subject to possible extension to compensate for any delays in the granting process.

33. WIPO was preceded by the Bureaux Internationaux Réunis pour la Protection de la Propriéte Intellectuelle, set up in 1893 to administer the Berne Convention for the Protection of Literary and Artistic Works (the first international agreement on copyright).

34. Linking one public policy issue (labor standards or intellectual property) to trade (and trade sanctions) is, naturally, called "linkage." The case for linkage is most compelling when what is at issue is the very well-being of everyone on the planet. In chapter 6, I argue that it does make sense to link trade agreements with the enforcement of global environmental agreements, such as the Kyoto Protocol on global warming.

35. Later that month, in Seoul, at a WIPO ministerial meeting of the least developed countries, I spelled out what such an intellectual property regime might look like. See "Towards a Pro-Development and Balanced Intellectual Property Regime," at www.gsb.columbia.edu/faculty/jstiglitz/download/2004_TOWARDS_A_PRO_DEVELOPMENT.htm.

36. Unfortunately, the United States in its many recently signed bilateral trade agreements has demanded, and received, a strengthening of intellectual property protection, a "TRIPs plus" agreement—the kind of agreement that gave rise to the protests in Morocco. Similar protests have marked bilateral trade negotiations elsewhere, such as in Thailand.

37. This is what is meant by knowledge being a public good.

38. See article by Donald G. McNeil Jr., "A Nation Challenged: The Drug; A Rush for Cipro, and the Global Ripples," *New York Times*, October 17, 2001, p. A1.

39. One seeming puzzle was why drug companies seem to care so much about generics in developing countries. After all, the profits they currently make from

developing countries is small, and therefore the loss in profits from generic production there is small. It is possible that they might even make more money
through licensing fees. The answer typically provided by the drug companies is
that they worry that the cheap drugs will be exported to the United States and
Europe, and this could affect their profits enormously. The argument, however,
is not totally persuasive. There are already huge price differences around the
world, and only limited circumvention, largely because this is a highly regulated
industry, with imports tightly controlled, and with most purchases paid by third
parties. The real reason, I suspect, has to do with the fear that if Americans (or
Europeans) were to see the discrepancy between what the drug companies are
charging and what the drugs could be purchased for, there would be enormous
pressures put on pricing.

40. Pharmaceutical companies filed a lawsuit against the government of South
    Africa to contest the government's ability to use WTO access provisions—in this
    case, compulsory licensing—to make HIV/AIDS drugs available there. The case
    was dropped in April 2001.

41. See paragraph 6 of the "Declaration on the TRIPS Agreement and Public
    Health" at Doha: "We recognize that WTO Members with insufficient or no
    manufacturing capacities in the pharmaceutical sector could face difficulties in
    making effective use of compulsory licensing under the TRIPS Agreement." In
    spite of the urgency of the matter, with thousands dying of AIDS, the Bush
    administration, under pressure from the drug companies, refused to go along
    with an agreement that had been reached by all of the other countries. Finally,
    in August 2003, an agreement was reached when the United States changed its
    position to allow least developed countries to import generic drugs from low-
    cost, non–patent holding producers in developing countries. But by then, the
    issue had already done considerable damage to the reputation of the WTO
    among developing countries.

42. In fact, since lifesaving medicines are a necessity, the drug companies' power to
    raise prices and increase profits from them is far higher than with cosmetic and
    lifestyle drugs.

43. New technologies may make it easier to trace where drugs are produced, mak-
    ing circumvention (sometimes called "parallel imports") even more difficult.

44. One study in *Lancet* found that "of 1393 new chemical entities marketed
    between 1975 and 1999, only sixteen were for tropical diseases and tuberculo-
    sis." P. Trouiller et al., "Drug Development for Neglected Diseases: A Deficient
    Market and a Public-Health Policy Failure," *Lancet*, vol. 359 (June 22, 2002),
    pp. 2188–94. Another noted, "Of the 137 medicines for infectious diseases in
    the pipeline during 2000, only one mentioned sleeping sickness as an indica-
    tion, and only one mentioned malaria. There were no new medicines in the
    pipeline for tuberculosis or leishmaniasis. PhRMA's current 'New Medicines in
    Development' list shows eight drugs in development for impotence and erectile

dysfunction, seven for obesity, and four for sleep disorders." Médecins Sans Frontières Access to Essential Medicines Campaign and the Drugs for Neglected Diseases Working Group, "Fatal Imbalance: The Crisis in Research and Development for Drugs for Neglected Diseases," Médecins Sans Frontières, September 2001, p. 12. A 2001 paper by Families USA Foundation entitled "Off the Charts: Pay, Profits and Spending by Drug Companies" shows that in 2000, eight of the nine largest drug companies in America spent over twice as much on marketing as on R&D. The one that didn't spent 1.5 times as much on marketing as R&D. For a more recent paper arguing that lower drug prices will not lower research, see Donald Light and Joel Lexchin, "Will Lower Drug Prices Jeopardize Drug Research? A Policy Fact Sheet," *American Journal of Bioethics*, vol. 4, no. 1 (January 2004), pp. W1–W4.

45. A third alternative is direct support for research, for instance, through the National Institutes of Health in the United States and similar research institutions in other countries.

46. There are, of course, other reforms that would reduce incentives to produce "me-too" drugs. The government, for instance, could disseminate information about the relative effectiveness and safety of drugs. Insurance companies might then be required to authorize the use of a more expensive drug only if it were shown to be significantly more effective or safer. Such a reform would encourage competition on product quality and price.

47. Knowledge is a global public good—a good from which everyone benefits. Private markets, by themselves, always provide an undersupply of public good. I have not addressed in the discussion here the question of the best location of research, whether in the public, private, or nongovernmental sector. Many of the most important innovations occur in government research labs and universities. It is clear that they have the ability to do first-rate research. The prize fund concept has been championed by James Love and the Consumer Project on Technology. Congressman Bernard Sanders introduced HR 417, the Medical Innovation Prize Act of 2005, to implement the idea.

48. Of course, prizes for diseases prevalent in developing countries would mostly benefit those in developing countries. These expenditures can be thought of as an important form of foreign assistance.

49. The first President Bush refused to sign the agreement, but on June 4, 1993, President Clinton did sign it. However, Congress has refused to ratify it.

50. Ruth Brand, "The Basmati Patent," in *Limits to Privatization: How to Avoid Too Much of a Good Thing*, ed. Ernst Ulrich von Weizäcker, Oran R. Young, and Matthias Finger (London and Sterling, VA: Earthscan Publications, 2005). Devinder Sharma, "Basmati Patent: Let Us Accept It, India Has Lost the Battle," June 22, 2005; available at http://www.eftafairtrade.org/Document.asp?DocID=150&tod=2112.

51. Interestingly, the scientists involved were South Asian.

52. The United States eventually rescinded all but five of the patent claims, and refused to allow RiceTec to market its rice as basmati rice. RiceTec may, of course, have improved the traditional variety; but critics claim that their patent at the same time tried to privatize a considerable body of traditional knowledge. See Brand, "The Basmati Patent," op. cit.; and John Madeley, "US Rice Group Wins Basmati Patents," *Financial Times*, August 24, 2001, Commodities & Agriculture section, p. 24.

53. See Vandana Shiva and Ruth Brand, "The Fight Against Patents on the Neem Tree," in *Limits to Privatization*, op. cit.

54. Qinghao, in English, is sweet wormwood, the active ingredient of which is called artemisinin. The Chinese had sought WHO approval for years before the Swiss got it fast-tracked. At the same time, Novartis shared patent rights with the Institute of Microbiology and Epidemiology of the Academy of Military Medical Sciences in Beijing. See Howard W. French, "Malaria Remedy Proves a Tonic for Remote China," *International Herald Tribune*, August 12, 2005, p. 1; and "A Feverish Response: Treating Malaria," *Economist*, November 20, 2004.

55. There are other reforms to patent procedures that are needed. For instance, see note 11 for this chapter.

56. The point, of course, is more general: they also need legal assistance, for instance, in fighting against the many nontariff barriers described in chapter 3.

57. In the entertainment industry, as in any other industry, domestic firms often try to engage in protectionist measures, and often use claims of the promotion of culture to defend such measures. I want to make it clear: I am not defending protectionism; but I am defending the rights of governments to promote their culture.

58. There was certainly never any open discussion within the White House about these provisions, and I was supposed to participate in all important meetings dealing with environmental matters. When, subsequently, I asked Mickey Kantor, the U.S. trade representative at the time, if he had been aware of this provision, in defense he pointed out that the agreement had been negotiated under the first President Bush; they simply took the agreement as it was, focusing on side agreements to placate labor and environmental groups, who also seemed to be unaware of its existence and potential importance.

Chapter Five

1. For a history of Azerbaijan oil, see Natig Aliyev, "The History of Oil in Azerbaijan," *Azerbaijan International*, vol. 2, no. 2 (Summer 1994), pp. 22–23.

2. See Terry Lynn Karl, *The Paradox of Plenty: Oil Booms and Petro-States*, Studies in International Political Economy 26 (Berkeley: University of California Press, 1997).

3. Xavier Sala-i-Martin and Arvind Subramanian, "Addressing the Natural

Resource Curse: An Illustration from Nigeria," Columbia University, Department of Economics, Discussion Paper Series 0203-15, May 2003, available at www.columbia.edu/cu/economics/discpapr/DP0203-15.pdf.

4. Even at the stringent $2-a-day standard used by the World Bank, one-third of the country is in poverty.

5. The phrase "resource curse" was first coined by Richard M. Auty in *Sustaining Development in Mineral Economies: The Resource Curse Thesis* (London and New York: Routledge, 1993).

6. The agreement that I helped forge at that meeting eventually led to the OECD Convention on Combating Bribery of Foreign Public Officials in International Business Transactions, which was signed on December 17, 1997, and entered into force on February 15, 1999. The signatories include all thirty OECD member countries and six nonmember countries (Argentina, Brazil, Bulgaria, Chile, Estonia, and Slovenia). Five years after our meeting, the OECD noted how slow governments have been to respond. France had just eliminated the provision grandfathering in tax deductibility of bribes for contracts signed before the OECD agreements, and New Zealand has still not fully complied. See Trade Compliance Center: OECD Antibribery Report 2001, "Laws Prohibiting Tax Deduction of Bribes," at www.mac.doc.gov/tcc/anti_b/oecd2001/html/ch04.html.

7. See the case of ExxonMobil in Kazakhstan, as reported in "Kazakhstan President Nazarbayev Accepted Bribes, U.S. Alleges," Bloomberg.com, April 16, 2004, at http://quote.bloomberg.com/apps/news?pid=10000087&sid=a_8QW26uoX_I&refer=top_world_news; Daniel Fisher, "ExxonMobil's Kazakstan Quagmire," *Forbes*, April 23, 2003, available at www.forbes.com/2003/04/23/cz_df_0423xom.html; Seymour M. Hersh, "The Price of Oil," *The New Yorker*, July 9, 2001, pp. 48–65; and Thomas Catan and Joshua Chaffin, "Bribery Has Long Been Used to Land International Contracts. New Laws Will Make That Tougher," *Financial Times*, May 8, 2003, p. 19. The *Financial Times* article notes, "In total, authorities have accused [Giffen] of taking more than $78 million in commissions and fees from Mobil and other western oil companies and then illegally funneling them to senior Kazakh officials." The trial is ongoing.

8. There eventually was an out of court settlement, in which the oil companies paid the state of Alaska more than $1 billion. Alaska was not the only state to encounter problems. So did Alabama—which succeeded in getting a large settlement from the oil companies.

9. Of course, governments may choose to shift some of the risk to others. They might sell the oil in futures markets—getting a certain price today rather than the uncertain price that might prevail two or three years from now. Contracts between oil companies and countries too may involve some risk shifting. If the government insisted that it get a larger share of the windfall profits when prices

soar, the oil companies might conceivably offer a smaller upfront bonus. There is, however, little evidence that this is the case, at least to any significant degree.

10. Formally, the value of the asset is the expected present discounted value of future profits (natural resource rents) that it generates. The huge profits earned by many of the privatized enterprises suggest that they got these assets for less than full value.

11. The United States under President Reagan engaged in rapid leasing of oil tracts—critics called it a fire sale; it resulted in a substantial reduction in the amount that the government received on average for each tract. See Jeffrey J. Leitzinger and Joseph E. Stiglitz, "Information Externalities in Oil and Gas Leasing," *Contemporary Economic Policy*, vol. 1, no. 5, pp. 44–57.

12. Borrowing by government-owned enterprises is treated as if it were borrowing by the government itself. This means that a country, like Brazil, that has committed itself to a certain level of government borrowing must cut back on other government expenditures—such as for education or health—if it wishes to invest more in these enterprises, no matter how high the return on those investments would be.

13. Dr. Mahathir bin Mohamad in an address to the Global Leadership Forum, Kuala Lumpur, September 7, 2005, entitled "The Past, Present and Future—Malaysia's Challenges in a Competitive Global Landscape," and in personal conversations with the author.

14. Chile has often been held up by the IMF as an exemplar of the success of the Washington Consensus model. But as former president Ricardo Lagos pointed out to me, Chile's policy differed from the Washington Consensus in several respects—including its refusal to fully privatize. It did not, for instance, fully liberalize its capital markets. Most important, it put considerable stress on education and fighting poverty—issues that were not part of the Washington Consensus.

15. Chrystia Freeland, *Sale of the Century: The Inside Story of the Second Russian Revolution* (New York: Crown, 2000).

16. Nigeria, for instance, had a Paris Club debt (that is, debt owed by the government) of more than $30 billion, before a write-down in October 2005.

17. Near the end of the IMF's Public Information Notice No. 01/73 (July 27, 2001), entitled "IMF Concludes 2001 Article IV Consultation with Chile," the IMF notes that its estimates of the Chilean government's fiscal balances are different (i.e., worse) than Chile's own estimates because of differing treatments of the revenues from the Copper Stabilization Fund (and capital gains from privatization).

18. See State oil fund of the Republic of Azerbaijan (SOFAZ) Web site at www.oil fund.az.

19. See Norges Bank Investment Management (NBIM) Web site at www.norges-bank.no/english/petroleum_fund/.

20. Similarly, while it is true that a successful market economy requires secure property rights, in a democracy property rights can only be secure if they are viewed

as legitimate. For a broader discussion of these issues, see, for example, Karla Hoff and Joseph E. Stiglitz, "After the Big Bang? Obstacles to the Emergence of the Rule of Law in Post-Communist Societies," *American Economic Review*, vol. 94, no. 3 (June 2004), pp. 753–76; and Karla Hoff and Joseph E. Stiglitz, "The Creation of the Rule of Law and the Legitimacy of Property Rights: The Political and Economic Consequences of a Corrupt Privatization," NBER Working Paper 11772, November 2005.

21. For further discussions of this major global initiative called the Extractive Industries Transparency Initiative (EITI)—sometimes called "publish what you pay" —see www.eitransparency.org/.

22. Arms sales, by supporting conflict, cause a major negative externality; and a standard way of responding to such externalities is to impose a tax. The leaders of several of the advanced industrial countries have called for such a tax. See "Action Against Hunger and Poverty: Report of the Technical Group on Innovative Financing Mechanisms," presented at the UN in September 2004, draft report authored by Anthony Atkinson et al., Technical Group on Innovative Financing Mechanisms, Brasilia, 2004, available at www.globalpolicy.org/socecon/glotax/general/2004/09innovative.pdf.

23. So far it seems to be relatively ineffective. In a quick canvas of diamond retailers in New York, I found that few knew about the issue, few cared, and most simply stated that the ban was impossible to implement.

24. Though in principle lumber is, like fish, a renewable natural resource—as distinguished from oil, gas, and minerals, which are depletable—hardwood forests take so long to grow that, in essence, they are a depletable resource.

25. See, for example, Sachs, *The End of Poverty*, op. cit.

26. And, in fact, much of the aid given to Russia quickly made its way into bank accounts in Cyprus and elsewhere. See Stiglitz, *Globalization and Its Discontents*, op. cit., p. 150.

27. The World Bank tried to play a positive role in norm setting—in ensuring that most of the money from the Chad-Cameroon oil project which it helped finance would go into development and not be spent on arms. The worry was that the oil money would simply strengthen Chad's military dictatorship. A complicated trust into which the oil money was supposed to go was established; but soon after oil started to flow, Chad's military government demanded that the trust be abrogated and that the money go directly to it, threatening to cut off the oil if this was not done. (As this book goes to press, the ultimate resolution is not certain.) The worst fears of the critics of the project have been realized. Why, they had asked, did ExxonMobil need World Bank assistance? If the project was a good project, it should have been able to get financing without the World Bank. Somewhat earlier, an independent review of World Bank lending in extractive industries had argued against the Bank lending in countries like Chad, where it

was unlikely that the money would help in poverty alleviation. The Bank sent away the review's recommendations for further study—a polite rejection.

Chapter Six

1. The term was popularized by Garrett Hardin in his classic article of that title in *Science*, vol. 162 (December 13, 1968), pp. 1243–48. There is a fundamental difference between the knowledge commons and the commons being discussed here. In the former, the use of the commons by one does not detract from what is available to others; the enclosure represents an inefficient restriction on usage. In the case of grazing land or fishing commons, usage by one reduces resources available to others. To use economists' jargon, in the former case the marginal cost of usage is zero; in the latter it is positive.

2. See chapter 3, p. 84, for a discussion of the concept of externality and the role of government in dealing with the inefficiences that result.

3. Greenhouse gases include not only carbon dioxide and methane (global average atmospheric concentrations of methane have increased 150 percent since 1750) but also such gases as nitrous oxide (N_2O). See Intergovernmental Panel on Climate Change, *Climate Change 2001: The Scientific Basis* (Geneva: United Nations Environment Programme, 2001).

4. The most comprehensive surveys of the science on global warming are provided by the Intergovernmental Panel on Climate Change (IPCC) in its periodic reports. See IPCC, *IPCC Third Assessment: Climate Change 2001* (Cambridge, UK: Cambridge University Press, 2001). The previous two assessments—IPCC, *IPCC First Assessment Report, 1990* (Cambridge, UK: Cambridge University Press, 1990); and IPCC, *IPCC Second Assessment: Climate Change 1995* (Cambridge, UK: Cambridge University Press, 1995)—can be found at www.ipcc.ch/pub/reports.htm.

5. The IPCC was established in 1988 by two United Nations organizations, the World Meteorological Organization and the United Nations Environment Programme, to assess the "risk of human-induced climate change." Since then, it has met almost continuously, reviewing new data and studies as they become available. I served on the Second Assessment (*IPCC Second Assessment: Climate Change 1995*, op. cit.).

6. The average reduction was 5.2 percent by 2012 compared to the year 1990. The average equals 5.2 percent because some countries, including Russia and Australia, have been permitted increases, or at least no reductions. While this may seem small, it represented a reduction of 29 percent compared to the emission levels that would be expected without the protocol. (The end date for the protocol itself is 2012; it is envisaged that tighter standards will be set going forward.)

7. Council of Economic Advisers, "The Kyoto Protocol and the President's Policies to Address Climate Change: Administration Economic Analysis," July 1998.

8. Energy Information Administration, *International Energy Annual 2003* (Washington, DC: U.S. Department of Energy, 2005), Table E.1G.

9. Ibid., Table H.1GCO2.

10. See Table H.1 of ibid.; and National Environmental Trust, *First in Emissions, Behind in Solutions: Global Warming Pollution from U.S. States Compared to 149 Developing Countries* (Washington, DC: National Environmental Trust, 2003); available at www.net.org/reports/globalwarming/emissionsreport.pdf.

11. See Table H.1GCO2 of Energy Information Administration, *International Energy Annual 2003*, op. cit. In chapter 2, I explained how GDP is an imperfect measure of living standards, and noted the more comprehensive measure used by the UN, called the "Human Development Indicator." By that indicator, the United States ranks tenth in the world in 2005.

12. In chapters 2 and 5, I explained why GDP is not a good measure of sustainable social well-being—which is why frequently heard industry arguments that restricting pollution has a GDP cost are not only self-serving but beside the point. Even if today's measured GDP were to decrease, if the result is that future losses from the effects of global warming are reduced, restricting emissions would be efficient, merely from the perspective of GDP, looked at from a long-term perspective. Equally irrelevant are arguments that jobs will be lost; if the fiscal authorities are doing their job, new jobs will be created elsewhere in the economy.

13. See the World Resources Institute's searchable database of "CO2 cumulative emissions, 1900–2002," located at http://earthtrends.wri.org/, which is compiled from various data published by the U.S. Department of Energy.

14. China's eleventh five-year plan, announced in March 2006, focuses on the environment, including increased energy efficiency. In the weeks following the announcement, the government raised taxes on gasoline and other oil products and announced other measures to discourage deforestation—including a tax on wooden chopsticks.

15. Energy Information Administration, *International Energy Outlook 2004* (Washington, DC: U.S. Department of Energy, 2004), Table 72. The comparison of developing to developed here ignores the category of Eastern Europe/FSU, which is projected to account for a relatively constant 12 percent of emissions through this entire period.

16. There are complex forces at play. Agriculture production, while it does not contribute to emissions as much as industrial production, still adds to greenhouse gases. Livestock, for instance, produce high levels of CH_4 (methane). Deforestation is a major problem, discussed more fully below. The developing countries also are very inefficient—that is, of course, almost a defining characteristic of being less developed; and it means that per unit of production, they have high levels of emissions. On the one hand, this means that as they industrialize, emissions grow rapidly; but it also means that there is enormous scope for emission reductions, as they become more efficient. In some of the developing countries,

like China, low energy prices contribute to this inefficiency. If the developing countries follow Europe's example, and not America's, of levying high taxes on oil, then their increases in emissions will be limited. Even without such taxes, China has shown that one can combine extremely rapid growth—7–9 percent per year—with only limited increases in emissions. For a discussion of projections of emissions, see Mustafa H. Babiker, John M. Reilly, Monika Mayer, Richard S. Eckaus, Ian Sue Wing, and Robert C. Hyman, "The MIT Emissions Prediction and Policy Analysis (EPPA) Model: Revisions, Sensitivities, and Comparisons of Results," MIT Joint Program on the Science and Policy of Global Change Report 71, February 2001. More recent projections published by the Energy Information Administration in *International Energy Outlook 2004*, op. cit., project the developing countries' emissions will exceed those of the developed countries in 2030, rather than 2025.

17. There are also some difficult problems associated with increases in population, which we do not have room to address here.

18. Based on the author's calculations, using UN data for emissions per capita (United Nations Millennium Indicators data series, "Carbon dioxide emissions (CO2), metric tons of CO2 per capita (CDIAC)," which can be found at http://unstats.un.org/unsd/mi/mi_series_list.asp.

19. This is one of several ways in which the playing field could be leveled. Europe could, for instance, impose a carbon tax (the Clinton administration actually proposed such a tax)—a tax on every commodity based on the magnitude of the emissions in its production, with a credit provided by energy taxes already paid. European producers would, of course, get a large credit, because of the high taxes already imposed on oil.

20. The Rainforest Coalition (see below) is *not* asking for compensation for this, partially because these "cleaning" services are hard to estimate, partially because analogous services are provided by the forests of the advanced industrial countries, including the United States, and these "negative emissions" have not been included in the carbon accounting for them.

21. There are a number of technical details in the implementation of avoided deforestation schemes: for instance, concerning monitoring. Modern technology makes this far easier today than even twenty years ago.

22. The Rainforest Coalition was announced on January 15, 2005, at Columbia University in New York, in a speech by Sir Michael Somare, the prime minister of Papua New Guinea. It has now garnered the support of at least twelve developing countries, including Costa Rica, Nigeria, Vietnam, and India. See www.rainforest coalition.org/eng.

23. While even the approach just outlined imposes different costs on different countries, the differences are small. Technically, the inefficiency cost of a tax is called the Harberger triangle, and is related to the elasticity of demand and supply.

Typically, these costs are small relative to GDP. The cost of a switch from taxing income to taxing pollution is the difference between the Harberger triangle associated with a pollution tax and, say, with an income tax—and this difference is likely to be truly small. Finally, the distributional impact is associated with the difference in this difference, a number that is also likely to be very small.

24. There is a reason to expect tax rates on oil, gas, and coal to rise over time if we are to continue reducing emissions: if we are successful in inducing energy reductions, the demand for these resources will fall and so too will the market price before tax. But as the price falls, so too will the incentive to reduce emissions.

25. Jared Diamond, *Collapse: How Societies Choose to Fail or Succeed* (New York: Viking, 2005), p. 498.

26. Just as I have argued that sanctions are justified as a way of ensuring compliance with global agreements, so too it makes sense to have assistance conditional on compliance with global agreements (including reductions in greenhouse gas emissions per unit of GDP and nonproliferation of nuclear weapons). Such conditionality would, I think, be both effective and enforceable.

Chapter Seven

1. In 2005 foreign direct investment (FDI) inflows to developing countries were $233 billion. See United Nations Conference on Trade and Development (UNCTAD), *World Investment Report 2005: Transnational Corporations and the Internationalization of R&D*, available at www.unctad.org/en/docs/wir2005_en.pdf.

2. Of course, some advertising is simply informative—like help-wanted ads, or ads letting consumers know what products are available at what prices.

3. Similar fortunes faced more than 100 other electric surface-traction systems in forty-five cities including New York, Philadelphia, St. Louis, Salt Lake City, and Tulsa. (Though market forces might have brought an end to rail systems on their own accord, GM and other firms dominant in the automotive industry thought it in their interest to hurry things along.) For a more extensive discussion, see Bradford C. Snell, *American Ground Transport: A Proposal for Restructuring the Automobile, Truck, Bus and Rail Industries*, report presented to the Committee of the Judiciary, Subcommittee on Antitrust and Monopoly, United States Senate, February 26, 1974 (Washington, DC: United States Government Printing Office, 1974), pp. 16–24.

4. As we noted in chapter 2, note 3, one of the main strands of research of modern economics has focused on the sense in which, and the circumstances under which, as per Adam Smith's argument, markets lead to efficiency. For our purposes, the subtleties on which so much attention has been focused are of little concern: it is clear that society suffers, for instance, when corporations pollute and do not pay the consequences.

5. On the other hand, campaign contributions to obtain a peerage may have limited economic consequences.

6. "Buy Now, Save Later: Campaign Contributions & Corporate Taxation," A Joint Project of the Institute on Taxation & Economic Policy, Citizens for Tax Justice, and Public Campaign, November 2001, available at www.itepnet.org/camptax.pdf.

7. M. Asif Ismail, "Prescription for Power: Drug Makers' Lobbying Army Ensures Their Legislative Dominance," Center for Public Integrity, April 28, 2005, available at www.publicintegrity.org/lobby/report.aspx?aid=685&sid=200. See also http://njcitizenaction.org/drugcampaignreport.html.

8. In recent years, there has been a growing recognition that for societies to function well—even for markets to function well—there has to be a certain level of trust, which is supported by a sense of community. The problem is that an unfettered market—especially in the context of globalization—may destroy, or at least weaken, trust. There is, by now, a large body of literature on the concept of social capital (which includes trust and other aspects of social cooperation) and the role that it plays in the functioning of society and markets. See, for instance, Robert D. Putnam with Robert Leonardi and Raffaella Y. Nanetti, *Making Democracy Work: Civic Traditions in Modern Italy* (Princeton: Princeton University Press, 1993); Robert D. Putnam, *Bowling Alone: The Collapse and Revival of American Community* (New York: Simon & Schuster, 2000); Partha Dasgupta, "Social Capital and Economic Performance: Analytics," in *Foundations of Social Capital*, ed. Elinor Ostrom and Toh-Kyeong Ahn (Cheltenham, UK, and Northampton, MA: Edward Elgar Publishing, 2003); Partha Dasgupta, "Economic Progress and the Idea of Social Capital," in *Social Capital: A Multifaceted Perspective*, ed. Partha Dasgupta and Ismail Serageldin (Washington, DC: World Bank, 2000); Partha Dasgupta, "Trust as a Commodity," in *Trust: Making and Breaking Cooperative Relations*, ed. Diego Gambetta (Oxford and New York: Basil Blackwell, 1988); Avner Greif, "Cultural Beliefs and the Organization of Society: A Historical and Theoretical Reflection on Collectivist and Individualist Societies," *Journal of Political Economy*, vol. 102, pp. 912-50.

9. Wal-mart has generated an enormous literature. See Andy Miller, "Wal-Mart Stands Out on Rolls of PeachCare; Sign-Up Ratio Far Exceeds Other Firms," *Atlanta Journal-Constitution*: February 27, 2004, available at www.goiam.org/territories.asp?c=5236.

10. The importance of the separation of ownership and control was emphasized in the 1930s by Adolf A. Berle and Gardiner C. Means. See Adolf A. Berle and Gardiner C. Means, *The Modern Corporation and Private Property* (New York: Macmillan, 1934). Earlier, the great Cambridge economist Alfred Marshall had identified the analysis of difference between the behavior of large corporations and the single proprietor firm as the most important problem to be tackled at the end of the nineteenth century. See Alfred Marshall, "The Old Generation of

Economists and the New," *Quarterly Journal of Economics*, vol. 11 (January 1897), pp. 115–35. By the 1960s, a large number of economists were arguing that the modern corporation could not be described by the simple profit or value maximization models beloved by standard economists. See, for example, William J. Baumol, *Business Behavior, Value and Growth* (New York: Macmillan, 1959); Robin Lapthorn Marris, *The Economic Theory of "Managerial" Capitalism* (London: Macmillan, 1968); and John Kenneth Galbraith, *American Capitalism: The Concept of Countervailing Power* (Boston: Houghton Mifflin, 1952).

Nobel Prize winner Herbert A. Simon continued the study of the behavior of firms as organizations, noting that it was not generally in the interests of those inside the organization to behave in a way that would have led the firms which they manage to behave in the way that the classical theory predicted. See Herbert A. Simon, "New Developments in the Theory of the Firm," *American Economic Review*, vol. 52, no. 2, Papers and Proceedings of the Seventy-Fourth Annual Meeting of the American Economic Association (May 1962), pp. 1–15; and James G. March and Herbert A. Simon, *Organizations* (New York: Wiley, 1958).

Subsequently, in work with Sanford J. Grossman, I showed that when information is imperfect and risk markets incomplete (as they always are), maximizing market value does not, in general, result in economic efficiency. See Sanford J. Grossman and Joseph E. Stiglitz, "On Value Maximization and Alternative Objectives of the Firm," *Journal of Finance*, vol. 32, no. 2 (May 1977), pp. 389–402, and "Stockholder Unanimity in the Making of Production and Financial Decisions," *Quarterly Journal of Economics*, vol. 94, no. 3 (May 1980), pp. 543–66; and Joseph E. Stiglitz, "On the Optimality of the Stock Market Allocation of Investment," *Quarterly Journal of Economics*, vol. 86, no. 1 (February 1972), pp. 25–60, and "The Inefficiency of the Stock Market Equilibrium," *Review of Economic Studies*, vol. 49, no. 2 (April 1982), pp. 241–61. Most important, I laid out the problems associated with what has since been called "corporate governance," and showed how the economics of information could be used to lay the foundations of a coherent theory of the modern corporation. See Joseph E. Stiglitz, "Credit Markets and the Control of Capital," *Journal of Money, Banking, and Credit*, vol. 17, no. 2 (May 1985), pp. 133–52, and "The Contributions of the Economics of Information to Twentieth Century Economics," *Quarterly Journal of Economics*, vol. 115, no. 4 (November 2000), pp. 1441–78; and Bruce Greenwald and Joseph E. Stiglitz, "Information, Finance and Markets: The Architecture of Allocative Mechanisms," *Industrial and Corporate Change*, vol. 1, no. 1 (1992), pp. 37–63.

11. The Bhopal episode has been extensively covered in the press and elsewhere. See, for example, Amnesty International, *Clouds of Injustice: Bhopal Disaster 20 Years On* (London: Amnesty International, 2004), available at http://web.amnesty .org/library/Index/ENGASA201042004?open&of=ENG-398.

12. These are not the only instances in which multinationals use politics; business

executives who talk about the importance of keeping government out of the way are quite willing to call upon governments for assistance when they need it. When Aguas Argentinas—in which France's Suez is a major stakeholder—found that it had overbid on a concession contract, it turned to the French government to put pressure on Argentina to renegotiate. Nor is this one-way traffic: when profits turn out to be excessively high and foreign governments try to renegotiate concessions, Western governments weigh in, talking about the sanctity of contracts.

13. For a discussion of this and other NAFTA Chapter 11 cases, see Public Citizen (a nonprofit organization), "Table of NAFTA Chapter 11 Investor-State Cases & Claims," at www.citizen.org/documents/Ch11cases_chart.pdf.

14. I have been impressed at the strength and diversity of the corporate responsibility movement. Hydro, a Norwegian firm working in a variety of areas including gas, has not only promoted transparency in the countries in which it operates but trumpets the UN's Declaration of Human Rights. ABN Amro, a major Dutch bank, not only talks about sustainability in its lending practices but has projects helping development in a number of countries. Many companies have gone to what is called the triple bottom line, focusing not only on profits but on impacts on the environment and broader issues of social responsibility.

15. *The Wealth of Nations* (New York: Modern Library, 1937), p. 128.

16. An additional level of complexity is added by international agreements that are supposed to deal with anti-competitive behavior. While the WTO allows countries to use dumping duties, as we saw in chapter 3, dumping, as traditionally defined, has little to do with anti-competitive behavior. Moreover, while dumping is concerned with firms that charge too little, the WTO seems unconcerned about the much greater danger of monopolization, of firms charging too much. In one instance, the United States did accuse Japan of anti-competitive behavior in film (Fuji outsold Kodak two to one, while in the United States, the ratios are reversed). But the U.S. position was not sustained.

17. See Stiglitz, *Globalization and Its Discontents*, op. cit., p. 173.

18. Some European countries have legal frameworks that recognize the obligations of corporations not only to shareholders but also to others affected by their policies.

19. The reason that America is the preferred venue is that it has traditionally had the strongest competition laws. The 2005 Supreme Court decision was in *F. Hoffman–LaRoche, Ltd.* (a Swiss-based multinational operating in more than 150 countries) *v. Empagran SA*, an Ecuadorean company injured by having to pay higher prices for vitamin C that it used in shrimp and fish farming. Hoffman-LaRoche and other producers of vitamin C had been found guilty of price-fixing, but they first settled claims by Americans who also had been injured. With American claimants out of the case, the Supreme Court ruled that Empagran and twenty other foreign companies could not seek redress in U.S.

courts. I thought the principles involved were so important for the preservation of global competition that I filed an *amicus curiae* (friend of the court) brief, describing the risks of global monopoly and what should be done. While the Court found against Empagran, its ruling did suggest an awareness of the problems posed by global monopolies.

20. There are innumerable dimensions to making a global legal regime that is both fair to the injured and incentivizes corporations to act responsibly. A more fundamental legal reform would separate out the issues of punishment and deterrence from the problem of just compensation. A claims board could establish, for instance, the magnitude of the damage suffered by each individual and provide compensation on that basis. A separate tribunal could establish the extent of the corporation's culpability, whether it took actions which caused harm—say, as a result of inappropriate environmental policies—and then assess, using a statistical model, appropriate penalties. Additional punitive damages might be assessed to provide further deterrence or in response to particularly outrageous behavior.

Chapter Eight

1. The ruble fell from R6.28 to the dollar before the crisis to R23 to the dollar in January 1999.

2. Argentina abandoned its long-standing foreign exchange regime, in which the peso was convertible to the dollar on a one-to-one basis, in December 2001. It was widely anticipated that this was a prelude to a default on its debt, which occurred early the next year. See Paul Blustein, *And the Money Kept Rolling In (and Out): Wall Street, the IMF, and the Bankrupting of Argentina* (New York: PublicAffairs, 2005).

3. Letter of Luis M. Drago, minister of foreign relations of the Argentine Republic, to Mr. Mérou, Argentine minister to the United States, December 29, 1902, Documents of American History, Durham Trust Library. Translations available at www.theantechamber.net/UsHistDoc/DocOfAmeriHist/DocOfAmeriHist3 .html. Drago also wrote: "The acknowledgement of the debt, the payment of it in its entirety, can and must be made by the nation without diminution of its inherent rights as a sovereign entity."

4. See Carlos Marichal, *A Century of Debt Crises in Latin America: From Independence to the Great Depression, 1820–1930* (Princeton: Princeton University Press, 1989).

5. The amount owed was nearly £100 million, or $11.12 billion today (using historical retail index, USD/GBP exchange rate on November 23, 2005). Source: EH.Net, at www.eh.net/hmit/ppowerbp/. See D. C. M. Platt, *Finance, Trade and Politics in British Foreign Policy 1815–1914* (Oxford: Clarendon Press, 1968).

6. I am indebted to David Hale for this example. See David Hale, "Newfoundland

and the Global Debt Crisis," *The Globalist*, April 28, 2003; available at www.theglobalist.com/DBWeb/StoryId.aspx?StoryId=3088.

7. The respected Pearson Commission (headed by former Canadian prime minister and Nobel Peace Prize winner Lester B. Pearson) made a similar point in its report to the World Bank almost twenty years ago, when it said, "The accumulation of excessive debts is usually the combined result of errors of borrower governments and their foreign creditors." Lester B. Pearson et al., *Partners in Development: Report of the Commission on International Development* (New York: Praeger, 1969), pp. 153ff.

8. But just as borrowers focus on the short-run gain—postponing to their successors the problems of repayment—so do lenders, leaving to successors the problems of collection.

9. Similar issues arise, of course, in developed countries. The problem is that those in developing countries, with less experience, may be more susceptible to the "advice" of an experienced Western bank, even if that advice is tainted by self-interest. Bribery and corruption (discussed more fully in earlier chapters) also sometimes play a role.

10. The argument is that by borrowing in dollars or euros, the country could establish a benchmark against which private borrowing interest rates could be set. Lending rates are often set by adding a company risk premium to a country risk premium. Thus, if Vietnam could borrow at, say, 8 percent, a lender thinking about lending to a relatively safe Vietnamese firm might charge 10 percent, a 2 percent company risk premium added to that of the country. But if Vietnam could only borrow at 10 percent, then lenders would want to lend only at 12 percent.

11. Firms should, to protect themselves against the risk, say, of going into bankruptcy, buy insurance against decreases in the exchange rate that would increase the value of what they owe. And often they do, though less often than standard economic theory would predict. But if firms feel that the government is going to prevent large exchange rate fluctuations, there may be less demand for such insurance; and if most firms do not have insurance, then exchange rate decreases may not provide much stimulus to the economy. This is because while a weaker currency leads to increased exports, it makes foreign debts more expensive and so the country becomes poorer; this in turn discourages consumption and investment. The result is that the IMF policies actually decreased the effectiveness of the exchange rate as part of the economy's adjustment process, increased countries' exposure to risk, and increased the cost of exchange rate volatility.

12. Actually, from the perspective of lenders, the system has not been working badly, because the creditors receive on average a higher than normal return on such loans, even when adjusted for risk.

13. With the so-called Brady plan, in which old bonds were exchanged for new bonds backed by U.S. T-bills.

14. The fiscal consequences of privatization of social security played a prominent role in the debate about partial privatization in the United States, where it was noted that it would lead in the first ten years alone to more than $1 trillion of increased deficits.

15. Of course, if there were a single global lender, he might want to punish the wayward country, to teach a lesson to any would-be defaulters. But in competitive financial markets, it is in no one's interest to provide that punishment.

16. As we noted in chapter 1, the debt of the developing countries by 2006 was roughly $1.5 trillion.

17. The list included Benin, Bolivia, Burkina Faso, Ethiopia, Ghana, Guyana, Honduras, Madagascar, Mali, Mauritania, Mozambique, Nicaragua, Niger, Rwanda, Senegal, Tanzania, Uganda, Zambia. Other countries may qualify in the future.

18. If the country would not have repaid the money in any case, in what sense is granting debt relief really providing additional assistance? Having the creditors off their backs may, of course, still be of considerable benefit to the developing countries. And it may be treated as assistance by the donors, who take it as a write-off on the debt in their books.

19. This is especially so because the interest rates on, say, World Bank loans are well below market; the loan is in fact largely (typically two-thirds) a grant. The grant element is calculated by taking the present discounted value of the difference between the "unsubsidized" interest rate and the interest rate the countries have to pay.

20. The Europeans are right, however, in insisting that there is still an important role for loans, for example, in financing electric power projects. Moreover, a country may take greater care in borrowing and spending money well when the money comes from a loan that has to be repaid, rather than when it comes simply as a gift.

21. There is a large and growing literature on odious debts. Patricia Adams, *Odious Debts: Loose Lending, Corruption, and the Third World's Environmental Legacy* (London: Earthscan Publications, 1991), provides a review of the historical literature. For a general discussion, including that of the application to Iraq, see Joseph E. Stiglitz, "Odious Rulers, Odious Debts," *Atlantic Monthly*, vol. 292, no. 4 (November 2003), pp. 39–45. For a discussion of the impact on legitimate lending, see Seema Jayachandran, Michael Kremer, and Jonathan Shafter, "Applying the Odious Debts Doctrine while Preserving Legitimate Lending," December 2005, available at http://post.economics.harvard.edu/faculty/kremer/webpapers/Odious_Debt_Doctrine.pdf.

22. Again, in the standard economics jargon, this is a classic case of an externality.

23. These policies are described in chapters 1 and 2.

24. See Barry Eichengreen and Ricardo Hausmann, eds., *Other People's Money: Debt Denomination and Financial Instability in Emerging Market Economics* (Chicago: University of Chicago Press, 2005).

25. The World Bank at one time or another has borrowed in more than forty currencies. (For a partial list, see World Bank Treasury, "List of Selected Recent World Bank Bonds," at http://treasury.worldbank.org/Services/Capital%2b Markets/Debt+Products/List+of+Recent+WB+Bond+Issuance.html.) It has helped serve as a catalyst for the creation of local bond markets.

26. John Maynard Keynes, *The Economic Consequences of the Peace* (New York: Harcourt, Brace and Howe, 1920).

27. There are two possible explanations for the U.S. position. One is that Wall Street wants to make sure that borrowers repay—it wants to make defaults as difficult as possible. The other is ideological: the Bush administration has consistently opposed efforts to create and strengthen multilateral institutions; an international bankruptcy court, which might naturally evolve as a result of an attempt to create a sovereign debt restructuring mechanism, would be seen as an anathema.

28. American bankruptcy law recognizes this difference; there is a separate chapter (Chapter 9) of bankruptcy law dealing with public bodies.

29. Andrei Shleifer, a professor at Harvard, and a close friend and associate of then undersecretary of the Treasury Larry Summers, was appointed to advise Russia on its privatization through an AID (America's development agency) contract with Harvard. (At the time, Treasury played a central role in designing economic policies toward Russia.) Amidst charges of the Harvard adviser using insider information for trading and inside connections to get a license for establishing a finance firm, AID suspended and then canceled the contract, and sued to recover what it had spent. The court sustained AID's position and the charges brought against Shleifer. After spending millions in legal bills, in an out of court settlement, Harvard paid more than $25 million and Shleifer more than $2 million. Summers, by then president of Harvard, resigned shortly thereafter, partially under pressure resulting from this incident, but as this book goes to press, Harvard has yet to mete out any punishment to Shleifer. For a detailed discussion of the incident, see David McClintick, "How Harvard Lost Russia," *Institutional Investor*, January 13, 2006; available at www.dailyii.com/print.asp? ArticleID=1039086.

30. See John Lloyd, "Who Lost Russia?," *New York Times Magazine*, August 15, 1999.

31. Higher interest rates may even increase overall efficiency, by reducing the disparity between social and private costs.

Chapter Nine

1. As this book goes to press, net capital has flowed away from newly industrialized countries for every year since 1997. For other developing countries, there has been net capital outflow for every year since 2000. See IMF, *World Economic*

Outlook, September 2004 (Washington, DC: IMF, 2004), Statistical Appendix, Table 25; available at www.imf.org/external/pubs/ft/weo/2004/02/pdf/statappx .pdf.

2. Gerard Caprio, James A. Hanson, Robert E. Litan, eds., *Financial Crises: Lessons from the Past, Preparation for the Future* (Washington, DC: Brookings Institution Press, 2005).

3. That is, countries should have the larger of the amount required to sustain imports and to cover the level of short-term dollar-denominated debt.

4. Thailand's July 2, 1997, crisis, for instance, occurred when it was recognized that the country didn't have enough reserves to sustain its currency.

5. See Dani Rodrik, "The Social Cost of Foreign Exchange Reserves," NBER Working Paper 11952, presented to the American Economic Association meeting, Boston, January 2006; available at www.nber.org/papers/w11952. Developed country reserves have not changed much as a percentage of GDP, remaining at slightly below 5 percent.

6. IMF, *World Economic Outlook*, September 2005 (Washington, DC: IMF, 2005), Statistical Appendix, Table 35.

7. In the early years of this decade, the T-bill interest rate fell to 1 percent. By mid-2006, it had risen to 5 percent. In real terms, accounting for inflation, the returns have been even more minuscule—ranging from –2 percent in 2003 to slightly above 1 percent in 2006.

8. Rodrik, "The Social Cost of Foreign Exchange Reserves," op. cit., presents a more conservative set of calculations. He focuses on the excess reserves—in excess of the traditional three-months-of-imports rule; and he presents calculations based on a spread of 3, 5, or 7 percent between the lending rate and the borrowing rates of sovereigns. Using the midpoint number, he calculates a cost of close to 1 percent of developing country GDP.

9. UNDP, *Investing in Development: A Practical Plan to Achieve the Millennium Development Goals* (London and Sterling, VA: Earthscan Publications, 2005), p. 57.

10. IMF, *Annual Report*, April 2005, Appendix I: International Reserves, Table I.2; available at www.imf.org/external/pubs/ft/ar/2005/eng/pdf/file7.pdf.

11. International Monetary Fund, International Financial Statistics, "Total Reserves 1s (w/gold at SDR 35 per oz)," accessed May 15, 2006 at http://ifs.apdi.net (using conversion factor of $1.5 per SDR).

12. Deflation is a symptom of inadequate aggregate demand; and with weak aggregate demand, output will be low and unemployment high. But deflation itself can be a problem, as borrowers have to pay back more in real dollars than they borrowed and than they anticipated paying back. The increased real debt burden (combined with a weak economy) often leads to high rates of default, leading in turn to problems in the banking system. The late nineteenth century and the Great Depression were periods of deflation. One of the great economists of

the first half of the twentieth century, Irving Fischer, analyzed the role of deflation and debt in the Great Depression; more recently, his theories have been revived and modernized in the works of Bruce Greenwald and myself. See, for instance, Joseph E. Stiglitz and Bruce Greenwald, *Towards a New Paradigm in Monetary Economics* (New York: Cambridge University Press, 2003).

13. In terms of supply and demand, the wish of others to hold these T-bills constitutes (part of) their demand, and it is easy for governments to respond to this demand, simply by borrowing money (issuing T-bills). Borrowing from abroad is frequently referred to as "capital inflows."

14. Total national savings is the sum of the savings of households, corporations, and government. The fiscal deficit—the difference between government's revenues and its expenditures—is simply negative savings. When the fiscal deficit goes up, then (unless household or corporation savings goes up) overall national savings is reduced. And if investment is unchanged, this means that there will be a shortfall of funds—the country will have to increase its borrowing from abroad. That is why the fiscal and trade deficits move in tandem *except* if investment or private savings changes simultaneously. In the 1990s, the fiscal deficit decreased, and investment increased, so the trade deficit remained large. Fiscal deficits mean the government is increasingly in debt. Trade deficits mean the country is increasingly in debt. Both can be a problem, especially when countries or governments spend what they borrow on consumption rather than investing it.

15. There is a risk of insufficiency of aggregate demand not just globally, but within the reserve currency country. Technically, we can express what is going on as follows: net imports subtract from aggregate demand. (There is another channel through which the demand for reserves abroad may depress aggregate demand at home. The increased demand for the reserve country currency, or T-bills, leads to currency appreciation in a flexible exchange rate system, and this in turn depresses exports and increases imports.)

16. No matter who is in government, given the insufficiency of aggregate demand there would be political pressures for expansionary fiscal policy. In this view, the trade deficit should be viewed as determined at least in part by the demand for the country's T-bills for reserves; the fiscal deficit adjusts to changes in the trade deficit. This contrasts with much of the standard analysis, which treats the fiscal deficit as determined by policy (such as tax cuts), with the trade deficit adjusting to reflect the resulting differences in domestic savings and investment.

17. Of course, there may come a time in the future when confidence in the euro too will erode, as the level of euro debt rises.

18. Moreover, Europe's Stability and Growth Pact prevents significant deficit spending by the member countries of the EU; with the deficit limits being regularly broken, there is some question whether, de facto, the pact is still in effect.

19. John Maynard Keynes, "Proposals for an International Clearing Union" (1942),

in *The Collected Writings of John Maynard Keynes*, vol. 25, *Activities 1940–1944*, ed. Donald E. Moggridge (London: Macmillan, 1980), pp. 168–95.

20. It has done so twice, for a total value of SDR 21.4 billion (as of June 14, 2006, an SDR is worth US$1.47). They are an asset held by central banks, convertible into any currency. In 1997 the IMF's board approved a further issuance doubling the SDRs, which will become effective when 60 percent of its membership (111 countries), with 85 percent of the voting power, accepts it. As of the end of August 2005, the United States, with 17.1 percent of the voting power, has exercised its effective veto.

21. The economic logic behind this proposal is spelled out more fully in Bruce Greenwald and Joseph E. Stiglitz, "A Modest Proposal for International Monetary Reform," paper presented to the American Economic Association, Boston, January 4, 2006; available at www.ofce.sciences-po.fr/pdf/documents/international_monetary_reform.pdf.

22. Just as central banks do not need full backing for the money they issue, the new Global Monetary Authority issuing the global greenbacks need not hold in its reserves an amount equal to the global greenbacks issued. Like the IMF or the World Bank, the member countries could agree to back up the global greenbacks, if necessary; such a guarantee would enhance confidence in the new global reserve system, but it is unlikely that these commitments would have to be drawn upon. (The global greenbacks would not be an ordinary medium of transaction; they would simply be a store of value, convertible, under specified conditions, into currencies that could be used to purchase goods and services.)

An important part of the proposal discussed below is that the issuance of global greenbacks need not be closely tied to the financial contributions made in helping establish the new global reserve system. While I have described how the system could work as a "pure exchange" between global greenbacks and each country's currency, the Global Monetary Authority could simply issue the global greenbacks (much as any other central bank issues fiat money). Below, I describe the principles that might guide the allocation of these annual emissions. Alternatively, those receiving the global greenbacks could agree to contribute a like amount to finance global public goods and development, along the lines described later in the chapter.

23. John Maynard Keynes, "Proposals for an International Clearing Union" (1942), in *The Collected Writings of John Maynard Keynes*, vol. 25, op. cit.

24. I argued earlier that the global reserve system encourages deficit spending. The ease of borrowing provides temptation to borrow recklessly, as America has done in the last few years. With the dollar no longer a reserve currency, this temptation would be reduced. Nor would the United States have the need to have huge fiscal deficits to stimulate the economy, to offset the effects of the trade deficit, which we have seen is just the flip side of the accumulation of U.S. T-bills in reserves.

25. Even a more limited reform than the one proposed in this section would be of

enormous benefit. Even if the global greenbacks went to various countries in proportion to their GDP, the reforms proposed in this chapter would enhance the strength and stability of the global economy.

26. Some disagreements may arise over the role of existing international institutions. Some critics of these institutions have less confidence in the capacity of international institutions than in national governments; and they argue that the almost inherent problems of governance and accountability make successful reform unlikely.

27. A UN report concluded that the cost of achieving those goals was modest—but substantially greater than current levels of expenditure on foreign assistance: it suggests that a plausible level of overall development assistance required for the attainment of the Millennium Development Goals during the coming decade will be $135 billion in 2006, rising to $195 billion in 2015. These figures are respectively equivalent to 0.44 percent and 0.54 percent of donor GNP. UNDP, *Investing in Development,* op. cit.

28. According to the International AIDS Vaccine Initiative, total expenditure on HIV/AIDS vaccine R&D as of 2002 has been between $430 million and $470 million, only between $50 million and $70 million of which has come from private industry. In contrast, total biopharmaceutical research and development expenditure has been about $50 billion a year. International AIDS Vaccine Initiative, "Delivering an AIDS Vaccine: A Briefing Paper," World Economic Forum Briefing Document, 2002.

29. See Shantayanan Devarajan, Margaret J. Miller, and Eric V. Swanson, "Goals for Development: History, Prospects and Costs," World Bank Policy Research Working Paper 2819, April 2002.

30. There is one argument for direct transfers: In many developing countries, the quality of publicly provided health and education services is deficient. With individuals purchasing the services directly (with the money provided by the transfers), the quality of services provided might increase substantially.

31. See George Soros, *George Soros on Globalization* (New York: PublicAffairs, 2002).

Chapter Ten

1. In January 2001, there were 17.1 million manufacturing jobs; by December 2004 this was down to 14.3 million. See Bureau of Labor Statistics (at www.bls.gov/), Employment, Hours, and Earnings from the Current Employment Statistics survey (National), Manufacturing employees (seasonally adjusted).

2. See Bureau of Labor Statistics (at www.bls.gov/), Employment, Hours, and Earnings from the Current Employment Statistics survey (National), Manufacturing employees and total nonfarm employees (seasonally adjusted). Probably more important than "outsourcing," however has been the tremendous increases

in productivity in manufacturing. Given this productivity increase, there would have been large job losses in manufacturing in any case.

3. World Bank, World Development Indicators, Manufacturing, Value Added (percent of GDP). World Bank, Development Data and Statistics; available by subscription at www.worldbank.org/data/onlinedatabases/onlinedatabases .html.

4. As we noted in Chapter 2, growth rates in India and China have been two to three times that of the Industrial Revolution, or of the golden age in America in the 1950s and 1960s. See Nicholas Crafts, "Productivity Growth in the Industrial Revolution: A New Growth Accounting Perspective," *Journal of Economic History*, vol. 64, no. 2 (June 2004), pp. 521–35.

5. OECD Observer, "China Ahead in Foreign Direct Investment," August 2003; available at www.oecdobserver.org/news/fullstory.php/aid/1037/China_ahead_ in_foreign_direct_investment.html.

6. Economic Policy Institute, "Hourly Wage Decile Cutoffs for All Workers, 1973–2003 (2003 Dollars)," at www.epinet.org/datazone/05/wagecuts_all.pdf.

7. Even the head of the IMF has recognized the problem, calling for a reallocation of voting rights at the spring 2006 meeting of the governors of the IMF. Mervyn King, the head of the Bank of England, the U.K.'s central bank, in a speech delivered in New Delhi on February 20, 2006, called for a broad reform of the IMF.

8. Wilson himself seems to have been more qualified in seeing the two interests as identical. He actually said, in his congressional testimony, "I used to think that what was good for our country was good for General Motors, and vice versa." See James G. Cobb, "G.M. Removes Itself from Industrial Pedestal," *New York Times*, May 30, 1999, sect. 3, p. 4.

9. This is an example of what is sometimes called the principle of subsidiarity— issues should be addressed at the lowest level at which effective action can be undertaken.

10. Just as, without national governments, there will be underprovision of national public goods. Economists refer to this as the "free rider problem"—since everybody benefits (and it may be impossible or costly to exclude anyone from the benefits), there is a tendency for each to free ride on the efforts of others.

11. In its spring 2006 meeting, the IMF's managing director proposed modest changes in voting rights in this direction, but, not surprisingly, such proposals encountered resistance from some of those whose relative voting rights would be reduced.

12. See the discussion in chapter 3.

13. We noted, however, in chapter 3, that the current system of trade sanctions is far more effective in inducing responses by developing countries to violations in WTO rules against developed countries than the converse.

14. See Table 1 of Robert C. Feenstra, "Integration of Trade and Disintegration of Production in the Global Economy," *Journal of Economic Perspectives*, vol. 12, no. 4 (Autumn 1998), pp. 31–50.

15. From this perspective, President Bush's unilateralism will, I hope, be just a temporary aberration of the first eight years of the twenty-first century.

INDEX